sec 5.9

chpt 5 - prob 14, 16, 18

$6^1 - 6.3$

chpt 6 - prob #3

A MODERN
INTRODUCTION TO
MECHANICS

The cover photograph shows a rare collision of a high-energy neutrino with a nucleus of the active liquid in the Gargamelle bubble
chamber at the European Laboratory for Particle Physics, CERN.
The projectile neutrino is not observed since it leaves no track, but
its existence can be inferred from a detailed analysis of the tracks
left by the charged particles created in the collision. The entire
chamber is immersed in a strong magnetic field that causes these
charged particles to move in circular paths. This photograph was
one of the first "events" to confirm new theoretical ideas about the
nature of weak forces. [CERN PHOTO]

A MODERN INTRODUCTION TO MECHANICS

Jonathan F. Reichert

STATE UNIVERSITY OF NEW YORK AT BUFFALO

Prentice Hall, Englewood Cliffs, New Jersey 07632

Library of Congress Cataloging-in-Publication Data

Reichert, Jonathan Freiberg.

   A modern introduction to mechanics / Jonathan F. Reichert.
       p.   cm.
   Includes bibliographical references.
   ISBN 0-13-596248-X:
   1. Mechanics.   I. Title.
OC125.2.R45   1991
531—dc20   90-34164
                                                              CIP

Illustration credits: The author and publisher acknowledge with appreciation those individuals and organizations that supplied illustrations for this book: American Physical Society (7); CERN PHOTO (10, 21, 32, 38, 44, 59, 132, 165, 289); Hale Observatories (13); Fermilab Visual Media Services (27, 53); Burndy Library, Norwalk, CT (35); Brookhaven National Laboratory (54, 70, 120, 158, 169); Lawrence Berkeley Laboratory (72); NASA (136); National Institute of Standards and Technology (87); *Scientific American*, Jan., 1925 (137); Webber, Inc. (235); Palm Press/Estate of H. E. Edgerton (255); Darian Lakes Rollercoaster (331); Empire State Games, Track and Field (364); Gasparini and Rhee, SUNY at Buffalo (473); *Beamline*, Nov. 1976 (Appendix C). Full-color insert (after page 76): Photos A, B, Stanford Linear Accelerator Center; Photo C, Brookhaven National Laboratory; Photo D, Fermilab Visual Media Services.

Production/interior design: Nicholas Romanelli
Cover design: Anne T. Bonanno
Manufacturing buyer: Paula Massenaro

© 1991 by Prentice-Hall, Inc.
A division of Simon & Schuster
Englewood Cliffs, New Jersey 07632

Printed in the United States of America
10   9   8   7   6   5   4   3   2   1

ISBN 0-13-596248-X

Prentice-Hall International (UK) Limited, *London*
Prentice-Hall of Australia Pty, Limited, *Sydney*
Prentice-Hall Canada Inc., *Toronto*
Prentice-Hall Hispanoamericana, S.A., *Mexico*
Prentice-Hall of India Private Limited, *New Delhi*
Prentice-Hall of Japan, Inc., *Tokyo*
Simon & Schuster Asia Pte. Ltd., *Singapore*
Editora Prentice-Hall do Brasil, Ltda, *Rio de Janeiro*

To My Parents

Louise and Victor Reichert

For a Lifetime of Encouragement

CONTENTS

15 HARMONIC OSCILLATION   455

Appendix A

Appendix B: ''Elementary'' Particle Properties   513

PREFACE

THE FARE usually served to freshman science and engineering students during their first course in college physics has, in my judgment, been something less than a rousing pedagogical success. Some 30 years of teaching experience leads me to this conclusion. This book was written to rectify our historical mistakes by introducing students to physics in a natural, modern, and even mathematically simpler way than done in the "standard" textbooks. Using this book will not change, to any significant degree, the *total* amount of material covered in the usual first semester of an introductory three- or four-semester sequence, but will distinctly modify the order and method by which that material is presented. First, let me explain the reordering and then the rationale behind this new approach.

The course begins with the introduction of conservation principles as fundamental laws of nature, with no attempt to derive them from other "laws," such as Newton's laws of motion. *At the very beginning* I use the relativistic expressions for momentum and energy so that we can discuss and solve significant problems involving the creation, annihilation, and decay of "elementary" particles. In the early chapters, modern high-energy physics is introduced by examining collision processes that occur in actual bubble and spark chambers. The first problems that are solved mathematically involve either scalar conservation laws or one-dimensional collisions.

The subject of vectors, familiar to some but not all students, is reviewed carefully. Two-dimensional collisions are considered in detail. A comprehensive collection of problems using *real* physical situations that are or were important in the study of elementary particles are provided. Students have an opportunity not only to learn modern physics, but also to test their ability to solve problems using vector and scalar conservation laws. Problems are presented for both the relativistic and nonrelativistic regimes. A brief discussion, as well as some problems, demonstrate to the student that the Newtonian conservation laws are a special case of the more general Einstein relations.

This is the "natural" approach to the subject. The traditional introduction left many of us with the notion that forces were the fundamental "things" in nature—that one had to get a feel for how objects pushed and pulled on each other. We were taught forces—only to be untaught them later. Modern physics is clearly in full retreat from this way of looking at nature. The concepts of conservation laws have not yet been shaken. In fact, the search for regularities in complex interactions is recognized to be one of the most useful and powerful methods of attacking new problems. By beginning with and emphasizing conservation laws, students will have less to unlearn in their later years and will gain a deeper understanding of physics as well.

It is not unusual for a majority of beginning physics students to be weak in calculus. The concept of a derivative, so necessary for all kinematic studies of accelerating bodies, is not clear to them. In this book I have avoided using calculus as long as possible while not sacrificing the sophistication of the physics. This is done in a natural way by introducing the conservation laws at the outset. You should notice that only high-school algebra and trigonometry are necessary to solve any problem in the first seven chapters. During the three to four-week period that students spend on Chapters 1 through 7, they undoubtedly will be studying calculus. Thus they will be far better prepared for the subsequent material on nonrelativistic classical mechanics, which does require a knowledge of differential and integral calculus. With this approach students do not have to learn calculus and particle kinematics and dynamics at the same time. All the material will be covered, but differential calculus and particle dynamics, both difficult to master, will be separated in time.

The first time calculus is needed is in Chapter 8. Here we begin the study of nonrelativistic particle motion in one dimension. Once again the book departs from the standard approach by introducing variable forces. Newton's second law is treated honestly. It is a differential equation and is examined as such. Most introductory books have students solve complicated constant force problems using the three algebraic equations that we have all memorized. Consequently, students often have the misconception that all particle motion problems can be solved with these magical equations. In this book we play down these kinds of problems. Instead, students are engaged in the study of very simple differential equations and are taught how to find the solutions for time-, position-, and velocity-dependent forces acting on a single particle. Students do exercises which require them to "read" differential equations and guess the solutions on physical grounds. Solutions to the differential equations are found by substituting the assumed solution. This method emphasizes the physics of the problem and does not require integration. An integral appears for the first time in Chapter 11, on work and energy.

Chapter 10 is also a departure from most introductory textbooks. I have included in one chapter the method of solving mechanics problems using coordinate transformation. Both the Galilean and Lorentzian transformation equations are used, but only the former is derived. The main emphasis in this chapter is on the center-of-mass coordinate system and its use in particle dynamics. We revisit elementary particle creation here and demonstrate the need for colliding beam apparatus in modern ultrahigh-energy accelerators. An optional section on pseudoforces and accelerating reference frames appears at the end of the chapter.

You will notice that the chapters are often longer than the chapters in most introductory texts. I believe that reading the short chapters contributes to students' notion that they understand physics when they have memorized a thousand different equations in fifty chapters. Most texts now evenly divide the problems and questions by sections within a chapter. The clear message to students is that the answer to the problem is somewhere in the appropriate section. I don't like that message. In this book there are only a few important principles to learn in mechanics, and they are discussed in the ten major chapters. The problems are placed so that they approximately track the material in the chapter, but they are not in perfect lock-step. The higher-numbered problems tend to be more difficult.

There is more material in this book than can be covered in a 14-week, one-semester course for freshmen. I have never taught more than 80% of what is in this book. The integrity of a one-semester course will not be adversely affected by eliminating up to one-third of the text. All of Chapter 7, much of Chapter 10, a third of Chapter 11, half of Chapter 12, a third of Chapter 13, half of Chapter 14, and half of Chapter 15 *could* be left out. The instructor has wide discretion in choosing topics in the latter part of the book. I have taught this course to large classes of engineering and science students, small honors courses with science majors, and medium-sized classes of liberal arts students over a period of ten years. Their response to this new approach inspired me to write this book: I would not have put myself through that ordeal if I didn't know that this way works!

The most important reason to begin the study of physics in this manner is that from the outset students are doing modern, current, living physics. We have all seen bright-eyed, eager students come into our classrooms looking for the excitement they experienced from a good high-school teacher or from reading *Scientific American* articles, only to turn glum and sullen because of the seemingly endless confrontation with $\mathbf{F} = m\mathbf{a}$ and sliding blocks down inclined planes. There is important classical physics to learn, but there is a no a priori reason to begin at this point. Nor should today's students study Newtonian mechanics without the clear recognition that it is a special case of a more general, more powerful, and more profound theory. It is a remarkable accident of physical theory that this new way of introducing physics can equip students to actually solve significant problems in modern physics, thereby giving them some insight into the research problems that confront physicists today. You can add topics in current activities in particle physics, the neutrinos from the sun, the proton's decay, or the Superconducting Supercollider. Physics *can be* interesting and exciting.

Jonathan F. Reichert

Buffalo, New York

ACKNOWLEDGMENTS

I NEVER THOUGHT I would write a textbook. Probably I wouldn't have done so if it had not had been for a chance conversation in 1978 with Abraham Pais, while I was a visiting faculty member at Princeton University. I'm sure he has long forgotten our discussion but I did not (we had not met before or since). I want to thank him for encouraging me to commit these new ideas for the introductory course to paper. His support at a time when few colleagues or publishers seemed interested in changing our approach to the subject proved pivotal in my decision to write this book.

A project of this magnitude cannot be carried out without the support and help of many people. I first tried beginning with particle physics in 1974, after long and extensive discussion with one of our bright graduate students, Barry Fell. The first full draft of the text was written while on sabbatical leave at Middlebury College in 1984–85. All the students who made it through that early version of the text deserve my apologies and my thanks.

I have benefitted greatly from all the reviewers of various drafts of the text: Cris Butler, Middlebury College; Frank Gasparini, SUNY Buffalo; Charles Hawkin, Northern Kentucky; Andy Langner, Rochester Institute of Technology; John McGervey, Case Western Reserve; Mike Naughton, SUNY Buffalo; Philip Pearle, Hamilton College; Uma Devi Venkateswanan and Richard Sentman, SUNY Buffalo. I especially want to thank Harvey Leff of California State Polytech who painstakingly read an early version of the manuscript and made many valuable suggestions and corrections.

I also had wonderful student help. John Newman and James Feigenbaum together solved every problem in the text. Paul Sokol, John Andersen, Scott Zelakiewikz and Ron Poling developed the two-dimension bubble chamber events. I also received help from Gordon Church, Anthony Chou, Laura McCarthy, John Chu and many other students.

I had essential editing help from Elizabeth Phillips, Sherry Mahady and Rose Glickman. Rose patiently read and improved the final version of the text. The early versions of the text were typed by Diane McCardle, Sherry Mahady and Marilyn Thorpe, and Darlene Miller expertly typed the entire final draft.

I would be remiss if I did not thank Holly Hodder, my editor, who believed in this idea when few others did. She encouraged and supported our work when such new ideas were not fashionable. Now that the physics community is considering new approaches to the introductory course, and many original suggestions are being considered by the profession, I hope her faith will be rewarded. Most of all, however, I must thank all of my students who have been my real inspiration over the years.

J.F.R.

1

THE ENTERPRISE CALLED PHYSICS

1.1 **The Method**

WHAT MOST DISTINGUISHES us from all other living creatures is probably our desire to understand and control the world in which we find ourselves. The struggle to accomplish this goal has taken many forms throughout history. We have tried magic, religion, indoctrination, cheating, guessing, experimenting, recording, observing, predicting, and more. Only very recently have we turned to science and the "scientific method" to help us understand the physical universe.

Had this book been published 300 years ago, I might have felt obligated to justify the heavy emphasis on physical sciences as *the* appropriate method for explaining natural phenomena. To do so now seems superfluous, for most of us readily accept science as the proper and indeed the only way to discover laws of nature. There is, however, a popular misconception that science proceeds according to orderly steps, commonly referred to as the "scientific method," that were first set down by Francis Bacon in the early seventeenth century. In his book *The Scientific Outlook*, Bertrand Russell gives us a modern statement of the scientific method as having "three main stages: the first consists of observing the significant facts; the second, in arriving at a hypothesis which, if it is true, would account for these facts; the third, in deducing from this hypothesis (further) consequences which can be tested by observation. . . ."* If the predictions of this new hypothesis turn out to be "incorrect," we must discard the hypothesis and formulate a new one. Science, it would seem, progresses in a neat and tidy ordered sequence.

Practicing scientists find such a description of their activities not so much incorrect as it is terribly misleading. Russell's definition leaves many open questions. Consider the first part of the statement: "observing the significant facts. . . ." How does one know, *in advance,* which are the significant facts? Does one record the

---

* B. Russell, *The Scientific Outlook* (New York: W.W. Norton, 1931).

temperature of the room, the time of day, the atmospheric pressure, the phase of the moon, the political party in power, and the distance to the center of the earth while studying the scattering of elementary particles? Clearly, the scientist's theoretical framework determines, to a large extent, the parameters that will be considered important. Without such a framework we would be forced to consider an endless number of variables in any single experiment.

Then, according to Russell, if the hypothesis is "true," it would "account" for these facts. In any given experiment there are limits to the accuracy of the measurements. It is impossible to eliminate all extraneous influences. For example, an experiment may yield data that were anticipated by the theory, but can we be sure that our theory has accounted for all forces and interactions? In fact, some theoretical conjectures have been made in the face of consistently contradictory data. One of the more remarkable examples was Galileo's proposal that bodies continue moving in a straight line, at a constant speed, unless acted on by external influences. Galileo came to this radical hypothesis in the face of endless data on terrestrial objects which did not seem to move like that. That is, if a ball is rolled on a flat plane, it eventually stops, no matter how hard and smooth the ball and surface are made. Galileo was able to extrapolate from the experiments that he could perform, to the case where friction would not be present, but he was never able to eliminate the frictional influences in his experiments. Did his hypothesis account for the "facts"?

Russell's compact statement gives the readers a false picture of science as a neat and orderly, cut-and-dried process of discovery. Nothing could be further from the truth. Science is a living and exciting enterprise full of fights, contradictions, discoveries, egos, cooperation, and bad guesses, with "very" human beings in the center of the storm. David Frisch, in his monograph on elementary particles, describes a bet of $500 between Hartland Snyder and Maurice Goldhaber over the existence (or nonexistence) of the antiproton.* (This was no small sum of money, considering the salaries paid to physicists in those days.) Fights at professional society meetings (of a verbal type, of course) are not as uncommon as you might think. Discoveries do not always come when or where one expects them, and only a few, a very few, change our entire way of looking at nature.† Some "discoveries" turn out to be fraudulent, others just sloppy work. As of this writing, the work of Ponds and Fleischmann on "cold fusion" seems to fall somewhere between these two limits.

---

* D. H. Frisch and A. M. Thorndike, *Elementary Particles* (Princeton, N.J.: D. Van Nostrand, 1964).

† T. Kuhn, *The Structure of Scientific Revolutions* (Chicago: Chicago Univ. Press, 1962).

Interplay: To exert influence reciprocally.

I WOULD LIKE to define science as "the quest for the understanding of the physical universe by a complex interplay of speculation, formal theoretical construction, and experimentation." Richard Feynman, in his lectures on introductory physics, states: "The principle of science, the definition almost, is the following: the test of all knowledge is experiment. Experiment is the sold judge of scientific 'truth.'"* While I agree with Feynman's statement, it gives the reader little insight into how actual scientific exploration is carried out.

In this interplay, which comes first? Does experimental evidence always precede a theoretical model? Or have there been well-developed theories before any experiments were performed? How many pure guesses have turned out to be correct, and how often do scientists misinterpret their measurements? Where does it all begin?

The starting point of this interplay is not always the same, as we observe from looking at some of physics' most significant advances. Sometimes theoretical ideas, indeed an entire theory, were well ahead of the experimental evidence. The premier example of this is the theory of special relativity, proposed by Einstein in 1905, followed about 10 years later by general relativity. Although it is still debated, the weight of the historical evidence suggests that there were few, if any, experimental data, of which Einstein was aware, that demonstrated Newton's laws to be incorrect (or incomplete). Einstein was not responding to unexplained experiments; rather, his remarkable theory came from his insight into the nature of the universe. Attempts to confirm the theory with experiments did not begin in earnest until thirty years after the original publication of the theory.

Experimental verification of the elementary particle, the neutrino, came many years after both the first speculation of the existence and the theory of its production had been worked out. Wolfgang Pauli originally proposed the existence of this unusual particle (which he called the neutron) in order to understand the puzzling data on the β-decay of radioactive nuclei.† Pauli wrote the

---

* R. P. Feynman, R. B. Leighton, and M. L. Sands, *Feynman Lectures on Physics*, Vol. I (Reading, Mass.: Addison-Wesley, 1963), p. 1.1.

† β-decay is electron emission from unstable nuclei which is accompanied by emission of neutrinos.

following letter to the Solvay Conference:

Zurich, December 4, 1930

Dear radioactive ladies and gentlemen:

I ask you most graciously to listen to the carrier of this letter. He will explain to you in some detail that in view of the "wrong" statistics of the N and ^{6}Li nuclei and of the continuous beta spectrum, I have taken a desperate step in order to save the conservation laws of statistics and energy. This is the possibility that electrically neutral particles might exist in the nuclei, which I shall call neutrons and which have spin $\frac{1}{2}$ and obey the exclusion principle. They are different from light quanta also as they do not travel with the velocity of light. The mass of the neutrons should of the same order as those of the electrons, in any case not more than 0.01 proton masses. The continuous spectrum would then be understandable if one assumes that in the beta decay an electron is emitted together with a neutron, such that the sum of the energies of neutron and electron is constant. . . .

At present I do not dare to publish something about this idea and I will first entrust myself to you, radioactive people, with the question about the potential for experimental detection of such a neutron, if it actually had a penetrating power about the same or ten times larger than gamma rays.

I admit that my remedy immediately may look unlikely because one probably would have seen these neutrons long ago if they were to exist. But only he who dares wins, and the seriousness of the situation with regard to the continuous beta spectrum is clearly shown in the remark of my predecessor, Mr. Debye, who recently said to me in Brussels: 'O, it is best not to think about it at all, just as with the new taxes.' Hence one should seriously discuss every path that may lead to a rescue.—Therefore, dear radioactive people, examine and judge.— Unfortunately I cannot come in person to Tübingen, because I am indispensable here in Zurich due to a ball which takes place on the night of December sixth.

Your most obedient servant,

W. Pauli

In 1938, Enrico Fermi formally worked out the theory of β-decay, including in it the process of emission of the elusive neutrinos. So physics had a coherent theory, reams of indirect experimental evidence pointing to the existence of this zero charge and extremely weakly interacting particle, but no direct experimental detection of the particle. It was not until 1953 that Reines and Cowan reported the first direct observation of the neutrino.* You will better understand the reason for this long delay if I tell you one remarkable property of the neutrino. It would take a lead shield *40 light years*

---

* F. Reines and C. L. Cowan, Jr., *Physical Review*, **90**, 492 (1953).

thick (10 times the distance to the nearest star) to stop all neutrinos from the typical β-decay process of radioactive nuclei. These particles interact *extremely* weakly with all types of matter. Generally, neutrinos pass through matter leaving no evidence that they have been there. This makes them very difficult to detect. Thus the theorists got well ahead of the experimentalists, developing a sophisticated theory some 15 years before direct experimental detection had been achieved. But without the theory, we would not have been able to design and construct apparatus capable of detecting these elusive particles.

The discovery of the omega-minus (Ω^-) particle at Brookhaven National Laboratory is another example of a "theory first" discovery. This was a dramatic moment in modern particle physics. There was a large body of experimental data and a brilliant theoretical framework to explain important features of the known particles. But there was a missing particle. The properties of this particle were predicted but never observed. Did the particle exist, or was the theory wrong? We shall return to this physics detective story later.

We do not have to look far to find examples of the reverse process: that is, the experimental data preceding theoretical understanding of their significance. One obvious example is the optical absorption and emission spectra of the elements that were known in the late nineteenth and early twentieth centuries, before the development of quantum mechanics. Entire books of various absorption and emission spectra of gases of these elements existed with no way to interpret them. Here is a fraction of one page of a table of such spectral lines from the *Chemical Rubber Company Handbook* to show you what such data look like.

II. EMISSION SPECTRA 2000–10,000 Å (Continued)

Cesium (Continued)

Wave length	Arc	Spark	Wave length	Arc	Spark	Wave length	Arc	Spark
4410.21	..	(20)	II 5096.60	..	(40)	6586.02	35	..
4425.66	..	(20)	II 5227.00	..	(200)	I 6586.51	500	(5)
I 4435.71	..	(20)	II 5249.37	..	(80)	I 6628.65	35	12
II 4501.52	..	(35)	II 5274.04	..	(40)	6723.28	500	6
II 4526.72	..	(35)	II 5306.61	..	(25)	I 6870.45	200	(5)
II 4538.94	..	(30)	5348.95	..	(25)	II 6955.52	..	(20)
I 4555.35	2000 r	100	II 5349.16	..	(25)	6973.29	500	..
I 4593.18	1000 r	50 r	II 5358.53	..	(500)	I 6983.49	25	..
II 4603.75	..	(60)	II 5370.98	..	(80)	7228.53	500	(2)
II 4623.09	..	(20)	5302.79	..	(40)	I 7229.01	35 1	..
II 4646.51	..	(25)	II 5419.69	..	(60)	I 7279.95	35 1	..
II 4670.28	..	(20)	Ii 5563.02	..	(125)	7609.01	500 1	..
II 4701.79	..	(25)	5566.7	..	(40)	7944.11	800	..
II 4732.97	..	(20)	II 5814.18	..	(25)	7990.68	100 s	..
4733.06	..	(20)	II 5831.16	..	(60)	I 8015.71	200	..
II 4739.66	..	(20)	5832.6	..	(25)	8053.35	100 s	..
4763.62	..	(25)	I 5844.7	30 w	..	I 8078.92	100	..
II 4830.16	..	(30)	5925.65	..	(60)	I 8079.02	1000	..
..	..	..	..	..	..	..	..	..
..	..	..	..	..	..	..	..	..
..	..	..	..	..	..	..	..	..

This is only one of about 180 pages of such data. Even the thought of making sense of and bringing some order to such a vast collection of numbers is enough to make one reconsider a physics career. From the experimenter's point of view, it was relatively easy to collect and record accurate data, but of what value was it?* Quantum mechanics, developed during the 1920s, gave us the theoretical tools to unravel this mess, but the story of this adventure would be the subject of a separate book.

The study of superconductivity is another example of experiment leading theory. For many years some physicists believed that new fundamental interactions would be discovered before superconductivity could be explained. They were unduly pessimistic. John Bardeen, Leon Cooper, and Robert Schrieffer, who shared the Nobel prize in physics for the theory of superconductivity, showed that the phenomenon can be understood by the proper application of quantum mechanics and well-known elecromagnetic interactions. But then Karl Müller and Johannes Bednorz observed that an oxide of barium, lanthanum, and copper had superconducting properties at temperatures higher than any theory had predicted. This work has set off a worldwide search for room-temperature superconductors and the practical application of these new materials. It has also left the theorists temporarily puzzled.†

Physicists often conjecture about possible regularities, symmetries, fundamental building "blocks," new forces, and other ideas without a complete mathematical or physical theory. One might even call this "physics gossip." Speculation about superweak interactions that might be escaping our detection, the graviton, the existence of magnetic monopoles, or of tachyons (particles that go faster than the speed of light—their *lower* limit is the speed of light) fall into this category. These leaps forward (or backward) are an essential part of the freewheeling nature of scientific thought and are important to its advancement. Many of the great discoveries have been made by young scientists, possibly because they are the only ones who are willing to break with the past and gamble on new ideas.

Science is alive and moving more rapidly today than in any time in recorded history. It is truly awesome to witness the attack by the physics community on an exciting new problem or discovery. During the closing days of 1974, two high-energy groups, one at Stanford and one at Brookhaven, simultaneously discovered the existence of a new particle called the J-particle by the east coast and the ψ-particle by the west coast group. The new particle had some novel and exciting properties that attracted wide attention in the press. To give you some insight into the activity that this discovery generated, I have

---

* These data have turned out to be of considerable value, with many applications. The design of modern optical lasers is one of the better known examples.

† For a description of the early work, see *Physics Today*, Apr. 1987, p. 17.

A Collage of Literature on Mechanics

Heavy Quarks and e^+e^- Annihilation*

Thomas Appelquist† and H. David Politzer‡
Lyman Laboratory of Physics, Harvard University, Cambridge, Massachusetts 02138
(Received 19 November 1974)

The effects of new, heavy quarks are examined in a colored quark-gluon model. The e^+e^- total cross section scales for energies far above any quark mass. However, it is much greater than the scaling prediction in a domain about the nominal two–heavy-quark threshold, despite $\sigma_{e^+e^-}$ being a weak-coupling problem above 2 GeV. We expect spikes at the low end of this domain and a broad enhancement at the upper end.

Possible Interactions of the J Particle*

Remarks on the New Resonances at 3.1 and 3.7 GeV*

C. G. Callan, R. L. Kingsley, S. B. Treiman, F. Wilczek, and A. Zee†
Jadwin Physical Laboratories, Princeton University, Princeton, New Jersey 08540
(Received 9 December 1974)

This is a collection of comments which may be useful in the search for an understanding of the recently discovered narrow resonances at 3.1 and 3.7 GeV.

Is Bound Charm Found?*

Possible Explanation of the New Resonance in e^+e^- Annihilation*

S. Borchardt, V. S. Mathur, and S. Okubo
University of Rochester, Rochester, New York 14623
(Received 18 November 1974)

We propose that the recently discovered resonance in e^+e^- annihilation is a member of the 15⊕1 dimensional representation of the SU(4) group. This hypothesis is consistent with the various experimental features reported for the resonance. In addition, we make a prediction for the masses of the charmed vector mesons belonging to the same representation.

Interpretation of a Narrow Resonance in e^+e^- Annihilation*

Julian Schwinger
University of California at Los Angeles, Los Angeles, California 90024
(Received 25 November 1974)

De Rújula
University, Cambridge, Massachusetts
and
Technology, Cambridge, Mass
- 1974)

We argue that the newly discovered narrow resonance at 3.1 GeV is a $^3S_1$ bound state of charmed quarks and we show the consistency of this interpretation with known systematics. The crucial test of this notion is the existence of charmed hadrons near 2 GeV.

Lyman Laboratory of Phys

*Center for Theoretical Physics, Massachusetts
(Received*

Model with Three Charmed Quarks*

R. Michael Barnett
Lyman Laboratory of Physics, Harvard University, Cambridge, Massachusetts 02138
(Received 25 November 1974)

The spectroscopy and weak couplings of a quark model with three charmed quarks are discussed in the context of recent results from Brookhaven National Laboratory, Stanford Linear Accelerator Center, and Fermi National Accelerator Laboratory.

H. T. Nieh
Institute for Theoretical Physics, State University of New York, Stony Brook, New York 11794

Tai Tsun Wu
Gordon McKay Laboratory, Harvard University, Cambridge, Massachusetts 02138

and

Chen Ning Yang
Institute for Theoretical Physics, State University of New York, Stony Brook, New York 11794
(Received 25 November 1974)

Are the New Particles Baryon-Antibaryon Nuclei?

Alfred S. Goldhaber
Institute for Theoretical Physics, State University of New York, Stony Brook, New York 11794*

and

Maurice Goldhaber
Physics Department, Brookhaven National Laboratory,† Upton, New York 11973
(Received 25 November 1974)

reprinted the abstracts and titles of several articles that appeared *two months* later in the prestigious *Physical Review Letters.* Even if you cannot fully understand the contents of these papers (something the reader shares with this author), it should be abundantly clear that these theoreticians did not agree on how to interpret the data.

As you may know, a superconductor is a substance that carries electric current without resistance and thus without loss of power due to heating. Such materials have obvious practical value in the electrical power industry. Early in 1973, a group of well-respected solid-state experimentalists at the University of Pennsylvania reported the existence of a long-sought "high-temperature" superconductor. Until this announcement, all known superconducting materials only became superconducting at very low temperatures, which presents serious technological problems for practical use. If indeed this material, called TTF-TCNQ (an abbreviation for the monstrously long chemical name, tetrathiafulvalene–tetracyanoquinodimethane, a type of organic charge-transfer salt), was superconducting at a considerably higher temperature, the implications would be profound for all science and industry.

Many groups immediately pounced on this area of research and tried to repeat the experiments. The Penn group kept claiming new results and other groups kept disputing them. Accusations, contradictory data, new theories, requests for samples, and personal challenges were commonplace. As it turned out, superconductivity does not exist in these materials at high temperatures. The high-temperature superconducting materials were to be discovered in oxide materials in 1987. Controversies go on, as they should. Some, such as the twin paradox in special relativity, never seem to end.

What is the scientific method? How does science progress? There is no pretty, compact answer. The pace of exploration and discovery is accelerating, the amount of knowledge is growing (at an alarming rate for those who attempt to keep up), the number of people engaged in this adventure is increasing, and the results are changing the way in which we view ourselves and our world. Whatever we call it, however it works, no matter how the interplay plays, science is exciting. My job is to let you in on the excitement— maybe even entice you into participating.

1.3 **A New Approach**

A QUICK GLANCE through almost all introductory physics texts reveals a common set of materials in a common sequence. One is tempted to say that physics education is in a rut. This book consciously departs from the standard approach. Rather than beginning with Newtonian mechanics as it was developed 200 years ago, we will learn more fundamental concepts which can be applied to contemporary research problems in physics. I believe this presentation is a significantly better way to begin a study of *the* basic physical science, physics. Let me explain my reasons for making these changes.

Conservation laws are introduced here as the *first* physics principles you will study. I present them at the very beginning because they are among the most important and universal ideas in all of physics. These laws are believed to be true for *all* known systems and interactions; they are not specialized concepts and equations that you will have to "unlearn" if you proceed further with your science or engineering education. These laws are introduced using the correct relativistic expressions for the dynamical variables. You may have studied Newtonian physics in high school, but you will soon appreciate that Newtonian physics is a special case of relativistic mechanics and useful only for a certain subset of mechanics problems.

There are real advantages to learning relativistic conservation laws at the very beginning. *They are easy*! You need only high school algebra to manipulate them! They are correct! They are fun! You can analyze modern physics experiments with them. In fact, using these principles, you will be studying the very frontiers of current elementary particle experiments. It is highly probable that sometime during this semester, a new discovery in the field of particle physics will be announced in the popular press. With this course as background you may be able to understand the experiment and the significance of the discovery.

Some of you may be apprehensive of this new approach. I can hear you say, "I'm an engineering student. Why should I learn relativistic particle physics? I'll never use it." Granted, the chances that you will apply the relativistic conservation laws to a problem you may encounter in your future job are very small, nearly zero. But you will use, and use often, various conservation laws. You will become an expert on the application of these principles. But beyond that, this book, like the standard textbooks, covers all the important Newtonian mechanics and some topics in even greater detail. I guarantee that you will not be shortchanged in your preparation for advanced engineering or physics courses when you have completed this book. So don't panic.

In just a few weeks you will be familiar with some of the important analytical tools used in elementary particle physics. In fact, you will soon feel familiar enough with elementary particles to set out on your own to follow current events in an area some believe to be the cutting edge of new physics. You can follow the "interplay," join the race, and share the excitement of scientific discovery. All this while you are learning physics!

2

THE KNOWN FORCES

2.1 INTRODUCTION

HAVE YOU EVER stopped to think about the light you are using to read this book? It probably comes from a fluorescent or an incandescent lamp. Does this light obey the same physical laws as those that pertain to the light from our sun? As you read, are you listening to the radio? What allows your tiny transistor radio to sense the radio waves that apparently surround you but which you cannot hear? What is a transistor? How does the sound from your radio travel to your ears? What is sound?

Curiosity about the incalculable number of objects and phenomena that crowd the world compels scientists to search for underlying reasons for all of nature's phenomena and to find the simplest explanations. Can the laws that govern the light from incandescent and fluorescent bulbs, the light from the sun, radio waves, radiation in microwave ovens, and x-rays somehow be related? Physicists look for the connections and hope to reduce the laws to a few "fundamental" principles. Indeed, the physicist's great dream is to reduce *all* natural phenomena to *one* all-encompassing, unified law. It was Einstein's fondest ambition to create a "unified field theory" to connect all of nature's forces. But he died without achieving his goal or seeing it accomplished by others.

Even if we knew every basic law of nature, we could not explain every phenomenon in all its details. One may know the laws of a few particle interactions, but many, if not most systems in nature consist of an enormously large number of interactions. In general, it is not feasible to describe the exact behavior of real systems. Consider a "simple" solid such as the electrical conductor copper. There are about 10^{23} atoms and about 10^{25} particles (electrons, protons, neutrons) in a piece of copper the size of a grain of sand. In principle, each particle interacts with *every* other particle in the copper grain. Even if some of these interactions are very weak, the forces among the electrons are not. It would take more than a human lifetime just to write down the mathematical equations that describe all these electronic interactions. Scientists may have to give up the idea that they can explain *all* phenomena and consider that once they know all the "rules," they "understand" nature.

Scientists have divided the phenomena of nature into arbitrary categories which, to this day, separate the study of modern science into specializations. We have university departments, great research centers, and industrial laboratories divided into biology, geology, chemistry, physics, astronomy, medicine, psychology, and many other smaller subdivisions. It is truly remarkable that only four fundamental "rules" describe the interaction of all matter and presumably govern all the fields and subfields noted above. Even with the enormous explosion in scientific knowledge since the turn of the century, we have discovered only four interactions. It may surprise and should comfort you, the young aspiring science student, to realize that despite enormous increases in scientific research (as witnessed by the ever-fatter and more frequently published journals, as well as the myriad of new textbooks and monographs), science has not continually discovered new basic interactions but has, in fact, reduced their numbers. What are the known interactions?

2.2 Gravity

IT MAY SEEM curious—almost an accident—that the first force to be discovered and "explained" turns out to be the weakest of all interactions in nature: GRAVITY. Gravity is truly an extremely weak interaction; it is approximately

$$.0000000000000000000000000000000000000001 \quad (10^{-40})$$

times weaker than the strongest known force. For example, if only gravity bound the hydrogen atom (instead of electrical forces), the radius of the hydrogen atom would be about 10^{13} light years! But it is no accident that gravity was discovered first. Later we explain why the stronger interactions remained hidden for such a long time.

The first law governing gravitational interaction was discovered by Newton. He proposed that *any* two objects exert on each other an *attractive* force that is inversely proportional to the square of the distance between them and directly proportional to the product of their masses. For the two particles shown in Fig. 2.1, Newton's universal law of gravitation is written

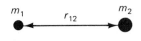

Fig. 2.1

$$F_{12} \text{ (gravity)} = G \frac{m_1 m_2}{r_{12}^2}, \qquad (2.1)$$

where $G = 6.67 \times 10^{-11}$ newton-meter^2/kilogram^2 (N-m^2/kg^2), the universal gravitational constant; m_1 and m_2 are the masses measured in kilograms; the separation r_{12} is in meters; and F_{12} is the gravitational force in newtons.

The deuteron is the nucleus of heavy hydrogen, containing one proton and one neutron packed very close together [about 10^{-15} meter (m) apart]. It is instructive to ask whether the gravitational attraction of these particles is responsible for holding them together to form this nucleus. It is a straightforward calculation using Newton's law (Eq. 2.1). The masses of the proton and neutron are approximately the same, as we see from Appendix A.III:

$$m_{p} \approx m_{n} = 1.67 \times 10^{-27} \text{ kg}$$

$$r_{12} = 10^{-15} \text{ m}$$

$$F_{\text{gravity}} = \frac{(6.67 \times 10^{-11} \text{ N-m}^2/\text{kg}^2)(1.67 \times 10^{-27} \text{ kg})^2}{(10^{-15} \text{ m})^2}$$

$$= 1.86 \times 10^{-34} \text{ N}.$$

Since a newton of force is about two-tenths of a pound, you can see that the gravitational force between the particles is unimaginably small. It could hardly be the source of the binding force.

This force depends on both the mass and the distance, so it is somewhat arbitrary to ascribe to it a "strength." Since we are only interested in giving you some feeling for the *relative* strengths of the various types of forces, we can assign to the gravitational force a strength of 1 for a standard distance.

We are continually aware of gravity in our daily lives: when we watch something fall to earth, when we attempt to lift a heavy object, when we try to jump over a fence. Scientists see this force at work in

Andromeda galaxy

the universe as well: in the motion of the moon around the earth, the earth around the sun, the sun around the galaxy, the galaxies expanding in the universe, and in hundreds of other ways. The subject of gravity is by no means a closed book in scientific research. Although most scientists accept Einstein's general theory of relativity (which has subsumed Newton's theory) as the "correct" theory of gravitation, there are competing theories that must be tested.* With the advent of space technology and modern ultrahigh-precision metrology, there has been a resurgence of experimental attempts to measure, with great accuracy, this weakest of all interactions.

Physicists believe that for each interaction in nature, there is a particle that is the propagator of the force. This means that some special particle transmits the force from one body to the other. The propagator of gravity is believed to be the graviton, which presumably moves at a velocity equal to the speed of light. No one has yet observed a graviton, but claims have been made for the experimental detection of large numbers of these propagators in the form of gravitational waves. The pioneering work in this field was done by Joseph Weber, but the validity of his measurements has been seriously challenged.† No one now claims to have detected gravitational waves directly.

Why is it so difficult to detect the graviton or a gravitational wave (presumably coming from a galactic explosion or an upheaval in space) given that scientists have long known of the existence of gravity? The answer lies in the very weakness of the interaction itself. It takes massive bodies to produce an appreciable gravitational effect, and the graviton's interaction is incredibly small. You should not, however, conclude that the forces of gravity are always weak. Large objects such as the sun or an average star produce enormous gravitational forces that confine the thermonuclear furnaces inside them. The recently discovered neutron stars are apparently mainly made up of neutrons packed closely together. They have almost unbelievable densities and produce colossal gravitational forces. Black holes are presumably objects of such incredible density and mass that the electromagnetic radiation that is emitted from them cannot escape and is drawn back into the black hole. Regrettably, black holes are not available for experimentalists to play with in the laboratory.

It is ironic that Newton explained a good deal of how gravity behaves in the seventeenth century, yet today it is the least under-

---

* A review of the competing theories of gravity can be found in the October 1972 issue of *Physics Today* and some interesting letters discussing this paper in the March 1973 issue of the same magazine. Some physicists have suggested that gravity must be modified by a new "fifth force" (see Section 9.10).

† Jonathan Logan, "Gravitational Waves, Progress Report," *Physics Today*, Mar. 1973. This is a most controversial subject, as you can easily see by reading the letters to the editor in the December 1974 issue of the same magazine. Joseph Weber's article in *Scientific American* (May 1971) is also recommended. For a more recent account, see Ref. 11 at the end of this chapter.

stood of the four fundamental forces. We understand pretty well what it does, but not how it works.

2.3 Electromagnetism

Viscosity - to non ability to flow w/out adhering to something

ALTHOUGH THE ancient Greeks had proposed the existence of atoms as the fundamental building blocks of living and nonliving matter, more than 1600 years passed before chemists began to isolate what were once thought to be indivisible "lumps" of matter. We now call these lumps the chemical elements. But what mechanisms hold these atoms together to form the molecules and compounds that science and nature have created? What makes carbon atoms form both soft coal and the hardest substance known, diamond? Why does hydrogen bond to oxygen to form water? Why does copper conduct electricity and liquid helium flow without viscosity? The gravitational forces are just too weak to explain these phenomena (see Problem 4).

These and many, many more questions about the strange behavior of matter demand that we introduce a second interaction, one that involves *two* types of particles: positive and negative. Unlike gravity, which is always an attractive interaction, the ELECTROMAGNETIC force can be *either* repulsive or attractive, depending on the charges of the particles. The laws of classical electromagnetism were completed by James Clerk Maxwell in 1873. The four equations of electromagnetism are too advanced to discuss at this point in your education, but you can begin to understand the strength of the electromagnetic interaction by considering the special case of the electrostatic force between two charged particles at rest. The equation for this force (Coulomb's law) has the same form as Newton's law of gravity, but the constants are different and charge replaces mass in the equation:

$$F_{12} \text{ (electrostatic)} = \frac{1}{4\pi\varepsilon_0}\frac{q_1 q_2}{r_{12}^2}, \qquad (2.2)$$

where $(4\pi\varepsilon_0)^{-1} = 8.99 \times 10^9$ newton-meter^2/coulomb^2, the charge on the two particles is measured in coulombs (C), and r_{12} is the separation of the two charges in meters. It is an attractive force if q_1 and q_2 have opposite signs and is repulsive if they are like signed charges.

Helium has two protons in its nucleus whose separation is about the same as the distance between the protons and the neutrons in the deuterium nucleus ($r_{12} \approx 10^{-15}$ m). It is instructive to calculate the electrostatic *repulsion* of the protons inside this nucleus. The charge of the proton, in the SI units of coulombs, is given in Appendix A.III:

$$q_{\text{proton}} = 1.6 \times 10^{-19} \text{ C}.$$

Thus

$$F_{12} \text{ (electrostatics)} = (8.99 \times 10^9 \text{ N-m}^2/\text{C}^2) \frac{(1.6 \times 10^{-19} \text{ C})^2}{(10^{-15} \text{ m})^2}$$

$$= 230 \text{ N}.$$

This is 36 orders of magnitude* larger than the gravitational attraction. So what holds the protons together with the neutrons to form the helium nucleus?

You will get some idea of the enormous strength of the electrostatic force by estimating the force between two grains of sand separated by 1 m, assuming that all the electrons have been stripped off both grains. (This would leave two tiny specks of positively charged matter.) The answer (which you should check) turns out to be about *3,000 million tons!* Impossible, you say. No, it is correct. We are unfamiliar with such forces because matter is essentially electrically neutral, having both protons and electrons in nearly equal numbers all the time. The strong electrical force between the positively charged protons and the negatively charged electrons is responsible for keeping ordinary matter electrically neutral. On our scale which puts gravity at a strength of 1, the electromagnetic force is 10^{38}. It is fair to say that 10^{38} is an incomprehensibly large number.

Let's return to the question we asked earlier: Why did gravity, rather than the much, much stronger electrical force, command the attention of early scientists? Gravitational forces have one very important and unique property that may not be apparent from a casual review of Eq. 2.1. Gravity is a *cumulative* force. All matter produces *attractive* gravitational forces; there is no "negative" mass, no repulsion. Thus the larger the mass, the greater the force. Gravity cannot be neutralized, canceled out, or shielded; it simply accumulates. We feel the gravitational force of the earth because of the earth's large mass and rarely observe its electric forces, because all the matter around us is electrically neutral. The electrical force of the protons is canceled by the electrical force of the electrons. It is the cumulative nature of gravity that is responsible for many of the exotic properties of large clusters of matter in the universe, such as neutron stars, black holes, and white dwarfs. If you hold two baseballs, one in each hand, you cannot feel an attraction between the two balls no matter how close you bring them together. The force is far too weak. However, you are always aware of the gravitational attraction to the nearby larger mass, the earth. This attraction is the

---

* An order of magnitude is a factor of 10. For example, if you say that the proton is about three orders of magnitude more massive than the electron, that means its mass is about 1000 times larger.

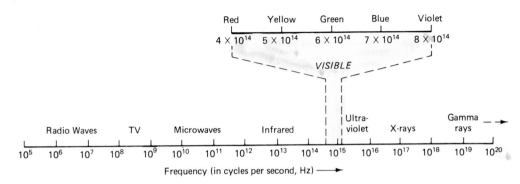

Fig. 2.2

"weight" of the object. Whenever I am carrying too heavy a load I keep telling myself "gravity is by far the weakest force"—but it doesn't help.

James Clerk Maxwell's theory of electromagnetism was one of the great syntheses of modern science. With four relatively simple equations he connected the phenomena of magnetism, electric current, charge, induction, light, x-rays, and many others. One of the remarkable aspects of Maxwell's theory was the discovery that the numerous examples of known propagating energy were all manifestations of electromagnetic fields, simply at different frequencies. For historical reasons, the various ranges of frequency have retained different names. They are, we now know, all propagating electromagnetic waves. Figure 2.2 shows the electromagnetic spectrum.

Quantum physics has penetrated even deeper into our understanding of electromagnetism. It explains electromagnetic radiation in terms of a particle force propagator that we call the photon. The photon has remarkable properties, which we study in greater detail in Chapter 6. Although it is the propagator of the electromagnetic field, it has *no electrical charge*, it has *no mass*, and it *always moves* at precisely the *same speed in empty space*. To emphasize the latter property, note that although photons' velocities have been measured experimentally over *energy* ranges spanning 20 orders of magnitude (10^{20}), all photons were found to move with the same speed within the experimental error.* In fact, the photon cannot exist if it is not moving. Thus the energy of a photon, unlike particles that have mass, is related to its associated frequency, *not* to its speed. Max Planck and Albert Einstein proposed the precise relationship to explain the frequency distribution of radiation coming from a hot radiating body and the electrons emitted from a metal surface as a result

---

* One precise measurement of the velocity of light at various frequencies was made by using the giant eruptions on the sun called solar flares [B. Lovell, F. L. Whipple, and L. H. Solomon, *Nature*, **202**, 377 (1964)].

of light striking the surface. This relation is

$$E = hf, \qquad (2.3)$$

where E is the energy of the photon, f is the frequency associated with the photon, and h is a constant whose value is 6.63×10^{-34} joule-second, (J-s), appropriately called Planck's constant. Of all the particles in nature, we probably understand the photon best of all. It may seem strange that a particle without mass can have energy and even stranger that it can have any value of energy, yet only one speed. Nature has plenty of other surprises in store for you.

2.4 Strong Interactions

NIELS BOHR proposed a model of atoms that were not solid lumps but a planetary system of particles held together by electrical forces. His first model led the way to our present understanding of the structure of atoms. The atom consists of a small, extremely dense nucleus and a surrounding cloud of electrons. But what holds the nucleus together? Is it a solid lump, or is it made up of neutrons and protons? What is the "nuclear glue" that holds these particles together in such a small volume?

Historically, these were the last questions to be answered (or partially answered, since physicists still do not have a comprehensive theory of nuclear interaction). The reason for the delay can be explained by analogy. Suppose that your golfing partner asked you to do experiments to learn the composition of a golf ball. Your first instinct would be to inspect it, look at it, and try to examine the inside. But what if you had only your bare hands? You might have to conclude that a golf ball is made of a hard white substance. But if your scientific curiosity got aroused, you might try harder to look inside. In fact, after much planning, you write a proposal to the National Science Foundation for the purchase of a "golf ball smasher" with the intention of cracking open the ball and examining the fragments.

You get the grant. Congratulations! With your smasher you hit the ball. All you observe is the ball flying away. You hit it again and again. Finally, a crack appears, but it is not large enough to allow you to look inside. So you write another proposal for a bigger machine, and because of the importance to science of this experiment, your second grant is funded. With your new apparatus you strike the ball a tremendous blow, and to the great excitement of the scientific world, it splits into fragments. You report to the *New York Times*, *Time* magazine, and the *Physical Review Letters* that the fundamental particles of the golf ball are silly putty, string, and hard rubber!

High-energy physicists do analogous experiments. To probe the structure of atoms or nuclei requires instruments capable of splitting apart these tightly bound systems. The very strength of the forces that hold the nucleons together to form a nucleus makes these forces difficult to observe. If the skin of the golf ball had been sufficiently tough, you might never have discovered the silly putty inside. The first human-made probes of matter were appropriately called "atom smashers." We now call them particle accelerators. But the early name is more descriptive of what physicists had in mind when they developed these machines. Even better, they might be called particle and nuclei smashers.

Hideki Yukawa, a Japanese theoretical physicist working in Kyoto, developed a theory to explain the extremely large forces that bind the particles together within the atomic nucleus, the STRONG FORCE. His theory predicted a force dependence on distance different from that of the inverse-square relationship of electrostatics and gravity. In fact, it suggested an extremely short range force with a rapid falloff in intensity outside the nucleus. Such a rapid degradation in the force is not described by an inverse power law, but it can be characterized by a distance called the RANGE. His theory also predicted the properties of the particle responsible for transmitting this short-range nuclear force. The range of the strong force and the mass of the particle propagator of the force are intimately connected by the approximate relationship

$$\text{RANGE} \approx \frac{\hbar}{mc}, \qquad (2.4)$$

where $\hbar$ (pronounced "h-bar") is Planck's constant divided by 2π $\left(\hbar \equiv \dfrac{h}{2\pi}\right)$, m is the mass of the particle propagator, and c is the velocity of light (3.0×10^8 m/s).

Even in the 1930s it was known that the nuclear force was effective over distances on the order of a proton's size (about 10^{-15} m), so Yukawa was able to predict the mass of its propagator. Rearranging Eq. 2.4 yields

$$m_{\text{nuclear force propagator}} \approx \frac{\hbar}{(\text{range})c}$$

$$\approx \frac{6.6 \times 10^{-34} \text{ J-s}}{2\pi \times 10^{-15} \text{ m} \times 3 \times 10^8 \text{ m/s}}$$

$$\approx 3.5 \times 10^{-28} \text{ kg*}.$$

The theory predicted a new particle whose mass was approximately 300 times the electron mass. In 1947, Cecil Powell, working at Bris-

---

* Joules have the units of energy or kg-m²/s² (see Appendix A.VI).

tol University in England, identified this particle in photographic emulsions that had been exposed to cosmic radiation on the tops of mountains in France. It is now called the pi-meson (π-meson). Both Yukawa and Powell were awarded the Nobel prize in physics for their work in unraveling some of the mysteries of the nuclear "glue."*

Equation 2.4 explains why electromagnetic forces can act over a very large (in principle, infinite) range. The particle responsible for propagating these forces has no mass, and thus the range is infinite. This relationship helps us to understand the even shorter range "weak" forces, whose particle propagators are even more massive than the π-mesons.

The forces that bind the nuclei together are also responsible for many of the particle interactions we examine later. These nuclear forces are the strongest known interactions, about 10^{40} times stronger than gravity. Because of their enormous strength and extremely short range, many strongly interacting particles have lifetimes on the order of 10^{-23} s. These particles cannot be directly observed, but their existence can be inferred from the particles into which they decay and the manner of their decay. The strong force, some *40* orders of magnitude stronger than the weakest force, was the last to be explored.

2.5 Weak Interactions

A FOURTH KNOWN interaction in nature, called the WEAK FORCE, seems to occupy a peculiar place in the grand design. Gravity is needed to hold together the galaxy, the sun, the solar system; electromagnetic forces are needed for all living and nonliving compounds; strong forces are needed to bind the nucleons in the nucleus; weak forces appear responsible only for some rather odd "games" that nature seems to be playing. You might think that we could get along quite well without this interaction. But closer examination proves that to be wrong.

The weak forces were observed first in natural radioactivity in a process called β-decay, in which the nucleus undergoes a transformation and an electron is ejected. Later, Enrico Fermi developed a theoretical picture of this process which also predicted the existence of the neutrino. The strength of this interaction is about 10^{25} times stronger than gravity and has an even shorter range than that of the strong forces.

---

*The theory of nuclear forces is currently undergoing major revisions. From the early 1950s until 1980, the π-meson was considered to be the propagator of the nuclear force. This theory could explain many but not all nuclear phenomena.

At first glance these shenanigans of nature seem relatively unimportant to the grand design. But to make such an assumption would be a monumental mistake. The rather small deviation from Newton's laws of motion at high velocities signaled not a small correction in a basic law, *but rather, an entirely new way of understanding nature.* The small rotation of the perihelion of Mercury's orbit around the sun, and the minute bending of a light beam from a star when it passes near the sun, were important clues to an entirely new way to understand gravitational interactions: that is, general relativity. The weak forces, for example, are responsible for preventing the sun from burning up its nuclear fuel so fast that life would never have formed on this planet. So it is fair to say that the weak forces are responsible for the sun shining.

Evidence has now accumulated that the weak force and the electromagnetic force are intimately related—that they are both aspects of what is called the "electroweak interaction." In 1979, the Nobel prize was awarded to Sheldon Glashow, Abdus Salam, and Steven Weinberg for their unified model of the weak and electromagnetic forces and for their prediction of the existence of the three carriers of the weak force: two charged particles, W^+ and W^-, and a neutral particle, Z^0. By the mid-1980s all three particles had been observed in high-energy experiments. This maverick force turns out to be related to one of the "big three." Someday, possibly with your help, we may be able to see all four forces as a part of a single grand plan.

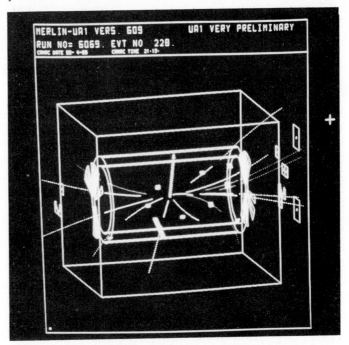

Detection of the W particle in the large UAI detector

2.6 **Summary of Important Equations**

$$F_{12} \text{ (gravity)} = G\,\frac{m_1 m_2}{r_{12}^2}$$

$$G = 6.67 \times 10^{-11} \text{ N-m}^2/\text{kg}^2$$

$$F_{12} \text{ (electrostatic)} = \frac{1}{4\pi\varepsilon_0}\,\frac{q_1 q_2}{r_{12}^2}$$

$$\frac{1}{4\pi\varepsilon_0} = 8.99 \times 10^9 \text{ N-m}^2/\text{C}^2$$

$$E = hf \text{ (for the photon)}$$

$$\text{Range} \approx \frac{\hbar}{mc}$$

$$h = 6.63 \times 10^{-34} \text{ J-s}$$

$$\hbar = \frac{h}{2\pi}$$

$$c = 3.00 \times 10^8 \text{ m/s}$$

2.7 **References**

The following books are written in a descriptive and interesting fashion, well within your grasp. You will find them of considerable value in obtaining a broad picture of elementary particle physics; you might even find them fun.

1. *The Particle Explosion*, Frank Close, Michael Marten, and Christine Sutton, Oxford Univ. Press, New York (1987). I highly recommend this book. Beautiful photographs, clean, straightforward explanations, and a good historical account of particle physics.
2. *The Hunting of the Quark*, Michael Riordan, Simon & Schuster, New York (1987). An accurate, yet engaging historical account of high-energy physics, by a scientist who participated in the work.
3. *The Particle Play*, J. C. Polkinghorne, W. H. Freeman, San Francisco (1979).
4. *The Nature of Matter*, J. H. Mulvey, Claredon Press, Oxford (1981).
5. "Particles and Fields," W. J. Kaufmann, in *Readings from Scientific American*, W.H. Freeman, San Francisco (1980). Reference within.
6. *Feynman Lectures on Physics*, Vol. I, R. P. Feynman, R. B. Leyton, and M. L. Sands, Addison-Wesley, Reading, Mass. (1963), Chapter 2.
7. *The World of Elementary Particles*, Kenneth W. Ford, Blaisdell, New York (1963).
8. *Elementary Particles*, C. N. Yang, Princeton Univ. Press, Princeton, N.J. (1962).

9. *Tracking Down Particles*, R. D. Hill, W.A. Benjamin, New York (1964).

10. *The Fundamental Particles*, Clifford E. Swartz, Addison-Wesley, Reading, Mass. (1965).

11. *Was Einstein Right*, Clifford Will, Basic Books, New York (1986). A prize-winning book on the exploration of general relativity.

12. *Quarks: The Stuff of Matter*, Harald Fritzsch, Basic Books, New York (1983).

13. *The Ideas of Particle Physics*, J. E. Dodd, Cambridge Univ. Press, Cambridge (1984). Somewhat more advanced treatment.

Chapter 2   QUESTIONS AND PROBLEMS

1. Pulsating stars (pulsars) are now believed to be dense rotating objects composed almost entirely of neutrons. The density of such objects is nearly beyond our comprehension. Calculate the approximate weight (on earth) of a 10-centimeter (cm) diameter sphere if it were composed entirely of closely packed neutrons. (*Note:* You may use 10^{-15} m for the approximate radius of the neutron.) (*Hint:* The weight of an object on the earth is the force exerted on it by the earth. This force can be calculated using Newton's law, assuming that the particle separation in the equation is the radius of the earth.)

2. Calculate the approximate diameter of the earth if the electrons were removed from all the atoms and the nuclei were closely packed. (*Note:* See Problem 1 for size of nucleus. You may assume that all nuclei are about the same size. A typical atom has a radius of 10^{-7} cm.)

3. Compute the strength of the gravitational attraction between two 1-millimeter (mm) diameter balls, composed entirely of protons, separated by 1 meter. Calculate the electrostatic repulsion for these two balls. Will they combine to form a stable single particle? Why are there no proton stars?

4. Could the gravitational attractive force hold together a simple solid material? Estimate the gravitational force between sodium and chlorine in common table salt, NaCl. The space between atoms is on the order of the atomic diameter, 10^{-7} cm. What characteristic of salt do you recognize which, along with the calculations above, enables you to answer this question?

5. Compare the electrostatic force with the gravitational force for two electrons separated by an arbitrary distance r. Compare these forces for the proton. How massive must a point particle become before the two forces are equal?

6. Why are high energy accelerators needed for research into elementary particles?

7. A new super-weak force has been discovered by a group of Tibetan scientists. They have determined the range of this force to be the shortest yet, 5×10^{-30} m. What is the approximate mass of the particle propagator of this new force?

THE TOOLS OF PARTICLE PHYSICS

3.1 In the Beginning

JOSEPH JOHN THOMSON opened the door to the physics of elementary particles in 1897. He was, by the way, my professor's professor.* His particle accelerator (Fig. 3.1) hardly resembles today's modern monsters. It was used to isolate and measure the charge-to-mass ratio of the first elementary particle discovered, the electron. The apparatus consisted of a section where the electrons were separated from their host atoms (ionization of the gas in the discharge region), an electric and magnetic force region, and the particle detector, consisting of a phosphor coating on the end of the glass bulb. The electrons, stripped from the gas atoms, were first accelerated and then subjected to known magnetic and electric forces before striking the phosphor screen. By studying the deflection of this electron beam observed on the phosphor screen, Thomson was able to measure the ratio of the electron's charge to its mass.† Physicists were searching for clues to the components and structure of atoms. Thomson's experiments were an important breakthrough in unraveling this puzzle, for he succeeded in splitting the atom as well as measuring a fundamental physical property of one of the fragments. For his measurements with this instrument, Thomson received the Nobel prize in physics in 1906.

In 1911, Ernest Rutherford, a New Zealand–born experimental genius, used nature's own particle accelerator to shoot α-particles at solid matter. Natural radioactivity, well known at the time, provided the first particle physicists with a reasonably high-energy "beam" of particles. The α-particle, a helium nucleus, consists of two protons

---

* Arthur L. Hughes, an expert in experimental atomic physics, particularly the photoelectric effect, was my teacher in graduate school at Washington University, St. Louis. He received his Ph.D. from Cambridge University working as J. J. Thomson's graduate student.

† This was one of the first "TV" pictures.

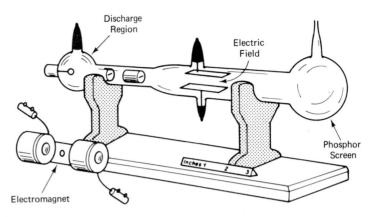

Discharge Region

Electric Field

Phosphor Screen

Electromagnet

Fig. 3.1

and two neutrons bound together by nuclear forces. By elementary particle standards, it is a rather massive projectile. In 1911, a 10 million-volt particle accelerator was well beyond the known technology, yet, remarkably, nature's own radioactivity did provide such an energetic beam. Although this particle was not able to fragment the nucleus, it could penetrate the surrounding electrons and get close to the nucleus. Rutherford used this penetrating beam to extract an important piece of information about atoms: that almost all of the mass of an atom was concentrated in an extremely small percentage of its volume. We now know this concentrated region to be the atomic nucleus. For a thorough account of the history of particle physics, including an up-to-date explanation of current high-energy physics, I suggest that you examine the beautiful book by F. Close, M. Marten, and C. Sutton, *The Particle Explosion* (New York: Oxford Univ. Press, 1987).

Today's experiments in particle physics require three types of apparatus: particle accelerators, particle detectors, and data processing systems. The purpose of the particle accelerator is to examine the small-scale structure of matter as well as to create the many "elementary" particles. The laws of quantum mechanics (which we do not examine in this book) require the use of high-energy particles in order to probe submicroscopic structures on the scale of nuclear dimensions, and the laws of special relativity (which we do examine) require high kinetic energy in order to create unstable particles that are not present in ordinary matter.* Originally, particle accelerators were built to split apart compound nuclei and examine their constituent parts. This required energy on the order of nuclear binding

---

* As the name implies, kinetic energy is the energy associated with the motion of the particle. It is usually measured in units of electron-volts. One electron-volt is the energy acquired by an electron moving through a difference of potential of 1 volt (eV). A million electron-volts, 10^6, is written MeV, and 10^9 electron-volts is written GeV. Machines capable of 10^{12} eV or 1 TeV are now operational. (See Appendix A.V for prefix definitions.)

energy, which turns out to be about 10 MeV for each particle in the nucleus. In the process of this search, physicists discovered a whole world of particles that require even higher energy to produce and are not part of the nuclear matter. They have a life of their own. The study of these new particles is called particle or high-energy physics.

In this chapter we take a brief look at the instruments used in high-energy experiments. By any measure, they are impressive and complicated. As a beginning student, I do not expect you to learn all the details of this equipment, but you can get some feeling for what all these measurements are about and some of the basic physical principles that are used. I would like this chapter to inspire you to visit Fermilab, SLAC, Brookhaven, CERN, or some other major high-energy physics laboratory and see these impressive machines for yourself. With what you learn from this course, you will impress the staff at these facilities. I guarantee it!

3.2 **Particle Accelerators**

THE LARGEST and most expensive component in a high-energy experiment is the particle accelerator. Development of these instruments has been rapid, from the first room-size, high-voltage linear accelerator, built by Cockcroft and Walton at the Cavendish Labs in England in 1931, and the first circular machine, a 30-centimeter cyclotron, built at Berkeley in California by Lawrence and Livingston in the same year, to the giant 6–7 km circumference accelerators at Fermilab in Batavia, Illinois and at CERN in Geneva, Switzerland and the 2-mile-long linear accelerator, SLAC, at Stanford University in California. In the course of these 50 years many important scientific and technological problems had to be solved to create each new generation of accelerators. This development has had important spin-offs in applied and pure science and technology.*

The earliest accelerators were either of the circular type, such as cyclotrons, or the linear type, such as the Cockcroft–Walton and Van de Graaff electrostatic accelerators. Later, more sophisticated accelerators, such as betatrons, synchrocyclotrons, synchrotrons, and alternating gradient synchrotrons, were developed. The modern superaccelerators use some of the early types of accelerators, such as the Van de Graaff accelerator, as their first stage. We cannot take the time to discuss each of these accelerators in detail, but we will briefly explore the basic physical principles of particle accelerators.

---

* For example, W. W. Hansen at Stanford University was interested in building a high-energy linear accelerator for electrons. Such a device required high-powered, high-frequency radio transmitting tubes which did not yet exist. Hanson developed such radio tubes, called klystrons, which are now the standard means of producing high-frequency power for radar as well as many other types of communications systems.

Cockcroft–Walton electrostatic generator used as a pre-accelerator at Fermilab to inject protons into main accelerator

No accelerator has yet been invented that can accelerate (add kinetic energy to) an electrically neutral particle. (A nuclear reactor can produce a beam of energetic electrically neutral neutrons, but it cannot add energy to the neutrons.) Energy is added to charged particles in *all* accelerators by an applied electric field. The force on a charged particle due to an electric field is

$$\mathbf{F}_{\text{electric}} = q\mathbf{E}, \quad \tag{3.1}$$

where $\mathbf{E}$ is the electric field and q is the charge of the particle.* It is this force that accelerates the charged particle and changes its kinetic energy.

---

* The SI units applicable to this equation are $\mathbf{F}$ in newtons, q in coulombs, and $\mathbf{E}$ in volts/meter.

Some "simple" machines, such as the Van de Graaff linear accelerator, use only this process. In these machines a very large static electric field is presented to the charged particle, which responds by accelerating down a straight evacuated tube. Because of the limitations of high-voltage breakdown, such accelerators can only produce particles with energy to about 20 to 30 MeV. Another type of linear accelerator, of which Stanford's SLAC is the largest and most important, also uses only an electric field. However, here the electric field is not static but in the form of an electromagnetic wave that propagates down a long tube, pushing a bunch of charged particles as it goes. To understand this type of accelerator, you may think of the surfer who rides a wave, gaining speed by staying on top. The charged particle and the electric field wave behave in an analogous way.

SLAC is 2 miles long, has 960 accelerating sections, can deliver 10^{12} electrons 360 times a second, and when it was completed in 1966 could accelerate electrons to an energy of 22 GeV (22×10^9 eV). With improvements in its radio-frequency power source, it now delivers 45-GeV electrons. In a sense, this enormous device is a monument to relativity, for *if* Newton's laws of motion were in fact correct, one could achieve the same *terminal speed* of an electron with a *1-inch-long* accelerator! A traveling electric field wave has to chase the electrons *2 miles* down the accelerator tube to give them a velocity of .9999 times the speed of light.

It became clear in the 1960s that the most practical way to produce the very highest energy particles is with circular cyclic machines. The first such device, the cyclotron, was designed by Ernest Lawrence and Stanley Livingston at Berkeley and used a magnetic

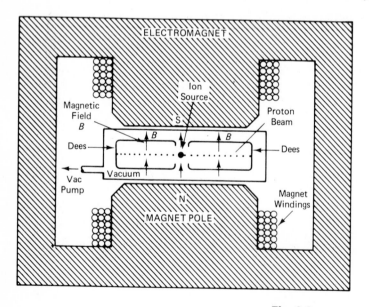

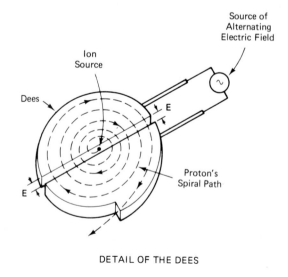

DETAIL OF THE DEES

Fig. 3.2

field to confine the charged particles to a circular orbit and an oscillating electric field to add kinetic energy periodically to the orbiting particles. Although in its original configuration (Fig. 3.2) it obtained a maximum particle energy of about 25 MeV (quite low by contemporary standards), its mode of operation illustrates many important principles that are incorporated into the new superaccelerator design.

The force produced by a magnetic field **B** on a charged particle moving with a velocity **v** is given by the mathematical expression

$$\mathbf{F}_{magnetic} = q(\mathbf{v} \times \mathbf{B}), \tag{3.2}$$

where the symbol ×, called the *cross product*, is a mathematical operation that we define in detail in Chapter 6. To simplify the discussion for the moment, let's consider the special case where the particle's velocity is perpendicular to the magnetic field (Fig. 3.3). For this important case, Eq. 3.2 reduces to

$$F_{magnetic} = qvB \quad (B \perp v). \tag{3.3}$$

How will a charged particle move when it is injected into a region of uniform magnetic field at right angles to the field? Consider the motion of a proton, shown in Fig. 3.4. To understand this motion, we arbitrarily divide the trajectory into small segments and examine the effect of the magnetic force over each segment. As the

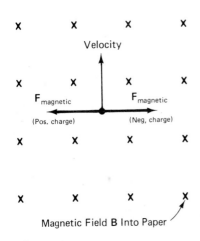

Fig. 3.3

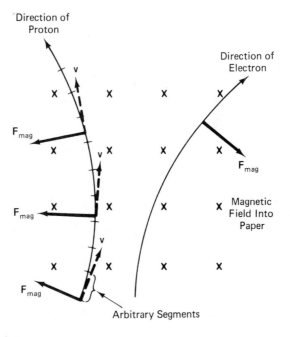

Fig. 3.4

proton enters the magnetic field the magnetic force accelerates the proton to the left. This changes the direction of the velocity (to the left), which in turn changes the direction of the force, and so on. Thus the particle appears to be moving on the surface of a many-sided polygon, always changing direction. If we divide the path into smaller and smaller segments, the path of the proton approaches a circle, which is, of course, the trajectory actually observed. This motion is analogous to twirling a ball around by a string, where the string exerts a force on the ball that keeps it moving in a circular path. The magnetic field acts like the string, providing a force, *always perpendicular to the motion*, which will make the charged particle move in a circle.

We show in Chapter 9 (but present it here without proof) that the radius of curvature r for a charged particle moving in a direction perpendicular to a uniform magnetic field is given by the equation

$$r = \frac{\text{linear momentum}}{qB}$$

or

$$p \equiv \text{linear momentum} = rqB. \tag{3.4}$$

This is an important equation to be used later. The radius of curvature is function *only* of the linear momentum of a given charged particle for the fixed magnetic field.* We will also show in Chapter 9 that the time it takes the charged particle to traverse one complete orbit of the circle is $2\pi m/qB$, which is *independent* of both the *radius* and the *speed* of the particle.

The cyclotron utilizes these properties in an elegant way. Consider Fig. 3.2 again. The particles to be accelerated circulate in a vacuum space between the poles of the electromagnet and pass through the gap between the hollowed-out conducting dees twice each complete revolution. An alternating electric field is applied in the gap between the dees. This electric field is synchronized to the motion of the charged particles in such a way that they experience an accelerating force each time they appear at the gaps. Thus the particles continue to gain kinetic energy and momentum as they spiral out to the edge of the magnet. They are ejected from the machine by a special electric field deflector placed at the outer edge of the magnet.

Protons can be accelerated to energies of about 25 MeV in these simple cyclotrons, but when the protons' velocity reaches about 20% of the speed of light, the synchronization begins to break down because of relativistic effects. The particles begin to act *as if* their mass

---

* Linear momentum has not yet been defined. For now, just consider momentum to be a dynamical property associated with a particle's motion.

were increasing, and take a longer time to complete an orbit. When this happens the protons no longer appear at the gaps at the correct moment to experience the accelerating electric field, and their energy cannot be increased.

This problem was circumvented ingeniously by Ed McMillan, who suggested that one could change the oscillation frequency of the electric field (slowing it down) as the particles increased their energy. The accelerator could no longer produce a continuous beam of particles; it had to be a pulsed beam, with the changing frequency synchronized to the motion of the particle bunch as they circled. These synchrocyclotrons were constructed and produced protons with about 800 MeV of kinetic energy. But now the size of the electromagnets limited the possibility of reaching higher energies.

The idea of spiraling orbits had to be abandoned in favor of fixed orbits with varying magnetic field strengths to achieve higher-energy particles. This type of machine is called a synchrotron. In these devices the frequency of the electric field is synchronized with the rising magnetic field so that the particle's radius remains constant. This must be done with high precision. It is accomplished by an automatic feedback system which senses the motion of the circulating bunch of charged particles and adjusts the oscillating electric field to keep the radius constant.

All modern superaccelerators are of the synchrotron design shown in Fig. 3.5 on page 32. They consist of two pole magnets, which bend the particles in a circular orbit, and quadrupole magnets, which are used to focus the particles into a very small pencil-like beam. The sections marked E are high-frequency resonant cavities that produce the accelerating electric field on the bunch of particles as they pass these sections.

When the particle speed is near the speed of light c, the maximum energy E_{max}, the magnetic field B, the particle's charge e, and the machine radius r are related by the simple relationship $E_{max} = ecrB$. To increase the maximum energy, two choices are available: increase B or increase r. Increasing r means purchasing more land, building more magnets, constructing a larger tunnel, building more cavity sections, and so on, all of which cost a great deal of money. The magnetic field strength can only be increased to about 2 tesla (20,000 gauss) with iron-core magnets, since the iron magnetically saturates at this field strength. Significantly higher magnetic fields require low-temperature superconducting magnets. This means special materials, low-temperature apparatus, liquid-helium refrigerant, and many more complicated cryogenic engineering difficulties. Such a superconducting magnetic system came on-line in 1985 at the Fermilab. This brought the proton beam up to the colossal energy of 1000 GeV (1 TeV), the tevetron. To operate this machine, a helium liquefier had to be built which *doubled* the world's production of liquid helium!

Fig. 3.5

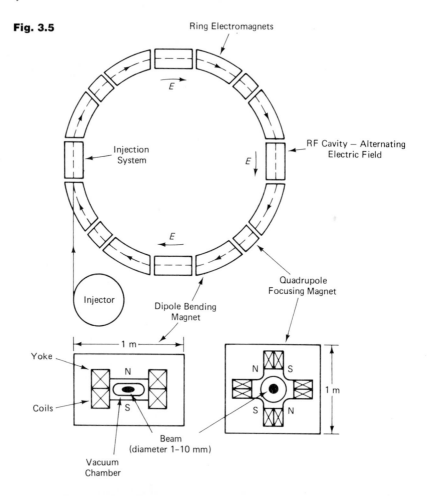

A section of the CERN super proton synchrotron

The SPS (super proton synchrotron) is currently the largest machine at the Center for European Nuclear Research (CERN), with a 7-km circumference and a capability of producing 450-GeV protons. It is divided up into periodic sequences called the "machine lattice." Each lattice consists of four bending magnets, a focusing quadrupole magnet, another four bending magnets, and another quadrupole. There are 108 of these lattice sections around the circumference. In all, there are 216 quadrupoles and 744 bending magnets. They are located in a tunnel bored 50 m underground in bedrock in the mountains on the border of France and Switzerland. The tunnel had to be bored to a tolerance of a few centimeters in a perfect circle. New engineering techniques were employed using gyroscopic guidance and laser beams to guide the underground boring equipment, with the earth's rotational axis as a reference. The machine is so large that allowances had to be made for the curvature of the earth in leveling the bending magnets. During their accelera-

What if we spend all these billions, and there just aren't any more particles to find?

tion to full energy, the protons travel more than 500,000 km (the moon is only 380,000 km away) in about 2 s, yet they do not deviate from their central orbit by more than a few millimeters. The SPS machine works at a frequency of 200 megahertz, rotating 4600 proton bunches around once every 23 microseconds. The total cost of this machine was about $\frac{1}{2}$ billion American dollars!

But Americans intend to upstage the Europeans, indeed the world. Twenty-five miles south of Dallas, Texas, a 53-mile-circumference, 20-trillion-electron-volt (20 TeV) Superconducting Supercollider accelerator is being constructed. It is estimated to cost over $5 billion, to employ 2500 scientists and technical staff as well as 500 visiting scientists, and to consume a $270 million annual budget. It will use 10,000 state-of-the-art superconducting magnets costing over $100,000 each. Not all physicists agree that this accelerator is a wise investment for the United States. P. W. Anderson (Nobel prize winner in physics) worries aloud that "science in the United States is dying of giantism."

3.3 Particle Detectors

IN THE SPAN of 50 years, physicists have increased the energy of particle accelerators by a factor of 10^7, but there have also been remarkable advances in the instruments used to detect and measure the physical properties of the new particles created in these high-energy collisions. Although the acclerators' advances are more dramatic, there have been no less spectacular devices designed and constructed for particle detectors. However, with all these advances, there is still no way to detect the presence of an electrically neutral particle without destroying or deflecting it.* All particle detectors use electromagnetic interaction in one form or another to signal the presence of a particle.

When charged particles pass through matter they exert an electrostatic force on the electrons bound to the atoms in the region. Undisturbed, the electrostatic force between the negatively charged electron and the positively charged nucleus is sufficient to keep the electrons bound in an atom. Occasionally (approximately one time in a thousand), the "intruder's" force is sufficient to break the electron's bond, freeing it and leaving behind a positively charged atom called an ion. The path of ionized atoms and free electrons is called the "ionization trail." Every particle detector utilizes the small amount of energy and charge in the ionization trail to locate, and in some cases identify, the charged particle that passed through.

---

* That is not to imply that there are no neutron detectors, but that all such detectors use nuclear interactions to convert the neutron plus nucleus to some electrically charged particle. High-energy photons (gamma rays) can be observed directly in some electrical detectors.

Particle detectors can be divided into two broad categories, COUNTERS and TRACK LOCATORS. The counters are electronic devices that record the presence of a charged particle and give information about the time the particle passed and, for some detectors, the kinetic energy of the particle. The track locators give spatial information and, in some cases, the time of passage of the charged particles. There is no single "best" detector for all high-energy experiments. Each detector has its advantages and its drawbacks. The experimenter must design the proper detector or the proper combination of detectors that is most appropriate for a given experiment. We will consider some of the most important types of detectors and discuss the basic physical principles on which they operate. Let's begin by examining various track locators.

A. Cloud Chamber

In 1911, C. T. R. Wilson developed the first track locator, called a cloud chamber, shown in Fig. 3.6. A supersaturated vapor, usually alcohol and water, will condense into droplets around free ions in the vapor. After a charged particle has passed through the chamber, the mechanical drive, triggered by external counters, expands the vapor, lowering its temperature below the dew point, and the droplets condense around the ions along the ionization trail.* In a cloud chamber the ions persist for about a second, giving sufficient time for

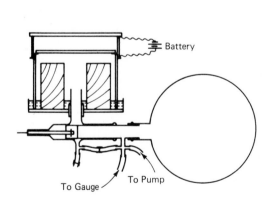

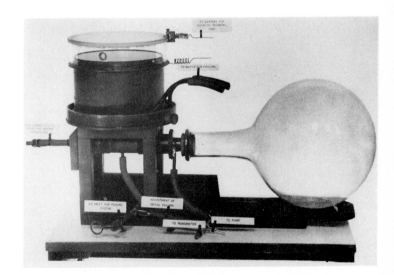

Fig. 3.6

---

* The counters are arranged in a "sandwich" geometry such as those shown for the spark chamber in Fig. 3.7.

the relatively slow mechanical system to respond to the electronic trigger. The experimenter reconstructs the path of the ionizing particle with stereoscopic photographs of the droplets in the chamber.

A major advantage of the cloud chamber is that it can remain quiescent for hours, waiting for an event to occur and then be triggered to record that event. This feature is particularly useful in cosmic ray research, where high-energy particles rarely "invade" the chamber.* Only when they pass through the counters and the chamber expands will the cameras record the event.

There are serious disadvantages to these detectors. They are inherently slow devices, taking about a second to trigger. Also, the ions from one event must be swept out to make the chamber ready for the next event. This severely limits the rate at which events can be recorded. This is not a handicap for cosmic ray experiments where events are rare, but it becomes a serious limitation for experiments on particle accelerators which produce high-energy particles in rapid repetitive bursts. A second problem is the limited materials that can be used in the chamber to serve both as condensing vapor *and* as targets for the high-energy particles. Since the active medium is low-density gas, the probability that a collision will take place with the nuclei in the chamber is small. The experimenter usually must observe the event at its vertex (the point of collision), since some particles created in the collision live only a short time. One can study the vertex of the event only if it occurs in the medium and not in a dense material above the chamber. Thus cloud chamber photographs rarely record an event that includes the point of collision.

Simple nuclear targets, such as hydrogen (which looks like a collection of protons to a high-energy particle) or deuterium (one proton and one neutron) do not form gases that work in the cloud chamber. Single nucleon targets are often necessary for some of the most important fundamental experiments so that physicists can study and unambiguously analyze the most basic interactions, such as a proton–proton collision. A proton collision with a compound nucleus, such as carbon, for example, is really a collision with 12 nucleons, 6 protons and 6 neutrons. That's a mess to analyze!

B. Bubble Chamber

In 1960, Donald Glaser received the Nobel prize in physics for his invention of the bubble chamber 10 years earlier. This was the most

---

* The first particle physics experiments were done on cosmic rays, the high-energy particles provided by nature from outer space. Combinations of electric and magnetic fields in our galaxy are apparently responsible for the production of the highest-energy particle we have yet observed, having energies of 2×10^{21} eV [see K. Suga et al., *Physical Review Letters*, **27**, 1604 (1971)]. Cosmic rays are still the source of the very highest energy particles, but the mechanism of their production remains a mystery.

important single tool in the high-energy physicist's arsenal for 20 years between 1955 and 1975. The principle of operation of a bubble chamber is as follows: a liquid is kept at a temperature and pressure very near (but below) its boiling point. The liquid is subjected to a sudden reduction in pressure and becomes superheated. That is, if it is allowed to stay at these conditions for a long time, the liquid will eventually boil. Glaser discovered that when a high-energy charged particle passes through the chamber in the superheated condition, the liquid will start to form its *first* bubbles of boiling along the path traversed by the charged particle. The growth of bubbles around the ionization trail is caused by local heating of the liquid. The ionized electrons transfer kinetic energy to the nearby liquid. The photographed bubbles are an image of a localized heat track which faithfully represents the paths the high-energy charged particles traversed in the liquid.

Liquid hydrogen, the ideal target material for many experiments (presenting one proton at a time to the incoming beam) can be used in these chambers. But experimenters are not limited to hydrogen; deuterium, helium, argon, or neon and a combination of liquids can also serve as both active media and targets. In bubble chambers the active media are liquids, which present a much higher density of targets to the incoming beam than a gas in a cloud chamber. This increases the probability of an interesting event occurring within the chamber and the observation of the vertex of the event. Bubble chambers are usually constructed inside a large electromagnet so that a uniform magnetic field exists over the entire active region of the chamber. This allows the unique determination of the linear momentum of all the charged particles that leave tracks. Bubble chamber tracks have the highest spatial definition of any detector, with the exception of tracks in photographic emulsions, permitting the most accurate measurement of scattering angles and momentum.

Bubble chambers do have their limitations. First, there is the practical difficulty of cost, size, and potential danger. The largest chamber built so far is about 4 meters in diameter and with its magnet occupies a 10-story building. Even this colossal chamber is too small to follow all the scattered particles in a very high-energy collision. Bubble chambers cannot be triggered after the particle has passed through because the heat deposited is dissipated in about 10^{-6} s, far too fast for a mechanical system to respond and expand the chamber. However, if you know in advance when the projectiles will arrive, as you do with a particle accelerator, the chamber can be expanded just *in advance* of the event. It is truly a remarkable scientific and engineering achievement to design and construct a 12-ft-diameter chamber, fill it with liquid hydrogen (at a temperature of $-249°C$), cycle it once a second, photograph 95% of its interior active regions, and have a uniform magnetic field of 2 tesla (20,000 gauss) over its active volume.

The large bubble chamber at CERN

C. Spark Chamber

Although the bubble chamber was the most versatile track detector used in high-energy physics, it had its limitations. The major drawback is size, which has become more important with super-high-energy collisions. These events simply cannot be recorded in a space of a few meters of liquid hydrogen or even liquid neon. To analyze them, such events must be tracked over tens and sometimes hundreds of meters. One device that has essentially no size limitation is the spark chamber, which was developed over a period of several years by many experimenters. The generation of a spark is an extremely complicated process. Fundamentally, it is chain reactions in a gas, started by the acceleration of the electrons in an electric field released in the initial ionization track. These accelerated electrons in turn ionize other gas atoms and cause an avalanche. The avalanche leads to an electric spark which is located at or near the path of the original ionizing high-energy charged particle. The spark chamber shown in Fig. 3.7 was the forerunner of the modern chambers. It consists of an array of thin, polished metal plates surrounded by an inert gas such as neon. The chamber is triggered by a logic circuit which turns on the high voltage to the plates. The high voltage is applied only when the logic circuits have determined that an interesting event has taken place in the chamber. The circuits do this by sensing the particles that pass through their detectors. Spark breakdown is then recorded on cameras for later analysis.

There is no practical limit to the size of these chambers (or rather, a series of chambers). They can also be stacked in any appropriate way for a given experiment. Cylindrical chambers have been built for the study of particles colliding along on the axis of the cylinder. The spatial resolution of the spark chamber cannot com-

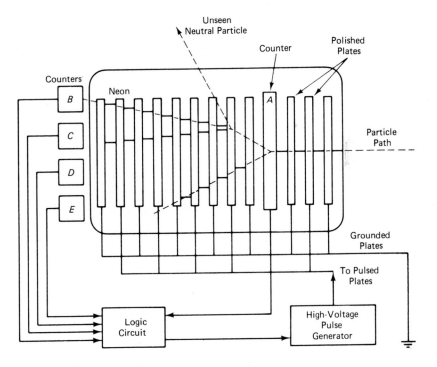

Fig. 3.7

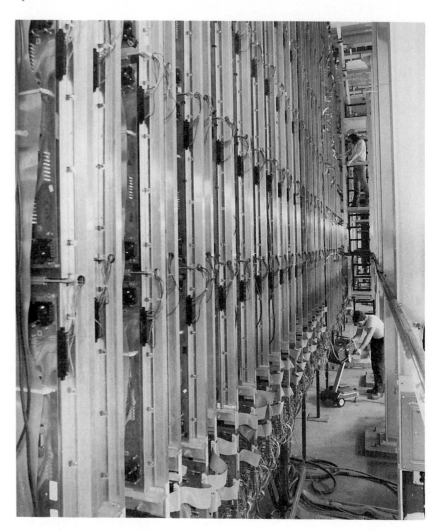

Large spark chamber used for neutrino experiments at Brookhaven National Laboratory

pete with the bubble chamber. It is also not practical to provide uniform magnetic fields for large chambers, so that the measurement of linear momentum must be estimated from other information, such as the penetration of the particle. This is not as precise a determination of momentum as the curvature in a magnetic field.

D. Drift Chamber

Several special detectors have now been developed. An important example of a modern detecting system combines many of the advantages of the bubble chamber and the spark chamber. It is called the multiwire drift chamber, one cell of which is shown in Fig. 3.8. As the charged particle passes through the chamber, it knocks electrons off the gas atoms. The electrons are attracted to the *nearest* anode

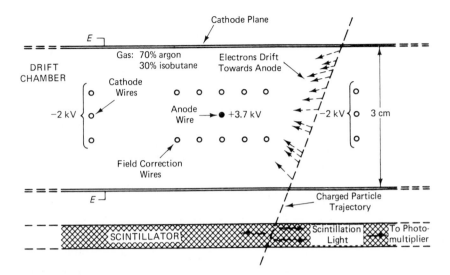

Fig. 3.8

wire that is positively charged. The electron drifts toward the wire (with a known speed of about 1 mm (millimeter) in 10^{-9} s) and creates a pulsed electrical signal on that wire. The shape and duration of this pulsed signal, along with the time of arrival of the particle in the chamber, can be used to determine the distance the original trajectory was from the wire. The time of arrival of the particle is obtained from the pulse of the scintillator detector below the cell. One such multiwire cell can only measure one coordinate of the beam; another cell, perpendicular to the first, can determine a second coordinate with high precision. Multiwire detectors are vast arrays of many such orthogonally stacked cells that are capable of measurements of position accurate to .1 mm! These multiwire detectors are connected to high-speed dedicated computers that analyze the complex electrical signals coming from the wires and scintillator detectors.

One advantage of these electronic detectors over bubble and spark chambers is that a data processing computer can select the events worth permanently recording. In bubble chamber experiments, literally tens of thousands of photographs must be scanned to find a few events of interest. The computer can reduce the amount of data stored by analyzing, *in real time*, each event fed into it from the drift detectors. The inherent disadvantage, however, is that the data reduction program may have subtle biases built in and may discard interesting and important events. The bubble and spark chamber photographs are always there to be reanalyzed for bias or new and unexpected phenomena. Often, a bubble chamber is placed in a region that is expected to be close to the vertex of the collision and a spark or drift chamber is used to measure the outgoing trajectories when they leave the bubble chamber.

E. Electronic Counters

A comprehensive discussion of just electronic particle detectors would consume several thick volumes. I will talk only briefly about these important tools for particle, nuclear, laser, and solid-state physics. These detectors serve two basic purposes, one for timing and the other for energy determination. Not all electronic detectors are capable of both functions. The earliest detector, the Geiger–Müller counter (Fig. 3.9), was invented in 1928. In these detectors the ionization of gas by the penetrating charged particle causes an electrical breakdown (such as the spark chamber) which produces an electrical pulse. The pulses are used to record the passage of a charged particle and the time of the event. A proportional counter was later developed with a similar configuration, but with the important additional ability to record the energy deposited by the penetrating particle in the counter. The pulse amplitude is proportional to the energy deposited in these counters.

The most widely used electronic detector in high-energy experiments is the scintillation counter. Here the ionization track occurs in a special crystal or plastic material where the ionization energy is converted to visible light. This light is monitored by a photomultiplier tube which converts the light to electrical pulses. The pulses contain information about both timing and energy of the original ionizing particle. Scintillation detectors are used widely in many parts of the high-energy experimental apparatus.

What kind of detectors are used in the the newest, highest-energy accelerators such as Fermilab, CERN, and SLAC? For reasons we discuss in Chapter 6, the highest-energy collision experiments are done with colliding beams. That is, instead of one projectile striking a stationary target, two particle beams are made to collide head-on. The detector must be designed to surround the point of intersection of these two beams of particle. The new detectors are hybrid collections of various electronic detectors and gigantic magnets, sometimes called an electronic bubble chamber. Take, for example, the UA1 detector at CERN, shown in the photo. It contains 2000 tons of iron, 100 tons of lead, 6300 m² of plastic scintillator,

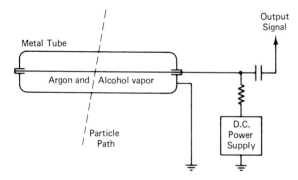

Fig. 3.9

The UA1 detector at CERN

7000 optical fibers, and cost over 20 million British pounds. It took over four years to build by a team of 52 scientists from all over the globe. By the time it had produced its first results in 1981 the team had expanded to 135 physicists. Headed by Carlo Rubbia, this team discovered the Z-particle, the propagator of the electroweak interactions, and collected a Nobel prize.

THE ONE PIECE of essential equipment that does not have to be "homemade" is the computer. The rapid progress in particle physics could not have been made without equivalent advances in high-speed computers. The amount of data collected, analyzed, and recorded is as mind-boggling as some of these giant machines and detectors. Literally millions of bubble chamber photographs and hundreds of billions of digital signals from various electronic detectors are created for a single experiment. The analysis of signals from one drift detector simply could not be done without an ultrafast computer. Computers are used in every phase of the experiment, from controlling and adjusting the accelerator, timing the detector, selecting significant events, and analyzing bubble chamber photographs, to projecting budgets for future grant proposals and sending out the paychecks. Many of the most sophisticated software programs ever written were developed by high-energy physicists to use in their experiments.

All this equipment—the accelerator, the detector, and the data analysis systems—must be kept operational and optimized so that no bottleneck prevents the entire experiment from functioning. If you look at a published paper in high-energy physics, you will often see 20 or more authors. Now you know why. It's big physics and getting bigger. Nature is such, as quantum mechanics tells us, that the finer the detail that we wish to study, the higher the energy of the probe. But the higher the energy of the probe, the larger the apparatus. To look small we have to build big!

CERN computers

3.5 Summary of Important Equations

$$\mathbf{F}_{\text{electric}} = q\mathbf{E}$$

$$\mathbf{F}_{\text{magnetic}} = q(\mathbf{v} \times \mathbf{B})$$

$$= qvB \qquad \text{for } v \perp B$$

$$r = \frac{\text{momentum}}{qB}$$

3.6 References

1. *The Particle Explosion*, Frank Close, Michael Marten, and Christine Sutton, Oxford Univ. Press, New York (1987).
2. Two comprehensive articles on SLAC can be found in *Physics Today*, April 1967, by R. B. Neal and J. Ballam. An informative article on the facility also appeared in *Scientific Research*, November 1966.
3. A good review of early particle accelerators can be found in *Fields and Particles: An Introduction to Electromagnetic Wave Phenomena and Quantum Physics*, by Francis Bitter and Heinrich Medicus, American Elsevier, New York (1973).
4. The celebration of the Batavia National Accelerator is recorded in *Physics Today*, May 1972, while the CERN 400-GeV accelerator is discussed in the December 1972 issue of the same journal.
5. *Particle Accelerators: A Brief History*, M. S. Livingston, Harvard Univ. Press, Cambridge, Mass. (1969).
6. "The Spark Chamber," G. K. O'Neal, *Scientific American*, August 1962.
7. *The Nature of Matter*, J. H. Mulvey, Ed., Clarendon Press, Oxford (1981).

Chapter 3   QUESTIONS

1. Several authors have noted a remarkable relationship between the maximum particle energy obtained in circular accelerators and the year this accelerator came "on line." The data are shown in the accompanying table. If this continues into the future, describe the characteristics of the accelerator that would be built in the year 2000. What would the diameter, energy, and cost be for the accelerator of the year 2000? Do you believe that it will be built? Explain. (*Hint:* Plot your data on semilog paper.) How does the Superconducting Supercollider fit this prediction?

Accelerator	Year of Completion	Diameter	Kinetic Energy	Approximate Cost
First cyclotron (Lawrence/Livingston)	1932	1 ft	1.2 MeV	?
Cyclotron (Columbia Univ.)	1938	3 ft	14 MeV	?
Synchocyclotron (Berkeley)	1946	15.3 ft	350 MeV	$ 3 million
Bevaton (Berkeley)	1954	100 ft	6 GeV	6 million
Brookhaven AGS	1960	840 ft	33 GeV	30 million
National Accelerator (Fermilab)	1972	1.24 mi	500 GeV	500 million

2. Why can't we directly detect neutral particles?

3. In the early days of particle physics, cosmic rays (coming into the earth from space) were used as a source of high-energy particles. The standard techniques used to detect them involved cloud chambers and associated Geiger counters. Can you explain the purpose of these counters?

4. How would you distinguish between the tracks of a positively and negatively charged particle in photographs of magnetic bubble chamber?

5. Why are modern bubble chambers and spark chambers increasing in size?

6. Liquid hydrogen is extremely dangerous and difficult to handle. Why isn't alcohol or some simple organic solvent used in bubble chambers?

7. How could the loss of particle energy along the particle's path in the bubble chamber (due to ionization of the hydrogen atoms in the liquid) be experimentally detected? Explain.

8. Why don't you find bubble chambers used as detectors on colliding beam experiments?

4

THE "ELEMENTARY" PARTICLES

4.1 History

AT ONE MOMENT in history, it must have seemed to physicists that they had discovered all the particles from which matter is built. J. J. Thomson isolated and identified the electron in 1897; in 1913, Rutherford analyzed Geiger and Marsden's α-particle scattering data and showed that positively charged matter was densely packed in a small nuclear volume. In that same year Niels Bohr set out his model of the atom, which pictured the electrons orbiting about a massive positively charged nucleus. In the period 1920–1930 the main elements of quantum mechanics were developed by many physicists, including Schrödinger, Heisenberg, and Dirac. This was one of the most triumphant periods in the history of science; the theory of the atom stood essentially complete, many of the known properties of ordinary matter could be deduced, and the major problems that chemistry had grappled with in the past century were solved in principle. It seemed just a matter of working out the details using the known particles and forces with the new quantum theory.

But in the early 1930s, just when things looked so good, physicists began to seriously probe the interior of the atomic nucleus. That is when their "troubles" began. They discovered that the nucleus is made up of both positively charged protons and particles that have no electrical charge, neutrons. Chadwick's discovery of the neutron in 1932 made it possible to construct models of atomic nuclei from both protons and neutrons. Hydrogen's nucleus was just a single proton; deuterium, a proton and a neutron; helium, two protons and two neutrons—all the way up to uranium, with 92 protons and 146 neutrons. The extremely short-range forces that held all these like-charged particles so closely together and that bound a charged particle to a neutral particle, however, were not understood at all. By 1933, the "mirror image" of the electron, the positron, had left its track in the cloud chamber of a 27-year-old Cal Tech physicist, Carl D. Anderson. But nature still appeared to be reasonably

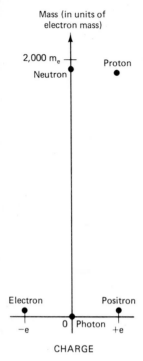

Fig. 4.1

uncomplicated with a short list of what were perceived as "elementary" particles. Figure 4.1 is an "organizational" chart of the particles known in 1933. Mass is plotted on the vertical axis in units of the electron's mass and charge in units of the electron's charge on the horizontal axis. The neutrino was still only a controversial theoretical conjecture. Counting the photon, the propagator of the electromagnetic field, only six "elementary" building blocks were available to construct all matter and explain every physical phenomena. Maybe we should have stopped here, because very soon the roof caved in.

Just 30 years later, in the early 1960s, the list of elementary particles had grown to about 30, and physicists were no longer confident that the name "elementary" made any sense. Indeed, in 1964, certain regularities began to show up which led Murray Gell-Mann and George Zweig independently to propose that a majority of these particles were composites of only three simpler particles. Today there is little doubt that the more than 100 known particles are *not* elementary or fundamental; however, their study has revealed many unexpected and exciting properties of nature.

I am getting way ahead of the story. I want to take you back to the 1950–1965 era, when new particles were being discovered at a feverish pace and new theoretical models were attempting to catalog, classify, and order them.

High-energy physicists discovered that these "elementary" particles seemed to fall into three categories: HADRONS, LEPTONS, and FIELD PROPAGATORS. This classification scheme is still used today, but the list of membership has grown considerably from the 30 or so particles known in the mid-1960s. The bulk of particles fall into the hadron category. They are almost all extremely short-lived, unstable particles, which are created in high-energy collisions and which decay into other particles in times ranging from 10^{-9} to 10^{-23} s. Yet stability alone cannot be used to classify an elementary particle. Other subtle and unfamiliar properties turn out to be more significant.

I invite you to come with me on a walk through the elementary particle zoo, where we will examine the various species of particles in three areas. A detailed "program" for our journey is provided for you in Appendix B. You should examine it now so that you are familiar with all the information it provides. You will use it extensively, particularly in Chapters 5 to 7.

Classification has always played an important role in scientific research, often providing the essential clues that lead to the development of a comprehensive theory. Mendeleeff's periodic table of elements, Rydberg's equation predicting the spectral lines of atomic hydrogen, and Darwin's examination of living species are a few prominent examples of the power of organization of raw data. It was in this spirit that physicists in the 1950s and 1960s attempted to discern patterns out of the myriad of particles being discovered. Let

the particle's *mass* and its *speed*. If the speed of the particle is zero, the kinetic energy is also zero, as one can easily see by examination of Eq. 5.1. Since the velocity of the particle appears in Eq. 5.1 as v^2, and since the velocity of a particle can never exceed the velocity of light in a vacuum ($v < c$), the kinetic energy of any particle must always be a *positive* quantity.

EXAMPLE 1   Find the kinetic energy of a proton that is moving at two-thirds the speed of light.

In the SI system of units, the mass of the proton is 1.67×10^{-27} kg, so

$$E_{KE} = \left[\frac{1}{\left(1 - \left(\frac{2}{3}\right)^2 \frac{c^2}{c^2} \right)^{1/2}} - 1 \right] (1.67 \times 10^{-27} \text{ kg})(3.0 \times 10^8 \text{ m/s})^2$$

$$= \left[\left(\frac{9}{5}\right)^{1/2} - 1 \right] (1.503 \times 10^{-10}) \text{ kg-m}^2/\text{s}^2 = 5.13 \times 10^{-11} \text{ J}$$

In high-energy physics, energy is more commonly measured in units of ELECTRON-VOLTS. One electron-volt is the energy acquired by an electron when it moves through a potential difference of 1 V. That definition is not of much help to those of you who have not studied electricity, since it does not explain the meaning of a volt or a potential difference. Do not let that concern you now, for the conversion factor from electron-volts to the SI unit of energy, joules, should be easy to remember (or look up). It is

$$\boxed{1 \text{ electron-volt} = 1.60 \times 10^{-19} \text{ joule}} \tag{5.2}$$

or its reciprocal,

$$\boxed{1 \text{ joule} = 6.25 \times 10^{18} \text{ electron-volts.}} \tag{5.3}$$

EXAMPLE 2   Calculate, in units of electron volts, the kinetic energy of the proton in Example 1.

$$E_{KE} = (5.13 \times 10^{-11} \text{ J})(6.25 \times 10^{18} \text{ eV/J})$$

$$= 3.21 \times 10^8 \text{ eV.}$$

High-energy physicists usually speak of millions of electron volts (MeV), sometimes billions of electron volts (GeV),* so that we

---

* Appendix A.V lists the commonly used prefixes.

may express our result for the kinetic energy of the proton as

$$E_{\text{KE}} = 321 \times 10^6 \text{ eV} = 321 \text{ MeV} = .321 \text{ GeV}.$$

One can see from Eqs. 5.2 and 5.3 that the electron volt is a small quantity of energy compared to the joule, but it is a more appropriate unit to use for high-energy particle physics.

The algebraic expression $\dfrac{1}{\left(1 - \dfrac{v^2}{c^2}\right)^{1/2}}$ will appear in many of the most important equations of particle physics. Because it is used so frequently, it is convenient to define a special symbol for this quantity. In most of the literature you will find the lowercase Greek letter gamma (γ), defined as

$$\gamma \equiv \frac{1}{\left(1 - \dfrac{v^2}{c^2}\right)^{1/2}}. \qquad (5.4)$$

Several comments should be made about γ. First, γ is always a positive quantity, since $c > v$. The smallest value of γ is 1, but γ has no upper bound. Gamma is dimensionless, since the particle's velocity v appears in the expression divided by the velocity of light in a vacuum. Finally, notice that gamma is *only* a function of the particle's *velocity*. Plotting γ versus v for velocities up to the speed of light is a useful exercise for the student.

The expression for the kinetic energy of a single particle Eq. 5.1 can now be written in an abbreviated form, using Eq. 5.4, as

$$E_{\text{KE}} = (\gamma - 1)mc^2. \qquad (5.5)$$

We have added nothing new to the discussion by making the algebraic definition of (gamma) γ, but as will soon be apparent, this shorthand way of writing kinetic energy will simplify some of our calculations.

B. Rest Energy

The second form of energy we must consider is the energy associated *only* with the mass of the particle, called the REST ENERGY or MASS ENERGY. Einstein first proposed the existence of such mass energy. He showed that mass energy was a direct consequence of his special theory of relativity. In the first papers on special relativity you will find the derivations of what surely must be the best known equation

PHOTO A. Carl Friedberg shown working on the SPEAR Mark I magnetic detector at the Stanford Linear Accelerator (SLAC). This complex detector for colliding-beams experiments is connected to processing and recording instrumentation by thousands of signal cables. Power and water cooling is also supplied to the large magnets and other instruments.

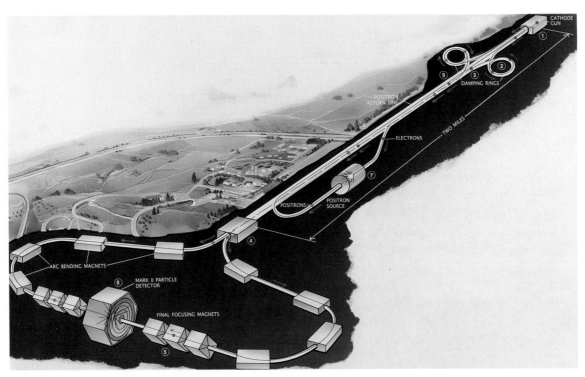

PHOTO B (*above*). The Stanford Linear Accelerator and Linear Collider addition viewed from the air. The original two-mile-long accelerator has been modified to create the position and electron colliding-beam apparatus shown below.

PHOTO C (*right*). The 80-inch liquid-hydrogen bubble chamber at Brookhaven National Laboratory. In 1964 it was the world's largest bubble chamber. This instrument recorded the first Ω^- events that confirmed the early theories of SU–3 symmetry.

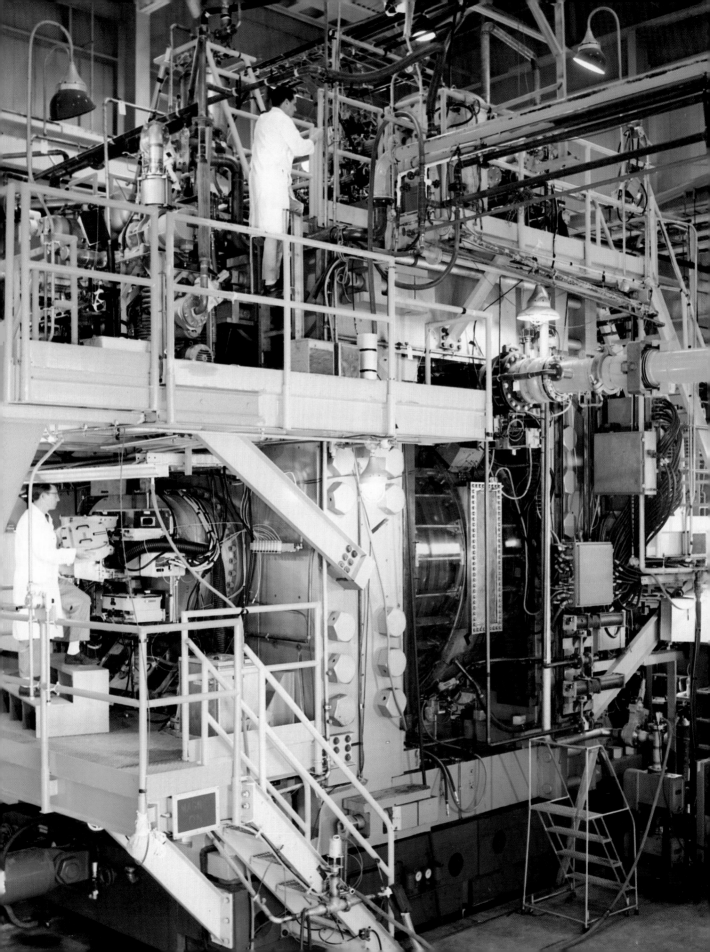

PHOTO D. The collider detector, commissioned in October, 1985 at Fermilab. The black arches removed to the sides contain modules of the central calorimeter which were built by a collaboration of scientists from the U.S., Japan, and Italy. The calorimeters contain photomultipliers and electronics used to measure particle energies from the collision. The calorimeters surround the region in the detector where the collisions of protons and antiprotons take place.

in all science:

$$E_{\text{Rest}} \equiv E_R = mc^2. \tag{5.6}$$

This is a truly remarkable equation. It relates the *intrinsic* energy of a particle to its mass. Einstein regarded the discovery of the connection between energy and mass to be of fundamental importance. In his own words:

> The most important result of a general character to which the special theory (of relativity) has led is concerned with the concept of mass. Before the advent of relativity, physics recognized two conservation laws of fundamental importance, namely, the law of the conservation of energy and the law of the conservation of mass; these two fundamental laws appeared to be quite independent of each other. By means of the theory of relativity, they have been united into one law.*

Take special note in this equation of the magnitude of the constant c^2 on the right side. The velocity of light squared is an enormous number: about 10^{17} in SI units. This means that a large—a very large—amount of energy is associated with even a tiny amount of mass!

EXAMPLE 1   Consider two small ball bearings about 2 mm in diameter, each with a mass of 10^{-3} kg (1 g). Imagine one made up of ordinary matter (protons, neutrons, and electrons), and the other made up of antimatter, (antiprotons, antineutrons, and positrons). If the two particles make contact (moving at arbitrarily low speed), suddenly all the mass disappears and the mass energy of the two balls is transformed into radiant energy of the γ-rays that are created. How much energy is carried away by the γ-rays?

$$E_R(1) + E_R(2) = m_1 c^2 + m_2 c^2 = (m_1 + m_2)c^2$$

$$= 2 \times 10^{-3} \text{ kg} \times 9 \times 10^{16} \ m^2/c^2$$

$$= 1.8 \times 10^{14} \text{ J}$$

$$= 1.8 \times 10^{14} \text{ J} \times 6.25 \times 10^{18} \text{ eV/J}$$

$$= 1.1 \times 10^{33} \text{ eV}.$$

Such a large quantity can be appreciated by considering the energy liberated in the explosions of 1 kilogram of TNT (quite a sufficient amount to destroy a large house).

$$1 \text{ kilogram of TNT} \approx 4 \times 10^6 \text{ joules.} \tag{5.7}$$

---

* A. Einstein, *Relativity* (New York: Crown, 1961).

The γ-ray energy that would be liberated from just 2 g of matter interacting in this manner is equivalent to 45,000,000 kg of TNT explosives, or about four fission bombs—all this energy stored in *two 1-g ball bearings.* Quite a compact storage unit!

Einstein's relationship permits us to measure mass in units other than kilograms or grams; it allows us to measure mass in energy units. In fact, energy units are the "natural" units to use for mass when one is studying high-energy decays and collisions. How do we measure mass in energy units? Consider, for example, a proton whose mass in SI units is 1.6×10^{-27} kg. From Eq. 5.6 one can see that the quantity

$$m_p c^2 = (1.6 \times 10^{-27} \text{ kg})(9 \times 10^{16} \text{ m}^2/\text{s}^2) = 1.5 \times 10^{-10} \text{ J}$$

$$= (1.5 \times 10^{-10} \text{ J})(6.25 \; 10^{18} \text{ eV/J}) = 9.39 \times 10^8 \text{ eV}$$

$$= 939 \text{ MeV}.$$

Multiplying the mass times the velocity of light squared gives the mass energy of the particle. High-energy physicists speak of the "mass of the proton as 939 MeV." A similar straightforward calculation would show that the "mass of the electron is .511 MeV." What is meant by this jargon, of course, is that the mass energy of the proton is 939 MeV and the mass energy of the electron is .511 MeV, but this is rarely stated. In fact, you will find in Appendix B all the "masses" of the particles listed in the table given in units of MeV. Actually, we have listed the mass energies, mc^2. As you become more familiar with problems in particle physics, it will become obvious why energy units are used for particle mass.

C. Total Energy

Since we are considering only particles in a force-free region of space (or where the forces are sufficiently weak so that we may ignore them), the total energy of a system of particles is simply the sum of the kinetic and rest energies:

$$E_{\text{Total}} = E_{\text{Kinetic}} + E_{\text{Rest}}. \tag{5.8}$$

Combining Eq. 5.6 with Eq. 5.4 gives us

$$E_{\text{Total}} = (\gamma - 1)mc^2 + mc^2$$

$$E_{\text{Total}} \equiv E_T = \gamma mc^2. \tag{5.9}$$

Equation 5.9 is remarkably simple. It states that the total energy of a single particle in a force-free region is equal to mass times the velocity of light (in a vacuum) squared times a dimensionless quantity γ, which is a function of velocity only. The total energy of a particle, like the rest and kinetic energy, is a positive scalar quantity.

The total energy for a *system* of n particles, where n is an integer greater than 1, is given by the algebraic sum of the total energies of the individual particles, or

$$E = \gamma_1 m_1 c^2 + \gamma_2 m_2 c^2 + \gamma_3 m_3 c^2 + \cdots + \gamma_n m_n c^2. \quad (5.10a)$$

Equation 5.10a can be written in a shorthand notation as

$$E = \sum_{i=1}^{n} \gamma_i m_i c^2. \quad (5.10b)$$

Obviously, the total energy of a system of particles is also a positive scalar quantity. If you calculate a negative total, rest, or kinetic energy either for an individual particle or for the entire system of particles, you know that you have made a mistake!

EXAMPLE 1 Find the total energy of an electron whose velocity is .99 times the velocity of light (.99c). What percentage of this total energy is kinetic?

$$E_T = \gamma m c^2 \qquad m_e = 9.1090 \times 10^{-31} \text{ kg}$$

$$\gamma = \frac{1}{\left[1 - \dfrac{(.99)^2 c^2}{c^2}\right]^{1/2}} = \frac{1}{(.0199)^{1/2}} = 7.088$$

$$E_T = 7.088 \times (9.109 \times 10^{-31} \text{ kg}) \times (3.00 \times 10^8 \text{ m/s})^2$$

$$= 5.81 \times 10^{-13} \text{ J} \times 6.25 \times 10^{18} \text{ eV/J}$$

$$= 3.63 \text{ MeV}.$$

Consider the rest energy of the electron:

$$E_R = m c^2$$

$$= 9.109 \times 10^{-31} \text{ kg} \times (3.00 \times 10^8 \text{ m/s})^2 \times 6.25 \times 10^{18} \text{ eV/J}$$

$$= .512 \text{ MeV}.$$

Rearranging Eq. 5.8 and solving for the kinetic energy, we obtain

$$E_{KE} = E_T - E_R = (3.63 - .512) \text{ MeV} = 3.12 \text{ MeV}.$$

The percentage of kinetic energy is

$$\frac{E_{KE}}{E_T} \times 100\% = \frac{3.12}{3.63} \times 100 = 86\% \text{ kinetic energy.}$$

The percentage of rest energy is

$$\frac{E_R}{E_T} \times 100\% = \frac{.512}{3.63} \times 100 = 14\% \text{ rest energy.}$$

A more direct way to calculate the percentage of kinetic energy is simply by forming the ratio of Eqs. 5.5 and 5.6:

$$\frac{E_{KE}}{E_T} = \frac{(\gamma - 1)mc^2}{\gamma mc^2} = 1 - \frac{1}{\gamma}.$$

If $\gamma \gg 1$, essentially all the energy is kinetic; if $\gamma \approx 1$, the energy is mainly rest.

EXAMPLE 2   Find the total energy of a 1-g BB shot from a gun at the velocity of sound (the velocity of sound is 330 m/s).

$$\gamma = \frac{1}{\left[1 - \left(\frac{330}{3 \times 10^8}\right)^2\right]^{1/2}} = \frac{1}{(1 - 1.2 \times 10^{-12})^{1/2}}$$

$$= 1.00000.$$

Thus $E_T = \gamma mc^2 \approx mc^2$, which is the rest energy of the BB. The error we made in calculating $\gamma = 1$ is about 1 part in 10^{-12}, or $10^{-10}\%$. A very small error indeed! Yet *if* the rest energy cannot be liberated, as in this ordinary lead BB, the small amount of kinetic energy is important. Try getting hit by one! Remember that we are all surrounded by an enormous amount of mass energy, but most of it, thank God, cannot be used!

Now that we know how to calculate the total energy of a system, we may write the conservation law associated with it:

> *The total energy of an isolated system of particles is constant, never changing in time, no matter how the system interacts or transforms itself.*

It is the TOTAL ENERGY of a system of particles that is a conserved quantity, *not* the rest or kinetic energy *separately*. If one calculates the sum of the kinetic energies and rest energies of all the particles in the system at an initial time, and then calculates the sum again at a later time (when there may be different particles with different rest and kinetic energies), conservation of energy requires that the two sums will be equal as long as the system remains isolated.

EXAMPLE 3   A π^+ meson decays at rest inside a bubble chamber into μ^+ and an unobserved particle. If the muon has a measured kinetic energy of 4.12 MeV, find the kinetic energy and identify the particle that was not observed.
The decay is

$$\pi^+ \rightarrow \mu^+ + x.$$

From Appendix B we have the following information:

$$E_R(\pi^+) = 139.58 \text{ MeV}$$

$$E_R(\mu^+) = 105.66 \text{ MeV}.$$

Then

initial total energy $= E_R(\pi^+) + $ (zero kinetic energy)

$$= 139.58 \text{ MeV}$$

final total energy $= E_R(\mu^+) + E_{KE}(\mu^+) + E_T(x)$

$$= 105.66 \text{ MeV} + 4.12 \text{ MeV} + E_T(x)$$

$$= 109.78 \text{ MeV} + E_T(x).$$

The conservation of energy requires equating the initial and final total energies of the system. Thus

$$139.58 \text{ MeV} = 109.78 \text{ MeV} + E_T(x)$$

or $\qquad E_T(x) = E_{KE}(x) + E_R(x) = 29.8 \text{ MeV}.$

Now we are stuck! However, conservation of charge and muon family membership tells us that the unknown and unobserved particle must be a muon neutrino.

Initial Muon Family Membership	$\pi^+ =$	0
Final Muon Family Membership	$\mu^+ =$	-1
	$x =$	?

Total must be zero.

Thus x must be an uncharged member of the muon family with family membership $+1$ and with a maximum mass of about 30 MeV. There is only one particle that fits the description; it is the muon's neutrino, ν_μ. The decay scheme is then

$$\pi^+ \rightarrow \mu^+ + \nu_\mu.$$

Since the neutrino has zero rest mass, all its energy is kinetic. Thus

$$E_T(x) = E_{KE}(\nu_\mu) = 29.8 \text{ MeV}.$$

EXAMPLE 4   Two identical protons, coming from opposite directions, collide head-on inside a bubble chamber. Each proton has a kinetic energy of 70.5 MeV. The result of this collision are three particles that move off with very low velocity. Two of these product particles are identified from their tracks as a proton and a π^+-meson, but the third particle leaves no track. Identify the third particle.

The reaction equation is

$$p + p \rightarrow p + \pi^+ + x.$$

Since the third (unknown) particle leaves no track, it is obviously a neutral particle. This is also apparent from charge conservation.

Conservation of baryon family membership:

Initial Baryon Number	p = 1
	p = 1
Total Baryons	+2.
Final Baryon Number	p = 1
	π^+ = 0
	x = ?
Total Baryons	+2.

So we are looking for a neutral particle whose baryon number is +1.

Conservation of energy:

initial total energy $= 2 \times [E_{KE}(p) + E_r(p)]$

$$= 2 \times (70.5 + 938.2) = 2017.4 \text{ MeV}$$

final total energy $= E_R(p) + E_R(\pi^+) + E_R(x) +$ neglected kinetic energy of all three product particles

$$= 938.2 \text{ MeV} + 139.6 + E_R(x).$$

Equating initial and final *total* energies yields

$$2017.4 = 1077.8 + E_R(x)$$

$$E_R(x) = 939.6 \text{ MeV}.$$

The rest energy of the neutron is 939.5 MeV. This clearly identifies the unknown particle as a neutron with kinetic energy of .1 MeV. It is a neutral baryon with a small kinetic energy.

5.5 Low-Velocity Limit

THE SPECIAL CASE of particles whose velocities are small compared to the velocity of light should be examined. The 1-g BB moving at 330 m/s (Example 2, Sec. 5.4C) was such a case. Recall we concluded that all the energy was in the form of rest energy, since γ was, for all practical purposes, equal to 1. Anyone who has ever been hit by such a BB knows that there is something very wrong with this calculation. Clearly the BB has kinetic energy, but we seem unable to calculate it. The reason we are having difficulty is that the rest energy of the BB is so much larger than the kinetic energy that our calculator lost the last digits in γ that are different from 1.0. We can, however, use a well-known mathematical trick to extract this small difference.

Consider γ, then, in the limit where $v \ll c$.

$$\gamma = \frac{1}{\left(1 - \frac{v^2}{c^2}\right)^{1/2}}.$$

But since $v \ll c$, it follows that $\frac{v^2}{c^2} \lll 1$. We can define α as the square of the fractional velocity, or

$$\alpha \equiv \frac{v^2}{c^2}$$

a very small number. Thus our expression for γ becomes

$$\gamma = (1 - \alpha)^{-1/2}.$$

To an arbitrary degree of accuracy, the binomial $(1 - \alpha)^{-1/2}$ can be approximated using the binomial expansion. The binomial expansion is written

$$(1 + x)^n = 1 + nx + \frac{n(n-1)}{2!} x^2$$
$$+ \frac{n(n-1)(n-2)}{3!} x^3 + \cdots .$$

(5.11)

This is a finite series if n is a positive integer. For other values of n, it is an infinite series that converges for $x < 1$. In our special case $n = -\frac{1}{2}$ and $x \lll 1$.

The numerical accuracy one can obtain from considering only a few terms in the series for the case of $n = -\frac{1}{2}$ is remarkable. Let us use the example of $x = .05$, which is a small number compared to 1, but not nearly as small as $\frac{v^2}{c^2}$ in our example.

Writing out the numerical values for the expansion term by term (for four terms) yields

$$(1 - .05)^{-1/2} = 1 - (-\tfrac{1}{2})(.05) + \frac{-\tfrac{1}{2}(-\tfrac{1}{2} - 1)}{2 \cdot 1} (.05)^2$$
$$- \frac{-\tfrac{1}{2}(-\tfrac{1}{2} - 1)(-\tfrac{1}{2} - 2)}{3 \cdot 2 \cdot 1} (.05)^3 + \cdots$$
$$= 1 + .025 + .00093 + .000039$$
$$= 1.025965.$$

The correct answer is 1.025978.

If we consider *only* the first *two terms* in the series, we obtain

$$(1 - .05)^{-1/2} \approx 1 + .025 = 1.025.$$

The error in this two-term estimate,

$$\frac{1.025978 - 1.025}{1.025978} \times 100\% = .095\%,$$

is an error of less than one-tenth of 1%. This approximation is extremely useful in many physics problems, since real experimental measurements always have a finite accuracy.

Now we consider the "low"-velocity expansion of γ, keeping *only* the first two terms of the series:

$$\gamma = \frac{1}{\left(1 - \dfrac{v^2}{c^2}\right)^{1/2}} \approx 1 + \frac{1}{2}\frac{v^2}{c^2}.$$

Substitution into Eq. 5.5 yields

$$E_{\text{KE}} = (\gamma - 1)mc^2 \approx \left(1 + \frac{1}{2}\frac{v^2}{c^2} - 1\right)mc^2 = \frac{1}{2}mv^2$$

$$E_{\text{KE}} \text{ (low-velocity limit)} = \tfrac{1}{2}mv^2. \qquad (5.12)$$

This Newtonian expression for the kinetic energy of a particle of mass m and velocity v may be familiar to some of you from your high school physics course. It is easy to see why this expression was

Its top speed is 186 M.P.H. That's 1/3,600,000 the speed of light.

considered correct for many years until the experimental techniques were developed to create and study particles with velocities near that of light.

For our 1-g BB moving at 330 m/s, we have

$$E_{KE} = \tfrac{1}{2}\,(.001\text{ kg})(330\text{ m/s})^2 = 54.45\text{ J}$$

and $\qquad E_R = (.001\text{ kg})(3 \times 10^8\text{ m/s})^2 = 90.0 \times 10^{12}\text{ J}.$

No wonder the kinetic energy got lost: it is over 12 orders of magnitude smaller than the rest energy. Nevertheless, for all Newtonian systems where mass cannot transform to energy, the kinetic energy is the *only* important quantity. The rest energy just remains a constant background. Now we have a straightforward way to calculate it for low-velocity systems (Eq. 5.12).

5.6 Vector Conservation Laws: Conservation of Linear Momentum in One Dimension

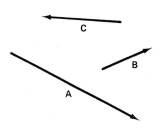

Fig. 5.4

THE NEXT ABSOLUTE conservation law is a VECTOR conservation law. A vector is an abstract mathematical quantity that has both magnitude and direction. It is usually presented pictorially as a line segment (an arrow in Fig. 5.4) whose length represents the magnitude of the vector and which points in the direction of the vector. It is written as a letter, both capital and lowercase, in boldface type, such as **A** or **a**. The more general two-dimensional vector problems are discussed in Chapter 6. For now, let's consider the simpler one-dimensional world where the vector quantity can have only two directions, which we will arbitrarily pick to be the plus or minus x-direction.

The correct relativistic expression for the LINEAR MOMENTUM of a single particle is

$$\mathbf{p} \equiv \text{linear momentum} = \gamma m \mathbf{v}, \qquad (5.13)$$

where **v** is the velocity of the particle whose rest mass is m. The first step in solving any problem that deals with a vector conservation principle is to establish a frame of reference—a coordinate system. For our one-dimensional problems we will arbitrarily assign all motion to the x-axis, defining motion to the right as positive and to the left as negative. The direction of the particle's velocity uniquely determines the direction of its linear momentum.

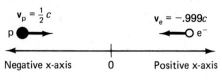

Fig. 5.5

EXAMPLE 1   Find the total momentum of the system of a colliding proton and electron moving in opposite directions along the x-axis, shown in Fig. 5.5.

First, choose a coordinate system. We define the positive axis to be to the right. Then*

$$\mathbf{P}\ (\text{proton}) = \gamma_p m_p v_p$$

$$= \frac{1}{\left[1 - \frac{(\frac{1}{2})^2 c^2}{c^2}\right]^{1/2}} (1.67 \times 10^{-27}\ \text{kg})(\tfrac{1}{2} \times 3 \times 10^8\ \text{m/s})$$

$$= 2.89 \times 10^{-19}\ \text{kg-m/s} \qquad (\text{to the right})$$

and

$$\mathbf{P}\ (\text{electron}) = \gamma_e m_e v_e$$

$$= \frac{1}{\left[1 - \frac{(.999)^2 c^2}{c^2}\right]^{1/2}} (9.11 \times 10^{-31}\ \text{kg})(-.999 \times 3 \times 10^8\ \text{m/s})$$

$$= -1.37 \times 10^{-19}\ \text{kg-m/s} \qquad (\text{to the left})$$

Note the minus sign! The electron is moving in the *minus* x-direction, with negative linear momentum.

$$\mathbf{P}_T = \mathbf{P}\ (\text{proton}) + \mathbf{P}\ (\text{electron})$$

$$= 2.89 \times 10^{-19} - 1.37 \times 10^{-19}$$

$$= +\ 1.52 \times 10^{-19}\ \text{kg-m/s}.$$

Isn't it interesting that although the electron is traveling at nearly the speed of light, most of the momentum of this *system* is carried by the more massive proton?

There is a system of units that is more convenient and more commonly used for high-energy particle physics than SI (kg, meters, seconds, etc.). For energy, MeV (million electron-volts) is the standard unit, and for momentum it is MeV/c. To understand this rather unusual unit, consider an example. What is the momentum of the proton in Example 1 in units of MeV/c?

Momentum has the dimensions of (mass)(distance)/(time) and energy has the dimensions of (mass)(distance)^2/(time)^2. Thus dividing energy by momentum gives

$$\frac{\text{energy}}{\text{momentum}} = \frac{(\text{mass})(\text{distance})^2}{(\text{time})^2} \cdot \frac{\text{time}}{(\text{mass})(\text{distance})} = \frac{\text{distance}}{\text{time}} = \text{velocity}.$$

---

* Both **P** and **p** are used for linear momentum. The capital letter is used when the proton's momentum is considered so as not to be confused with the symbol for the particle itself.

Rearranging the relationship, we obtain the dimensional relationship

$$\text{momentum} = \frac{\text{energy}}{\text{velocity}}$$

or

$$\text{velocity} \cdot \text{momentum} = \text{energy}. \qquad (5.14)$$

If we choose a *standard* velocity by which we agree to multiply all the values of linear momentum, the linear momentum times this velocity will be in units of energy. The obvious choice for this universal velocity multiplier is the velocity of light in a vacuum, c. Let's take as an example the proton in Example 1:

cP (proton) $= 3 \times 10^8$ m/s $\times 2.89 \times 10^{-19}$ kg-m/s $\times 6.25 \times 10^{18}$ eV/J

$\qquad = 542$ MeV,

and for the electron,

$\qquad cP$ (electron) $= 3 \times 10^8$ m/s $\times -1.37 \times 10^{-19}$ kg-m/s $\times 6.25$

$\qquad\qquad \times 10^{18}$ eV/J $= -257$ MeV.

This is, however, usually written

$\qquad P$ (proton) $= 542$ MeV/c

$\qquad P$ (electron) $= -257$ MeV/c.

In the final result the division by c is *not carried out arithmetically*. The units MeV/c or MeV/(3×10^8 m/s) are the units of momentum. To return to the original SI units, you need only convert MeV into joules and carry out the division by the velocity of light. You will find, however, that these strange units are very convenient when solving high-energy physics problems.

Now that you know how to calculate the linear momentum of a particle and how to add the momentum of several particles (at least in one dimension), we can examine the conservation law. This fundamental principle is written as follows:

> *The **total** linear momentum of an isolated system does not change with time, no matter how the system interacts or evolves.*

When the total linear momentum of a system (which is the quantity that is conserved, *not* the individual particle's momentum) is considered, it is necessary to add the individual particle's momentum by the rules of vector addition. The general rules for vector addition are discussed in Chapter 6, but for the one-dimensional case we are only required to keep track of the *sign* of the individual particle's momentum. Unlike energy, you must remember that *momentum can be either positive or negative in one dimension.*

EXAMPLE 2

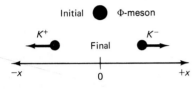

Initial ● Φ-meson

K^+ K^-
●————   Final   ————●

−x   0   +x

Fig. 5.6

A phi-meson (Φ) decays at rest into two K-mesons, by the reaction Φ → K⁺ + K⁻ (Fig. 5.6). Find the momentum of each meson.

In this example we must use both conservation of energy and conservation of momentum principles. From Appendix B we know the following:

$$m_\Phi c^2 = 1020 \text{ MeV} \qquad m_{K^+} c^2 = m_{K^-} c^2 = 493.8 \text{ MeV}.$$

Conservation of energy requires

$$E_R(\Phi) = E_R(K^+) + E_{KE}(K^+) + E_R(K^-) + E_{KE}(K^-)$$

$$1020 = 493.8 + E_{KE}(K^+) + 493.8 + E_{KE}(K^-) \qquad \text{(in MeV)}$$

Rearranging, we have

$$E_{KE}(K^+) + E_{KE}(K^-) = 32.4 \text{ MeV}.$$

Conservation of momentum requires that

$$\mathbf{P}(\Phi) = \mathbf{P}(K^+) + \mathbf{P}(K^-).$$

But $\mathbf{P}(\Phi) = 0$, thus $\mathbf{P}(K^+) = -\mathbf{P}(K^-)$. The two decay products have the same momentum but in opposite directions. Since the two K-mesons have the same mass and the same magnitude of momentum, they must also have the same kinetic energy. That allows us to further simplify the energy conservation equation (a) to read

$$2E_{KE}(K^\pm) = 32.4 \qquad \text{or} \qquad E_{KE}(K^\pm) = 16.2 \text{ MeV}.$$

But from the definition of kinetic energy,

$$E_{KE}(K^\pm) = (\gamma - 1)m_K c^2 = 16.2 \text{ MeV}$$

or

$$\gamma - 1 = \frac{16.2 \text{ MeV}}{493.8 \text{ MeV}} = .0328$$

$$\gamma = 1.0328.$$

To calculate the momentum of the decay products, we need three quantities: mass, velocity, and γ. Actually, only two are needed, since γ is a function of v only. There is a straightforward method for calculating the velocity, given the value of γ. Since this calculation is often needed in solving particle physics problems, it is worth our time to go through the algebraic steps.

$$\gamma = \frac{1}{\left(1 - \dfrac{v^2}{c^2}\right)^{1/2}}$$

Square both sides and cross-multiply to obtain

$$\gamma^2\left(1 - \frac{v_2}{c^2}\right) = 1 \quad \text{or} \quad \gamma^2 - \frac{\gamma^2 v^2}{c^2} = 1.$$

Rearrange:

$$\frac{v^2}{c^2} = \frac{\gamma^2 - 1}{\gamma^2}$$

and take the square root:

$$\frac{v}{c} = \left(\frac{\gamma^2 - 1}{\gamma^2}\right)^{1/2} \tag{5.15}$$

For our problem

$$\frac{v_{K^\pm}}{c} = \left[\frac{(1.0328)^2 - 1}{(1.0328)^2}\right]^{1/2} = .250.$$

You will note that I have not calculated the velocity in meters/second, but rather, left our result as the ratio of the velocity of the particle to the velocity of light, v/c. This is standard practice in particle physics since it is the velocity of light that sets the appropriate scale for these problems. Returning to our calculation of the linear momentum of the K^+ and K^- particles, we recall that

$$P(K^\pm) = \gamma_K m_K v_K.$$

Multiplying by c to get it in energy units, and multiplying the right side by c/c, we have the expression in a convenient form:

$$cP(K^\pm) = \gamma_K(m_K c^2)\frac{v_K}{c}$$

$$= (1.0328)(493.8 \text{ MeV})(.250) = 127.5 \text{ MeV}$$

or $\qquad P(K^\pm) = 127.5 \text{ MeV}/c.$

EXAMPLE 3   In a bubble chamber a lambda-nought (Λ^0) with 6.38 MeV of kinetic energy decays in flight into a proton and an unknown particle. If the proton remains at rest in the chamber, find the momentum, rest mass, and total energy of the unknown particle. Identify the unknown particle ($\Lambda^0 \rightarrow p + x$).

Conservation of charge requires the unknown particle to be negatively charged. Since all charged particles have nonzero rest mass, we are looking for a particle *with* rest mass. Conservation of baryon number demands that the particle is *not* a baryon. From Appendix B we know that

$$m_{\Lambda^0}c^2 = 1115.6 \text{ MeV} \qquad m_p c^2 = 938.26 \text{ MeV}.$$

Conservation of energy is written as

$$E_R(\Lambda^0) + E_{KE}(\Lambda^0) = E_R(p) + E_T(x)$$

$$1115.6 + 6.38 = 938.26 + E_T(x) \quad \text{(in MeV)}.$$

Rearranging yields

$$E_T(x) = 183.72 \text{ MeV}.$$

Conservation of momentum requires that

$$\mathbf{P}(\Lambda^0) = \mathbf{P}(p) + \mathbf{P}(x) \quad \text{but} \quad \mathbf{P}(p) = 0 \quad \text{so} \quad \mathbf{P}(\Lambda^0) = \mathbf{P}(x).$$

The magnitude and direction of the momentum of Λ^0 is the same as that of the unknown particle. To calculate the momentum of particle x, we first need to calculate the momentum of the initial Λ^0. To calculate γ_{Λ^0}, we use

$$E_T(\Lambda^0) = \gamma_{\Lambda^0} m_{\Lambda^0} c^2 \quad \text{or} \quad \gamma_{\Lambda^0} = \frac{E_T(\Lambda^0)}{m_{\Lambda^0} c^2}$$

$$= \frac{1115.6 \text{ MeV} + 6.38 \text{ MeV}}{1115.6 \text{ MeV}}$$

$$= 1.00572$$

and $\quad \dfrac{v_{\Lambda^0}}{c} = \left(\dfrac{\gamma_{\Lambda^0}^2 - 1}{\gamma_{\Lambda^0}^2}\right)^{1/2} = \left[\dfrac{(1.00572)^2 - 1}{(1.00572)^2}\right]^{1/2} = .10649.$

Now we can calculate the momentum of Λ^0:

$$cP(\Lambda^0) = \gamma_{\Lambda^0} m_{\Lambda^0} c^2 \frac{v_{\Lambda^0}}{c}$$

$$= (1.00572)(1115.6 \text{ MeV})(.10649) = 119.48 \text{ MeV}$$

$$cP(x) = 119.48 \text{ MeV} \quad \text{(since the momentums are the same)}.$$

We have two equations: one for energy, one for momentum:

$$cP(x) = c\gamma_x m_x v_x = 119.48 \text{ MeV}$$

$$E_T(x) = \gamma_x m_x c^2 = 183.72 \text{ MeV}$$

If we divide the momentum equation by the energy equation, we obtain an expression for the velocity of the unknown particle:

$$\frac{v_x}{c} = \frac{119.48 \text{ MeV}}{183.72 \text{ MeV}} = .6503.$$

Then $\quad \gamma_x = \dfrac{1}{[1 - (.6503)^2]^{1/2}} = 1.3164,$

but $\quad cp(x) = \gamma_x m_x c^2 \dfrac{v_x}{c} = 119.48 \text{ MeV}$

or
$$m_x c^2 = \frac{119.48 \text{ MeV}}{\gamma_x \dfrac{v_x}{c}}$$

Substituting in the values gives

$$m_x c^2 = \frac{119.48 \text{ MeV}}{(1.3164)(.6503)} = 139.57 \text{ MeV}.$$

The rest energy of the π^- is 139.58 MeV. Given experimental error in the measurements, we can identify the unknown particle as a π^-. It has all the right properties; it is negatively charged, is not a baryon, and has the correct rest mass.

5.7 Low-Velocity Limit: Momentum

AS WAS THE CASE for the kinetic energy of a particle, the linear momentum will also reduce to the Newtonian form in the limit of low velocities. For the case where $v \ll c$, one can expand γ in Eq. 5.13:

$$\mathbf{p} = \frac{m\mathbf{v}}{\left(1 - \dfrac{v^2}{c^2}\right)^{1/2}}$$

Once again using the binomial expansion for the polynomial, we obtain

$$\left(1 - \frac{v^2}{c^2}\right)^{-1/2} \approx 1 + \frac{1}{2}\frac{v^2}{c^2} + \cdots.$$

Substitution yields

$$\mathbf{p} \approx m\mathbf{v} + \frac{1}{2} m \frac{v^3}{c^2} + \text{higher-order terms.}$$

Neglecting the term in $\dfrac{v^3}{c^2}$ and all higher-order terms, the expression for linear momentum reduces to the Newtonian formulation:

$$\boxed{\mathbf{p} = m\mathbf{v} \quad \textit{low-velocity limit.}} \tag{5.16}$$

It is important to emphasize again that relativity is much more than a set of equations that are more accurate than Newton's laws in describing particle motion at high velocities. It is, in fact, an entirely new understanding of nature. The concept of rest energy, not present in any form in Newtonian mechanics, is an essential insight that allows us to understand particle creation, annihilation, and radioactive decay. It is reassuring, however, that this more general theory does reduce, under certain conditions, to the older Newtonian mechanics which served physics so well for many years.

EXAMPLE 1

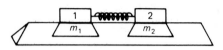

Fig. 5.7

Two gliders compress a spring between them as they rest on a frictionless air track (Fig. 5.7). The string that holds the gliders together is burned and the two gliders fly apart. Find the velocity of glider 1 given that $m_1 = 0.3$ kg, $m_2 = 0.13$ kg, and $v_2 = .037$ m/s. Clearly, this is a Newtonian system and we would be foolish to analyze the problem using the relativistic equations—although they *will* work. It is true that gravitational forces act on the gliders, but the air track exerts an equal and opposite force on them so that the *net* force is zero and the gliders may be considered an isolated system.

Before we can use the momentum conservation law, we must establish a coordinate system. Define the positive x-direction to be to the right along the direction of the air track. Then

$$\mathbf{P}_i = \mathbf{P}_f \qquad \text{or} \qquad 0 = m_2 v_2 - m_1 v_1.$$

Solving for v_1 gives

$$v_1 = \frac{m_2}{m_1} v_2 = \left(\frac{.13 \text{ kg}}{.3 \text{ kg}}\right)(.037 \text{ m/s}) = .016 \text{ m/s}.$$

5.8 Magnitude Equation

One more relationship is useful. This relationship, called the "magnitude equation," relates the three fundamental dynamical parameters of a given free particle: the total energy E_T, the mass m, and the magnitude of the linear momentum p. It is left as a problem at the end of the chapter for the reader to prove, using the equations for these quantities, that

$$E_T^2 = m^2 c^4 + p^2 c^2. \qquad (5.17)$$

The magnitude equation 5.17 holds *only* for a *single* particle. In practical terms, this equation means that if one knows two of the three dynamical parameters of any single free particle, one can always calculate the third parameter.

EXAMPLE 1

An electron has a total energy of 5 MeV. Find its linear momentum.

There are two ways to solve this problem; we will calculate the momentum both ways. First we use the magnitude equation:

$$E_T^2(e) = p_e^2 c^2 + m_e^2 c^4.$$

From Appendix B

$$m_e c^2 = .511 \text{ MeV}.$$

We next rearrange the magnitude equation,

$$c^2 p^2(e) = E_T^2(e) - m_e^2 c^4 = (5 \text{ MeV})^2 - (.511 \text{ MeV})^2$$

$$= 24.74 \text{ MeV}^2,$$

which becomes

$$cp(e) = 4.97 \text{ MeV} \quad \text{or} \quad p(e) = 4.97 \text{ MeV}/c.$$

The second method involves directly calculating the velocity from the equations for the total energy.

$$E_T = \gamma mc^2 \quad \text{or} \quad \gamma = \frac{E_T(e)}{m_e c^2}.$$

Substitution of the known values yields

$$\gamma = \frac{5 \text{ MeV}}{.511 \text{ MeV}} = 9.785.$$

Now we can use Eq. 5.15 to solve for the electron's velocity:

$$\frac{v_e}{c} = \left(\frac{\gamma^2 - 1}{\gamma^2}\right)^{1/2} = \left[\frac{(9.785)^2 - 1}{(9.785)^2}\right]^{1/2} = .9948.$$

Substituting the values for γ, $m_e c^2$, and $\frac{v_e}{c}$ into the definition for momentum gives us

$$cp(e) = \gamma m_e c^2 \frac{v_e}{c} = (9.785)(.511 \text{ MeV})(.9947)$$

$$cp(e) = 4.97 \text{ MeV} \quad \text{or} \quad p(e) = 4.97 \text{ MeV}/c.$$

Both methods give the same result (as they must), but for this problem the calculation is considerably simpler using the magnitude equation in units of MeV.

You may have noticed that we often use a slightly altered form of the expression for linear momentum. The equation used in the last few examples was

$$cp = \gamma(mc^2)\frac{v}{c}$$

which is obviously the same as the defining equation

$$p = \gamma mv.$$

By now the reason for this algebraic manipulation should be clear. The first term of the equation, written in energy units, contains mc^2 (the rest energy of the particle tabulated for the elementary particles in Appendix B) and the second term uses v/c, the fractional (dimensionless) velocity of the particle. All these terms are commonly used and are convenient to obtain when solving elementary particle physics problems.

EXAMPLE 2   Consider a stationary radioactive nucleus of mass $4m_0$. It decays spontaneously into two identical fragments, each of which has rest mass m_0 (Fig. 5.8). Find the final kinetic energy, velocity, and linear momentum of each of the fragments in terms of m_0.

Let's start with momentum.

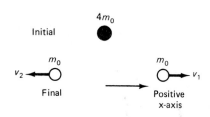

Fig. 5.8

$$\text{initial momentum} = 0$$

$$\text{final momentum} = \gamma_1 m_0 v_1 - \gamma_2 m_0 v_2.$$

Conservation of linear momentum requires that

$$0 = \gamma_1 m_0 v_1 - \gamma_2 m_0 v_2$$

or

$$\gamma_1 v_1 = \gamma_2 v_2.$$

Since γ is a function of velocity *only*, it must be that $\gamma_1 = \gamma_2$ and $v_1 = v_2$.

For simplicity we now write

$$\gamma_1 = \gamma_2 = \gamma \qquad v_1 = v_2 = v,$$

$$\text{initial energy} = E_R = 4m_0 c^2$$

$$\text{final energy} = \gamma_1 m_0 c^2 + \gamma_2 m_0 c^2 = 2\gamma m_0 c^2.$$

Conservation of energy requires that $4m_0 c^2 = 2\gamma m_0 c^2$, which reduces to $\gamma = 2$:

$$\text{kinetic energy of each particle} = (\gamma - 1)m_0 c^2 = (2 - 1)m_0 c^2$$

$$= m_0 c^2.$$

To find the velocity of each particle, we recall Eq. 5.14:

$$\frac{v}{c} = \left(\frac{\gamma^2 - 1}{\gamma^2}\right)^{1/2} = \left(\frac{4 - 1}{4}\right)^{1/2} = \frac{\sqrt{3}}{2}.$$

The magnitude of the momentum of each particle is then

$$p = \gamma m_0 v = (2)(m_0)\frac{\sqrt{3}}{2}c = \sqrt{3}\, m_0 c.$$

EXAMPLE 3   A particle of mass $2m$ and kinetic energy mc^2 hits and sticks to a stationary particle of mass m. Find the rest mass M of the composite particle.

You might be tempted to jump to the conclusion that the rest mass of the composite particle is simply the sum of the two rest masses of the original particles, that is, $2m + m = 3M$. This is wrong! Mass is *not* conserved. The solution to this problem can be obtained by a straightforward application of the two conservation

Fig. 5.9

laws. The variables for this problem are all defined in Fig. 5.9. We start with energy:

$$\text{initial total energy} = \underbrace{2mc^2 + mc^2}_{\substack{\text{rest + kinetic} \\ \text{projectile}}} + \underbrace{mc^2}_{\substack{\text{rest} \\ \text{target}}} = 4mc^2.$$

The final total energy of the composite particle $= \gamma_f Mc^2$. Conservation of energy requires that

$$4mc^2 = \gamma_f Mc^2 \quad \text{or} \quad 4m = \gamma_f M. \tag{a}$$

The magnitude equation for *projectile* particle can be used to calculate the momentum of the projectile.

$$m^2 c^4 = E_T^2 - p_1^2 c^2.$$

But the total energy of the projectile is

$$E_T = E_R + E_{KE} = 2mc^2 + mc^2 = 3mc^2.$$

Substituting into the magnitude equation, we obtain an expression for the projectile's momentum:

$$4m^2 c^4 = 9m^2 c^4 - p_1^2 c^2 \quad \text{or} \quad p_1^2 c^2 = 5m^2 c^4. \tag{b}$$

The magnitude equation for the *final composite* particle is

$$E_f^2 = P_f^2 c^2 + M^2 c^4 \quad \text{or} \quad E_f = (P_f^2 c^2 + M^2 c^4)^{1/2}.$$

Linear momentum conservation requires that $p_1 = P_f$, since all the initial momentum of the system was in the projectile particle. Energy conservation requires that the initial total energy of the system $4mc^2$ must equal the final total energy of the new composite particle E_f.

$$4mc^2 = E_f = (P_f c^2 + M^2 c^4)^{1/2}.$$

Substituting in from Eq. (b) yields

$$4mc^2 = (5m^2 c^4 + M^2 c^4)^{1/2}.$$

Squaring and rearranging gives us

$$16m^2 c^4 = 5m^2 c^4 + M^2 c^4$$

or

$$M^2 c^4 = (16 - 5)m^2 c^4 = 11m^2 c^4$$

$$M = \sqrt{11}\, m.$$

We never needed equation $4m = \gamma_f M$, but we can use it now to calculate the final velocity of the composite particle. Rearranging it as

$$\gamma_f = \frac{4m}{M} = \frac{4\,m}{\sqrt{11}\,m} = 1.206,$$

we have

$$\frac{v}{c} = \left(\frac{\gamma_f^2 - 1}{\gamma_f^2}\right)^{1/2} = .559.$$

SOME TEXTBOOKS imply that the special theory of relativity is important only for understanding particles that move at extremely high velocities. If that were so, the subject would be rather specialized and, in fact, somewhat esoteric. Nothing could be further from the truth. Relativity is a new view of our universe. Its consequences or predictions are many, but none is more important than the relationship between energy and rest mass. As we have discussed before, this relationship holds, and can in fact be experimentally observed, with particles moving at arbitrary *small* velocities.

Among relativity's so-called "paradoxes" is the postulate that light will have exactly the same speed no matter what the speed of the observer relative to the source. Perhaps even more remarkable is the theoretical prediction that *no* particle of nonzero rest mass can travel at a velocity greater than the speed of light in a vacuum. Is there experimental evidence to support this theory?

Indeed there is! Every large accelerator attests to the uncanny accuracy of this prediction. Let's consider some experimental data that demonstrate this.* Figure 5.10 shows a schematic diagram of the apparatus used to test the prediction. It consists of a high-voltage electrostatic accelerator (Van De Graaff, which injects electrons into the system with kinetic energies up to 1.5 MeV) in series with a linear accelerator (LINAC) capable of accelerating the electrons further, to 15 MeV. A bunch of electrons pass a set of plates used to measure the time of flight between two fixed points, A and B. The electrons are injected in short bursts (of 3×10^{-9} s duration) from a negative cathode in the electrostatic generator.

As these bunches of electrons pass the pickup electrodes, the voltage they induce is displayed on a fast oscilloscope (Fig. 5.10b). The oscilloscope measures the time of flight of the electrons between the pickup electrodes. The kinetic energy of the electron beam is measured separately. A calorimeter, which measures the heat input in a given time, is used to determine, *independently*, the kinetic energy of the electrons. When the electrons hit the calorimeter, they transform essentially all their kinetic energy into internal energy of the target, causing its temperature to rise. Thus the apparatus can separately measure the velocity of the electron bunch $v =$ distance/time and kinetic energy. From the experimental data we can study their relationship.

Newtonian mechanics makes a rather simple prediction of the results for this experiment. It predicts

$$\text{kinetic energy (Newtonian)} = E_{\text{KE}} = \tfrac{1}{2}mv^2$$

---

* A detailed description of these experiments can be found in an article by W. Bertozzi, *American Journal of Physics*, **32**, 551 (1964).

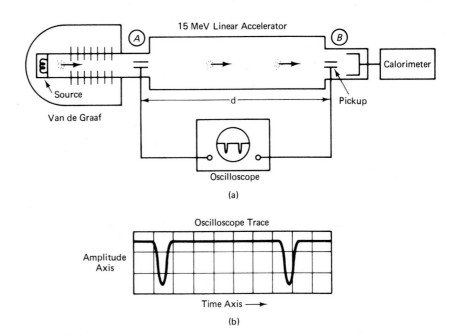

Fig. 5.10

or $v^2 = \dfrac{2E_{KE}}{m} = (3 \times 10^{17})E_{KE}$ (E_{KE} in MeV if v is in m/s).

If we plot the measured values of v^2 versus E_{KE}, we should obtain a straight line whose slope is 3×10^{17}. Figure 5.11 shows both the Newtonian theoretical plot and the experimental data. The results are striking. Only at *very low* values of kinetic energy does the classical theory predict the experimental results. According to Newton, an electron would acquire a velocity greater than the speed of light when it has less than .5 MeV of kinetic energy!

The data clearly do *not* support classical mechanics. Notice that the experimental value of the electron's velocity squared appears to be asymptotically approaching the value of 9×10^{16} m²/s², or

$$v = 3 \times 10^8 \text{ m/s.}$$

This, of course, is the velocity of light. What seems "strange" is that the apparatus can impart an arbitrarily large amount of kinetic energy to the electron, but it cannot give the electron a velocity greater than the speed of light.

I have illustrated only one of the many important breaks with the past that Einstein provided with his special theory of relativity. Many more await you. We cannot simply patch up Newtonian mechanics to solve high-velocity problems. We must recognize fundamental difficulties with Newtonian mechanics and understand the "correct" way to describe the motion of bodies with relativity.

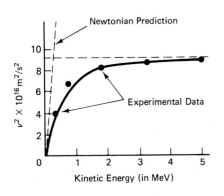

Fig. 5.11

We have restricted ourselves to one dimension in this chapter in order to simplify the problems. The basic techniques needed to use conservation laws have been presented. In this discussion we have paid no attention to the type or nature of particle interactions. We can do this for the rather miraculous reason that conservation principles transcend the details of the interactions; they are truly universal laws.

Obviously, the conservation laws are not the entire story. If one asks such questions as, "If I fire a π-meson at a proton, will it stick or will gamma rays come out?", or "What is the effective target size of the proton?", the conservation laws cannot give the answer. To deal with these important questions, one must understand the details of the forces between the particles. Nevertheless, the conservation principles are fundamental to physics and are deeply rooted in our understanding of the character of space and time. They are also an essential and powerful tool for unraveling complicated collision processes and identifying new particles.

5.10 Summary of Important Equations

Scalar conservation laws:

1. Family membership
   Baryons
   Electron
   Muons
   Tau-leptons
2. Total electrical charge
3. Total energy

$$E_{\text{Kinetic}} = E_{\text{KE}} = (\gamma - 1)mc^2$$

$$E_{\text{Rest}} = E_R = mc^2$$

$$E_{\text{Total}} = E_T = \gamma mc^2$$

$$\gamma = \frac{1}{\left(1 - \dfrac{v^2}{c^2}\right)^{1/2}}$$

Total linear momentum:

$$\mathbf{p} = \gamma m\mathbf{v}$$

Magnitude equation:

$$E_T^2 = p^2c^2 + m^2c^4$$

Low-velocity limit; Newtonian formulation:

$$E_{\text{KE}} = \tfrac{1}{2}mv^2 \qquad \mathbf{p} = m\mathbf{v}$$

Conversion factors:

$$1 \text{ J} = 6.25 \times 10^{18} \text{ eV} \quad \text{or} \quad 1 \text{ eV} = 1.6 \times 10^{-19} \text{ J}$$

Binomial expansion:

$$(1 + x)^n = 1 + nx + \frac{n(n-1)}{2!} x^2 + \frac{n(n-1)(n-2)}{3!} x^3 + \cdots$$

If n is a positive integer, the series consists of $(n + 1)$ terms; otherwise, the series is infinite.

Useful relationship for velocity:

$$\frac{v}{c} = \left(\frac{\gamma^2 - 1}{\gamma^2} \right)^{1/2}$$

Chapter 5   QUESTIONS

1. Discuss some consequences for living and nonliving forms of matter if the charge on the electron were not exactly equal to the charge of the proton.

2. A high-energy positron is directed into a bubble chamber, leaving a single track that suddenly disappears. No trace can be found of it anywhere. Doesn't this conflict with the laws of the conservation of charge, energy, and momentum? Explain what has happened.

3. If a K^0 decays into a π-meson, an electron, and a positron neutrino, does this mean that the K^0 is composed of these particles? Explain.

4. A photon converts into a positron–electron pair. Doesn't this imply that electron family membership is *not* conserved, since initially there was no electron and afterward there was an electron? Explain.

5. If physicists eventually discover that the proton does decay, will baryon conservation be a law of nature? Will this conservation law be useful in the laboratory? Explain.

6. Recently there has been speculation that the neutrino does have rest mass. If that is so, what implication might that have for astrophysics and cosmology?

7. If the liquid in a bubble chamber were neon instead of hydrogen, how would that affect the analysis of bubble chamber events?

Chapter 5   PROBLEMS

1. For each of the following hypothetical collisions or decays, state and explain the conservation laws that are violated.

    (a) $\mu^+ \rightarrow e^+ + \nu_e + \gamma$

    (b) $e^- \rightarrow \nu_e + \gamma$

    (c) $p + p \rightarrow p + \Lambda^0 + \Sigma^+$

    (d) $\pi^- + p \rightarrow \pi^- + n + \Lambda^0 + K^+$

2. Supply the single missing particle in the following neutrino events.

    (a) $\nu_\mu + n \rightarrow p + ?$

    (b) $\nu_e + n \rightarrow p + ?$

    (c) $K^0 \rightarrow \bar{\nu}_\mu + \pi^+ + ?$

    (d) $\mu^- \rightarrow e^- + \nu_\mu + ?$

    (e) $\pi^- \rightarrow \bar{\nu}_\mu + ?$

3. Calculate the following values.

    (a) The kinetic energy of a Σ^0 whose total energy is 1400 MeV.

    (b) The velocity of a π^0 whose total energy is twice its rest energy.

    (c) The name of the neutral particle whose total energy is 1400 MeV and whose velocity is .2648c.

(d) The rest mass in kilograms of a particle whose rest mass (or rest mass energy) is 548.8 MeV.

(e) The kinetic energy of a .006-kg bullet traveling at 420 m/s.

(f) The total energy of a proton whose velocity is 2.9978×10^8 m/s.

(g) The value of γ for a π^--meson whose total energy is 350 MeV.

(h) The velocity in meters/second of a Σ^0 whose kinetic energy is 78 MeV.

4. Calculate the total kinetic energy (in MeV) of the particles produced in the following decays (assuming that the initial particle was at rest).

(a) $\mu^- \rightarrow e^- + \nu_e + \nu_\mu$

(b) $\Lambda^0 \rightarrow p + \pi^-$

(c) $n \rightarrow p + e^- + \nu_e$

5. For each of the following proposed collisions or decays, state the conservation laws that are violated.

(a) $n \rightarrow p + p + \pi^0$

(b) $\mu^- \rightarrow e^- + e^+ + \nu_\mu$

(c) $p \rightarrow \Sigma^+ + K^0$

(d) $\mu^+ \rightarrow \Lambda^0 + e^+$

6. A K^0-meson with kinetic energy equal to half of its rest energy spontaneously decays into a π^--meson whose kinetic energy is 200 MeV, an electron's neutrino (ν_e) with a total energy of 350 MeV, and a third, unknown particle (x).

$$K^0 \rightarrow \pi^- + \nu_e + x$$

(a) What is the electric charge of x?

(b) To what family must x belong?

(c) What is x?

(d) What is the speed of x?

7. An Ω^- traveling with a speed of .65c spontaneously decays into a Ξ^0 particle with a velocity of .75c and a second unknown particle (x).

$$\Omega^- \rightarrow \Xi^0 + x.$$

(a) What is the charge of x?

(b) What is the total energy of x?

(c) What is x?

(d) Find the velocity of x as a fraction of the velocity of light.

8. Make a plot of γ as a function of the dimensionless velocity v/c for values of v/c ranging from .1 to .99. Indicate on the plot the "regions" where Newtonian physics is applicable and the regions where rela-tivistic effects become dominant. What happens to the plot when $v/c > 1$?

9. A proton and an antiproton collide head-on, resulting in a complete transformation into four particles, all of the same family, remaining at rest. If both the proton and the antiproton have 2.8 GeV of kinetic energy, find and identify the four resulting particles.

10. Calculate the following values.

(a) The linear momentum of an electron with a total energy of 5 MeV.

(b) The kinetic energy of a proton whose total energy is 939 MeV.

(c) The total energy of a π^+-meson whose linear momentum is 5 MeV/c.

(d) The rest mass of an unknown particle of total energy 531 MeV and velocity .98c.

(e) The total energy of a Σ^0 particle moving with half the velocity of light.

(f) The velocity of a K^+-meson whose total energy is 500 MeV.

(g) The rest mass (in kilograms) of a particle whose rest energy is 105 MeV.

(h) The name of the particle whose total energy is 829.7 MeV and whose velocity is $\frac{3}{4}c$.

11. Consider the magnitude equation.

(a) From the definition of linear momentum and total energy, prove that the magnitude equation $E^2 = p^2c^2 + m^2c^4$ holds true for a single particle.

(b) Show that for a collection of particles (even in one dimension), the equation

$$E_T^2 = p_T^2 c^2 + m_T^2 c^4$$

where E_T, P_T, and m_T are the total energy, total momentum, and total mass of the system of particles, is *not* true.

12. A K^0-meson at rest decays into two π^0-mesons via the reaction $K^0 \rightarrow \pi^0 + \pi^0$. Find the kinetic energy and velocity of each π^0. Find the momentum of each π^0.

13. A Ξ^- at rest decays into a Λ^0 and a π^--meson. What is the sum of the kinetic energy of both decay products? How is the kinetic energy shared among the two particles?

14. Two tracks of opposite curvature emerge from a single point on a bubble chamber photograph. Analysis shows that these tracks are the result of a decay of unstable particle at rest. One track is identified as a proton with kinetic energy of 5.37 MeV and the sec-

ond track is a π^--meson moving in the opposite direction whose kinetic energy is 32.29 MeV. Identify the original unstable particle.

15. A neutron whose kinetic energy is equal to its rest-mass energy strikes another neutron, whose total energy is three times its rest energy, coming from the opposite direction. After a head-on collision, the two neutrons briefly form a new composite particle. Find the velocity and the rest mass of this new composite particle.

16. A particle of rest mass m_0 moves at a velocity $\frac{2}{3}c$ before colliding with an identical particle (the same rest mass) at rest. The two particles form a composite single particle.

    (a) Find the final velocity, rest mass, and kinetic energy of the final composite particle.

    (b) What happened to the initial kinetic energy of the projectile particle?

17. A positron with a kinetic energy of 1.02 MeV collides with an electron at rest and forms a bound system called positronium. Find the velocity and the mass of the recoiling positronium "atom."

18. Using conservation of energy and momentum, show that it is dynamically impossible for a particle of arbitrarily large kinetic energy and rest mass m_0 to decay spontaneously into a single particle of smaller rest mass m.

19. Three identical particles of rest mass m are going to collide. They all move along the x-axis. The particle on the left is moving with velocity .7c toward the right, the particle in the middle is stationary, and the particle on the right is moving with velocity .95c toward the left. If all three particles stick together after the collision:

    (a) Find the final velocity of the composite particle.

    (b) Find the rest mass of the composite particle.

20. A Σ^+ traverses a short distance through a bubble chamber before it decays into a neutron n and a π-meson (π^+). If the π^+ is left at rest in the chamber, find the initial energy of the Σ^+.

21. Find the velocity of the recoil nucleus $_{60}Nd^{148}$ and the initial mass of the $_{62}Sm^{152}$ parent nucleus in the α-decay process $_{62}Sm^{152} \rightarrow {}_{60}Nd^{148} + {}_2He^4$ if the α-particle has 2.0 MeV of kinetic energy and the $_{60}Nd^{148}$ has a mass of 137,781.75 MeV. (For historical reasons the $_2He^4$ particle is called the α-particle. Its rest-mass energy is 3728.35 MeV.) Is it necessary to use the relativistic expression for momentum for the $_{60}Nd^{148}$ recoiling nucleus? Explain.

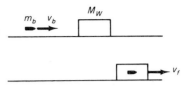

Fig. 5.12

22. A hunter shoots a rifle bullet into a block of wood resting on a smooth ice surface. The bullet hits the block and lodges permanently in the wood (Fig. 5.12). The hunter's daughter, standing nearby, measures the speed of the block after the bullet had lodged inside (v_f). Using the conservation of linear momentum, her data, and subsequent measurements of the mass of the block M_w and bullet M_b, the father and daughter determine the velocity of the bullet when it left the gun. They also had to assume that the block slid across the ice without any frictional resistance.

    (a) What is the velocity of the bullet v_b?

    (b) Show that the ratio of the kinetic energy of the original bullet to the final kinetic energy of the bullet–block system is given by

$$\frac{KE_{initial}}{KE_{final}} = \frac{M_b + M_w}{M_b}.$$

    (c) Does the result in part (b) surprise you? What has happened to the original kinetic energy of the bullet? Where has it gone? Do you still believe in the conservation of energy principle? Why don't bullets conserve energy as elementary particles do?

23. Two gliders compress a spring between them on a horizontal air track as shown in Fig. 5.13. The two gliders are moving with a velocity of 10 cm/s in the positive x-direction when the string breaks. Find the new velocity of m_2 if $m_1 = 5$ kg and $m_2 = 15$ kg and m_1 is left at rest.

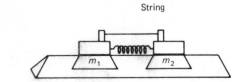

String

Fig. 5.13

24. A bullet weighing 10 g is fired from a rifle with a muzzle velocity of 600 m/s. If the rifle had a mass of 1.5 kg:

(a) Find the recoil velocity of the rifle.

(b) Find the ratio of the kinetic energy of the rifle to the kinetic energy of the bullet assuming that the rifle is free to recoil.

25. A toy popular in the 1960s (and still available) that can teach some physics consists of a set of balls of equal mass hung by a string so that they lie in a horizontal plane (Fig. 5.14). One can easily perform various collision experiments with this "apparatus" to demonstrate important properties of collisions that conserve *kinetic energy*.

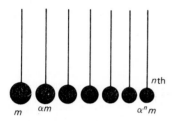

Fig. 5.14

(a) Consider a modification of the standard setup, in which the balls have different masses. These masses have a simple mathematical relationship to each other: the first being m, the second αm, and the third $\alpha^2 m$. If the projectile ball has mass m and strikes the first ball with velocity v, show that the last mass $\alpha^n m$ will shoot off with a velocity

$$v_n = v \left(\frac{2}{1 + \alpha} \right)^n.$$

(b) Consider a specific example. If $\alpha = .95$ and $n = 15$, calculate the ratio of the kinetic energy of the last ball to that of the incident projectile.

(c) Show that the general expression for the velocity of the last ball, if $\alpha \ll 1$, is

$$v_n \approx 2^n (1 - \alpha)^n.$$

(*Hint:* Use the binomial expansion.) How does the kinetic energy of the last ball compare to the kinetic energy of the incident ball?

6

COLLISIONS
AND DECAYS
IN TWO DIMENSIONS

6.1   **Conservation in Two Dimensions**

NOW THAT YOU have been introduced to the conservation laws, we will apply them to high-energy events that are observed in bubble, spark, and drift chambers. All high-energy collisions and decays occur in three-dimensional space. Both the spark and bubble chamber tracks are recorded photographically on a two-dimensional film surface using special stereoscopic optical techniques. These techniques allow spatial reconstruction of the tracks into their original three-dimensional configuration and permit the measurement of both scattering angles and track curvature.

High-energy experiments demand accurate measurements of track curvatures and of the relative angles between the tracks on each photographed event. To do this in three dimensions requires long and tedious solid geometry calculations, which are now done routinely by high-speed electronic computers. Drift chamber data are directly recorded, reconstructed, and analyzed "on-line" by computers. To understand the basic physics of such analysis, it serves no useful purpose to analyze three-dimensional collisions. *All the physics can be understood by studying events that occur in a plane.* Such two-dimensional collisions and decays do occur infrequently in a certain fraction of the events recorded, and we will select these for analysis. Because the geometry is *so much* easier, we can focus our attention on the fundamental physics of the processes.

The analysis of a two-dimensional bubble chamber event proceeds by determining the momentum, charge, and if possible, the energy of each particle that leaves a track. Using the conservation laws discussed in Chapter 5, the energies and rest masses can be calculated and the individual particles identified. But how do we apply these conservation principles in a two-dimensional world? Before we actually analyze an event, we must develop the mathe-

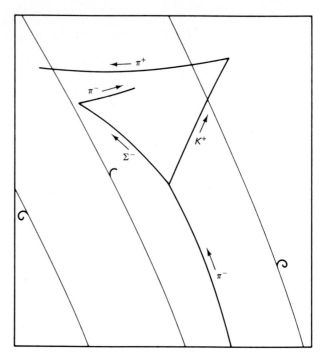

Fig. 6.1

matical machinery necessary to use the conservation laws in more than one dimension.

Let's reconsider the first bubble chamber photograph that we studied in Chapter 5 (Fig. 5.1), which is reconstructed in Fig. 6.1. In this event a high-velocity negatively charged π^--meson collides with the proton of a hydrogen atom (which is at rest in the bubble chamber) to produce a baryon, the Σ^-, and a K^+-meson:

$$\pi^- + p \rightarrow \Sigma^- + K^+.$$

What can we say about the motion of the particles that take part in these events? First we note that each particle travels in the direction indicated by the arrows. The arrows, of course, do *not* appear on the photograph, but the experimenter knows the direction of the projectile particle and usually knows its energy.*

As we discussed in Chapter 5, the total momentum of the system is conserved. For a *single particle*, the momentum is

$$\mathbf{p} = \frac{m\mathbf{v}}{\left(1 - \dfrac{v^2}{c^2}\right)^{1/2}}. \tag{6.1}$$

But what is the *total momentum*? We are faced with the problem of

---

* The arrows are drawn here to aid you in visualizing the collision process.

how to add vector quantities such as momentum in more than one dimension. We can assign momentum to each of the particles that appear in one event. They are, for this event:

$$\mathbf{P}_{\pi^-} = \frac{m_{\pi^-}\mathbf{v}_{\pi^-}}{\left[1 - \left(\dfrac{v_{\pi^-}}{c}\right)^2\right]^{1/2}} \qquad \mathbf{P}_{\Sigma^-} = \frac{m_{\Sigma^-}\mathbf{v}_{\Sigma^-}}{\left[1 - \left(\dfrac{v_{\Sigma^-}}{c}\right)^2\right]^{1/2}}$$

$$\mathbf{P}_{p} = \frac{(m_{p}) \cdot 0}{\left[1 - \left(\dfrac{v_{p}}{c}\right)^2\right]^{1/2}} \qquad \mathbf{P}_{K^+} = \frac{m_{K^+}\mathbf{v}_{K^+}}{\left[1 - \left(\dfrac{v_{K^+}}{c}\right)^2\right]^{1/2}} \tag{6.2}$$

Now what do we do? You notice that in the expression for linear momentum, velocity appears in *two* places and in *two different ways:* in the numerator as **v** and in the denominator as *v*. Velocity is a vector quantity with both magnitude and direction. When it appears as *v*, it is a scalar quantity with magnitude *only*, and when it appears as **v** it has both magnitude and direction. The momentum of each of these particles has some magnitude or numerical value associated with it, as well as a particular orientation or direction. The π^- has a momentum that is directed "up," the K^+ is directed to the "right," and the Σ^- momentum is directed to the "left." The specification of a particle's momentum involves the assignment of *both* a *direction* and a *magnitude*. As we shall see, the conservation of linear momentum requires *not only* the magnitude but also the *direction* of the *total momentum* of the system to be conserved in any collision or decay. Both are required!

That's all very well, but we have not addressed the problem of how to calculate "total momentum"; that is, how do we add (or subtract) momentum? We must develop the mathematical tools for handling vector quantities: how to add them, subtract them, find their magnitude and direction, multiply, and divide them. Vector algebra is probably as important a mathematical tool as you will learn in physics. You will use it in all kinds of engineering and science for a wide variety of problems. Let's examine this new algebra.

6.2 Vector Algebra

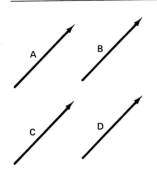

Fig. 6.2

A QUANTITY that has direction as well as magnitude is called a VECTOR QUANTITY. To specify a vector we need a unit with which to measure its magnitude, as well as some way to specify its direction. A geometric representation of a vector is a *directed line segment*, drawn as an arrow whose length specifies its magnitude. The arrow and straight-line segment specify the direction. Figure 6.2 shows four *equal* vectors **A**, **B**, **C**, **D**. If two vectors point in the same direction *and* have the same length, they are equal. The parallel translation of vectors does *not* affect the equality of the vectors.

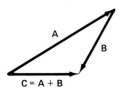

Fig. 6.3

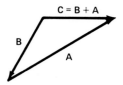

Fig. 6.4

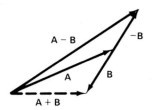

Fig. 6.5

A. Vector Addition

Addition of two vectors **A** + **B** is represented geometrically in Fig. 6.3. The geometric rule of vector addition is to place the tail of **B** to the head of **A**. The sum of **A** and **B** is a new vector drawn as a line segment from the tail of **A** to the head of **B**. This operation is written mathematically as

$$\mathbf{A} + \mathbf{B} = \mathbf{C}. \tag{6.3}$$

The order of the addition does not change the result (Fig. 6.4):

$$\mathbf{A} + \mathbf{B} = \mathbf{B} + \mathbf{A} \quad \text{(commutative law)}. \tag{6.4}$$

The associative law also holds for vectors: that is,

$$\mathbf{A} + (\mathbf{B} + \mathbf{C}) = (\mathbf{A} + \mathbf{B}) + \mathbf{C}. \tag{6.5}$$

The negative of a vector is defined as a vector of the same magnitude but opposite direction. Thus the subtraction of vectors can be defined in terms of the addition of a negative vector:

$$\mathbf{A} - \mathbf{B} \equiv \mathbf{A} + (-\mathbf{B}), \tag{6.6}$$

which is represented geometrically in Fig. 6.5. I have deliberately chosen two vectors so that the magnitude of **A** + **B** is *less* than the magnitude of **A** − **B**. This is to emphasize that this is truly a new kind of algebraic manipulation and that your old prejudices (and habits) will have to be modified when you think about vector addition.

One can easily show that the distributive law

$$\alpha(\mathbf{A} + \mathbf{B}) = \alpha\mathbf{A} + \alpha\mathbf{B}$$

and

$$(\alpha + \beta)\mathbf{A} = \alpha\mathbf{A} + \beta\mathbf{A}, \tag{6.7}$$

where α and β are scalar quantities, holds for vectors.

B. Vector Multiplication

Two distinctly different kinds of multiplications exist in vector algebra, the DOT PRODUCT and the VECTOR PRODUCT.

1. Dot Product. This multiplication procedure, also known as the SCALAR PRODUCT, results in a scalar quantity. The dot product of the two vectors **A** and **B** is defined as

$$\mathbf{A} \cdot \mathbf{B} \equiv |\mathbf{A}||\mathbf{B}| \cos \theta, \tag{6.8}$$

where $|\mathbf{A}|$ is the magnitude of the vector **A**, $|\mathbf{B}|$ is the magnitude of **B**, and θ is the *smallest* angle between the two vectors when their tails are brought together. The commutative law also holds for scalar multiplication: namely,

$$\mathbf{A} \cdot \mathbf{B} = \mathbf{B} \cdot \mathbf{A}. \tag{6.9}$$

The proof is left as a problem.

If **A** and **B** are perpendicular, *then*

$$\mathbf{A} \cdot \mathbf{B} = 0 \qquad \text{for } \mathbf{A} \perp \mathbf{B}$$

since the cosine of 90° is zero. The dot product of a vector with *itself* yields

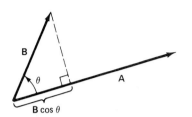

Fig. 6.6

$$\mathbf{A} \cdot \mathbf{A} = |\mathbf{A}||\mathbf{A}| = |\mathbf{A}|^2 \qquad \text{(magnitude of } \mathbf{A} \text{ squared)} \qquad (6.10)$$

or

$$|\mathbf{A}| = (\mathbf{A} \cdot \mathbf{A})^{1/2}. \qquad (6.11)$$

The dot product can also be used to determine the projection of one vector on another. The quantity

$$\mathbf{A} \cdot \mathbf{B} = (\text{projection of } \mathbf{B} \text{ on } \mathbf{A})(\text{magnitude of } \mathbf{A})$$

so (Fig. 6.6)

$$\text{the projection of } \mathbf{B} \text{ on } \mathbf{A} = \frac{\mathbf{A} \cdot \mathbf{B}}{|\mathbf{A}|} = |B| \cos \theta \qquad (6.12)$$

$$\text{the projection of } \mathbf{A} \text{ on } \mathbf{B} = \frac{\mathbf{A} \cdot \mathbf{B}}{|\mathbf{B}|} = |A| \cos \theta. \qquad (6.13)$$

2. *Cross Product.* The CROSS PRODUCT or VECTOR PRODUCT is an *entirely different* mathematical operation. To begin, the cross product of two vectors *is itself a vector*, not a scalar like the dot product. The cross product is written

$$\mathbf{A} \times \mathbf{B} = \mathbf{C} \qquad (6.14)$$

and the *magnitude* of this new vector **C** is found by the equation

$$|\mathbf{C}| = |\mathbf{A}||\mathbf{B}| \sin \theta, \qquad (6.15)$$

where θ is again the *smallest* angle between the vectors **A** and **B** with their tails drawn together (Fig. 6.7). In the case of the cross product, the new vector **C** has zero magnitude when the two vectors are parallel (i.e., when $\theta = 0$).

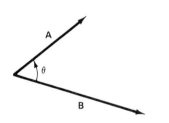

Fig. 6.7

The direction of the new vector **C** is a little more complicated to remember. Any two vectors determine a plane. The vector product **C** is always *perpendicular to the plane* of **A** and **B**. For example, if **A** and **B** are in the plane of this paper, **C** must be perpendicular to the plane of the paper. But there are *two* perpendicular directions, one out of the paper toward you and the other into the book. Since a vector must have a *unique* direction, only one of these perpendicular directions is permissible. Which one?

The convention used to determine the unique perpendicular direction is called the RIGHT-HAND RULE. There are many tricks to remembering this rule. (The one I use is shown in Fig. 6.8a.) Imagine that the vectors **A** and **B** are attached to an ordinary wood screw (which, by the way, is a right-hand screw). If **A** is a "handle" on this screw, imagine turning it toward the second vector **B** and ask: Which way would the screw advance if it were embedded in a piece of wood? The direction of the advance is the direction of the cross

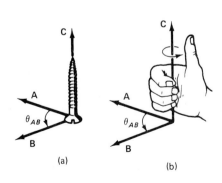

Fig. 6.8

product **C**. If the screw in the diagram was pointing down, then **A** was turned into **B**, the screw would *still advance* in the upward direction. A second way to remember this is shown in Fig. 6.8b, where the right hand is used with the thumb giving the correct perpendicular direction. (*Warning!* It is very easy to make a mistake. When you use the cross product, check your answer several times!)

The cross product is *not* commutative. You can easily show that

$$\mathbf{A} \times \mathbf{B} = -\mathbf{B} \times \mathbf{A}. \tag{6.16}$$

The order of the multiplication *is* important. Finally, the cross product of any vector with itself must be zero:*

$$\mathbf{A} \times \mathbf{A} = \mathbf{B} \times \mathbf{B} = 0. \tag{6.17}$$

C. Components

So far we have discussed vector algebra without referring to a coordinate system. In principle, vector multiplication can, using geometric construction, be carried out without reference to a coordinate system. But the geometric method of adding, subtracting, and multiplying vectors is not very practical, particularly for dealing with two- and three-dimensional problems. The most common analytical method of manipulating vectors involves the resolution of each vector into its COMPONENTS. The components of a vector *depend* on the particular coordinate system chosen.

To simplify the discussion, we will first consider a vector **C** in the two dimensional x–y plane, shown in Fig. 6.9. This figure shows the x and y axes mutually perpendicular and the components of **C**, C_x and C_y. This is called a CARTESIAN coordinate system and is by far the most commonly used system.†

The components of a vector are not unique to the vector itself but depend on the coordinate system used. Consider the vector **R** shown in Fig. 6.10. This vector can be written in terms of its components r_x and r_y as an ordered array:

$$\mathbf{R} = (r_x, r_y).$$

But in terms of the second rotated coordinate system x', y',

$$\mathbf{R} = (r_{x'}, r_{y'}).$$

The *components* of the vectors are *not* the same in the x, y and x', y' coordinate systems:

$$r_x \neq r_{x'} \qquad r_y \neq r_{y'} \qquad \text{(in general)}.$$

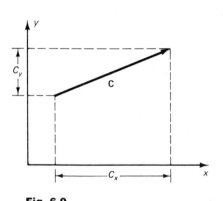

Fig. 6.9

---

* Division by vectors **A**/**B** is undefined. One can divide a vector by a scalar α, such as $\mathbf{A}/\alpha = (1/\alpha)\mathbf{A}$, which means that the magnitude of the vector **A** is multiplied by the quantity $1/\alpha$.

† Vectors may be represented by any coordinate system that spans space, even those that are *not* perpendicular.

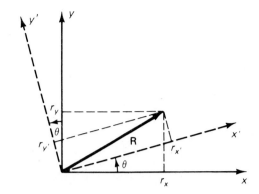

Fig. 6.10

In fact, there is an algebraic relationship between the components, which is

$$r_{x'} = r_x \cos \theta + r_y \sin \theta$$

$$r_{y'} = r_y \cos \theta - r_x \sin \theta. \qquad (6.18)$$

The proof of Eq. 6.18 is left as a problem.

If we restrict ourselves to *one* coordinate system, it is true that:

> *Two vectors are equal if and only if all their components are equal.*

For example, if **A** = **B** = **C**, then, as shown in Fig. 6.11, it must be true that

$$A_x = B_x = C_x \qquad A_y = B_y = C_y \qquad A_z = B_z = C_z. \qquad (6.19)$$

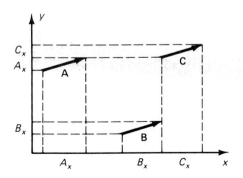

Fig. 6.11

The single equation $\mathbf{A} = \mathbf{B} = \mathbf{C}$ represents these *three* scalar equations. The law of vector addition,

$$\mathbf{A} + \mathbf{B} = (A_x + B_x, A_y + B_y, A_z + B_z) \qquad \text{a vector,} \qquad (6.20)$$

the dot product

$$\mathbf{A} \cdot \mathbf{B} = (A_x B_x + A_y B_y + A_z B_z) \qquad \text{a scalar} \qquad (6.21)$$

$$\mathbf{A} \cdot \mathbf{A} = A_x^2 + A_y^2 + A_z^2 \qquad (6.22)$$

$$|\mathbf{A}| = |\mathbf{A} \cdot \mathbf{A}|^{1/2} = (A_x^2 + A_y^2 + A_z^2)^{1/2}, \qquad (6.23)$$

and multiplication by a scalar quantity α,

$$\alpha \mathbf{A} = (\alpha A_x, \alpha A_y, \alpha A_z) \qquad \text{a vector.} \qquad (6.24)$$

EXAMPLE 1   Consider two vectors,

$$\mathbf{A} = (4.5, -1.0, 1.6) \qquad \mathbf{B} = (-3.0, -2.0, 3.5).$$

The magnitudes of these vectors can be calculated using Eq. 6.23:

$$|\mathbf{A}| = [(4.5)^2 + (-1.0)^2 + (1.6)^2]^{1/2} = 4.88$$

$$|\mathbf{B}| = [(-3.0)^2 + (-2.0)^2 + (3.5)^2]^{1/2} = 5.02.$$

The sum and difference of the two vectors are

$$\mathbf{A} + \mathbf{B} = [4.5 + (-3.0), -1.0 + (-2.0), 1.6 + 3.5]$$

$$= (1.5, -3.0, 5.1)$$

$$\mathbf{A} - \mathbf{B} = [4.5 - (-3.0), -1.0 - (-2.0), 1.6 - 3.5]$$

$$= (7.5, 1.0, -1.9).$$

To find the angle between the two vectors $\mathbf{A}$ and $\mathbf{B}$, we note that

$$\cos \theta = \cos (\mathbf{A}, \mathbf{B}) = \frac{\mathbf{A} \cdot \mathbf{B}}{|\mathbf{A}||\mathbf{B}|}.$$

Now, by Eq. 6.21,

$$\mathbf{A} \cdot \mathbf{B} = (4.5)(-3.0) + (-1.0)(-2.0) + (1.6)(3.5)$$

$$= -5.9.$$

Thus

$$\cos \theta = - \frac{5.9}{(4.88)(5.02)} = -.241$$

$$\theta = 104°.$$

I invite you to draw these two vectors on a three-dimensional projection and attempt to visualize them.

D. Unit Vectors

A more useful way of using components is to express the vectors in terms of UNIT VECTORS rather than in an ordered array. As their name implies, unit vectors are vectors of unit length in the coordinate

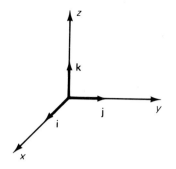

Fig. 6.12

system chosen. For the Cartesian coordinate system, these vectors (shown in Fig. 6.12) are called **i**, **j**, and **k**. One may write the vector quantity **A** in terms of unit vectors as

$$\mathbf{A} = a_x\mathbf{i} + a_y\mathbf{j} + a_z\mathbf{k}$$
$$\mathbf{B} = b_x\mathbf{i} + b_y\mathbf{j} + b_z\mathbf{k},$$

(6.25)

where a_x, a_y, and a_z are numerics and the **i**, **j**, and **k** are the unit vectors in the respective direction: x, y, and z. In this notation the addition of two vectors **A** + **B** is

$$\mathbf{A} + \mathbf{B} = (a_x + b_x)\mathbf{i} + (a_y + b_y)\mathbf{j} + (a_z + b_z)\mathbf{k},$$

(6.26)

and the subtraction of two vectors is

$$\mathbf{A} - \mathbf{B} = (a_x - b_x)\mathbf{i} + (a_y - b_y)\mathbf{j} + (a_z - b_z)\mathbf{k}.$$

(6.27)

The dot product

$$\mathbf{A} \cdot \mathbf{B} = (a_x\mathbf{i} + a_y\mathbf{j} + a_z\mathbf{k}) \cdot (b_x\mathbf{i} + b_y\mathbf{j} + b_z\mathbf{k})$$
$$= a_xb_x(\mathbf{i} \cdot \mathbf{i}) + a_xb_y(\mathbf{i} \cdot \mathbf{j}) + a_xb_z(\mathbf{i} \cdot \mathbf{k})$$
$$+ a_yb_x(\mathbf{j} \cdot \mathbf{i}) + a_yb_y(\mathbf{j} \cdot \mathbf{j}) + a_yb_z(\mathbf{j} \cdot \mathbf{k})$$
$$+ a_zb_x(\mathbf{k} \cdot \mathbf{i}) + a_zb_y(\mathbf{k} \cdot \mathbf{j}) + a_zb_z(\mathbf{k} \cdot \mathbf{k})$$

and

$$\mathbf{i} \cdot \mathbf{j} = 0 \quad \mathbf{i} \cdot \mathbf{k} = 0 \quad \mathbf{j} \cdot \mathbf{k} = \mathbf{j} \cdot \mathbf{i} = 0 \quad \text{etc.,}$$

(6.28)

because the cosine of 90° is zero. The long expression above collapses to

$$\boxed{\mathbf{A} \cdot \mathbf{B} = a_xb_x + a_yb_y + a_zb_z \quad \text{(as before).}}$$

The cross products of the unit vectors are easily shown to be

$$\mathbf{i} \times \mathbf{j} = \mathbf{k} \qquad \mathbf{j} \times \mathbf{k} = \mathbf{i} \qquad \mathbf{k} \times \mathbf{i} = \mathbf{j}$$
$$\mathbf{i} \times \mathbf{k} = -\mathbf{j} \qquad \mathbf{j} \times \mathbf{i} = -\mathbf{k} \qquad \mathbf{k} \times \mathbf{j} = -\mathbf{i}$$
$$\mathbf{i} \times \mathbf{i} = \mathbf{j} \times \mathbf{j} = \mathbf{k} \times \mathbf{k} = 0.$$

(6.29)

The cross product, in terms of unit vectors, is

$$\mathbf{A} \times \mathbf{B} = (a_x\mathbf{i} + a_y\mathbf{j} + a_z\mathbf{k}) \times (b_x\mathbf{i} + b_y\mathbf{j} + b_z\mathbf{k})$$
$$= a_xb_x(\mathbf{i} \times \mathbf{i}) + a_xb_y(\mathbf{i} \times \mathbf{j}) + a_xb_z(\mathbf{i} \times \mathbf{k})$$
$$+ a_yb_x(\mathbf{j} \times \mathbf{i}) + a_yb_y(\mathbf{j} \times \mathbf{j}) + a_yb_z(\mathbf{j} \times \mathbf{k})$$
$$+ a_zb_x(\mathbf{k} \times \mathbf{i}) + a_zb_y(\mathbf{k} \times \mathbf{j}) + a_zb_z(\mathbf{k} \times \mathbf{k}).$$

Using Eqs. 6.29, the expression above reduces to

$$\boxed{\mathbf{A} \times \mathbf{B} = (a_yb_z - a_zb_y)\mathbf{i} + (a_zb_x - a_xb_z)\mathbf{j} + (a_xb_y - a_yb_x)\mathbf{k}.}$$

(6.30)

For those of you who remember determinants from high school algebra (used for solving simultaneous linear algebraic equations), expression (6.30) can be written as a three-row determinant:

$$\mathbf{A} \times \mathbf{B} = \begin{Vmatrix} \mathbf{i} & \mathbf{j} & \mathbf{k} \\ a_x & a_y & a_z \\ b_x & b_y & b_z \end{Vmatrix} \tag{6.31}$$

EXAMPLE 1 Consider a particle with momentum $P(1) = 6$ MeV/c which is traveling at an angle $\theta_p = -30°$ with respect to the positive x-axis as shown in Fig. 6.13. The components of this vector are

$$P_x(1) = P \cos \theta_p = (6 \text{ MeV}/c) \cos (-30°)$$
$$= 5.2 \text{ MeV}/c$$
$$P_y(1) = P \sin \theta_p = (6 \text{ MeV}/c) \sin (-30°)$$
$$= -3 \text{ MeV}/c.$$

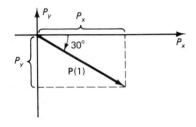

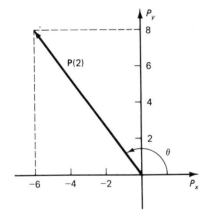

Fig. 6.13

A second particle's momentum is described by its components, $P_x(2) = -6$ MeV/c and $P_y(2) = 8$ MeV/c (Fig. 6.14). The magnitude of the momentum $|\mathbf{P}(2)|$ is found from Eq. 6.23:

$$|\mathbf{P}(2)| = (P_x^2 + P_y^2)^{1/2}$$
$$= [(-6)^2 + (8)^2]^{1/2} = 10 \text{ MeV}/c.$$

The direction is found by

$$\tan \theta = \frac{P_y}{P_x} = \frac{8 \text{ MeV}/c}{-6 \text{ MeV}/c} = -1.33 \text{ in the second quadrant}$$
$$\theta = 180° - \tan^{-1} 1.33 = 180° - 53° = 127°.$$

Now let's imagine that these two particles collide. What is the initial *total momentum* of the system before they collide? Using unit vectors, we find that the

$$\text{total momentum} = \mathbf{P}_T = \mathbf{P}(1) + \mathbf{P}(2)$$
$$= [5.2 + (-6)]\mathbf{i} + (-3 + 8)\mathbf{j}$$
$$\mathbf{P}_T = -0.8i + 5j \qquad (\text{MeV}/c).$$

Fig. 6.14

The magnitude of the initial total momentum is

$$|\mathbf{P}_T| = [(-0.8)^2 + (5)^2]^{1/2} = 5.06 \text{ MeV}/c$$

and the total momentum is directed at the angle θ_T, given by

$$\tan \theta_T = \frac{P_y}{P_x} = \frac{5 \text{ MeV}/c}{-.8 \text{ MeV}/c} = -6.25$$
$$\theta_T = 180° - \tan^{-1}(-6.25) = 180° - 80.9 = 99.1°.$$

If for some reason you want to find the angle between $\mathbf{P}(1)$ and $\mathbf{P}(2)$ (we shall call ϕ) you can find it as follows:

$$\cos \phi = \frac{\mathbf{P}(1) \cdot \mathbf{P}(2)}{\mathbf{P}(1)\|\mathbf{P}(2)\|} = \frac{5.2 \times -6 + (-3) \times 8}{6 \times 10} = -.92$$

$$\phi = 157°.$$

The direction of the momentum of particle 1 was given to be $-30°$ and we determined the direction of the momentum of particle 2 to be $127°$; thus the angle between them is $127° + 30° = 157°$, which checks.

6.3 Analysis of Bubble Chamber Photographs

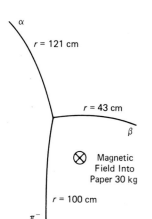

α

$r = 121$ cm

$r = 43$ cm

β

⊗ Magnetic
Field Into
Paper 30 kg

$r = 100$ cm

π^-

Fig. 6.15

NOW THAT you have mastered the algebra of vectors, it is time to use it, together with the conservation laws, to identify the tracks recorded in a photograph of two-dimensional high-energy collisions. We begin with an uncomplicated event shown in Fig. 6.15. This experiment used an incident beam of π^--mesons that entered a liquid-hydrogen chamber from below and struck a stationary proton. The collision process is

$$\pi^- + \text{p} \rightarrow \alpha + \beta.$$

How does one know that the π^--meson strikes a proton? Is it not possible that the π^--meson decays into two other particles, or possibly even more particles, some of which have no charge and therefore leave no track?

It is possible that our initial guess was wrong. Only a *complete* analysis can verify our assumptions. In other words, there is no cookbook procedure to analyzing bubble chamber events in real research; one must *guess* what has happened and see if it holds up under analysis. Other information on a bubble chamber photograph, such as the bubble density, the length of the track, and the variations in radius with distance, is often essential in unscrambling complicated events. We will not consider them in our discussion and they are not needed to analyze the events presented in this book. For this first example, I have given you the correct interpretation.

A brief examination of the event gives us the sign of the charge of both α and β. As we discussed in Chapter 3, the direction of the magnetic force on a charged particle depends on the sign of the particle. For the magnetic field into the paper, negatively charged particles curve clockwise and positively charged particles curve counterclockwise. Obviously, β has a negative charge and α a positive charge. Since almost all free elementary particles have the same amount of charge (e) as the electron, we will assume that α and β

have $+e$ and $-e$, respectively. Does our collision process obey conservation of charge?

$$\pi^- + \text{p} \rightarrow \alpha + \beta$$

$$-e + (+e) \rightarrow +e + (-e)$$

$$0 \rightarrow 0.$$

The answer is obviously yes.

What can we learn about α and β from the conservation of energy? Since we know the incident particle is a π^--meson, its rest energy tabulated in Appendix B is

$$E_{\text{rest}}(\pi^-) = 139.58 \text{ MeV}.$$

But how do we obtain its total energy, which is the sum of its rest energy and its kinetic energy? There are two ways of calculating the total energy, but both require a calculation of the magnitude of the linear momentum of the particle. We learned from Eq. 3.4 that there is a unique relationship between the momentum of a singly charged particle moving perpendicular to a magnetic field and its radius of curvature in that magnetic field. Equation 3.4 can be written in a more useful form as

$$\boxed{\text{MAGNITUDE OF LINEAR MOMENTUM} \equiv |\mathbf{p}| = (.3)\,rB,} \qquad (6.32)$$

where $|\mathbf{p}|$ is measured in MeV/c, r is in centimeters, and B is the magnetic field strength in units of kilogauss (kG).* This is an important practical relationship to use in your analysis of bubble chamber data.

From the data given in Fig. 6.15, one can easily calculate the magnitude of the momentum of the π^-:

$$|\mathbf{P}(\pi^-)| = .3 \times 10^2 \text{ cm} \times 30 \text{ kG} = 9 \times 10^2 \text{ MeV}/c. \qquad (6.33)$$

One way to calculate the total energy is to recall that

$$E_T = \gamma mc^2 \qquad \text{and} \qquad p = \gamma mv,$$

and by solving the second equation for γ, and substituting into the first, one can obtain the total energy. However, the mathematically

---

* The units in this equation are crazy; they are certainly not units that one uses in basic physics equations. The constant (.3) must have units of $(\text{MeV}/c)(\text{kG-cm})^{-1}$ for Eq. 6.20 to make sense dimensionally. This equation is introduced at this point in our discussion to make it easy for you to convert a measurement on a two-dimensional bubble chamber photograph to a determination of the magnitude of the particles' linear momentum. This is the type of equation one might give to a technician who is making measurements on the bubble chamber photographs but who does not have a physics background. In Chap. 9 we derive Eq. 6.32 from the fundamental relationship, Eq. 3.2.

simpler method is to use the magnitude equation in units of MeV:

$$E_T^2 = p^2c^2 + m^2c^4$$

or, substituting,

$$E_T^2 = (9 \times 10^2 \text{ MeV}/c)^2c^2 + (139.58 \text{ MeV})^2 = 82.95 \times 10^4 \text{ MeV}^2$$

$$E_T(\pi^-) = 910.76 \text{ MeV}.$$

Now we have all of the important parameters of the incident projectile particle. We know its charge, rest mass, total energy, and linear momentum. To obtain these parameters for α and β, we must use conservation of linear momentum and energy. But to use conservation of momentum, we must know the directions of the momenta at the point of impact for all four particles. The target particle, the proton, is easiest, because it is at rest and has no momentum. But what about the other three? Because these charged particles are moving in a magnetic field, the direction of their momentum is changing along the track. Consider the single curved track shown in Fig. 6.16. The direction of the momentum is changing from position A to B to C, but the *magnitude* of the vector momentum does *not* change. If we are going to use the linear momentum conservation law, we must consider the momentum of each particle *at the point of impact*, called the VERTEX of the collision.

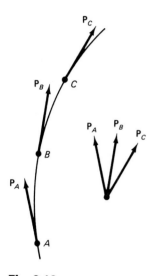

Fig. 6.16

The direction of the linear momentum at the vertex can be constructed geometrically by drawing the *tangent* to the track *at that point*. This must be done extremely carefully because an error in this construction may lead to serious errors in the determination of the properties of the particle. Figure 6.17 shows the geometrical construction of such tangents. Note that we have intentionally set the y-axis along the direction of the incident π^--meson's momentum. This is *not* essential, but it simplifies the calculation.

From the radius of curvature we can determine the magnitude of the momentum for α and β:

$$|\mathbf{P}(\alpha)| = (.3) \times 121 \times 30 = 1089 \text{ MeV}/c$$

$$|\mathbf{P}(\beta)| = (.3) \times 43 \times 30 = 387 \text{ MeV}/c.$$

Let us see if we can account for all the initial momentum of the system in the final products α and β, or whether there are uncharged particles (not observed in the photograph) which we must include in our calculations. *We will examine each component of the momentum separately.*

The x-direction:

$$P_x \text{ (total initial)} = 0$$

$$P_x \text{ (total final)} = 387 \text{ MeV}/c \text{ (cos 19.2°)} - 1089 \text{ MeV}/c \text{ (cos 70.4°)}$$

$$= 365.47 - 365.30 = .17 \text{ MeV}/c.$$

Only .17 MeV/c of momentum in the x-direction is *not* accounted for. Does this mean that a neutral particle carried away .17 MeV/c

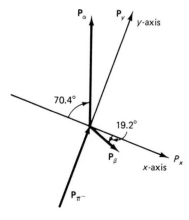

Fig. 6.17

momentum in the positive x-direction? Not necessarily. The observed final momentum of the two particles is about 365 MeV/c; the .17 MeV/c excess is $.17/365 \times 100\% = .05\%$ of the *measured* momentum. The measurements of radius and angle are typically accurate only to 2%. They are not sufficiently precise to justify believing that this .17 MeV/c represents a particle rather than reflecting the errors inherent in the measurement of radius and collision angles. We assert that there is no x-component of momentum that is unaccounted for.

The y-direction:

$$P_y \text{ (total initial)} = 900 \text{ MeV/}c$$

$$P_y \text{ (total final)} = 1089 \sin 70.4° - 387 \sin 19.2°$$

$$= 1025.9 - 127.2 = 898.6 \text{ MeV/}c.$$

Again the agreement is remarkable, leaving only .15% of the initial y-momentum unaccounted for. We are clearly justified in stating that all the momentum of the final particle is accounted for by α and β and that no third, neutral particle participated in the collision.

Now we turn to the principle of conservation of total energy. For this process, the scalar equation can be written

$$E_T(\pi^-) + E_R(\text{p}) = E_{\text{KE}}(\alpha) + E_R(\alpha) + E_{\text{KE}}(\beta) + E_R(\beta). \quad (6.34)$$

Substituting in what is determined, we have

$$910.76 + 938.26 = 1849.02 \text{ MeV}$$

$$= E_{\text{KE}}(\alpha) + m_\alpha c^2 + E_{\text{KE}}(\beta) + m_\beta c^2. \quad (6.35)$$

The magnitude equations for each unknown particle are

$$\begin{aligned} E_\alpha^2 &= (1089 \text{ MeV/}c)^2 c^2 + m_\alpha^2 c^4 \\ E_\beta^2 &= (387 \text{ MeV/}c)^2 c^2 + m_\beta^2 c^4. \end{aligned} \quad (6.36)$$

Now we are stuck! We have more unknowns than we have independent equations. Mathematically, the problem *cannot* be solved. But this is *not* just a problem in mathematics. It is a physics problem, and we can try to use other physics information. What I am saying is that you are free to *guess*; yes, guess the answer (or part of the answer) and see if it works. It is now time to assume that you know the rest mass of particle α and β; that is, you know the identity of these collision products. Since we already know that no other neutral particles are involved in this event (from momentum conservation), the most logical assumption is that α and β are the particles that were involved in this collision; that is, the proton and the π^--meson just "bounce" off each other. Still, it is not clear which is which. We recall that the direction of curvature tells us that

α-positive charge $\rightarrow$ proton β-negative charge $\rightarrow$ π^--meson.

Now Eqs. 6.35 and 6.36 reduce to a solvable form: conservation of total energy

$$1849.02 \text{ MeV} = E_T(\alpha) + E_T(\beta)$$

and from the magnitude equation,

$$E_T(\alpha)^2 = (1089 \text{ MeV})^2 + (938.26 \text{ MeV})^2 \qquad E_T(\alpha) = 1437.44 \text{ MeV}$$

$$E_T(\beta)^2 = (387 \text{ MeV})^2 + (139.58 \text{ MeV})^2 \qquad E_T(\beta) = 411.40 \text{ MeV}.$$

Adding the last two yields

$$E_T(\alpha) + E_T(\beta) = 1848.85 \text{ MeV}.$$

The agreement is fortuitous, within .01% (better than we had any right to expect).

We have demonstrated that this is an elastic collision of a π^--meson with a stationary proton. You might want to make another guess for this collision and see if other particles, with other rest masses, will satisfy all the conservation laws. In this collision, as your intuition might lead you to expect, the light particle bounces backward and the proton goes nearly straight forward, carrying with it most of the kinetic energy. Whoops, I didn't show the last statement. Let's do it.

$$E_{\text{KE}}(\alpha) = E_T(\alpha) - E_R(\alpha)$$

$$= 1437.44 - 938.26 = 499.18 \text{ MeV}$$

$$E_{\text{KE}}(\beta) = 411.40 - 139.58 = 271.82 \text{ MeV}$$

total kinetic energy of final particles $E_{\text{KE}}(\alpha) + E_{\text{KE}}(\beta) = 771.0 \text{ MeV}$

$$E_{\text{KE,Initial}}(\pi^-) = E_T(\pi^-) - E_R(\pi^-) = 910.76 - 139.58 = 771.18 \text{ MeV}.$$

We see that kinetic energy (as well as total energy) is conserved. Collisions that conserve *kinetic energy* are called ELASTIC COLLISIONS. In particle physics, an elastic collision is one where no new particles are created and none are destroyed. All elastic collisions conserve kinetic energy as well as total energy.

That was a long-winded account of a rather straightforward problem. I hope such verbosity will clarify, at the outset, the details of this kind of analysis.

We should examine a different kind of track, one that is the result of an unstable particle's decay. In Fig. 6.18 we have a replica of the tracks left by the decay of a Σ^- particle. We see only two tracks, one left by the Σ^- and a second, the result of a negatively charged particle moving to the right, which we will label C^-.

Obviously something is happening at the vertex A, since the decay products that leave a *visible* track do *not* conserve momentum. To analyze this event, we will postulate the existence of a second decay product which has no charge, and call it "0." If we construct the proper tangents to the curved tracks at the vertex and

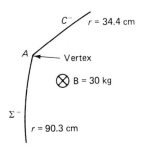

Fig. 6.18

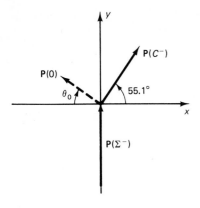

Fig. 6.19

sketch in the postulated neutral particle, the event appears as in Fig. 6.19. Consider conservation of momentum first. The measured radii give the magnitude of the linear momentum as

$$|\mathbf{P}(\Sigma^-)| = (.3)(90.3 \text{ cm})(30 \text{ kG}) = 812.7 \text{ MeV}/c$$

$$|\mathbf{P}(C^-)| = (.3)(34.4 \text{ cm})(30 \text{ kG}) = 309.6 \text{ MeV}/c.$$

Momentum conservation can be applied by setting the initial momentum in the x-direction equal to the final momentum in the x-direction, and doing the same for the y-axis.

x-comp.:

$$0 = 309.6 \cos 55.1° - P(0) \cos \theta_0$$

or

$$P(0) \cos \theta_0 = 177.1 \text{ MeV}/c. \tag{6.37}$$

y-comp.:

$$812.7 = 309.6 \sin 55.1° + P(0) \sin \theta_0$$

or

$$P(0) \sin \theta_0 = 558.8 \text{ MeV}/c. \tag{6.38}$$

Dividing Eq. 6.38 by Eq. 6.37 gives us

$$\tan \theta_0 = 3.155 \qquad \theta_0 = 72.4°.$$

Substituting this value of θ_0 into Eq. 6.38, we obtain the magnitude of $P(0)$, or

$$P(0) = \frac{558.8 \text{ MeV}/c}{\sin 72.4°} \qquad \text{or} \qquad P(0) = 586.2 \text{ MeV}/c. \tag{6.39}$$

That is as far as we can go using only momentum conservation. Now we must apply energy conservation principles. Using the magnitude equation, one can determine the initial total energy of the Σ^- particle ($m_{\Sigma^-}c^2 = 1197.4$ MeV):

$$E_T(\Sigma^-) = [(812.7)^2 + (1197.4)^2]^{1/2} = 1447.2 \text{ MeV}.$$

Energy conservation requires that

$$E_T(\Sigma^-) = 1447.2 \text{ MeV}$$

$$= E_R(C^-) + E_{KE}(C^-) + E_R(0) + E_{KE}(0). \tag{6.40}$$

But we cannot solve this equation without knowing the rest mass of 0 or C^-, both of which are unknown. Again we must guess the result. Let us assume that

$$C^- \text{ is a } \mu\text{-meson} \qquad m_\mu\text{-}c^2 = 105.7 \text{ MeV}.$$

Since we already know the momentum of this particle (from its curvature), we can determine its total energy from the magnitude equation.

$$E_T(\mu^-) = [(309.6)^2 + (105.7)^2]^{1/2} = 327.1 \text{ MeV}.$$

Substituting into Eq. 6.40, we can determine $E_T(0)$:

$$E_T(0) = 1447.2 - 327.1 = 1120 \text{ MeV}.$$

The rest energy of an unseen particle can be calculated from the magnitude equation, since we know the magnitude of the momentum of the unseen particle.

$$E_R(0) = m_0c^2 = [(1120)^2 - (586.2)^2]^{1/2} = 954.4 \text{ MeV}.$$

But there is no particle with such a rest mass! Try again! This time assume that C^- is a π^--meson, $m_{\pi^-}c^2 = 139.6$ MeV.

$$E_T(\pi^-) = [(309.6)^2 + (139.6)^2]^{1/2} = 339.6 \text{ MeV}$$

$$E_T(0) = 1447.2 - 339.6 = 1107.5 \text{ MeV}$$

$$m_0c^2 = [(1107.5)^2 - (586.2)^2]^{1/2} = 939.7 \text{ MeV}.$$

The neutron has a mass of 939.6 MeV, which is within .02% of the rest mass of our unknown particle. Thus we identify the neutral particle as a neutron moving off at an angle of 72.4° with respect to the x-axis and the charged particle, which left its track as a π^--meson.

6.4 Zero-Rest-Mass Particles

ZERO-REST-MASS particles require a little special attention because some of the equations we have developed so far are not appropriate for these particles. The obvious question is: Does a zero-rest-mass particle have momentum, since momentum is defined as γmv? The answer is yes, they do have both energy and momentum; but no, the expression γmv does not apply to zero-rest-mass particles. We examine two examples of important zero-rest-mass particles, the photon and the neutrino, and learn how to analyze high-energy interactions where these particles are involved.

A. Photons

To remind you of our discussion of the photon in previous chapters, let me list its basic characteristics:

1. Zero rest mass
2. Zero charge
3. Energy related to its frequency by Planck's equation: $E = hf$
4. Always has velocity (in vacuum): $c = 2.9979 \times 10^8$ m/s
5. Family (membership) *not* conserved
6. Can be absorbed by and emitted from other particles

How do we handle such a creature in the type of collision processes we are considering? The answer comes directly from the magnitude equation:

$$E_T^2 = p^2c^2 + m^2c^4.$$

If the rest mass is identically zero, this equation takes the particularly simple form

$$E_T \text{ (photon)} = pc, \tag{6.41}$$

where p is the magnitude of the linear momentum. This and

$$E_T \text{ (photon)} = hf$$

An example of an electron–positron pair production in the 80-inch bubble chamber at the 33 BeV Alternating Gradient Synchrotron at Brookhaven National Laboratory. The electron–positron pair was created from the decay of a neutral particle which left no track in the chamber.

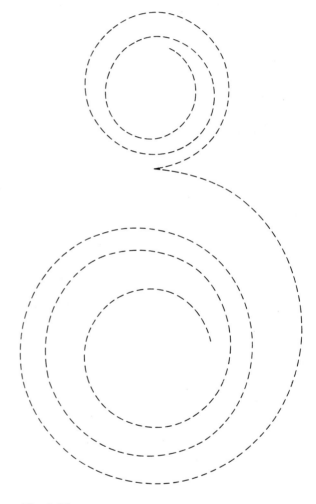

Fig. 6.20

are the only "new" equations you will need to solve collision dynamic problems involving photons. Do not confuse them! One relates the energy of a photon to its linear momentum, the other to its frequency of oscillation. They are independent relationships.

EXAMPLE 1 Consider the simplified drawing of the bubble chamber Photo 6.1 shown in Fig. 6.20. The two spirals are the signature of a positron–electron pair (called pair production), produced by a high-energy photon (gamma ray) which leaves no track in the chamber. The tracks form a spiral (a circle with an ever-decreasing radius) because both the positron and the electron interact with the atoms in the liquid and rapidly lose energy and momentum while moving through the chamber. Nevertheless, by measuring the radius *near the vertex*, one can accurately determine the initial momentum of both the positron and electrons.

It might appear that this event can be written as $\gamma \rightarrow e^+ + e^-$, but this is *not* correct. To demonstrate that this reaction violates the conservation laws, we will *assume* that it is correct and proceed with the analysis. To simplify the mathematics we will consider the electron and positron to have the same angle with respect to the direction of the initial photon's momentum ($\theta_{e^-} = \theta_{e^+} = \theta$) Fig 6.21.

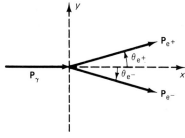

Fig. 6.21

Conservation of momentum:

x-comp.: $\qquad P(\gamma) = P(e^-) \cos \theta + P(e^+) \cos \theta$

y-comp.: $\qquad 0 = P(e^+) \sin \theta - P(e^-) \sin \theta.$

The y-component shows that

$$P(e^+) = P(e^-).$$

The magnitude of the electron's momentum equals the magnitude of the positron's momentum. That fact reduces the first equation to

$$P(\gamma) = 2P(e) \cos \theta.$$

Conservation of energy:

$$E_T(\gamma) = E_T(e^+) + E_T(e^-) = 2E_T(e).$$

But for a photon

$$E_T(\gamma) = P(\gamma)c \qquad \text{or} \qquad P(\gamma) = \frac{E_T(\gamma)}{c}.$$

Combining energy and momentum conservation equations, we obtain

$$\frac{E_T(\gamma)}{c} = 2P(e) \cos \theta \qquad \text{or} \qquad E_T(\gamma) = 2cP(e) \cos \theta = 2E_T(e)$$

or

$$E_T(e) = cP(e) \cos \theta.$$

This equation is nonsense! Consider the maximum value of the $\cos \theta$, which is 1. Then the equation becomes $E_T(e) = P(e)c$. But for any nonzero-rest-mass particle, the magnitude equation states that

$$E_T(e) = [(P(e)c)^2 + m_e^2 c^4]^{1/2}.$$

These two equations disagree! Our original assumption must be incorrect.* What we have forgotten is frequently not observable in bubble chamber pictures: a stationary particle in this case, a nucleus. The photon can only undergo such a transformation near a material object (a nucleus or atom) that will take up the recoil momentum. The real process is

$$\gamma + N \rightarrow N + e^+ + e^-,$$

where N stands for the nucleus. However, a measurement of the total energy of e^+ and e^- gives a very precise measurement of the initial γ-ray's energy, since the relatively massive nucleus generally takes up very little energy in the recoil.

EXAMPLE 2
The Compton Effect

In 1923 at Washington University in St. Louis, Missouri, Arthur Holly Compton discovered that x-rays (or γ-rays) appear to behave like classical particles in certain scattering experiments. This came as a great shock to the physics community, since well-known experiments clearly demonstrated that x-rays behave like classical waves.† Compton measured the energy of x-rays scattered from a carbon target as a function of the scattering angle. Figure 6.22a shows a schematic diagram of his apparatus. Figure 6.22b is the vector diagram of the scattering processes that Compton observed.

The photon enters the carbon target and scatters off one of the many essentially free electrons in the carbon, losing some energy to the electron and glancing off at the angle θ. The analysis of this collision involves a straightforward application of the conservation principles. The "trick" in this photon–electron scattering problem is to arrive at a mathematical expression that involves the experimental variables that Compton was able to measure. He was unable to measure *any* dynamical variables of the electron, but *could* measure the incident and reflected photons' energies as well as the scattering angle θ: that is, $E(\gamma)$, $E(\gamma')$, and θ; nothing else was measurable. Figure 6.22b defines all the variables in this scattering process.

Conservation of momentum:

x-comp.: $\qquad P(\gamma) = P(\gamma') \cos \theta + P(e) \cos \phi$ $\qquad$ (a)

y-comp.: $\qquad 0 = P(\gamma') \sin \theta - P(e) \sin \phi.$ $\qquad$ (b)

---

* When we learn about coordinate transformations in Chapter 10, we will show a more elegant proof that the spontaneous decay of a photon into any particle of nonzero rest mass is impossible.

† This is now called the wave–particle duality, which can be understood from the perspective of quantum mechanics.

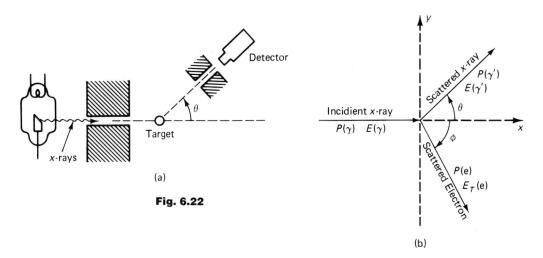

Fig. 6.22

Conservation of energy:

$$E(\gamma) + m_e c^2 = E(\gamma') + E_{\text{KE}}(e) + m_e c^2$$

or

$$E(\gamma) - E(\gamma') = E_{\text{KE}}(e). \tag{c}$$

To reduce Eqs. (a), (b), and (c) to a form that does *not* contain the dynamical variables of the electron, namely, $E(e)$, $P(e)$, and ϕ (since they are not measurable), we transpose and square Eq. (a):

$$P(\gamma)^2 - 2P(\gamma')P(\gamma) \cos \theta + P(\gamma')^2 \cos^2 \theta = P(e)^2 \cos^2 \phi. \tag{d}$$

Transposing and squaring (b) yields

$$P(\gamma')^2 \sin^2 \theta = P(e)^2 \sin^2 \phi. \tag{e}$$

Adding (d) and (e) gives us

$$P(\gamma)^2 - 2P(\gamma')P(\gamma) \cos \theta + P(\gamma')^2(\cos^2 \theta + \sin^2 \theta)$$
$$= P(e)^2(\cos^2 \phi + \sin^2 \phi). \tag{f}$$

Recall that for any angle α, $\sin^2 \alpha + \cos^2 \alpha = 1$, so Eq. (f) reduces to

$$P(\gamma)^2 - 2P(\gamma)P(\gamma') \cos \theta + P(\gamma')^2 = P(e)^2. \tag{g}$$

Now consider the magnitude equation for the *scattered* electron

$$(E_{\text{KE}}(e) + m_e c^2)^2 = c^2 P(e)^2 + m_e^2 c^4$$

or

$$\frac{E_{\text{KE}}(e)^2}{c^2} + 2E_{\text{KE}}(e)m_e = P(e)^2. \tag{h}$$

But from conservation of energy [Eq. (c)], we can substitute in $E_{\text{KE}}(e)$ into (h) and get

$$\frac{[E(\gamma) - E(\gamma')]^2}{c^2} + 2[E(\gamma) - E(\gamma')]m_e = P(e)^2. \tag{i}$$

Now we have two expressions for $P(e)^2$, Eqs. (i) and (g). They can be equated and $P(\gamma)$ and $P(\gamma')$ can be eliminated by substituting the relationship between energy and momentum for a photon:

$$E(\gamma) = P(\gamma)c \qquad \text{and} \qquad E(\gamma') = P(\gamma')c.$$

One arrives at

$$\frac{E(\gamma)^2}{c^2} - 2\frac{E(\gamma')E(\gamma)}{c^2}\cos\theta + \frac{E(\gamma')^2}{c^2}$$

$$= \frac{[E(\gamma) - E(\gamma')]^2}{c^2} + 2[E(\gamma) - E(\gamma')]m_e,$$

which reduces to

$$E(\gamma') = \frac{E(\gamma)}{\left[1 + \dfrac{E(\gamma)}{m_e c^2}(1 - \cos\theta)\right]}. \qquad (6.42)$$

This is the equation of the Compton scattering of x-rays off electrons. It shows that the energy of the scattered photon $E(\gamma')$ is a *unique* function of the scattering angle θ. Figure 6.23 shows some experimental data with the theoretical line calculated from this equation. The agreement is excellent. This experiment, with its explanation, produced a dramatic moment in physics, since it established some particlelike properties of the photon. The wave–particle "paradox" was to be an issue of hot debate for the next 10 years in physics. The interested reader can find many well-written works on this subject.*

B. Neutrinos

The July 1, 1962 issue of the *Physical Review Letters* reported the "divorce" of one of the famous families of the physical world, the lepton family. This family separation was declared by the high-energy physics group headed by Leon Lederman working on the large AGS accelerator at Brookhaven National Laboratories. They discovered that the lepton family was indeed two families, the electron family and the muon family, with separate memberships.

The experiments demonstrated that in nature there are at least *two* types of neutrinos: the electron neutrino, ν_e, and the muon neutrino, ν_μ (along with their antiparticles, $\bar{\nu}_e$ and $\bar{\nu}_\mu$). The names "electron" and "muon" come from the fact that these distinct particles are associated with different decay products. The decay of the free neutron has an electron and an electron antineutrino,

$$n \rightarrow p + e^- + \bar{\nu}_e$$

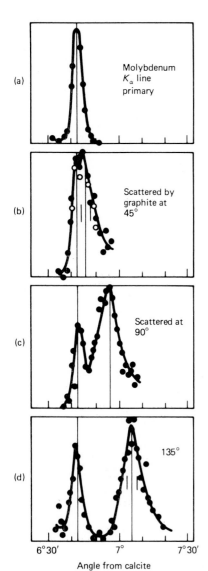

(a) Molybdenum K_α line primary

(b) Scattered by graphite at 45°

(c) Scattered at 90°

(d) 135°

6°30'   7°   7°30'

Angle from calcite

Fig. 6.23

---

* I particularly recommend the *Feynman Lectures on Physics*, Vol. II, Chaps. 37 and 38 (repeated in Vol. III), R. P. Feynman, R. B. Leighton, and M. L. Sands (Reading, Mass.: Addison-Wesley, 1963).

and the decay of the π-meson has a muon and a muon antineutrino,

$$\pi^+ \to \mu^+ + \bar{\nu}_\mu.$$

Like the photon, the neutrino is apparently a zero-rest-mass particle. The analysis of its collision dynamics involves the same methods that were used for the photon. Let us consider the example of a free stationary neutron decaying to a proton, an electron, and an electron antineutrino. If such an event were observed in a bubble chamber, it might appear as in Fig. 6.24a. The only particles observed would be the electron and the proton, since they are charged. The vector diagram representing the momentum of the particles is shown in Fig. 6.24b with the x-axis constructed parallel to the momentum of the electron.

You may ask: Where do the proton, electron, and neutrino get their kinetic energy, since the neutron was initially at rest? The answer is simply that it comes from the rest mass of the neutron, since the combined rest mass of p and e$^-$ is less than that of the neutron.

Initial   *Final*

$m_n c^2 = 939.55$ MeV $m_p c^2 = 938.26$ MeV

$m_e c^2 = .511$ MeV

$m_\nu c^2 = 0$

$m_T c^2 = 938.77$ MeV

$$m_n c^2 - m_T c^2 = .779 \text{ MeV} \qquad \text{(available energy)}.$$

This can also be seen by directly applying conservation of energy:

$$E_R(n) = E_R(p) + E_{KE}(p) + E_R(e^-) + E_{KE}(e^-) + E_{KE}(\bar{\nu}_e)$$

or

$$m_n c^2 - (m_p c^2 + m_e c^2) = .779 \text{ MeV} = E_{KE}(p) + E_{KE}(e^-) + E_{KE}(\bar{\nu}_e).$$

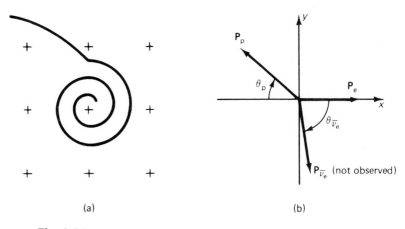

(a)   (b)

Fig. 6.24

We will not put numbers in, but rather, just set up the useful equations. Measurements of the bubble chamber photograph would establish the numerical values of $P(p)$, $P(e)$, and θ_p. The unknowns are

$$E_{KE}(p) \qquad E_{KE}(e^-) \qquad E_{KE}(\bar{\nu}_e).$$

This can be determined from the following:

Conservation of momentum:

x-comp.: $\qquad P(e) + P(\bar{\nu}_e) \cos \theta_{\bar{\nu}_e} - P(p) \cos \theta_p = 0$

y-comp.: $\qquad P(p) \sin \theta_p - P(\bar{\nu}_e) \sin \theta_{\bar{\nu}_e} = 0.$

Magnitude equation:

Proton: $\qquad E_T(p)^2 = P(p)^2 c^2 + (938.26 \text{ MeV})^2$

Electron: $\qquad E_T(e)^2 = P(e)^2 c^2 + (.511 \text{ MeV})^2$

Neutrino: $\qquad P(\bar{\nu}_e) = \dfrac{E(\bar{\nu}_e)}{c}.$

6.5 **Particle Creation**

WE HAVE EXAMINED particle decays, as in the last example, where the mechanism for decay is the weak interactions, and decays such as $\Lambda^0 \rightarrow p + \pi^-$, where the mechanism is the strong interactions. In these examples, the energy of the product particles, both rest and kinetic, is supplied by the parent particle. In both cases the product particles necessarily had less mass than the initial particle that decayed. But how does one create heavier particles starting with lighter ones? Conservation of energy requires that if more massive particles are to be created, the projectile must have at least an amount of kinetic energy equivalent to the additional rest energy of the new particles.

So much for the generalities. We must be much more precise, especially if we are about to design a new multibillion-dollar accelerator to find some predicted but undetected particles. The Snyder–Goldhaber $500 bet (discussed in Chapter 1) on the existence of the antiproton is a good example. In the early 1950s O. Chamberlain and E. Segrè headed a research group that some have claimed "constructed a Nobel prize." Using a particle accelerator, they set about to design an experiment to detect the existence (or nonexistence) of the antiproton. You can be quite confident that they first calculated the minimum particle energy that the accelerator would need to produce if the antiprotons were to be observed. Using a proton accelerator, one might hope that a simple reaction, such as

$$p + p \rightarrow \bar{p} \qquad \text{or} \qquad p + p \rightarrow \bar{p} + p$$

might work, but clearly they cannot; these reactions violate charge and baryon conservation. The reaction that creates the fewest particles and conserves both charge and baryon family number is

$$p + p \rightarrow \bar{p} + p + p + p,$$

which has four particles in the final state. In order to conserve baryon family number, it is necessary to create a proton and an antiproton in pairs.

The question remains: What is the minimum energy of the projectile proton for this reaction to take place? This is called the THRESHOLD ENERGY for the reaction. One might guess that a collision process, like the one shown in Fig. 6.25, is the most efficient use of the kinetic energy of the projectile proton. In this hypothetical collision, the initial projectile proton strikes a stationary target proton and four new particles emerge, *all at rest*. Since these new particles have no kinetic energy, no energy is "wasted"; it presumably all went into creating the rest mass of the products. Indeed, it is the most efficient; unfortunately, it is *not* a possible process. The reason is simply that linear momentum is not conserved. If all the product particles are at rest after the collision, the initial momentum of the system (all in the projectile proton) has disappeared in the final state of the system, violating linear momentum conservation. How then do we find the *most efficient* of all *possible* processes?

Imagine that you could view this proton–proton collision from a spaceship moving with a constant velocity in the same direction as the projectile proton. Imagine also that the ship is moving at such a velocity that the target proton *and* the projectile proton appear to an observer on board to have *equal* but oppositely directed momentum. In such a REFERENCE FRAME the total initial momentum of the system $\mathbf{p} + \mathbf{p} = p - p$ is zero. In *this* frame of reference, the four product particles could be at rest after the collision. This is shown schematically in Fig. 6.26.

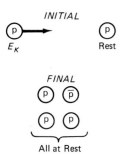

INITIAL

E_K Rest

FINAL

All at Rest

Fig. 6.25

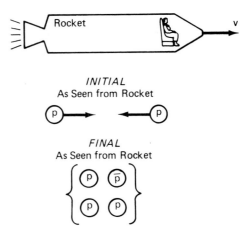

INITIAL
As Seen from Rocket

FINAL
As Seen from Rocket

Fig. 6.26

LABORATORY FRAME

INITIAL

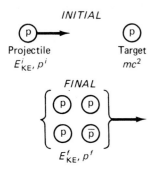

Projectile
E^i_{KE}, p^i

Target
mc^2

FINAL

E^f_{KE}, p^f

All particles have the
same velocity

Fig. 6.27

The observer in the rocketship would see the two protons colliding and the products at rest as the *most efficient* particle creation process. The reason should be clear. If any product particle had some velocity after the collision, there would have to be another particle moving with the same speed in the opposite direction (to make the total momentum add up to zero) and these particles would carry off kinetic energy. This kinetic energy could *not* be used to create the new particles.

To calculate the necessary initial kinetic energy of the projectile proton for this "minimum" particle creation process, we return to the laboratory reference frame. From the point of view of the lab observer, the minimum process involves all four product particles not at rest but moving off with the *same* velocity and thus the same kinetic energy (Fig. 6.27). We define the following variables for our collision process:

$$m = \text{mass of proton or antiproton}$$
$$K^i_{KE} = \text{initial kinetic energy of projectile proton}$$
$$E^f_{KE} = \text{final kinetic and of } each \text{ product particle}$$
$$P^i = \text{initial momentum of the projectile}$$
$$P^f = \text{final momentum of } each \text{ product particle.}$$

Conservation of energy requires that

$$E^i_{KE} + mc^2 + mc^2 = 4(E^f_{KE} + mc^2) \quad \text{or} \quad E^f_{KE} = \frac{E^i_{KE}}{4} - \frac{mc^2}{2}. \quad \text{(a)}$$

Conservation of linear momentum requires (one-dimensional problem)

$$P^i = 4P^f \quad \text{or, in energy units,} \quad cP^i = 4cP^f. \quad \text{(b)}$$

The magnitude equation for the projectile becomes

$$c^2 P^{i2} = (E^i_{KE} + mc^2)^2 - m^2c^4$$

or
$$cP^i = (E^{i2}_{KE} + 2mc^2 E^i_{KE})^{1/2}.$$

For each product particle the magnitude equation also becomes

$$cP^f = (E^{f2}_{KE} + 2mc^2 E^f_{KE})^{1/2}.$$

Substituting into the momentum conservation equation (b) yields

$$(E^{i2}_{KE} + 2mc^2 E^i_{KE})^{1/2} = 4(E^{f2}_{KE} + 2mc^2 E^f_{KE})^{1/2}.$$

Squaring and writing E^f_{KE} in terms of E^i_{KE} from Eq. (a) gives

$$E^{i2}_{KE} + 2mc^2 E^i_{KE} = 16\left(\frac{E^i_{KE}}{4} - \frac{mc^2}{2}\right)^2 + 32mc^2\left(\frac{E^i_{KE}}{4} - \frac{mc^2}{2}\right)$$

or
$$E^i_{KE} = 6mc^2.$$

Since the rest energy of the proton is about 1 GeV, we see that to create the antiproton the accelerator must be capable of giving the projectile particle 6 GeV of kinetic energy. This seems quite inefficient since the antiproton we seek has only about 1 GeV of rest energy.

Where does the energy go? The final kinetic energy of each product particle E_{KE}^f can easily be calculated from Eq. (a):

$$E_{KE}^f = \frac{E_{KE}^i}{4} - \frac{mc^2}{2} = \frac{6mc^2}{4} - \frac{mc^2}{2} = mc^2.$$

The energy balance is as follows:

$$E_{KE}^i(\text{proj.}) + E_R(\text{proj.}) + E_R(\text{target}) = 4E_R^f(\text{product}) + 4E_{KE}(\text{product})$$

$$6 \text{ GeV} + 1 \text{ GeV} + 1 \text{ GeV} = 4 \text{ GeV} + 4 \text{ GeV}.$$

Chamberlain and Segrè [*Physical Review*, **100**, 947 (1955)] designed and used a 6-GeV accelerator at Berkeley, called the bevatron, to find the antiproton and win the Nobel prize. Snyder won the $500 bet!

Although there is no simple equation or rule for all kinds of particle production, since each new particle creation process involves different products, the problem of "efficiency" becomes increasingly acute as one attempts to create more massive particles. Already accelerator dimensions are of the order of several miles and projectile kinetic energies are thousands of GeV. Still, the appetite and curiosity of high-energy physicists make them seek events of higher and higher energy. Recent technological advances have made it possible to study these processes in a more efficient way. The basic idea is to use *two* projectiles in a *head-on* collision instead of directing one particle at the target particle at rest.

Consider the antiproton example we just discussed. If two projectiles of equal kinetic energy E_{KE}^i make a head-on collision, the final products of the collision *could* be four particles at rest (Fig. 6.28). The initial momentum of the system is zero *in the laboratory frame of reference,* and the final momentum in this frame of reference must also be zero. For a head-on collision of particles with equal momentum, it is not necessary for the product particles to have any final kinetic energy to conserve linear momentum.

The analysis is trivial, requiring only conservation of energy.

$$2E_{KE}^i + 2mc^2 = 4mc^2$$

or
$$E_{KE}^i = mc^2 \quad (1 \text{ GeV}).$$

Our accelerator need only accelerate *both* particles to 1 GeV of kinetic energy, not the 6-GeV threshold of the single-particle projectile. A considerable savings!

Of course, there is a fly in the ointment. It is not easy to get two particles to collide head-on, particularly when each particle has an effective diameter on the order of 10^{-15} m. In the standard experi-

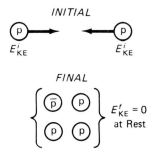

INITIAL

FINAL

$E_{KE}^f = 0$ at Rest

Fig. 6.28

ment, the target (liquid hydrogen) had a rather high density of particles (10^{24} protons/m^3) in the path of the projectile beam, as well as a long collision length (several meters in the larger chambers) over which the projectile could find a target. In the case of two colliding particles, once they have passed each other there is no chance for a collision to occur.* The number of collisions per unit time in a given volume is proportional to the number of target particles in that volume. In the standard bubble chamber experiment, with stationary proton targets, the product of the collision rate and the particle density in the beam, called LUMINOSITY, is on the order of 10^{27}.

This characteristic luminosity produces about one interaction per second for each centimeter of target thickness, assuming that the particles act as though they cast a "shadow" 10^{-27} cm^2 in area (πr^2). If one just aimed the beams of two typical accelerators at each other, the luminosity would be only 10^{18}. Although 10^{18} seems like a large number, such a crude approach would produce 10^9 *fewer* collisions per unit time than the standard fixed-target experiment! That would be roughly equivalent to waiting for a collision to take place between two 1-mm spheres thrown at each other every second by two blind men at opposite ends of a football field. It is damned hard to get much information about high-energy physics while waiting for the collision to take place in that kind of setup.

How does one devise an experiment with colliding particles and make up this factor of 10^9? Three modifications greatly improve the situation: (1) increase the number of particles in the beam, (2) focus the beam to an even smaller diameter, and (3) give each particle in the beam *many* chances to collide. These have been achieved by storing the beam in ringlike devices with special focusing magnets and keeping them confined in the storage ring for long periods of time.† All major accelerators now have an associated storage ring in operation. A schematic diagram of the early CERN instruments is shown in Fig. 6.29.

Storage rings play a major part in current experimental high-energy physics. If we plot (Fig. 6.30) the available energy (the kinetic energy of the projectile particle that could be used for creation

---

* A typical particle accelerator produces about 10^{13} particles per pulse in the beam. If the beam can be focused to a 1-mm-diameter cylinder, the ratio of the "shadow" area case by the particles (protons) to the area of the beam is simply

$$\frac{\text{area of proton}}{\text{area of beam}} = \frac{\pi r_{\text{proton}}^2}{\pi r_{\text{beam}}^2} = \frac{(10^{-13} \text{ cm})^2}{(.1 \text{ cm})^2} = 10^{-24}.$$

Although it might seem that a 1-mm focused beam of protons is a rather high concentration of particles, almost all of the space in the beam is *empty*. Now try to get two "empty" beams to collide.

† Two excellent *Scientific American* articles explain different types and uses of storage rings: Gerald K. O'Neill, "Particle Storage Rings" (Nov. 1966) and Sidney D. Drell, "Positron Annihilation and the New Particles" (June 1975).

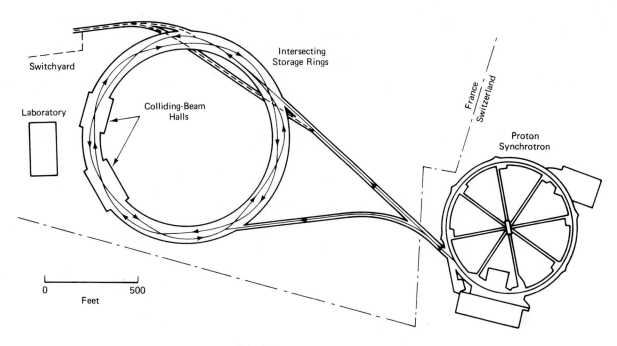

Fig. 6.29

of a new particle) for a stationary target experiment versus the projectile's energy, the results are indeed discouraging. Using relatively modest particle accelerators in conjunction with storage rings, one obtains *available* energies larger than the existing or projected large accelerators can produce on fixed targets.

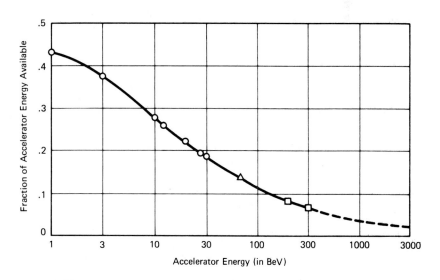

Fig. 6.30

An experimental station at CERN's Intersecting Storage Ring (ISR). Although protons fed into the ring from the accelerator will continue circulating for hours, only a very few will collide. There are eight intersecting points around the ring where special detectors, such as the one shown, observe high-energy collisions.

6.6 Newtonian "Collisions"

MOST OF YOU probably will never have to use relativistic dynamics. But as you learn to solve high-energy physics problems, you also develop important skills in using conservation principles that can also be applied to nonrelativistic classical physics and engineering problems. Some differences should be kept in mind for Newtonian mechanics, so let's consider them now.

We start with the example of a highly idealized game of pool. These idealizations greatly simplify the mathematics and physics. We will make two assumptions:

1. The billiard balls *slide* and do not roll (like hockey pucks).
2. The collision between two balls is perfectly elastic (i.e., kinetic energy is conserved in the collision).

The first idealization is necessary since a rolling or spinning ball has kinetic energy associated with the rotational motion, which at the moment we are not ready to treat. The assumption of elastic collision is quite reasonable for a good pool ball. If this assumption is not made, some of the cue ball's initial kinetic energy would be converted into random *internal* motion of both the cue ball and the target (the eight ball). This internal motion is known as INTERNAL ENERGY. The collision process would cause a temperature rise in both balls as well as a transfer of kinetic energy from cue ball to eight ball. We will not include these processes in our analysis.

The problem to be solved is a two-body collision shown in Fig. 6.31. Two sliding identical balls collide in such a way that the stationary ball rebounds at an angle of 30°. The problem is to find both the angle of recoil of the projectile particle [if the initial speed of the first particle $v^i(1) = 10$ m/s] and the final velocity $\mathbf{v}^f(1)$ and $\mathbf{v}^f(2)$ of each ball.

We could attack this problem using the relativistic expressions for momentum and energy, but this could require unnecessary work. Because this is a low-velocity collision, with no creation of new particle or mass–energy conversion, the problem can be solved using the Newtonian expressions for kinetic energy and linear momentum. To remind you, these equations are $\mathbf{p} = m\mathbf{v}$ and $E_{\text{KE}} = \frac{1}{2}mv^2$. Conservation of kinetic energy (elastic collision) requires that

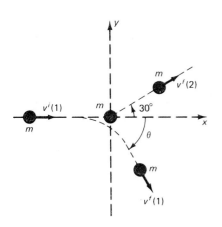

Fig. 6.31

$$\tfrac{1}{2}mv^i(1)^2 = \tfrac{1}{2}mv^f(1)^2 + \tfrac{1}{2}mv^f(2)^2$$

or

$$v^i(1)^2 = v^f(1)^2 + v^f(2)^2. \qquad \text{(a)}$$

Conservation of linear momentum requires:

  x-comp.:

$$mv^i(1) = mv^f(1)\cos\theta + mv^f(2)\cos 30°$$

or

$$v^i(1) = v^f(1)\cos\theta + (.866)v^f(2) \qquad \text{(b)}$$

  y-comp.:

$$0 = mv^f(2)\sin 30° - mv^f(1)\sin\theta$$

or

$$\tfrac{1}{2}v^f(2) = v^f(1)\sin\theta. \qquad \text{(c)}$$

Equations (a), (b), and (c) are three independent equations with three unknowns. They can be solved in the straightforward manner: transposing (b), squaring, and adding to (c) squared (the same mathematical procedure we used in the Compton effect) gives

$$[v^i(1) - (.866)v^f(2)]^2 = v^f(1)^2 \cos^2\theta$$

$$\tfrac{1}{4}v^f(2)^2 = v^f(1)^2 \sin^2\theta.$$

Adding yields

$$v^i(1)^2 - (1.73)v^i(1)v^f(2) + v^f(2)^2 = v^f(1)^2.$$

Substituting from (a) for $v^f(1)^2$, we obtain

$$v^f(2) = .866v^i(1) = 8.66 \text{ m/s}.$$

The final velocity of the projectile, from Eq. (a),

$$v^f(2) = [v^i(1)^2 - v^f(1)^2]^{1/2} = 5 \text{ m/s}.$$

The scattering angle θ from Eq. (c) is

$$\sin \theta = \frac{v^f(2)}{2v^f(1)} = .866 \qquad \theta = 60°.$$

A second example is an inelastic Newtonian collision. Consider a 15,000-kg (about 15 tons) truck traveling at 25 m/s (about 63 mi/hr) on a two-lane ice-covered highway that makes a head-on collision with a 2000-kg automobile moving in the opposite direction at a speed of 40 m/s (100 mi/hr). If the two vehicles stick together (as might very well happen in such a collision), find the final velocity of both just after the collision. What fraction of the total initial kinetic energy has been lost in the collision? Where has this energy gone?

Although *kinetic* energy is *not* conserved in such an inelastic collision (collisions that either create new particles or in which some of the kinetic energy is transformed into internal energy of the colliding bodies), linear momentum is *always conserved* if the net external force that acts on the system is zero. As long as we ignore the friction with the road and the air, it is a good approximation to assume that this system is isolated. Is this a relativistic process? The "litmus" test for relativity is the magnitude of $\dfrac{v^2}{c^2}$ compared to 1:

$$\frac{v^2}{c^2} = \frac{(40 \text{ m/s})^2}{(3 \times 10^8 \text{ m/s})^2} = 1.8 \times 10^{-14}.$$

It is clearly *not* a relativity problem. Not even close! We define

$$m_T = \text{mass of truck} \qquad v_c = \text{velocity of car}$$

$$m_c = \text{mass of car} \qquad v_T = \text{velocity of truck}.$$

$$v_f = \text{velocity of truck and car together}$$

Conservation of linear momentum requires that

$$m_T v_T - m_c v_c = (m_T + m_c)v_f$$

or

$$v_f = \frac{m_T v_T - m_c v_c}{m_T + m_c} = \frac{(1.5 \times 10^4)(25) - (2 \times 10^3)(40)}{2 \times 10^3 + 15 \times 10^3} = 17.4 \text{ m/s}.$$

The initial total kinetic energy of the system before the collision was

$$E^i_{KE} = \tfrac{1}{2}m_T v_T^2 + \tfrac{1}{2}m_c v_c^2$$

$$= 4.7 \times 10^6 + 1.6 \times 10^6 = 6.3 \times 10^6 \text{ J}.$$

The final kinetic energy was

$$E_{\text{KE}}^{f} = \tfrac{1}{2}(m_T + m_c)v_f^2 = 2.57 \times 10^6 \text{ J}.$$

The percentage change in kinetic energy is then

$$\frac{E_{\text{KE}}^{i} - E_{\text{KE}}^{f}}{E_{\text{KE}}^{i}} \times 100\% = \frac{6.3 \times 10^6 - 2.57 \times 10^6}{6.3 \times 10^6} \times 100\% = 59\%.$$

Fifty-nine percent of the initial kinetic energy has been converted into the internal energy of the smashed vehicles.

Historically, physics has divided collision processes into two categories, labeled elastic and inelastic. To correct a wrong impression in some students' minds, I wish to point out quite emphatically that *energy* is conserved for *both* kinds of collisions. What, then, is the essential difference? In elastic collisions *kinetic energy is conserved*.

If we think about ordinary everyday events, the collision of two billiard balls is to a very high degree an elastic collision, whereas the collision of two automobiles is inelastic. In the billiard ball problem we can use an equation such as

(KINETIC ENERGY)_{BEFORE} = (KINETIC ENERGY)_{AFTER}   *elastic*.

This is clearly not applicable in the automobile wreck. What happened to the initial kinetic energy of the automobile?

The answer is simply that the energy has gone ultimately into elastic energy, and possibly light, and internal vibrational energy in a complicated way. The neatly ordered form of kinetic energy before the collision has been transformed in too many "disorganized" forms of energy. *But no energy is lost!* All the energy can, in principle, be accounted for. For particle physics, "inelastic collision" means *the creation of new particles after the collision*. We can account for the total energy in all particle collisions, whether they are elastic or inelastic.

6.7 **Rocket Travel**

NASA DEPENDS on momentum conservation for its very existence (although the U.S. Congress would take exception to this statement). Automobiles, planes, and boats depend on friction for their propulsion along the earth's surface. But rockets propel themselves *not* by pushing against air, but literally by breaking themselves into pieces and throwing away their parts with as great a velocity as they can. Of course, some of its pieces were designed to be thrown away, such as the fuel, which is usually exhausted in the form of a high-velocity stream of gas at the rear of the rocket. The basic principles

Launch of the space shuttle by NASA

Before "Burning"

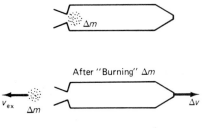

After "Burning" Δm

v_{ex} Δm Δv

Fig. 6.32

of rocket motion can be explained quite easily by using conservation of linear momentum.

Consider a rocket of mass M at rest in outer space where there is no gravitational force. (We might imagine that the rocketship is being observed from a nearby asteroid, and that the rocketship is at rest with respect to that asteroid.) If the astronaut wants to get the rocket moving, he cannot grab, push, or pull on anything. In fact, to move the rocket, he must discard part of it. By burning the fuel in the engine, he takes an amount of mass Δm and gives it a velocity $v_{exhaust}$ with respect to the rocket out the rear (Fig. 6.32). Conservation of momentum for this isolated system requires the *total* momentum to remain the same before and after the burning, or

$$\mathbf{P}_{Before} = \mathbf{P}_{After}$$

or

$$0 = -\Delta m \, \mathbf{v}_{ex} + (M - \Delta m) \, \Delta \mathbf{v}$$

and thus

$$\Delta \mathbf{v} = \frac{\Delta m \, \mathbf{v}_{ex}}{M - \Delta m}$$

Since $m \gg \Delta m$,

$$\Delta \mathbf{v} \approx \frac{\Delta m \, \mathbf{v}_{ex}}{M}.$$

So the rocket now has less mass $(M - \Delta m)$ but is moving with a velocity $\Delta \mathbf{v}$, presumably toward home. This principle is also used to stabilize the orbits of space satellites to prevent them from tumbling and gyrating in space. Small jets of gas are mounted in various positions on the vehicle and the astronauts can change the motion of their capsule by ejecting streams of high-velocity gas through the proper jet.

In a real accelerating rocket, fuel is continuously exhausted from the engines and the rocket continuously changes its velocity. This is readily understood by considering the next moment in our example. After the first puff of gas is released, the rocketship changes its velocity from zero to $\Delta \mathbf{v}$; if a second burst is emitted, its mass will again decrease by an amount Δm, and its velocity will also increase. Thus the engine will cause a continual *change in the velocity* of the rocket (called acceleration) until the desired velocity is obtained (or it runs out of fuel).

We have, as you recall, also ignored such things as the gravitational field and the friction of the air when the rocket is near Earth. These will be considered later in the book. Still, we have outlined the basic principle of rocket propulsion, conservation of linear momentum for isolated systems.

Newton would have no trouble understanding NASA's program. Yet the literature reveals that the conservation of linear momentum was not well understood even by some "scientists" 65 years ago. The *Scientific American* of January 1975 had an amusing note about one of the world's pioneers in rocket propulsion. Read it and weep for the teaching of physics 65 years ago!

JANUARY, 1925: "Critics of Professor Robert H. Goddard's project for sending a rocket to the moon, who have taken the position that in a vacuum such as exists in space between the earth and the moon the gases emitted by the rocket would have nothing to impinge upon in order to propel the rocket, will be silenced by recent experimental proof that this criticism is a fallacy. Professor Goddard answered his critics by placing a revolver in a vacuum and firing a blank cartridge from it. Although the hot gases of the burning powder had no air to impinge against, the revolver recoiled on its prepared axis just as if it had not been fired in vacuum. Further experiments along this line were made with a small model of the rocket which it is planned to send to the moon. Professor Goddard never expected his rocket to get its kick from the impact of its escaping gases against air. On the contrary, its propulsion depends on Newton's Third Law, which states that, 'To every action, there is an equal and opposite reaction.' "

6.8 **Summary of Important Equations**

Vector algebra:

$$\mathbf{A} + \mathbf{B} = \mathbf{B} + \mathbf{A}$$

$$\mathbf{A} - \mathbf{B} = \mathbf{A} + (-\mathbf{B})$$

$$\alpha(\mathbf{A} + \mathbf{B}) = \alpha\mathbf{A} + \alpha\mathbf{B}$$

$$(\alpha + \beta)\mathbf{A} = \alpha\mathbf{A} + \beta\mathbf{A}$$

$$\mathbf{A} \cdot \mathbf{B} = |\mathbf{A}||\mathbf{B}| \cos \theta = a_x b_x + a_y b_y + a_z b_z = \mathbf{B} \cdot \mathbf{A}$$

$$\mathbf{A} \cdot \mathbf{B} = (\text{projection of } \mathbf{B} \text{ on } \mathbf{A}) \cdot (\text{magnitude of } \mathbf{A})$$

$$|\mathbf{A} \times \mathbf{B}| = |\mathbf{A}||\mathbf{B}| \sin \theta$$

$$\mathbf{A} \times \mathbf{B} = -\mathbf{B} \times \mathbf{A} = (a_y b_z - a_z b_y)\mathbf{i} + (a_z b_x - a_x b_z)\mathbf{j}$$
$$+ (a_x b_y - a_y b_x)\mathbf{k}$$

$$\mathbf{i} \times \mathbf{j} = \mathbf{k} \qquad \mathbf{j} \times \mathbf{k} = \mathbf{i} \qquad \mathbf{k} \times \mathbf{i} = \mathbf{j}$$

$$\mathbf{p} = \frac{m v_x \mathbf{i} + m v_y \mathbf{j} + m v_z \mathbf{k}}{\left(1 - \dfrac{v_x^2 + v_y^2 + v_z^2}{c^2}\right)^{1/2}}$$

$$E_{\text{photon}} = pc \qquad E_{\text{neutrino}} = pc$$

Special equation for bubble chamber analysis determination of the magnitude of the linear momentum:

$$p = (.3)rB \qquad p \text{ in MeV}/c, \ r \text{ in cm}, \ B \text{ in kG}$$

Chapter 6 QUESTIONS

1. Can two vectors of different magnitude be combined to give a zero resultant vector? Explain.

2. Which of the following are scalar, vectors, or neither quantities: (a) mass; (b) shape; (c) velocity; (d) brightness; (e) force; (f) magnetic field; (g) charge; (h) family membership; (i) momentum; (j) baryon number; (k) volume.

3. Three vectors sum to zero, $A + B + C = 0$. If A and B have equal magnitude, what are the minimum and maximum magnitudes of the vector C?

4. If $A \cdot B = 0$ and neither A nor B is zero, does it follow that A and B are perpendicular to each other? Explain.

5. Can the sum of two vectors be a scalar quantity?

6. Explain how a vector equation contains more information than a scalar equation.

7. Consider two vectors A and B. We form two sums

(1) $|A| + |B|$ and   (2) $|A + B|$.

Is sum (1) greater than, less than, or equal to sum (2)?

8. A boy walks 3 miles south, 6 miles east, 2 miles west, and turns north to walk 1 mile. What is his final vector position? How far did he walk?

9. Can any of the Cartesian components of a vector

have a magnitude greater than the magnitude of the vector?

10. Is time a vector quantity?

11. Why is it common to observe spiraling electron tracks in bubble chamber photographs and not spiraling proton or other baryon tracks?

12. How can you measure the scattering angles of the charge particles in a magnetic field if the tracks are curving?

13. A proton–proton collision has the following consequences:

$$p + p \rightarrow p + \text{pions} + \text{baryons}$$

If there is 700 MeV *available* energy for the creation of new particles, write four possible collision processes that conserve all known quantities.

14. A proton strikes an antineutron. Write down several possible products for this collision that conserve all known quantities.

15. An astronaut finds herself far away from the space shuttle vehicle and out of rocket fuel. However, she discovers that she has two loaded pistols on her person ("star wars" weapons). How can she use these to return herself to the ship? Explain.

Chapter 6 PROBLEMS

1. Two vectors have the property that

$$a + b = 3i + 2j$$
$$a - b = 5i + 8j.$$

Find the vectors a and b.

2. If I walk 4 m east, 6 m south, and you walk 6 m west and 10 m north, how far are we away from each other if we started at the same origin?

3. Given $A = 2i - 5j + 6k$, $B = 3i + 8j + k$, $C = -i - 5j - 5k$. Find the following sums and products.

(a) $A + B$
(b) $A + B - C$
(c) $B - A$
(d) $(A + B + C)_{x\text{-comp.}}$
(e) $A \cdot B$
(f) $i \cdot A$
(g) $j \times (A - B)$
(h) $(A \cdot C)B - (A \cdot B)C$
(i) $(i + j) \times (A - C)$
(j) $A \times C$
(k) $|A||C|$

4. Find the projection of the vector A on the vector B and the projection of C on A using the vectors defined in Problem 3.

5. Consider the following vectors in a two-dimensional space—the x–y plane. Note that the notation (a, b) means that a is the x-component and b the y-component of the vector.

$$A = (7, 3) \qquad C = (10, 4)$$
$$B = (2, 0) \qquad D = (0, 6)$$

(a) By geometric construction on graph paper, find the x and y components of the vector S, where

$$S = 3B - A + C - 2D.$$

From your construction, determine the length of S and the angle that S makes with the positive x-axis. Make you construction as carefully as possible so that you can check it with the more accurate analytical calculation. Solve this problem analytically. Discuss the comparison of your two results. What is the source of error?

(b) Graphically determine $\mathbf{A} \cdot \mathbf{C}$ and $\mathbf{C} \cdot \mathbf{A}$. Analytically compute $\mathbf{A} \cdot \mathbf{C}$ and $\mathbf{C} \cdot \mathbf{A}$. Compare these two results. What conclusion can you draw about the relationship of $\mathbf{A} \cdot \mathbf{C}$ to $\mathbf{C} \cdot \mathbf{A}$? Can you prove this relationship for any two vectors in general?

6. Consider the three vectors shown in Fig. 6.33. The "law of cosines" is a relationship between the magnitude of these vectors $\mathbf{A}$, $\mathbf{B}$, $\mathbf{C}$ and the angle θ, and is written

$$C^2 = A^2 + B^2 - 2|\mathbf{A}||\mathbf{B}| \cos \theta.$$

Derive this expansion. (*Hint:* Form the vector relationship $\mathbf{A} + \mathbf{B} = \mathbf{C}$ and "square" it. Note that $C^2 = |\mathbf{C}||\mathbf{C}|$, etc.)

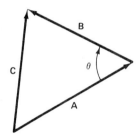

Fig. 6.33

7. Find the unit vector in the x–y plane that is perpendicular to the vector $\mathbf{R} = 4\mathbf{i} + \mathbf{j} + 6\mathbf{k}$. Is your answer unique? Explain.

8. Two vectors have unit magnitude. Find the magnitude of their *sum* and *difference* when (a) they are parallel; (b) they are at right angles. How can these vectors be arranged so that their sum is also a unit vector?

9. An airplane is heading in the easterly direction with an airspeed of 400 mi/hr. On that day the wind is out of the northwest (exactly) blowing at 75 mi/hr. What is the ground speed of the plane? Draw vector diagrams representing the plane and the wind. Solve the problem both graphically and analytically.

10. Prove Eq. 6.18, the transformation equation for two rotated Cartesian coordinate systems with common origin.

11. Consider a vector $\mathbf{R}$ represented in two rotated Cartesian coordinate systems (see Fig. 6.10) whose transformation equation is given in Eq. 6.14.

For the unprimed: $\mathbf{R} = (R_x, R_y)$

For the primed: $\mathbf{R} = (R_{x'}, R_{y'})$

Show that

$$R_{x'}^2 + R_{y'}^2 = R_x^2 + R_y^2,$$

which proves that the magnitude of the two vectors is the same in both coordinate systems. In physics jargon this is an example of INVARIANCE, a quantity that does not change in two different reference frames.

12. Prove that if three vectors sum to zero, they must all lie in the same plane.

13. A student can row a boat with a velocity V in still water. Show that if he rows his boat in a stream that flows uniformly with velocity v, the round trip time t it will take him to travel upstream a distance D and return is

$$t = \frac{2VD}{V^2 - v^2}.$$

What is the significance of the two cases $v > V$ and $v = V$? Explain.

14. A truck is traveling south at a speed of 55 mph in a wind that is blowing due west. The smoke from the exhaust streams behind at an angle of 42° with respect to the line of the truck. How fast is the wind blowing?

15. Find the component of the vector $\mathbf{A} = 5\mathbf{i} + 6\mathbf{j}$ along the direction of the unit vector $\mathbf{u} = (1/\sqrt{5})(\mathbf{i} + 2\mathbf{j})$.

16. Consider the two vectors $\mathbf{A}$ with $|\mathbf{A}| = 5$ and $\mathbf{B}$ with $|\mathbf{B}| = 4$. If $|\mathbf{A} - \mathbf{B}| = 3$, find the angle between the two vectors.

17. The law of sines is written

$$\frac{a}{\sin \alpha} = \frac{b}{\sin \beta} = \frac{c}{\sin \gamma},$$

where a, b, and c refer to the sides of the triangle and α, β, and γ are the included angles (Fig. 6.34). Prove

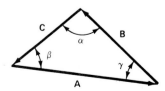

Fig. 6.34

this relationship using vectors and the vector cross product. (*Hint:* Consider the vector $\mathbf{A} + \mathbf{B} + \mathbf{C} = 0$.)

18. Prove the relationships between the unit vectors in Eq. 6.29.

19. Prove the expression for the cross product of two vectors $\mathbf{A} \times \mathbf{B}$ in Eq. 6.31.

20. Consider a unit cube with the two vectors $\mathbf{A}$ and $\mathbf{B}$, as shown in Fig. 6.35.

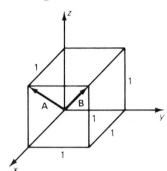

Fig. 6.35

   (a) Calculate $\mathbf{A} \times \mathbf{B}$. Call this vector $\mathbf{C}$.

   (b) Calculate $\mathbf{C} \cdot \mathbf{A}$ and $\mathbf{C} \cdot \mathbf{B}$.

   (c) Find the angle between $\mathbf{A}$ and $\mathbf{B}$ and between $\mathbf{A}$ and $\mathbf{C}$.

21. In the first example of a bubble chamber event that was analyzed in this chapter, a proton and π^--meson were assumed to be the product particles. Show that if other particles are assumed, which obey charge and family conservation laws, the energy–momentum relationships will not work out.

22. Explain what absolute conservation laws are violated by the following hypothetical decay schemes.

   (a) $\pi^0 \rightarrow \gamma$ (e) $e^- \rightarrow \nu_e + \gamma$

   (b) $\pi^+ \rightarrow \gamma + \gamma$ (f) $n \rightarrow \mu^+ + e^- + \gamma$

   (c) $\pi^+ \rightarrow \mu^+ + \gamma$ (g) $\Lambda^0 \rightarrow p + e^+$

   (d) $p \rightarrow e^+ + \nu_e$

23. Complete the following reactions.

   (a) $K^- + p \rightarrow \Xi^- + \cdots$

   (b) $\Xi^0 + p \rightarrow \Lambda^0 + \cdots$

Complete the following decays.

   (c) $\Sigma^0 \rightarrow \Lambda^0 + \cdots$

   (d) $\pi^+ \rightarrow \nu_\mu + \cdots$

   (e) $K^+ \rightarrow \pi^0 + \cdots$

24. An unknown particle at rest decays into two gamma rays which travel in the opposite direction. If each gamma ray has a total energy of 269.94 MeV, find the rest mass and the name of the unknown particle.

25. Consider a photon of energy 10 MeV colliding head-on with a moving π^--meson. What initial velocity must the π^--meson have if the photon is to recoil backward with the same total energy that it had before the collision?

26. A particle of rest mass M decays into three identical particles of mass m while at rest in the laboratory. The velocities and direction of two of the particles are given in Fig. 6.36.

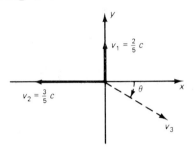

Fig. 6.36

   (a) Find the velocity (the direction and speed) of the third particle.

   (b) Find the momentum of each of the particles.

   (c) Find the ratio M/m.

27. A π^0-meson traveling at three-fifths the speed of light decays in flight into two γ-rays (photons) whose direction of motion makes equal angle θ with respect to the π^0-meson's original path (Fig. 6.37).

   (a) Prove that the γ-rays have equal energy.

   (b) Find the energies of the γ-rays.

   (c) Calculate the angle θ.

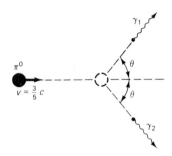

Fig. 6.37

28. For an extreme relativistic particle, the total energy E_T is much greater than the particle's rest energy. Prove that the velocity of such a particle is approximately $v \approx c\left(1 - \dfrac{m^2c^4}{2E_T^2}\right)$.

[*Hint:* You will need to use the binomial approximation $(1 + \alpha)^{1/2} \approx 1 + \frac{1}{2}\alpha$ and the magnitude equation. For example, note that

$$E = (p^2c^2 + m^2c^4)^{1/2} = pc\left(1 + \frac{m^2c^4}{p^2c^2}\right)^{1/2} \approx pc\left(1 + \frac{m^2c^4}{2p^2c^2}\right).$$

29. A K^- disintegrates in flight into three particles, as shown in Fig. 6.38. The K^- has a radius of 52 cm and a particle comes off along the same line as K^- with a radius of curvature of 36.6 cm in a 5-kG magnetic field normal to the direction of motion. Particles 2 and 3 move off at equal angles θ at the vertex, with respect to the path of the K^-.

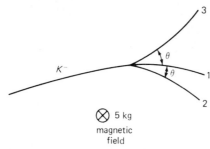

Fig. 6.38

(a) What are the particles 1, 2, and 3?

(b) Find the energy and momentum of particle 1, 2, 3. (*Hint:* Look up possible decay products of the K^- particle.)

(c) Find the angle θ.

30. A photon strikes a stationary proton. The proton is observed to recoil with a kinetic energy of 50 MeV at an angle of 60° with respect to the line of flight of the incident photon.

(a) Calculate the frequency of the incident photon.

(b) Calculate the direction of the recoil photon.

31. An unknown particle strikes a neutron at rest and creates a μ^--meson and a proton. The μ^- has a radius of 60.4 cm in a 27-kG field and leaves at a 1.93° angle, while the proton is observed to leave at an angle of 55.6° (Fig. 6.39). Find the energy, momentum, rest mass, and the name of the unknown particle. (*Hint:* You must guess the names of the unseen and unnamed particles and use energy and momentum conservation in the same way as Problem 30 was solved.)

32. An antiproton of 1.5 GeV kinetic energy collides with a proton in a hydrogen bubble chamber. The result is complete annihilation of both material particles and the creation of two photons. The two photons travel in opposite directions along the path of the incident projectile. Find the energies of each photon. Use 1 GeV for the rest energy of the proton.

33. A Σ^0 initially at rest, decays into lambda-nought (Λ^0) and a photon γ ($\Sigma^0 \to \Lambda^0 + \gamma$).

(a) Show that the ratio of the masses $M_{\Sigma^0}/M_{\Lambda^0}$ is given by $\gamma_{\Lambda^0}(1 + v_{\Lambda^0}/c)$, where γ_{Λ^0} and v_{Λ^0} refer to the lambda decay product.

(b) Using the expression derive in part (a), show that the velocity of the lambda is given by

$$\frac{v_{\Lambda^0}}{c} = \frac{\left(\dfrac{M_{\Sigma^0}}{m_{\Lambda^0}}\right)^2 - 1}{\left(\dfrac{M_{\Sigma^0}}{m_{\Lambda^0}}\right)^2 + 1}.$$

34. Prove that the following processes violate absolute conservation principles. (None of these processes has been observed, so theory and experiment are in agreement.)

(a) A photon spontaneously transforms itself into a π^+ and a π^- meson. (Assume that both mesons emerge at equal angles with respect to the original direction of the photon.)

(b) A photon collides with an electron at rest and gives all its energy to the stationary electron.

(c) A positron and an electron collide in a storage ring (each with the same magnitude of momentum, but with opposite direction) and produce only one photon.

35. Prove that a photon cannot transform itself spontaneously into a positron and an electron, where the product particles can be moving in any direction (not the symmetric transformation shown in Section 6.4).

36. A proton whose kinetic energy is equal to its rest energy strikes another proton, whose total energy is three times its rest energy, coming from the opposite

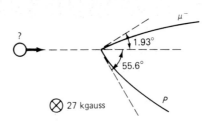

Fig. 6.39

direction. After a head-on collision, they form a new composite particle. Find the velocity (direction and magnitude) of the new composite particle.

37. Consider an atom of mass M in an excited energy state. If the atom (initially at rest) deexcites, returns to a lower-energy state by emission of a photon of frequency f, and recoils, show that the frequency of the emitted photon is given by

$$f = \frac{M^2 c^2 - m^2 c^2}{2Mh},$$

where m is the mass of the atom in the lower-energy state and h is Planck's constant.

38. A K^- disintegrates in flight into a π^--meson and an unseen π^0 particle. The radius of curvature of the K^- is 28.8 cm in a 9-kG field and the π^- comes off at an angle of 79°, as shown in Fig. 6.40. The radius of the track is not measured.

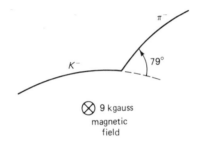

Fig. 6.40

(a) Using conservation of energy and momentum and the magnitude equation, show that the following equation holds for the magnitude of the momentum of the π^--meson (p_{π^-}):

499.88 MeV $= [c^2(77.76 \text{ MeV}/c - p_{\pi^-} \cos 79°)^2$

$+ P_{\pi^-}^2 c^2 \sin^2 79° + m_{\pi^0}^2 c^4]^{1/2} + [p_{\pi^-}^2 c^2 + m_{\pi^-}^2 c^4]^{1/2}$

(b) Although the equation arrived at in part (a) contains only one unknown (p_{π^-}), it is difficult to solve analytically. However, with your calculator you may either program the right side of this expression and substitute in values for p_{π^-} or simply go through the algebraic steps each time to find the solution.

(c) Find the angle of the flight of the π^0 as well as its energy and momentum, using the results obtained in part (b).

39. A particle moving at a relativistic velocity and having rest mass m strikes an identical particle at rest, and elastically scatters, as shown in Fig. 6.41. If the

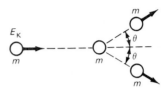

Fig. 6.41

relative scattering angle for this symmetric scattering is θ, show that

$$\cos 2\theta = \frac{E_{KE}}{4mc^2 + E_{KE}},$$

where E_{KE} is the initial kinetic energy of the projectile particle.

40. An antiproton with kinetic energy 1 GeV interacts with a stationary proton to form a new atom, "protonium," which is bound by electrostatic forces for a measurable lifetime. Eventually, the two particles annihilate each other and produce two high-energy γ-rays.

(a) Find the velocity of the freely recoiling protonium.

(b) Calculate the total energy of the protonium.

(c) Calculate the rest mass of the protonium.

(d) Calculate the energy of each of the photons produced in the annihilation if both protons are emitted along the line of the incident antiproton.

41. A particle of mass m moving with a speed v_1 collides with a particle of mass $2m$, which was at rest. The initial projectile particle rebounds at an angle of 45° and with a speed of $v_1/2$. If $v_1 \ll c$,

(a) Find the kinetic energy and direction of the target particle after the collision.

(b) Was kinetic energy conserved in the collision?

42. Prove that for elastic scattering of two particles of the same mass in the nonrelativistic limit, the recoil angle between the two particles is always 90°. You may solve this problem by considering one particle to be at rest.

43. A puck slides along the ice (without friction) at a speed of 2.2 m/s and strikes a glancing blow on an identical puck. After the collision, the projectile puck is found moving off at an angle of 60° with respect to the original line of the puck, at a speed of 1.1 m/s.

(a) Find the velocity of the other puck.

(b) Is this an elastic collision?

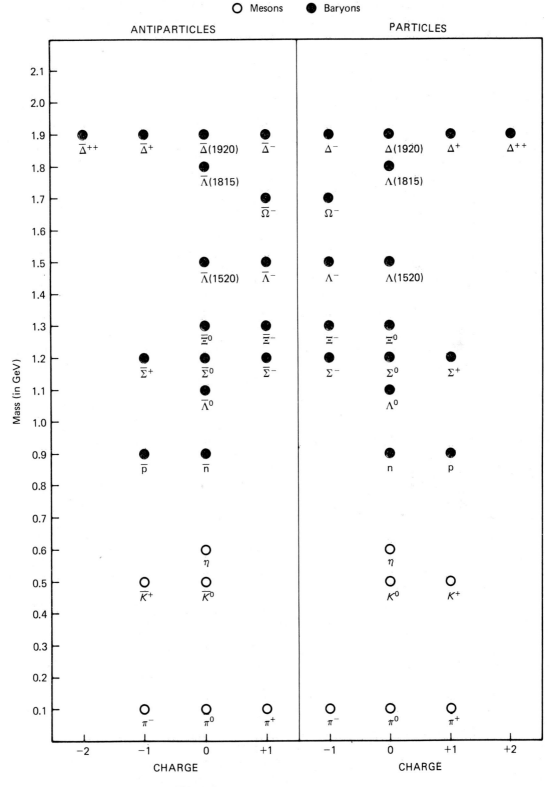

Fig. 7.2

of the deuteron. The binding energy is then

$$\text{deuteron binding energy} = (m_T - m_D)c^2.$$

The general equation for the binding energy of any compound nucleus is*

$$\boxed{\text{binding energy} = \left(\sum_i m_i - m_{\text{composite}}\right)c^2.} \tag{7.1}$$

Both C^{14} (carbon) and O^{14} (oxygen) are nuclei whose mass has been accurately measured. If the electrostatic repulsive energy of the two additional protons is subtracted from the O^{14}'s binding energy, the binding energy of the two nuclei is identical. This tells us that the binding of the p–p and the n–n to the nuclear core $_6C^{12}$ are the same. Many other mirror nuclei exist in nature (for example, Be^7–Li^7, B^9–Be^9) and all confirm the charge independence of the nuclear forces. Thus, *as far as strong forces are concerned*, the proton and the neutron are the same; they are interchangeable.

Stable nuclei can only give us information about the nature of the strong forces between the lowest-energy baryon multiplet, the proton and neutron. The strong interactions among the Σ triplet, Ξ doublet, and so on, have to be surmised from other data. The experimental results indicate that:

> *All hadrons within a given multiplet state have the same strong interactions.*

A useful mathematical model, proposed by Heisenberg in 1932, has been worked out to describe the multiplet particle states. This formalism was the prototype of both particle classification schemes and modern dynamical theories of the fundamental forces. The mathematical description of the multiplet quantum number is directly analogous to the mathematics that describe the spin angular momentum of stable particles in quantum physics. It is the mathematics of a *quantized* vector quantity. The analogy is carried even to the extent of naming this multiplet structure quantum number ISOTOPIC SPIN or I-SPIN, but do *not* be misled into thinking that isospin has *anything to do with a physical rotation of the particle*. The name derives from the similarity in the *mathematical structure* to

---

* This is an important equation in nuclear physics. We could use it to understand many things about nuclear forces and the stability of nuclei, but this would divert us from our subject. It is interesting to note that both high-energy physics and nuclear physics (medium range) was carried out with a *very* low energy "scale": the mass spectrometer.

quantized rotational motion. If you go on to study quantum mechanics, you will learn this mathematical formalism. I will just show you the results and explain how to use it.

All the particles in a particular multiplet are assigned a total isotopic spin **I**, which can easily be calculated from the following expression:

$$\text{isotopic spin multiplicity} = 2\mathbf{I} + 1. \qquad (7.2)$$

For example, the proton and neutron doublet has a multiplicity of 2; therefore, $2 = 2\mathbf{I} + 1$ or $\mathbf{I}$ (proton, neutron) $= \frac{1}{2}$. The π-mesons form a hadron triplet, having a multiplicity of 3; therefore, $3 = 2\mathbf{I} + 1$ or $\mathbf{I}$ (π-mesons) $= 1$. Let me repeat: all members of a hadron multiplet have the same total isotopic spin. Figure 7.3 lists the values of the isotopic spin and charge of the "common" baryons and mesons.

But Fig. 7.3 also lists the ISOTOPIC SPIN PROJECTION (I_3) for *each* particle. For example, the π^+ has $I_3 = +1$, the π^0 has $I_3 = 0$, and the π^- has $I_3 = -1$. Each hadron has a value of the total isotopic spin **I** and its own value of the isotopic spin projection I_3. If you invest a few moments of careful study of Fig. 7.3, the rules for these assignments will become clear. The assignments can also be depicted graphically as vectors where I_3 is a quantized component of the total isotopic spin vector **I**.

Now we return to the problem of assigning strangeness quantum numbers to the strongly interacting particles. First consider what is called the CENTER OF CHARGE of the NORMAL multiplets of both the baryons and mesons. The "normal" multiplet comprises the lowest-energy particles of a given hadron family. To be specific, the normal

	Singlets $n = 1$			Doublets $n = 2$					Triplets $n = 3$						Quadruple $n = 4$			
Isotopic Multiplets	η^0	Λ^0	Ω^-	K^+	K^0	p	n	Ξ^0	Ξ^-	π^+	π^0	π^-	Σ^+	Σ^0	Σ^-	Δ^{++}	Δ^+	Δ^0 Δ^-
Electric charge	0	0	-1	$+1$	0	$+1$	0	0	-1	$+1$	0	-1	$+1$	0	-1	$+2$	$+1$	0 -1
Isotopic projection I_3	0	0	0	$+1/2$	$-1/2$	$+1/2$	$-1/2$	$+1/2$	$-1/2$	$+1$	0	-1	$+1$	0	-1	$+3/2$	$+1/2$	$-1/2$ $-3/2$
Isotopic spin I	0	0	0	1/2		1/2		1/2		1			1			3/2		

Fig. 7.3

multiplet of the baryon is the proton–neutron cluster, and the normal multiplet for the mesons is the π^+, π^0, π^- triplet. The assignment of strangeness quantum number depends on the deviation of the center of charge of the particle's multiplet from the center of charge of the normal multiplet of that particle's family. That's a lot to swallow in one sentence, so let's take it one step at a time.

The rules for assigning strangeness quantum number to strongly interacting particles are as follows:

1. Identify the multiplet of the particle under consideration.
2. Find the center of charge of that multiplet.
3. Calculate how many units of charge the center of charge of the particle's *multiplet* is displaced from the "normal" center of charge of that family.
4. Multiply the units of displacement of the center of charge times 2. This is the magnitude of the strangeness.
5. If the displacement of charge is to the left, the sign of the strangeness is negative; if it is to the right, it is positive.
6. All particles with the same total isotopic spin (all members of the same multiplet) have the same strangeness. This is as it should be, since it is experimentally observed that particles with the same isotopic spin have the same strong interactions.

EXAMPLE 1   Find the strangeness quantum number of the K^0-meson.

The K^0 resides in a doublet at about .5 GeV in Fig. 7.2. Its center of charge of the multiplet is $+\frac{1}{2}$; the mean between zero and $+1$. The normal multiplet of the mesons (the π-meson triplet) has a center of charge at zero, so the displacement of the center of charge is $\frac{1}{2}$. Multiplying times 2 gives K^0 a strangeness $S = +1$ since the displacement is to the right. The K^+ has a strangeness quantum number of $+1$.

There is an equation for hadrons that relates the charge of the particle Q, the strangeness S, the third component of the isotopic spin I_3, and the baryon number B. It is

$$Q = e\left(I_3 + \frac{B}{2} + \frac{S}{2}\right), \qquad (7.3)$$

where e is the charge on the electron. Another system of nomenclature for strongly interacting particles uses the quantum number HYPERCHARGE, which is defined as

$$Y \equiv \text{HYPERCHARGE} = B + S. \qquad (7.4)$$

Since hypercharge is simply the algebraic sum of the baryon and strangeness quantum number, no new physics has been added. Equation 7.3 becomes

$$Q = e\left(I_3 + \frac{Y}{2}\right).\tag{7.5}$$

EXAMPLE 2   Calculate the third component of the isotopic spin I_3 for the K^+-mesons using Eq. 7.3.

We know the following quantum numbers for K^+:

$$\text{charge of } K^+ = +e$$

$$\text{baryon number } B = 0$$

$$\text{strangeness } S = +1$$

Equation 7.3:

$$Q = e\left(I_3 + \frac{B}{2} + \frac{S}{2}\right)$$

Substitution yields

$$+e = e(I_3 + 0 + \tfrac{1}{2}) \quad \text{or} \quad I_3 = +\tfrac{1}{2}.$$

It is possible to use any combination of the three quantum numbers in this equation to determine the fourth.

The isotopic spin projection I_3 (also called the third component) is conserved in *strong* interactions. This should be evident from Eq. 7.3, where charge, baryon number, and strangeness are all conserved in strong interactions. The formulation of strangeness by Gell-Mann and Nishijima in terms of isotopic spin was a significant theoretical advance because now one could account for particles carrying more than one unit of strangeness. The Pais conjecture that strange particles are always created in pairs cannot explain why the reaction

$$\pi^- + p \rightarrow \Xi^0 + K^0 \tag{7.6}$$

does *not* occur. The reaction products are a pair of strange particles, but I_3 is not conserved. An examination of Fig. 7.2 shows that K^0 has $I_3 = -\tfrac{1}{2}$, Ξ^0 has $I_3 = +\tfrac{1}{2}$, π^- has $I_3 = -1$, and the proton's $I_3 = +\tfrac{1}{2}$. The isotopic spin projection equation reads

$$I_3\text{-component:} \quad -1 + \tfrac{1}{2} = \tfrac{1}{2} - \tfrac{1}{2}$$

and does not balance. A reaction that does balance and which, in fact, is observed involves the creation of two K^0 particles:

$$\pi^- + p \rightarrow \Xi^0 + K^0 + K^0 \tag{7.7}$$

$$I_3\text{-component:} \quad -1 + \tfrac{1}{2} = \tfrac{1}{2} - \tfrac{1}{2} - \tfrac{1}{2}.$$

Although the weak interactions are not bound to conserve strangeness, they do have some "respect" for it. Experiments have shown that strangeness *changes* as little as possible in weak interactions. Even though strangeness is *not* conserved in weak decay processes, it never changes by more than one unit. Mathematically, this is written

$$\Delta S \text{ (change in strangeness)} = \pm 1 \text{ in weak decays.} \qquad (7.8)$$

7.4 Resonances

THE MOST numerous types of particles were observed last. This seems like a paradox. How can so many particles be staring physicists in the face and not be seen?

The answer lies in one physical characteristic of these particles: their extremely short lifetime. All these particles interact strongly in *both* their production and their decay. Some are baryons and some are mesons. They all decay by strong interactions, with no conservation principle to prevent it, and hence live only on the order of 10^{-22} s. They are called RESONANCES. Even if such a particle were moving almost at the speed of light, it could only travel a nuclear radius in its *entire lifetime*. How can one observe such a short-lived particle? Surely there is no hope of seeing their tracks in a bubble chamber.

Resonance particles are detected in two different types of experiments: RESONANCE FORMATION and RESONANCE PRODUCTION. Although resonance formation is easier to explain, resonance production is the more common method of detecting these short-lived particles. In resonance formation two relatively stable particles collide with sufficient energy and momentum to form a new intermediate (or resonance) particle, which decays almost immediately into "long"-lived particles. An example is the formation of the Δ^{++} particle in a π^+ collision with a proton. Experimentally, this formation is detected by measuring the probability that the reaction

$$p + \pi^+ \rightarrow \Delta^{++} \rightarrow p + \pi^+$$

will occur.

The probability that a collision will occur, which is called the CROSS SECTION, is measured in millibarns.* When the projectile parti-

---

* The concept of "cross section" comes from a classical model of the collision process which measures probability in terms of the geometric area that a target presents to the incoming projectile; the larger the area, the greater the probability of a collision. The units of barn = 10^{-28} m² or millibarn (10^{-3} barn) was derived from the old expression, "You can't hit the side of a barn." Who says physicists have no sense of humor!

cle has sufficient energy to create the intermediate particle, the probability of the collision, or cross section, increases. A plot of the measured cross section for the p + π^+ collision as a function of the π-meson's momentum is shown in Fig. 7.4. Something very unusual is happening when the π^+ beam has about .32 GeV/c of momentum. It is also clear that this dramatic increase in the cross section is not a sharp spike at exactly one value of momentum but occurs over a range of energies of about .1 GeV on the mass-energy scale. Is this width due to the apparatus, experimental error, or is there some important physics to be gleaned from the line shape?

The answer is the latter. Quantum mechanics provides the correct interpretation of the line width. If the particles were absolutely stable (infinite lifetime), quantum physics tells us that the line would be a sharp single spike of zero width. Because the particle has a finite lifetime (in this case very short) there is a range of values of the mass energy of the particle which can be calculated from the Heisenberg uncertainty principle:

$$\Delta E \; \Delta \tau \geq h \qquad (7.9)$$

or

$$\Delta \tau \geq \frac{h}{\Delta E}$$

where $\Delta \tau$ is the lifetime of the particle and ΔE is the uncertainty in the rest energy (or rest mass) of the particle.* Equation 7.9 shows that the longer the lifetime of the particle, the smaller the uncertainty in the mass energy of the particle. In practice, the analysis is usually inverted; that is, the measured width of the line (ΔE) is used to determine the lifetime of the intermediate or resonance particle.

We can use Eq. 7.9 to estimate the lifetime of the resonance particle Δ^{++}. It is apparent from Fig. 7.4 that the "uncertainty" in the mass energy of the Δ^{++} is about .1 GeV = 10^8 eV. Thus

$$\Delta \tau \geq \frac{6.6 \times 10^{-34} \text{ J-s}}{10^8 \text{ eV} \times 1.6 \times 10^{-19} \text{ J/eV}} \approx 10^{-23} \text{ s}.$$

This is a typical life expectancy for a strongly interacting particle that decays by strong interactions.

In resonance production experiments, the presence of an intermediate short-lived particle is inferred when the outgoing particles emerge with a particular value of combined mass. Consider the ex-

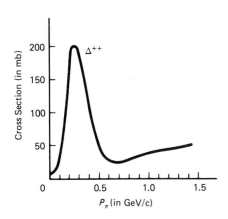

Fig. 7.4

---

* The word "uncertainty" can be misleading. I prefer to use the more cumbersome (but more precise) statement: "ΔE is the *range of probable values* of the rest energy." The experimenter is not uncertain about the measurement of the particle's mass; nature actually creates these particles with a range of values of their rest mass. It is another of those wonderful but puzzling realities of nature that is described by quantum physics.

ample of the ρ^--meson, which is produced in the reaction

$$\pi^- + p \rightarrow \rho^- + p$$
$$ \hookrightarrow \pi^- + \pi^0.$$

The bubble chamber observations of this process only show the π^- interacting in the chamber, and the p, π^-, and π^0 (π^0 found from conservation laws, since it leaves no track) emerging from the collision. An analysis of the energies of the emerging particles demonstrates that the reaction was really a *two-stage* process, with a particle formed as an intermediate state. How does the analysis reveal this?

Consider the case when the ρ^- is *not* formed. Then the total momentum of the π^- projectile would be shared by the three product particles π^-, π^0, and p, and the energies of these products would be smoothly distributed among all three. But in the actual reaction, the ρ^- is formed and the energy as well as the momentum of the emerging proton are uniquely determined by the energy-momentum conservation laws applied to this *two-body* (rather than three-body) reaction. The decay products will reflect that the ρ^- is also uniquely determined by the *two*-body collision. If one plots the particle production rate for many experiments as a function of its energy distribution, a peak is observed that can be used to determine the mass and the lifetime of the intermediate particle. Figure 7.5 shows such a plot for a closely related resonance particle, the ρ^0.

By now, hundreds of these resonance particles have been detected. The word *resonance* may leave you with the impression that these are not "really" particles. After all, you don't "see" them in bubble, drift, or spark chambers; they are merely peaks on a plot. Such an interpretation would be a serious mistake. These particles are every bit as "real" and as "elementary" as any of the other hadrons; they just have unimaginably short life spans. It seems obvious that with so many particles in our collection, all of them cannot be "elementary." It is highly unlikely that nature has built all matter out of so many parts.* As new particles were discovered and catalogued, interesting regularities appeared that gave theorists clues to a new underlying "simple" structure for all particles. These resonance particles, which at first appeared to muddy the theoretical waters, played an important, if not essential role in revealing a hidden inner structure of matter.

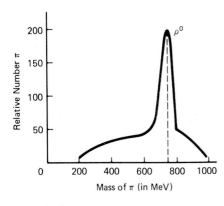

Fig. 7.5

7.5 Internal Symmetry: "Order Out of Chaos"

HISTORY, even scientific history, has a way of repeating itself. Mendeleyev's periodic table was the ordering scheme that provided an essential clue to understanding the composition of atoms. By the

---

* This is hardly a scientific statement. It is far closer to an article of scientific faith, whatever that is.

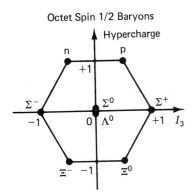

Fig. 7.6

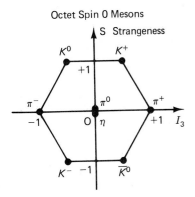

Fig. 7.7

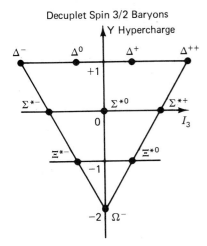

Fig. 7.8

early 1960s physicists realized that they would never understand the Pandora's box of particles until they could classify them. Once again Murray Gell-Mann (and Yuval Ne'eman) came to the rescue with a classification scheme called THE EIGHTFOLD WAY because it involves the operation of eight quantum numbers and because it recalls an aphorism attributed to the Buddha about the appropriate path to Nirvana: "Now this, O monks, is the noble truth that leads to the cessation of pain: this is the noble Eightfold Way: namely right views, right intention, right speech, right action, right living, right effort, right mindfulness, right concentration."

The basis of the eightfold way classification scheme is the mathematics known as Lie (pronounced "lee") groups, developed by the Norwegian mathematician Sophus Lie in the nineteenth century. We have neither the time nor the space to indulge ourselves in an extensive discussion of this complicated group theory, but we can examine a few of the most important examples. The groups are called SPECIAL UNITARY groups, and the one proposed by Gell-Mann and Ne'eman was called the SU(3) group. Figure 7.6 shows the octet of spin-$\frac{1}{2}$ (angular momentum, not isotopic spin) baryons. The identification of the correct symmetry group was perhaps the most important single theoretical advance in particle physics up to that moment, since it pointed the way to understanding the substructures responsible for this symmetry. The octet of spin-0 mesons is shown in Fig. 7.7 and the famous decuplet of spin-$\frac{3}{2}$ baryons in Fig. 7.8. It should now be apparent to you that isotopic spin plays as central a role in the classification scheme as do the other more familiar quantum numbers, such as charge and spin angular momentum.

There is an exciting story associated with the discovery of one of the particles, Ω^-, shown in Fig. 7.8 at the bottom of the decuplet. The theory predicted not only the existence of the Ω^- but also, as you can see from the symmetry, many important physical properties of this particle, which was supposed to sit at the "bottom" of the triangular array. This was one of those dramatic moments in science when the theory was ahead of the experiments, and the theorists were able to tell the experimentalists what to search for. A tremendous amount of effort was spent looking for this particle. If it existed and had the "correct" properties, it would strongly support the ideas of SU(3) symmetry.*

All the particles in the spin-$\frac{3}{2}$ baryon decuplet had been observed except the Ω^- with strangeness -3. Not only can the Lie group symmetry predict the particle's strangeness quantum number, its isotopic spin-3 component, and its charge, but from a careful examination of the properties of this decuplet, one can estimate the new

---

* A classmate of mine from graduate school, Medford S. Webster, was one of the discoverers of this particle. He was in charge of the beam designed at Brookhaven and told me some of the personal details about the experiments. V. E. Barnes et al., "Observation of the Hyperon with Strangeness Minus Three," *Physical Review Letters*, **12**(8), Feb. 24, 1964.

particle's mass. The average mass of the Δ is 1232 MeV, the Σ* is 1385 MeV, and the Ξ* is 1530. The difference in mass between the neighboring hypercharge levels can easily be calculated:

$$\text{between } \Delta \text{ and } \Sigma \rightarrow 1385 - 1232 = 153$$

$$\text{between } \Xi \text{ and } \Sigma \rightarrow 1530 - 1385 = 145,$$

an average difference of about 150 MeV. A respectable guess, then, for the mass of the Ω^- would be $1530 + 150 = 1680$ MeV. The mass of the Ω^- was measured to be 1672, very close to the prediction. The discovery of the Ω^- gave physicists confidence in this abstract classification scheme. It appeared to represent the complex structure of matter.

The decay mode of the Ω^- is interesting. The simplest way to imagine the Ω^- breaking up in a strangeness-conserving strong inter-

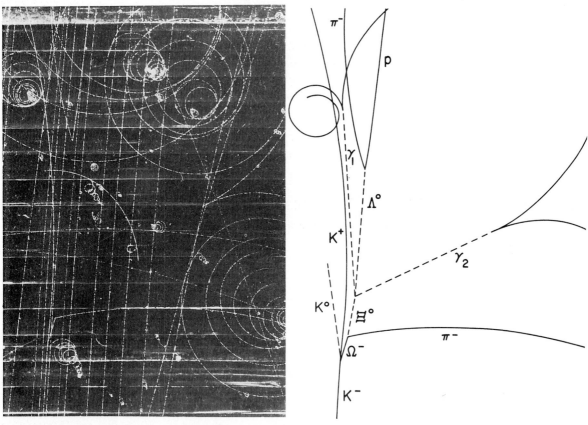

Fig. 7.9

Liquid hydrogen bubble-chamber photograph showing the production of a negatively charged omega baryon (Ω^-) by the interaction of a K^--meson with a proton in the chamber. The 5 BeV/c K^--mesons were produced from a proton beam of the Brookhaven AGS accelerator in 1964.

action would be to produce a doubly strange Ξ^- and an antikaon K_0 (-1 strangeness):

$$\Omega^- \rightarrow \Xi^- + K_0.$$

For this decay to conserve energy, the mass of the Ω^- must be greater than the sum of the masses of the two decay products. But that doesn't work because the sum of the masses of the product particles is greater than the mass of the Ω^- ($1321 + 497 = 1818$). That rules out the strong (and very fast) decay. The Ω^-, *unlike all the other members* of this SU(3) decuplet, which are "resonances," is long-lived and decays by weak interactions. It should therefore be directly observable in a bubble chamber photograph. Because it is triply strange, it cascades in three unit steps (since strangeness can change by only one unit in weak decays), leaving its telltale signature, as shown in Photo 7.1 and diagrammed in Fig. 7.9. An abstract mathematical theory of groups, which had found no practical application for 100 years, became the cornerstone of a new physical theory.

7.6 Quarks

THE DISCOVERY of the Ω^- set off a frenzy of activity in the theoretical physics community. The next breakthrough came from taking seriously the "3" in "SU(3)." The mathematics of Lie groups is too complex for the reader (and the author) to understand in detail, but it is fair to say that SU(3) is concerned with shuffling operations performed on three objects. It is natural to assume that these three objects have a basic role to play in the theory. Again Gell-Mann and Zweig independently and simultaneously (although both were at Cal Tech) proposed a surprisingly economical model of the building blocks responsible for all the observed hadrons. Zweig called them aces, but that name has disappeared and Gell-Mann once again supplied the name: QUARKS.* The three varieties of quarks, or QUARK FLAVORS as they are now called, are the up, the down, and the strange quarks. The quantum numbers assigned to these quarks are given in Table 7.1, where "up" and "down" refer to the orientations of the isotopic spin and "strange" refers to the -1 strangeness quantum number of the s quark. Along with these quarks are their antiquarks,

---

* The name for these particles came from a poem in James Joyce's *Finnegan's Wake*:

Three quarks for Muster Mark!
Sure he hasn't got much of a bark
And sure any he has it's all beside the mark
But O, Wreneagle Almighty, wouldn't un be a sky of a lark
To see that old buzzard whoopin about for uns shirt in the dark
And he hunting round for uns speckled trousers aroundly Palmers town Park?

TABLE 7.1

Quark	Symbol	Spin	Charge	I	I_3	S	B
Up	u	1/2	+2/3	1/2	+1/2	0	1/3
Down	d	1/2	−1/3	1/2	−1/2	0	1/3
Strange	s	1/2	−1/3	0	0	−1	1/3

making a total of six quarks. The combination of quarks that gives
the occupied representations of SU(3) are three quarks for the baryon
multiplets (qqq) (Figs. 7.10 and 7.12) and quark–antiquark for the
meson multiplets (qq) (Fig. 7.11). Quarks are not believed to be the
constituent particles of leptons and field propagators.

SU(3) fixes several properties of the quarks. A particularly significant consequence of the SU(3) symmetry scheme is that if three
quarks make up every baryon, the quarks then must have a $\frac{1}{3}$ baryon
number, since the baryons have unity baryon number. If, as theorists believe, Eq. 7.3, $Q = e(I_3 + B/2 + S/2)$, is applicable to quarks
as well as hadrons, because they are both strongly interacting parti-
cles, quarks must have a fractional electronic charge of $\frac{2}{3}$ and $−\frac{1}{3}$. The
existence of these unusual fractional quantum numbers is the most
celebrated property of the quarks.

But no one had ever observed a particle with fractional electronic charge. Moreover, physicists firmly believe that charge is an
absolutely conserved quantity and can never get lost or destroyed.
Therefore, if an isolated fractionally charged quark exists, it must be
stable, for it cannot decay into any known particle. For the fractionally charged quark to decay into an integral charged particle or a

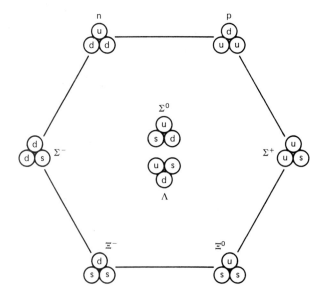

Fig. 7.10

collection of integral charged particles) would violate charge conservation.

That gave the experimentalists an idea. These fractionally charged and stable particles might be lying around somewhere, everywhere, waiting to be detected. This produced another flurry of activity, sometimes almost comical. One investigator developed the argument that quarks would accumulate in the muscles of bivalve creatures. He collected a large quantity of oysters to test his theory (or to fill his stomach). Some scientists reasoned that quarks would be at the bottom of the oceans, others that they would exist in the cosmic ray flux. A major effort was made to repeat Millikan's oil-drop experiment, which had first measured the charge of the electron. The new method used low-temperature superconducting techniques and, in fact, reported the observation of $\frac{1}{3}$ fractionally charged particles on superconducting metal spheres. Other investigators, using similar techniques, could not reproduce those results on the same metal spheres. All this activity has not yet yielded one confirmed fractionally charged *free* particle.

Well, that's that—another good idea gone down the drain! Fractional charge was not the only "problem" that the quark theory presented. Why weren't other quark configurations, such as qqqq, four quark states, detected? Experimenters searched for evidence of such particles but none has been discovered. A third serious problem existed that can best be explained by an example. In the spin-$\frac{3}{2}$ baryon decuplet the Δ^- is supposed to be composed of three identical quarks (ddd), all with their spin aligned in the same direction. This appears to violate one of the basic laws of quantum physics, the Pauli exclusion principle. This principle states that no two identical

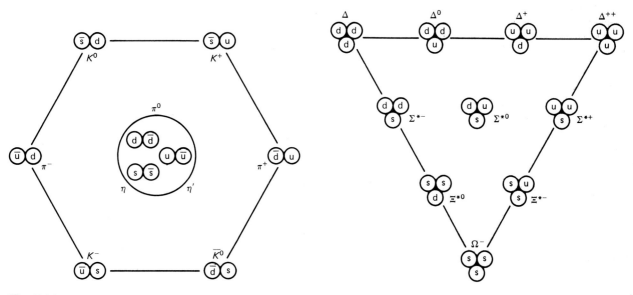

Fig. 7.11   **Fig. 7.12**

half-integer spin particles (called *fermions*) can exist in the same state. The d quarks are all identical, all have their spin in the same direction, all are fermions (spin $\frac{1}{2}$), and all appear in the same state. Some theorists proposed that quarks don't obey the Pauli exclusion principle—they seem so odd, why not add one more outlandish characteristic to them? But for others, the idea of abandoning the Pauli principle was too bitter a pill to swallow.

Particle physicists, although at first skeptical, began to accept the quark model as its successes continued. Not only did it give the correct groups for the mesons and baryons, it was able to show that mesons cannot be decuplets, and none have been found that are. It explains in a straightforward way why SU(3) is not an exact symmetry. By "exact" we mean that mass difference does exist in the spin multiplets. Exact symmetry in the multiplets means that all the particles in the multiplet would have the same mass, and clearly that is not what is observed. By giving the s quark a different mass from that of the u and d quarks, we can reproduce the mass differences (called splittings) between the particles of different hypercharge. Let's see how this works.*

If we assume that the s quark has a mass greater than the u and d quarks (which have the same mass), this can be written

$$m_s = m_0 + \delta,$$

where $m_0 = m_u = m_d$, the mass of the u or d quark.

The masses of the various particles in the spin-$\frac{3}{2}$ decuplet should be split by an amount δ. The separation between the various levels should be proportional to their hypercharge. This theoretical prediction can be compared to the experimental results in Table 7.2. Notice how we obtain a reasonably consistent value of δ for such a simple model, where the details of the binding energy of the quarks has been ignored. This evidence is indeed compelling.

TABLE 7.2

	Observed Mass	δ (experimental)
Mass of $\Delta = 1$ $= m_0$	1238	
Mass of $\Sigma = 0$ $= m_0 + \delta$	1385	147
Mass of $\Xi = -1 = m_0 + 2\delta$	1530	145
Mass of $\Omega = -2 = m_0 + 3\delta$	1674	144

The concept of quark mass must be tempered by the fact that no quark has been observed as a free particle. Recall that the mass of any composite particle is always smaller than the sum of the masses of its constituents by the amount of the binding energy that holds the composite together. We discussed this concept when we examined mirror nuclei (Eq. 7.1). Since we have been unable to isolate quarks,

---

* Rosenfeld et al. *Reviews of Modern Physics*, **39**, 1, 1967.

perhaps they are very massive particles bound by enormous forces. However, the foregoing argument still demonstrates a kind of "effective mass" of the quarks.

7.7 **Deep Inelastic Scattering**

THE FIRST *dynamical* evidence for the existence of quarks (in contrast to the "static" evidence that we have been discussing) came from scattering of very high-energy electrons off protons. The 2-mile-long linear accelerator at Stanford University (SLAC) was designed and constructed primarily to carry out these experiments. If fractionally charged entities existed inside the proton, the scattering of an electron off a proton should reflect the localization of these objects. Electrons were chosen as projectiles for these experiments because they do *not* interact by the strong forces, but do "feel" the electromagnetic forces of the proton. These "forces" can be accurately calculated by quantum electrodynamics (QED), probably the best understood and best tested theory in physics.

Why is a 2-mile-long accelerator needed? To answer this question, I must pull out one more equation from quantum physics.* De Broglie showed that the wavelength λ associated with a particle is related to its linear momentum p by the equation

$$\lambda = \frac{h}{p}.$$

(7.10)

To obtain useful information about the internal structure of the proton, it is reasonable to suggest that the electron's wavelength must be small compared to the size of the proton. This allows the electron to interact with only a small portion of the nucleon.† The idea for the experiment had its roots in Rutherford's experiments in 1911 when he detected the hard core of the atom with a "high-energy" α-particle beam. For those experiments the beam had to probe within dimensions of 10^{-8} cm. To probe the proton *itself* requires a seven-order-of-magnitude higher resolution. Equation 7.10

---

* As hard as I try, there simply is no way to avoid quantum mechanics when describing the fundamental processes in nature. Although this book deals principally with classical physics, we have already discussed Planck's constant, Einstein's relation for the photon $E = hf$, the uncertainty principle $\Delta E \, \Delta \tau \geq h$, Pauli's exclusion principle, quantized angular momentum, and now the de Broglie relationship.

† As an analogy, suppose that we wish to determine the shape of a rock in the center of a pond by examining the water waves that bounce off it. If we use long-wavelength waves, they would merely wash over the rock and we could see little detail of the rock's shape. Only when the waves are smaller than the dimensions of the rock will the reflections be characteristic of the shape of the rock.

shows that the larger the momentum, the smaller the wavelength of the electron and the more precise the probe. The results of the SLAC experiments using 20-GeV and higher-energy electrons startled the physics community. They showed that the proton has a distinct internal structure, with localized fractional electric charges and three localized spin-$\frac{1}{2}$ point particles. In short, it behaves exactly the way the quark model had predicted!

However, yet another surprise appeared in the analysis of the scattering data. Quarks do *not* account for all the matter in the proton; they appeared to carry only 50% of the proton's momentum. The other half of the momentum had to be carried by some other particles that did *not* interact with the electromagnetic forces and thus were "invisible" in these electron scattering experiments. Theorists conjectured that the missing momentum is carried by the exchange particles, which are the propagators of the force that holds the quark together. Appropriately they were called GLUONS (no explanation needed). Now we had a theory to explain the "static" properties of the many free particles and the first dynamic evidence that the baryons were themselves composite particles with internal localized fractionally charged centers.

7.8 Confinement: Freedom

STILL THE problem remained: Why don't quarks turn up as free particles, as the protons and neutrons had the courtesy to do for physicists in the 1930s? Why don't these high-energy electrons knock the quarks out of the proton? What is going on? Some theorists began to suspect that the very elusiveness of quarks was the key to understanding their behavior. Maybe quarks have "no life of their own," as we have come to expect of all the other constituent parts of matter. Maybe the quarks are always "confined." Models began to emerge that formally proposed such permanent confinement. Two of these are the "bag" and "elastic string" models of quark confinement. The latter concept is analogous to a string that is normally slack so that the quarks respond freely to the deep inelastic scattering, but as the quarks recoil the string becomes taut. If the force becomes great enough, the string snaps, but in so doing it creates two new quarks, like a bar magnet "creates" two new "poles" when you cut it in two pieces, leaving behind two new magnets with two poles each. It is a nice idea, a vivid picture, but it is not based on fundamental theoretical concepts of forces and fields; rather, it was an ad hoc proposal.

If the electrons are so useful in exploring the interior of the proton, maybe the neutrinos could be used in a similar way to explore the structure of hadrons. Neutrinos, of course, do not interact electromagnetically since they have no charge, but they do feel the weak forces of the proton's core. Are there centers of the weak forces, like there are centers of the electromagnetic forces inside the hadrons?

Neutrino beams were constructed at Fermilab and at CERN to study the high-energy interaction of neutrinos with the proton. In 1973 the Gargamelle bubble chamber at CERN had recorded a sufficient number of neutrino events (Photo 7.2) to answer this question. The proton also had a fractional structure for the weak forces, and

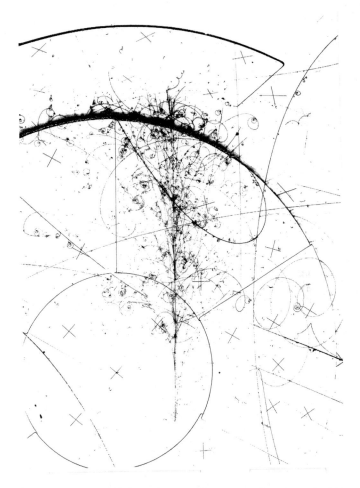

One of the first neutrino interactions in BEBC (Big European Bubble Chamber) after the start-up of the SPS and the 200-GeV narrow-band neutrino beam in the CERN West Area. An electromagnetic shower, showing up in a great number of $e^{\pm}$ pairs curling in the 35 kG magnetic field of BEBC. In spite of the shower, the details of the event at the interaction point are easily recognized.

these force centers appeared to be pointlike particles. That is, these quarks also appeared to be the center for weak interactions (as well as the electromagnetic forces) and to have no structure of their own. Again, as the SLAC experiments had suggested, the CERN neutrino results indicated that the quarks were almost free point particles as long as they remained inside the proton. This property is called ASYMTOTIC FREEDOM. What a crazy world! The string model at least gives us a picture of what a hadron must be like: a composite object with almost free pointlike internal particles, if they don't stretch their bonds. The binding is called INFRARED SLAVERY (infrared light is long wavelength, referring to slavery at larger distances); the binding force increases as the quark separation increases. Quarks are "free" like the black people of South Africa.

7.9 Particles with Charm

IN THE MOST sensational experiment of the early 1970s, a new and highly unusual resonance peak was discovered simultaneously by two groups, one at SLAC and the other at Fermilab. Burton Richter at SLAC examined the debris from electron–positron collisions, and Samuel Ting's group at Brookhaven studied proton–proton collisions within the energy range 3 to 4 GeV.* Richter's experiment used a colliding beam from the SPEAR storage ring of electrons and positrons in the newly designed MARK I detecting system (see full color insert, Photo A). The experiment measured the cross section for the interaction

$$e^+ + e^- \rightarrow \text{hadrons}$$

as a function of the center of mass energy.† Two startling characteristics appeared in the original data shown in Fig. 7.13. There was an enormous increase, by a factor of 70, in the cross section (typically, the cross section increases by a factor of 2 to 5 for strongly interacting resonant particles), and an extremely narrow line width, indicating a long lifetime. Something *very* different was going on here! After a brief period of speculation, an agreed-upon interpretation of the data emerged. At the new increased energy for the e^+–e^- collision in SPEAR, sufficient energy had become available to create a quark of another flavor. The new heavier quark had been anticipated by some theoreticians and given the name CHARM. Of course, the discovery of the charmed quark automatically implies a hoard of new particles, some in the form of excited states of the qq-meson configuration, and others in the qqq-baryon configurations with the u, d, and s quarks.

---

* Richter and Ting shared the 1976 Nobel prize in physics for their discovery.
† The concept of center of mass is discussed in detail in Chapter 10.

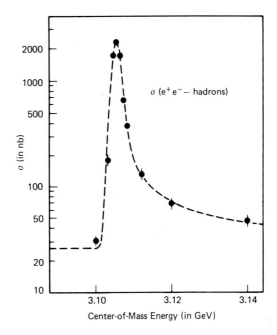

Fig. 7.13

The J/ψ, as this particle is now called, has the cc configuration, which goes under the name "charmonium".* Electrons and positrons, orbiting around one another form a bound state, called "positronium." The bound pair lives for about 10^{-6} s before the electron and positron annihilate each other. The cc charmonium is the analogous quark bound system, which decays much faster (10^{-20} s), since the quarks are strongly interacting particles. The new quantum number charm is important not only because it defines a fourth quark, but because it has associated with it a conservation principle in strong interactions. In fact, it is charm's conservation principle that makes the J/ψ particle live *100 times longer* than one expects for strongly interacting particles. This is the reason the line width is so narrow in Fig. 7.13.

Charmonium also exists in higher-energy states (think of two charmed quarks rotating around one another at higher and higher speeds) (Fig. 7.14). Studies of the properties of these states give important information about the nature of the strong forces between the charmed quarks and have convinced most physicists of the reality of quarks. You can get some idea of the drama of experimental physics from reading Gerson Goldhaber's account of the discovery of the ψ-particle at SLAC, reprinted in Appendix C.

---

* The Brookhaven group proposed the letter J, the Chinese character for Ting, for the new particle, but the SLAC group had already named it ψ. They compromised with J/ψ.

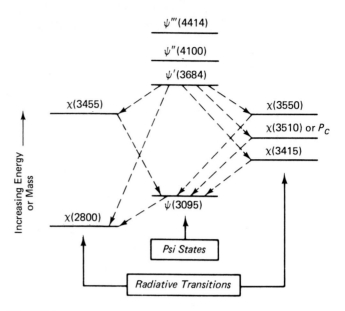

Fig. 7.14

7.10 **Red, Green, and Blue**

THE EVIDENCE mounted to support the quark hypothesis as the charmed baryon Λ_c (discovered in 1979, Fig. 7.15), and other excited states of charmed quarks appeared. Yet two problems plagued the theory: the fact that quarks do not exist as independent particles, and the apparent violations of the Pauli principle in, for example, the Δ^{++} particle. Several proposals attempted to explain the Δ^{++} puzzle, but apparently only one works. To get around the problem of the Pauli exclusion principle it was assumed that each quark has associated with it *another* quantum number, called COLOR. Again physicists were having fun with words, so you must not think that this so-called color charge (or color quantum number) has *anything at all to do with color* as we ordinarily think of it. It was proposed that quarks come in different colors: red, green, and blue.

Now one can construct the Δ^{++} out of three differently colored quarks, all of the same flavor, without violating the Pauli principle. The Pauli principle only applies to *identical* particles, and since each quark in the Δ^{++} has a different color, they are *not* identical. That was easy. But as you might imagine, the new color concept has other profound consequences. One of them is that all observed particles are COLOR SINGLETS. The concept of a color singlet is new. It means that there is a kind of "democracy" of color. Each color has the same "rights" as any other; there is no color preference. There is a "threeness" of the color quantum number; all baryons are made up

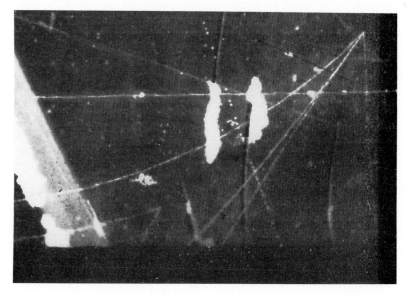

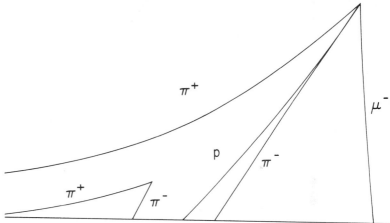

Fig. 7.15 An example of the production of a charmed baryon utilizing the 7-foot bubble chamber at Brookhaven. One half million pictures were required in order to obtain two events. The event is interpreted as an interaction between a neutrino of energy 3.9 GeV (unseen in the bubble chamber) and a neutron which is a constituent of the deuterium target in the bubble chamber. The resultant products of the collision are a μ^--meson and the Λ_e^+ (2260) charmed baryon which is not directly observed but is seen to decay into a proton, a positive pion, a negative pion, and a neutral kaon. The neutral kaon is observed to decay some distance from the event into a positive and a negative pion.

of three colors. What do we do with mesons, which have only two quarks? Here we must invoke a concept of quantum physics that I cannot explain. I beg your indulgence and simply state that mesons are superpositions of three different quark–antiquark terms: one a red quark and its antiquark, one a green quark and its antiquark, and finally a blue quark and its antiquark, written

$$q\bar{q} = \frac{1}{\sqrt{3}}\,(r\bar{r} + g\bar{g} + b\bar{b}).$$

Color singlets, composed of all three colors, are called "white."

Not only did the color conjecture explain the Pauli puzzle in the quark composition of hadrons, but for reasons that we shall not discuss, it resolved a long-standing mystery of the lifetime of the π^0-meson and the unexplained rate of production of hadrons in electron–positron collisions. So far, so good. But the real glory for the color hypothesis is that it points the way toward the development of a *fundamental* theory of hadrons and strong interactions, QUANTUM CHROMODYNAMICS (QCD). Oh, my—we are getting in deep!

7.11 QCD

NOW THAT you are all experts in classical mechanics, relativity, quantum mechanics, quantum electrodynamics, and Lie group algebra, I need hardly explain quantum chromodynamics. I'm only kidding—don't panic. No problems will be assigned! For fun, let's take a peek at this new theoretical framework of matter. Even if it sounds like it, it is *not* the mathematical theory of oil painting! It is a promising new theoretical framework which some theorists believe will ultimately explain the complete zoo of hadrons, the confinement of quarks (both asymtotic freedom and infrared slavery), the properties of gluons, the color force, the spectrum of states of charmonium, and the nature of the nuclear forces. It is not a simple theory with which to do calculations, at least that's what my theorist friends tell me. Unlike the sort of ad hoc assumptions and conjectures that have been made along the way, quantum chromodynamics (QCD) is a theory of interactions among quarks which may some day take its place along with QED (quantum electrodynamics) as one of the basic theories of the forces of nature. It has already shown promise of attaining that lofty goal. From my vantage point as an experimentalist, I will just sit back and admire; but maybe some of you will be inspired to dig in and help nurture this infant to adulthood, a mature theory of strong interactions.

Where are we today? By today I mean the day I am writing this book, because things are happening so fast that by the time you read this, some of it is likely to be out of date. The table of "truly" fundamental particles, that is, leptons, quarks, and field propagators, consists of 18 particles: six flavors of quarks, six leptons, and six field propagators—a beautiful symmetry. All but the graviton and the top quark have been observed. I have listed them in Table 7.3 with their known properties.

TABLE 7.3

LEPTONS

Particle Name	Symbol	Mass at Rest (MeV)	Electric Charge
Electron neutrino	ν_e	About 0	0
Electron	e^-	0.511	-1
Muon neutrino	ν_μ	About 0	0
Muon	μ^-	106.6	-1
Tau neutrino	ν_τ	Less than 164	0
Tau	τ^-	1784	-1

QUARKS

Particle Name	Symbol	Mass at Rest (MeV)	Electric Charge
Up	u	310	$+\frac{2}{3}$
Down	d	310	$-\frac{1}{3}$
Charm	c	1500	$+\frac{2}{3}$
Strange	s	505	$-\frac{1}{3}$
Top/truth	t	22,500; hypothetical particle	$+\frac{2}{3}$
Bottom/beauty	b	About 5000	$-\frac{1}{3}$

Force	Range	Strength at 10^{-13} cm Compared with Strong Force
Gravity	Infinite	10^{-38}
Electromagnetism	Infinite	10^{-2}
Weak	Less than 10^{-16} cm	10^{-13}
Strong	Less than 10^{-13}	1

Carrier	Mass at Rest (GeV)	Spin	Electric Charge	Remarks
Graviton	0	2	0	Conjectured
Photon	0	1	0	Observed directly
Weak-sector bosons				
W^+	81	1	$+1$	Observed directly
W^-	81	1	-1	Observed directly
Z^0	93	1	0	Observed directly
Gluons	0	1	0	Permanently confined

The theory of weak interactions is in good shape, the particle propagators have been discovered, and the connection between the weak interactions and the electromagnetic forces has been worked out reasonably well. Experimentalists continue to develop new, more sophisticated multicomponent detectors and ultrahigh-energy colliding beam systems to examine new quarks and leptons and the nature of the strong interactions. Theorists once again speculated about a grand unified theory of all forces, this time with some cautious optimism. But the proton still refuses to decay (as some theorists tell us it must), not everyone agrees that quarks can never exist as free particles, no one has seen the graviton, and. . . . There are many questions not yet answered! Plenty left for you!

Chapter 7 QUESTIONS and PROBLEMS

1. What would the world be like if strangeness were an absolute conservation law? What particles would be stable?

2. Why does the Ω^- leave a track in the bubble chamber when all the other spin $-\frac{3}{2}$ baryons in its decuplet do not?

3. What wavelength is associated with the following particles, whose kinetic energy is listed? (Remember that the units of wavelength are meters—the length of a wave.)

    (a) 1.2 GeV, proton

    (b) 30 GeV, electron

    (c) 400 MeV, π-meson

    (d) $\frac{1}{30}$ eV, neutron

4. Calculate the strangeness and isotopic spin quantum number of the $\bar{\Delta}$ (1920-MeV) spin particles. Explain your calculation.

5. Calculate the strangeness of the $\bar{K}^0$-meson. Explain your calculation.

6. The rest mass of the He^4 nucleus is 3728.43 MeV. Calculate the total binding energy of the four nucleons in this nucleus (two protons and two neutrons).

7. Find the mean lifetime of a resonance particle whose energy width for its production cross section is 1300 MeV.

8. If a particle has an average lifetime of 10^{-10} s, what is the minimum range of possible values of its mass?

9. The following are reactions or decays of "elementary" particles. Which of the following could not occur as strong interactions? Which could not occur at all? Explain your reasons for each case.

    (a) $\pi^+ + p \rightarrow \Sigma^+ + \pi^+$

    (b) $\Lambda^0 \rightarrow p + \pi^+$

    (c) $\Sigma^- \rightarrow \pi^- + n$

    (d) $\Sigma^0 \rightarrow \begin{cases} \pi^0 + n \\ \pi^- + p \end{cases}$

    (e) $p + p \rightarrow \Xi^0 + K^+ + K^+$

    (f) $\pi^0 \rightarrow e^+ + e^- + \nu_e$

    (g) $\Lambda^0 \rightarrow \pi^- + p$

    (h) $p + p \rightarrow \Lambda^0 + \Lambda^0$

    (i) $p + p \rightarrow n + \Lambda^0 + K^0$

    (j) $\pi^- + p \rightarrow \Lambda^0 + K^0$

    (k) $\mu^0 \rightarrow \gamma + \gamma$

    (l) $p + n \rightarrow \bar{\Xi}^- + \Sigma^0$

    (m) $n + n \rightarrow \Lambda^0 + \Lambda^0$

<table>
<tr><td>

8

</td><td>

MOTION WITH EXTERNAL FORCES: ONE DIMENSION

</td></tr>
</table>

8.1 Introduction

I HOPE you found the "stratosphere" of quarks and chromodynamics exhilarating, but the time has come to return the spaceship to Earth and begin our study of more familiar objects. The physics of everyday phenomena can also be fascinating, sometimes counterintuitive, and certainly challenging. You are already armed with the powerful conservation laws but we will add many new principles to the arsenal that you will use to analyze the motion of familiar tangible objects. Welcome back; now let's get started!

8.2 Details of the Motion

WE HAVE BEEN examining systems of particles isolated from external forces before, during, and after their interactions.* In fact, we have not examined these particles during the actual collision, only before and after the interaction had taken place. Observant students may also have noticed that we "cheated," since we never precisely defined some of the dynamical variables that were involved, such as velocity. Our discussion of the conservation principles relied on your past knowledge and intuition of these parameters. To study the details of particle motion, how their position and velocity change when they are subjected to external forces, we must explore all these concepts in greater depth.

Consider, for example, a beam of electrons extracted from a particle accelerator and directed into a set of charged metal plates, shown

---

* A more accurate statement would be that we have considered systems as though they were isolated, since the effects of the external forces were undetectable by the experimental apparatus. The one exception was the magnetic field used to determine the charged particle's momentum by measuring the curvature of its path.

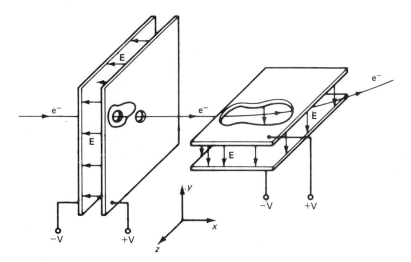

Fig. 8.1

in Fig. 8.1. The electrons are traveling in a high-vacuum region where, for this discussion, we can disregard their infrequent collisions with residual gas atoms.

If you could see this electron beam, you would notice that the electrons emerge from the first set of metal plates with a greater velocity than when they entered. If the polarity on the power supply were reversed, you would observe a decrease in the velocity of the electrons when they emerge. In both cases the beam has a velocity only in the x-direction. When the beam passes the second set of plates, the x-component of velocity remains the same, but the electrons acquire a y-component of velocity. Clearly, these devices produce a *change* in the particles' velocity.

The cathode ray tube (CRT) should be familiar to all of you; it is the picture tube of your television set. It is a device that displays the time dependence of electrical signals. For the electrical engineer, experimental physicist, and even the physician, the CRT is the heart of the oscilloscope, a device for observing time-dependent electrical signals in all kinds of electrical and electronic circuits. Figure 8.2 shows a simplified representation of its essential elements.

The accelerating anode gives the electron beam its kinetic energy. The vertical and horizontal plates deflect the beam. When the voltage on these plates changes, the beam strikes a different spot on the fluorescent screen on the front of the tube. When the electrons strike the phosphor atoms on this surface, they give up their kinetic energy to the atoms in an inelastic collision. In a complicated way, the atoms convert this kinetic energy into visible light. It is this light that you see on your TV or oscilloscope.

Our concern here is to understand the mechanism that causes the electrons to *change* their velocities. Of course, we recognize that the electric field between the charged metal surfaces is responsible for

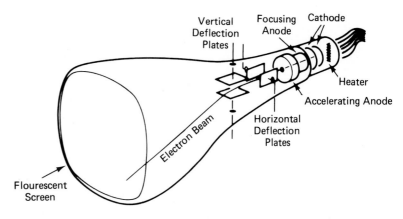

Fig. 8.2

these velocity changes, but how do we develop a more quantitative description of the motion? Can we calculate the change in velocity? Is the exact path of the particle predictable? Given the time when the electrons enter, when will they emerge from the accelerator plates?

Are there general laws to describe the motion of particles subjected to various kinds of forces, as well as to predict the motion of larger objects, such as bullets, billiard balls, airplanes, spacecraft, and celestial bodies in the presence of spring forces, electrical forces, gravitational forces, viscous forces, and others? This book would end here if the answer were anything but *yes*. Indeed, one of the great triumphs of eighteenth- and nineteenth-century physics was to develop a theoretical framework to explain such motion. This work is no less important today, since much modern engineering is based on these principles. Today, these principles can be applied, in conjunction with high-speed computers, to solve problems that were once believed to be intractable. But we are getting ahead of the story.

8.3 **Velocity and Speed**

THE CONCEPT of speed is familiar to everyone, but to be precise we will define the average speed of a particle as follows:

$$\text{AVERAGE SPEED} = \frac{\text{total distance traveled}}{\text{total time}}. \tag{8.1}$$

In the SI units, meters/second (m/s) is used to measure speed. Average speed is indeed a crude measurement of a body's motion, for it tells you little about the speed at any particular moment. For example, when I travel to Middlebury, Vermont, from Buffalo, New York, a distance of some 400 miles, it usually takes me 8 hours; an average speed of 50 mi/hr (I will get back to the SI units when we do real physics, I promise). However, I rarely drive 50 mi/hr. I usually make two 15-minute stops and drive somewhat over the 55-mi/hr speed limit on thruways. None of that information could be gleaned from knowing my *average* speed.

Although everyday language does not distinguish between *speed* and *velocity*, it is essential for the student of mechanics to know that physicists make an important distinction between them. The average velocity is defined as

$$\text{AVERAGE VELOCITY} = \frac{\text{total displacement}}{\text{total time}}, \qquad (8.2)$$

where average velocity is a *vector* quantity with the same units as average speed. The distinction between DISPLACEMENT and DISTANCE is obviously the key. If I drive to Vermont and back to Buffalo in 16 hours, I have traveled a distance of 800 miles, but my displacement, a vector quantity, is *zero* since I ended up where I began. My average *speed* is still 50 mi/hr, but my average *velocity* is zero! The second important distinction is that velocity is a vector and speed a scalar quantity.

A useful theory of dynamics (the study of the motion of bodies under the influence of forces) must be able to describe the moment-by-moment position and velocity of the system under consideration. Average parameters are not sufficient to describe the motion. To do the job we must define "instantaneous" parameters. Let's focus on one new idea at a time, and begin our analysis with the study of motion in one dimension.

Suppose that a particle moves along the x-axis near a meterstick so that the distance from an arbitrary origin is described by the mathematical equation

$$x(t) = 1 + 2t - 3t^2, \qquad (8.3)$$

where $x(t)$ is the position of the particle in meters, at the time t, measured in seconds.* There are at least two ways to represent this motion. Figure 8.3a is an artist's conception of the one-dimensional

---

* Students will note that Eq. 8.3 has a problem with units. Distances $x(t)$ are measured in meters and time in seconds, and thus the equation apparently equates seconds to meters, a no-no! We can get around this problem by considering the constants to have units: namely, 1 has units of meters; 2, units of m/s; 3, units of m/s²; which makes the equation dimensionally correct. This is strictly a kinematic exercise, since we have not specified the force or forces that cause this motion.

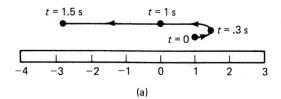

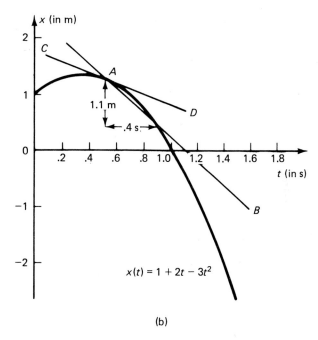

(b)

Fig. 8.3

motion. We see that the particle starts off 1 m from the origin, moves in the positive *x*-direction for about .3 s, stops, reverses direction, and begins moving in the minus *x*-direction, crossing the origin 1 s after the start. Plotting the position $x(t)$ versus time t, as in Fig. 8.3b is, however, a more useful way to represent the motion described by Eq. 8.3. The reason for this will soon become apparent.

It is clear from the plot of $x(t)$ versus t in Fig. 8.3b that the particle travels different distances in successive time intervals. The velocity is apparently always changing. An approximate measurement of the velocity during some particular time interval can be obtained using our definition of average velocity. Consider a time interval τ. The displacement during the interval τ is the difference between the particle's position at time $(t + \tau)$, namely $x(t + \tau)$, and the position of the particle at some earlier time t, $x(t)$; or

DISPLACEMENT (during time interval τ) $= x(t + \tau) - x(t)$, (8.4)

and thus the average velocity over this interval is

$$\langle v(t, \tau) \rangle = \frac{x(t + \tau) - x(t)}{\tau}, \tag{8.5}$$

where the symbol $\langle \ \rangle$ indicates average value. The average velocity depends on both the time t from which the displacement is measured and the time interval τ over which the average is taken. Thus both t and τ appear in the argument of the velocity v. One way to conceptualize this average velocity is to imagine another particle traveling at a constant velocity equal to $\langle v(t, \tau) \rangle$. Such a particle, starting out at $x(t)$ at time t, would also reach $x(t + \tau)$ at the time $t + \tau$.

The algebraic quantity $\langle v(t, \tau) \rangle$ can also be given a graphical interpretation which may aid the student. Suppose, for example, that you wished to find $\langle v(.5, .4) \rangle$. From Eq. 8.3 we see that

$$x(t) = 1 + 2t - 3t^2 \qquad x(.5) = 1 + 2(.5) - 3(.5)^2 = 1.25 \text{ m}$$

$$x(t + \tau) = x(.5 + .4) = x(.9) = 1 + 2(.9) - 3(.9)^2 = .37 \text{ m}.$$

Thus from Eq. 8.5 we have

$$\langle v(.5, .4) \rangle = \frac{.37 - 1.25}{.4} = -2.2 \text{ m/s}.$$

The straight line AB in Fig. 8.3b, drawn through points ($x = 1.25$, $t = .5$) and ($x = .37$, $t = .9$), has a slope equal to $\langle v(.5, .4) \rangle$. Note the minus sign. Velocity is a vector quantity, and for our one-dimensional example the minus sign indicates that the average displacement is in the minus x-direction.

The calculation gives a crude idea of how this particular particle is moving during a .4-s time interval, starting at $t = .5$ s. But suppose that we want to know more precisely what the velocity was at or near $t = .5$ s. To achieve this we will calculate the average velocity over smaller time intervals, for example, .2 s or .1 s or even .0001 s. In Table 8.1 the displacement from $x(.5) - 1.25$ m and the average velocity $\langle v(.5, \tau) \rangle$ for a sequence of decreasing values of the time interval τ have been calculated.

TABLE 8.1

τ	.4	.2	.1	.05	.01	.005	.001	.0001
$x(.5 + \tau) - x(.5)$	$-.88$	$-.32$	$-.13$	$-.0575$	$-.0103$	$-.0051$	$-.001003$	$-.00010003$
$\langle v(.5, \tau) \rangle$	-2.2	-1.6	-1.30	-1.15	-1.03	-1.02	-1.003	-1.0003

This table demonstrates a remarkable phenomenon. Even though the time interval τ is sequentially reduced by four orders of magnitude, the average velocity over the interval is apparently "settling down" to a constant value of about -1.00 m/s. The reason this occurs, of course, is that we are calculating a *ratio of two* quantities, displacement and time interval. Although the displacement decreases in the numerator, the corresponding time interval also decreases in the denominator, keeping the ratio constant. Mathematicians are cautious people. They have developed cautious procedures (originally discovered by Newton), called "taking a limit," which tell us how to deal with the problem of making the time interval arbitrarily small, even zero. This procedure is formally written as

$$\mathbf{v}(t) = \langle \mathbf{v}(t, 0) \rangle = \underset{\tau \to 0}{\text{limit}}\ \frac{x(t + \tau) - x(t)}{\tau}, \tag{8.6}$$

which means that we define the instantaneous velocity at the time t as the ratio of the displacement from $x(t)$ to $x(t + \tau)$ divided by the time interval τ as we let τ get smaller and smaller. It is important to understand that the division by τ is executed *first*, then the time interval τ is allowed to go to zero.

Mathematicians assure us that this ratio does not blow up, even if the time interval goes all the way to zero. If you are still puzzled, you can adopt an experimentalist's attitude, noting that in an actual measurement of any of these dynamical quantities, finite time and displacement intervals are always necessary. That is, one can never measure the velocity of a particle without allowing the particle to travel some finite physical distance in a small but finite time interval. The ratio of these two finite experimentally measurable quantities must also be a finite quantity. In any case, we can now use this limiting process to calculate the velocity of the particle described by Eq. 8.3 as follows:

$$v(t) = \lim_{\tau \to 0} \frac{[1 + 2(t + \tau) - 3(t + \tau)^2] - (1 + 2t - 3t^2)}{\tau}$$

$$= \lim_{\tau \to 0} \frac{2\tau - 6t\tau - 3\tau^2}{\tau} = \lim_{\tau \to 0} (2 - 6t - 3\tau).$$

Now, taking the limit, letting $\tau \to 0$, we have

$$v(t) = 2 - 6t.$$

Notice the order of the limiting process. First we substituted the expression for the position as a function of time (Eq. 8.3) and carried out the indicated algebraic manipulation. Second, we divided by τ

and then allowed $\tau \to 0$. I'm sure you all recognize this process as taking the DERIVATIVE of x with respect to t. It is formally written

$$v(t) = \frac{dx(t)}{dt} \equiv \lim_{\tau \to 0} \frac{x(t + \tau) - x(t)}{\tau},$$ (8.7)

where the symbol $\equiv$ means "defined as equal to."

Equation 8.7 is remarkable, for it allows us to calculate the velocity of any particle that is described by a mathematical expression for the position $x(t)$ at *any* time t. For example, consider the particle we have been playing with, described by Eq. 8.3. What is its velocity at .5 s? We have already calculated $v(t)$ (the velocity at any time t) above, and now it is just a trivial matter of substitution, namely,

$$v(t) = 2 - 6t$$

or $$v(.5) = 2 - 6(.5) = 2 - 3 = -1 \text{ m/s},$$

which agrees with the calculations done in Table 8.1. But unlike the rather laborious job we performed in constructing this table, we can now calculate instantaneous velocities *at any time*, simply by the appropriate algebraic substitution into our expression for the velocity, $v(t)$.

Figure 8.4 is a plot of the velocity as a function of time of the particular particle described by Eq. 8.3. The equation $v(t) = 2 - 6t$ represents the slope of Eq. 8.3 at any time t. This particle starts out at $t = 0$ traveling in the positive x-direction, slows down, stops at .333 s, reverses direction, and moves in the minus x-direction at a speed that increases linearly with time.

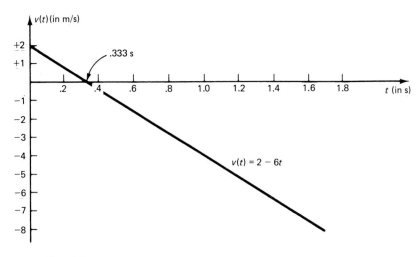

Fig. 8.4

8.4 Acceleration

WHEN THE instantaneous velocity itself is changing with time, as in our example, the particle is said to be ACCELERATING. The formal definition of acceleration is

$$\mathbf{a}(t) = \frac{d\mathbf{v}(t)}{dt} \equiv \lim_{\tau \to 0} \frac{\mathbf{v}(t + \tau) - \mathbf{v}(t)}{\tau}, \tag{8.8}$$

Like velocity, acceleration is a vector quantity. Students often confuse velocity and acceleration. A particle may be on the positive x-axis and move in the positive x-direction, but this does not necessarily mean that its acceleration is positive, since acceleration is the *rate of change* of the particle's *velocity*. Consider a ball tossed vertically into the air. As the ball travels upward, it slows down, and when it comes down, it speeds up, but it is *always accelerating in the downward direction*.

Returning to our example, where the particle's velocity is given by $v(t) = 2 - 6t$

$$a(t) = \lim_{\tau \to 0} \frac{(2 - 6t - 6\tau) - (2 - 6t)}{\tau} = \lim_{\tau \to 0} -6 = -6 \text{ m/s}^2.$$

The acceleration of this particle is constant in the negative x-direction. Particles that exhibit constant acceleration form an important class of problems, but the student should not attach "religious" significance to them and assume that all motion can be described by equations of constant acceleration. Such is not the case!

8.5 Derivatives

THE CALCULATIONS of velocity and acceleration are two specific examples of the important calculus operation of differentiation. In the general case, when y is a function of the independent variable x,

$$y = f(x),$$

the derivative of y with respect to x is defined as

$$\frac{dy}{dx} = \frac{df(x)}{dx} \equiv \lim_{\varepsilon \to 0} \frac{y(x + \varepsilon) - y(x)}{\varepsilon}.$$

Examples:

(a) $y = x^{-1}$

$$\frac{dy}{dx} = \frac{d}{dx}(x^{-1}) = \lim_{\varepsilon \to 0} \left(\frac{\frac{1}{x + \varepsilon} - \frac{1}{x}}{\varepsilon} \right) = \lim_{\varepsilon \to 0} -\left(\frac{1}{x^2 + \varepsilon x} \right) = -\frac{1}{x^2}.$$

(b) $y = \sin ax$

$$\frac{dy}{dx} = \lim_{\varepsilon \to 0} \frac{\sin a(x + \varepsilon) - \sin ax}{\varepsilon}.$$

But

$$\sin (a + b) = \sin a \cos b + \cos a \sin b$$

$$\frac{dy}{dx} = \lim_{\varepsilon \to 0} \frac{\sin ax \cos a\varepsilon + \cos ax \sin a\varepsilon - \sin ax}{\varepsilon}$$

$$= \lim_{\varepsilon \to 0} \left(\sin ax \frac{(\cos a\varepsilon - 1)}{\varepsilon} + \cos ax \frac{\sin a\varepsilon}{\varepsilon} \right).$$

For small θ, where θ is expressed in radians,* one can represent the trigonometric functions by the following Taylor series expansions:

$$\cos \theta = 1 - (\tfrac{1}{2})\theta^2 + (\tfrac{1}{24})\theta^4 + \text{higher-order terms in } \theta$$

$$\sin \theta = \theta - (\tfrac{1}{6})\theta^3 + (\tfrac{1}{120})\theta^5 + \text{higher-order terms in } \theta$$

Substituting these series expansions into the expression above for $\cos a\varepsilon$ and $\sin a\varepsilon$ (neglecting terms of higher order than θ^3) gives

$$\frac{dy}{dx} = \frac{d}{dx} \sin ax = \lim_{\varepsilon \to 0} [\sin ax(-\tfrac{1}{2}a^2\varepsilon) + \cos ax(a - \tfrac{1}{6}a^3\varepsilon^2)].$$

Taking the limit as $\varepsilon \to 0$ yields

$$\frac{d}{dx} \sin ax = a \cos ax$$

(c) $y = x^n$ (n is a positive integer)

$$\frac{dy}{dx} = \lim_{\varepsilon \to 0} \frac{(x + \varepsilon)^n - x^n}{\varepsilon}.$$

The binomial series allows us to expand the $(x + \varepsilon)^n$ term as

$$(x + \varepsilon)^n = x^n + nx^{n-1}\varepsilon + \frac{n(n - 1)}{2!} x^{n-2}\varepsilon^2$$

$$+ \frac{n(n - 1)(n - 2)}{3!} x^{n-3}\varepsilon^3 + \cdots,$$

where there are $(n + 1)$ terms in the *finite* series. Substituting in the expression above and dividing by ε, we have

$$\frac{dy}{dx} = \frac{d}{dx} (x^n)$$

$$= \lim_{\varepsilon \to 0} \left[nx^{n-1} + \frac{n(n - 1)}{2!} x^{n-2}\varepsilon + \frac{n(n - 1)(n - 2)}{3!} x^{n-3}\varepsilon^2 \right]$$

$$\frac{d}{dx} (x^n) = nx^{n-1}.$$

---

* See Chapter 9 for the formal definition of radians. Conversion from degrees to radians is given by 1 radian = 57.3 degrees.

Differentiation yields a mathematical expression for the *rate of change of the function differentiated* with respect to the variable indicated. An extensive list of important and commonly used derivatives is given in Appendix A.IX.

8.6 The Exponential Function

THE EXPONENTIAL function e^x, where e is a particular number (namely, 2.718 . . .), describes the behavior of many important physical systems that you will be studying, and therefore is worthy of special attention. It has the unusual property that its derivative is equal to itself. That is,

$$\frac{d}{dx} e^x = e^x. \tag{8.9}$$

Equation 8.9 expressed in words is:

> *The rate of change of the exponential function at any point x is equal to the value of the function itself at x.*

No other mathematical function acts like this.

We can determine the value of the constant e from the property expressed in Eq. 8.9. Consider a number which we shall call α. Can we find a number α such that

$$\frac{d}{dx} \alpha^x \overset{?}{=} \alpha^x ?$$

From the definition of the derivative of any function, we have

$$\frac{d\alpha^x}{dx} \equiv \lim_{\varepsilon \to 0} \left(\frac{\alpha^{x+\varepsilon} - \alpha^x}{\varepsilon} \right) = \lim_{\varepsilon \to 0} \left(\alpha^x \frac{\alpha^\varepsilon - 1}{\varepsilon} \right)$$

$$= \alpha^x \lim_{\varepsilon \to 0} \left(\frac{\alpha^\varepsilon - 1}{\varepsilon} \right).$$

Now it is clear that *if*

$$\lim_{\varepsilon \to 0} \left(\frac{\alpha^\varepsilon - 1}{\varepsilon} \right) = 1,$$

we will have proved what we started out to discover, namely that

$$\frac{d\alpha^x}{dx} = \alpha^x.$$

Is there a special value of α such that

$$\lim_{\varepsilon \to 0} \left(\frac{\alpha^{\varepsilon} - 1}{\varepsilon} \right) = 1? \tag{8.10}$$

Equation 8.10 would be satisfied if

$$\alpha^{\varepsilon} \approx 1 + \varepsilon + \text{higher-order terms in } \varepsilon \tag{8.11}$$

since we can substitute Eq. 8.11 into Eq. 8.10 and obtain

$$\lim_{\varepsilon \to 0} \left[\frac{(1 + \varepsilon + \cdots) - 1}{\varepsilon} \right] = \lim_{\varepsilon \to 0} (1 + \text{terms in } \varepsilon, \varepsilon^2, \ldots, \text{etc.})$$

$$= 1.$$

Equation 8.11 can be used to obtain the special numerical value of α. Taking the logarithms of this equation, we obtain

$$\varepsilon \log_{10} \alpha \approx \log_{10}(1 + \varepsilon)$$

or

$$\log_{10} \alpha = \frac{1}{\varepsilon} \log_{10}(1 + \varepsilon). \tag{8.12}$$

You can use your calculator now to obtain α, by substituting various small values for ε in Eq. 8.12. Table 8.2 contains the results for three values of ε. It is clear from Table 8.2 that we can find a particular value for α that will satisfy Eq. 8.10. That value is written as e and is

$$e = 2.71828183. \ldots$$

Other properties of the exponential function are

$$\frac{d}{dx} e^x = e^x$$

as well as

$$\frac{d}{dx} e^{ax} = ae^{ax}$$

$$\frac{d}{dx} ae^x = ae^x.$$

The exponential function may also be expressed as an infinite power series

$$e^x = 1 + x + \frac{x^2}{2!} + \frac{x^3}{3!} + \cdots .$$

TABLE 8.2

ε	$\log_{10}(1 + \varepsilon)$	$\frac{1}{\varepsilon} \log_{10}(1 + \varepsilon)$	$\text{antilog}\left[\frac{1}{\varepsilon} \log_{10}(1 + \varepsilon) \right] = \alpha$
.01	.00432137	.432137	2.704814
.0001	.00004327	.434272	2.7181459
.000001	.000000432943	.4342943	2.718280

NEWTON'S THREE LAWS of motion, deceptively simple to write down, provide a sophisticated and powerful theoretical framework for understanding both individual particle motion and the complicated motion of mechanical systems. True, these laws have their limitations. Newtonian mechanics cannot describe particles moving with velocities comparable to that of light, interacting particles where mass and energy are exchanged, and systems on the scale of atoms. Still, there is an enormous array of familiar and unfamiliar systems for which Newtonian mechanics is applicable. Mechanical, civil, and aeronautical engineering students in particular will discover these equations to be the basis of much of their later engineering studies. These laws have withstood the test of time. They have challenged students for several hundred years to master their subtleties. Now it's your turn.

Without in any way diminishing Newton's contribution to the study of dynamics, remember that it was Galileo, the first modern scientist, who first clearly stated what is commonly called Newton's first law of motion. The first law states that:

> ***I.*** *Every body persists in its state of rest or uniform motion in a straight line (constant velocity) unless compelled to change that state of motion by an external force.*

I consider this law to be the greatest single intellectual "leap" ever achieved by a scientist (and I am not Italian!). If this judgment surprises you, let me argue the case. Consider the experimental data available to Galileo in 1638. Imagine yourself examining the motion of bodies; let a ball drop, it accelerates; roll it along a smooth floor, it slows down and stops; watch a pendulum swing, it slows down and stops. Nothing Galileo or anyone else had ever observed seemed to obey the first law. Some law! Why didn't the experimental data fit the law? Was the law wrong?

A moment's reflection gives you the answer that we can now see with 20/20 hindsight. No one had been able to remove the effects of the external forces acting on the system. Galileo realized that the "natural" state of a particle is to move with constant velocity (including the special case of zero velocity). He was the first to develop the concept we call INERTIA.* It does not require angels to flap their wings, forces from distant stars, or fields from unseen objects for

---

* I urge you to read R. Feynman's delightful discussion of the concept of inertia, *The Character of Physical Law* (Cambridge, Mass.: MIT Press, 1965), Chap. 1.

bodies to behave this way: it is simply the natural state of matter. No other explanation has ever been discovered.

Both Galileo and Newton developed the so-called second law. It gives a mathematical formulation to the effect of applied forces on the motion of a particle.

> **II.** *The vector sum of the forces acting* **on** *a particle is equal to the time rate of change of the linear momentum of the particle.*

or

$$\sum_i \mathbf{F}_i = \frac{d}{dt}\,\mathbf{p},$$

(8.13)

where $\sum_i \mathbf{F}_i$ is the vector sum of all the forces $\mathbf{F}_i$ acting *on* the particle, and $\mathbf{p}$ is the linear momentum of the particle.

The way I have written Newton's second law, it is valid for particles even at velocities near that of light. In other words, the equation *is relativistically correct.* However, to use it at large velocities requires a deeper understanding of relativistic kinematics than we have time to develop. You might guess from anecdotal knowledge of relativity that part of the problem in using Eq. 8.13 at high velocities comes from the differentiation with respect to time. Whose time? What time are you using when you take the limit? For the interested student, a series of excellent references is given at the end of chapter 10. You have all the mathematical tools necessary to understand the concepts of kinematics in special relativity, so don't be afraid to look into it. Relativistic kinematics examines the questions of the twin paradox, simultaneity, the red shift, space "wars," the "pole in the barn," and many more fascinating subjects.

We will restrict ourselves to the study of Newtonian dynamics, low velocity, and conserved mass. For this type of problem the second law can also be written

$$\sum_i \mathbf{F}_i = \frac{d}{dt}\,m\mathbf{v},$$

(8.14)

or, for a *single force* $\mathbf{F}$ acting on a *single particle* of mass m,

$$\mathbf{F} = m\mathbf{a}.$$

(8.15)

Let's think about the meaning of Eq. 8.15. Since we have already discussed acceleration (at least in one dimension), it is mass and force that need our attention. Imagine an experimental situation where the same force is applied to two different particles. For example, a proton and an electron are separated and allowed to drift into the same region of constant electric field in a CRT. If the acceleration of the proton is compared with the acceleration of the electron, we discover that the acceleration of the proton is about 1/1833 times the electron's acceleration and in the opposite direction. The same experiment could be done with a μ-meson, an oxygen-positive ion, or other charged particles. Since the same force is applied to all three particles, Newton's second law states that*

$$F = m_e a_e = m_p a_p = m_\mu a_\mu = m_0 a_0 \qquad (8.16)$$

where the subscripts refer to the electron, proton, μ-meson, and the oxygen ion. Rearranging Eq. 8.16, we have

$$\frac{m_e}{m_p} = \frac{a_p}{a_e} \qquad \frac{m_e}{m_\mu} = \frac{a_\mu}{a_e} \qquad \frac{m_e}{m_0} = \frac{a_0}{a_e}. \qquad (8.17)$$

This series of experiments can only be used to determine the ratio of masses of two objects from their measured accelerations. It is necessary to establish an arbitrary standard mass and call it one mass unit. It would have been nice if physicists had defined the lightest stable particle (the electron) as one mass unit, but they didn't. Originally, the kilogram (SI mass units) was defined to be the mass of 1000 cm^3 of water at a temperature of 4°C, but this proved to be a poor standard because exact volume measurements are exceedingly difficult; water absorbs gases and evaporates. The standard kilogram is one particular cylindrical hunk of platinum–iridium alloy kept under inert atmosphere at the International Bureau of Weights and Measures in Sèvres, France. All masses are measured by comparison to that standard.

If inertial mass is indeed a property of the object alone, and if we repeat the experiments described above for a different force (gravitational, magnetic, etc.), the ratio given in Eq. 8.17 will remain the same. If we connect the two particles so that their mass is now the scalar sum $(m_1 + m_2)$, the measured acceleration produced by the force F will be reduced by the amount

$$a_{1+2} = \frac{1}{m_1 + m_2} F.$$

This has been observed over a wide range of velocities.†

Photo 8.1 *The standard kilogram at the Bureau of Standards in Washington, D.C. It is the cylinder on the right. It is being compared to a lower standard.*

---

* The electrical force depends only on the electric field **E** and the charge of the particle q (Eq. 3.1).

† Here we are neglecting relativistic effects. If one combined a proton and a neutron to form a deuteron, the mass of the deuteron is less than the sum of the mass of the two constituent particles. This mass loss has been transformed into "binding" energy of the deuteron. Such effects were not known to Newton. The binding energy for everyday large-scale objects is insignificant compared to their rest energy and thus can be neglected in our calculations.

Having defined the mass scale in kilograms and acceleration in meters/(second)^2, the units of force are defined to be in newtons. A newton (N) is that force which, when acting on a 1-kg mass, will produce an acceleration of 1 m/s^2, or

$$1 \text{ N} = 1 \text{ kg m/s}^2. \qquad (8.18)$$

Now we turn to the most difficult task of all, the definition of force. Many introductory textbooks give the impression that defining "force" is straightforward, by stating that "if a body is accelerating, there is a force acting on it"—or an equivalent statement. In other words, use the second law to define force. But the thoughtful reader recognizes this to be a circular argument. For Newton's first law to have any meaning, forces must have *independent properties* in addition to those described by $\mathbf{F} = ma$. Such properties do exist, but it is difficult, if not impossible, to define them in a few neat sentences. For the sake of simplicity, let us rely on our experience and knowledge to explain forces. A spring, for example, produces forces that are related to its elongation from its equilibrium configuration and the stiffness (or elastic modulus) of the material from which it is fabricated. Electric and magnetic forces have properties related to charges, metal shapes, electric currents, and number of windings, while viscous drag forces depend on surface texture, viscosity, size, and so on. Pushes and pulls are the most familiar forces, yet at the microscopic atomic scale these are the most complicated of all forces! The essential point is that forces have their own properties, independent of the fact that they produce acceleration when they act on a particle.

This discussion may not satisfy the mathematical purists among you, but in my defense I quote the great Dutch physicist H. A. Kramers, who remarked: "My own pet notion is that in the world of human thought generally, and in physical science particularly, the most important and most fruitful concepts are those to which it is impossible to attach a well-defined meaning."*

Newton's third law is often stated: "For every action there is an equal and opposite reaction." A more complete and precise definition will help you to understand this law.

> **III.** *When two bodies interact, the first exerts a force $\mathbf{F}_{12}$ on the second, and the second body exerts a force $\mathbf{F}_{21}$ on the first, such that $\mathbf{F}_{12} = -\mathbf{F}_{21}$. These two forces act along the same line.*

---

* *Physical Science and Human Values* (Princeton, N.J.: Princeton Univ. Press, 1947).

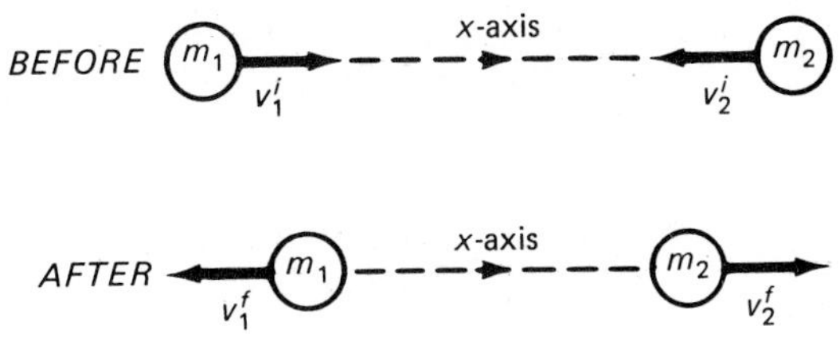

Fig. 8.5

Newton's third law makes it clear that a single isolated force cannot exist; forces must always come in pairs. Some examples may help to clarify this law. Consider two different hockey pucks colliding head-on as shown in Fig. 8.5. (again, for simplicity, a one-dimensional problem). During the collision, puck 1 exerted a force on puck 2 called $\mathbf{F}_{12}$, and conversely puck 2 exerted a force $\mathbf{F}_{21}$ on puck 1. We assume what appears to be obvious—that the two forces act on their respective pucks for an equal amount of time. Now to simplify the discussion, we will assume that the deformation force remains constant during the time of the collision Δt.* From Newton's second law,

$$\mathbf{a}_1 = \frac{\mathbf{F}_{21}}{m_1} \qquad \mathbf{a}_2 = \frac{\mathbf{F}_{12}}{m_2}.$$

Each puck experiences a constant acceleration for a time Δt, resulting in a change in the velocity of the puck. Any particle that executes constant accelerated motion changes its velocity according to the expression $\mathbf{v}^f = \mathbf{v}^i + \mathbf{a}t$.† Then substituting for the acceleration, we have

$$\mathbf{v}_1^f = \mathbf{v}_1^i + \frac{\mathbf{F}_{21}}{m_1}\Delta t \qquad \mathbf{v}_2^f = \mathbf{v}_2^i + \frac{\mathbf{F}_{12}}{m_2}\Delta t.$$

Multiplying the first equation by m_1 and the second by m_2, and adding them together yields

$$m_1\mathbf{v}_1^f + m_2\mathbf{v}_2^f = m_1\mathbf{v}_1^i + m_2\mathbf{v}_2^i + (\mathbf{F}_{21} + \mathbf{F}_{12})\,\Delta t. \qquad (8.19)$$

We can look at Eq. 8.19 in two ways. Recalling our discussions of the great conservation principles, you should recognize the term $m_1\mathbf{v}_1^i + m_2\mathbf{v}_2^i$ as the initial total nonrelativistic momentum of the two-puck system, and $m_1\mathbf{v}_1^f + m_2\mathbf{v}_2^f$ as the total final momentum of the system. Momentum conservation requires that they be equal since

---

* A more realistic model of the deformation forces, such as one where the force depends on the amount of deformation, used in this derivation will not alter its conclusions.

† This equation will be derived from general kinematic consideration later in this chapter (Eq. 8.30).

there is no net *external* force on the system. Thus Eq. 8.19 reduces to

$$(\mathbf{F}_{21} + \mathbf{F}_{12})\,\Delta t = 0 \qquad \text{or} \qquad \mathbf{F}_{21} = -\mathbf{F}_{12},$$

demonstrating Newton's third law. Another way to interpret Eq. 8.19 is to use Newton's third law, which requires that $\mathbf{F}_{21} = -\mathbf{F}_{12}$, and then this example shows that the linear momentum of the system must be conserved. Both are valid interpretations.

There is, however, a subtle problem with this discussion, which becomes important in collisions involving long-range electromagnetic forces. For such events, momentum conservation remains a valid principle, but Newton's third law is not directly applicable. The problem comes about because of the assumptions of simultaneity of the two forces, the action and reaction. Consider two charged particles colliding. The force of particle 1 on particle 2 does *not* occur instantaneously since the electromagnetic field is propagated with a finite velocity, the velocity of light. This time, delay has been ignored in the problem. That is not all. It can be shown that the electromagnetic fields themselves carry momentum as well as energy that must be taken into account when one analyzes the collision. All this is by way of warning the reader to be skeptical of "obvious" assumptions. Having said all this, we shall assume in the rest of this book that the third law *is* valid and use it extensively.

Here is a demonstration that I use in my classes. I urge you to try it. You need only one other student, a piece of rope, and a reasonably smooth, clean floor. Both students sit on the floor and hold on to the ends of the rope, as shown in Fig. 8.6. One of the students (*a*) braces her feet on the floor; the other (*b*) lifts his feet up so that he can slide on his fanny. (Role reversal certainly is acceptable.) Student *a* begins to pull on the rope slowly while asking student *b* whether he feels a force. The answer, of course, is yes, if he holds on. In fact, if we gave each student a spring scale inserted between the rope and their hands, and asked each to read the scale, they would each observe the same elongation, the same force. If student *a* pulls hard enough and is braced well enough, student *b* will accelerate. Simple enough. Why get your pants dirty doing it? You do it to demonstrate a paradox!

The paradox is this: student *a* pulls with force $\mathbf{F}_a$ on student *b*, but the third law tells us that student *b* pulls on student *a* with exactly the same force $|\mathbf{F}_b|$ (in magnitude) but opposite in direction, $\mathbf{F}_a = -\mathbf{F}_b$. Since the sum of these forces, $\mathbf{F}_a + \mathbf{F}_b$, is zero, there is *no net force* and no acceleration, according to the second law. Try the experiment. You will confirm that both students feel forces in their arms. There is no doubt about it: these forces are in the opposite direction, yet student *b* does accelerate! How can this be? How do we resolve this paradox?

There is no real paradox here. The apparent paradox comes from the way I have analyzed the problem. Look carefully at Newton's

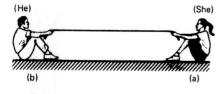

(He)   (She)

(b)   (a)

Fig. 8.6

second law and notice the critical preposition *"on"* in the law. It is the vector sum of the forces *on* a body, *not* the vector sum of the force *on and by*, that cause a body to accelerate. It is irrelevant that student *b* produces a force $\mathbf{F}_b$, at least it is irrelevant to the motion of student *b*. What is relevant, of course, is the force student *a* exerts *on* student *b*, and any other forces *on* student *b*. Think of the electrons in the accelerator of the CRT. We never mentioned that the electric field created by the electrons produces a force *on* the plates of the accelerator, since this force does not affect the motion of the electron. Never forget this! In analyzing the many problems you will shortly encounter, always ask and think carefully; what are the forces *on* the particular body I wish to study? Those forces are the ones that affect the motion of the body.

8.8 Single-Particle Motion with Constant Forces

NOW WE HAVE the three classical laws of motion. Given the position and velocity of a particle at a certain time and the forces on the particle, how can we calculate the position and velocity of the particle at a later time?

The answer seems to be, use Newton's second law,

$$\sum_i \mathbf{F}_i = m \frac{dv}{dt}.$$

But this is a DIFFERENTIAL EQUATION, *not* an algebraic equation such as those you are used to. It may very well be the first differential equation you have ever encountered, so we must pause and talk about this new kind of equation. Differential equations describe how things *change*, not how things are. A differential equation always involves derivatives in some form or other and, as you know, derivatives express how mathematical functions change with respect to some independent variable. In this discussion of the motion of single particles, the differential equations we will deal with express how position or velocity change with time. These equations might tell us, for example, that the velocity of a particular particle is increasing at a rate proportional to the time squared, or that the position of another particle is oscillating back and forth with regular periodicity, or that a particle's velocity does not change with time. Differential equations are not mathematical expressions that give you the velocity at any time or location you choose. Such information comes from the *solutions* to these differential equations. The solutions to these differential equations are the functions $r(t)$ and $v(t)$ in algebraic form.

The first task then is to set up differential equations for the particle using Newton's second law and the forces that act *on* the body.

After the correct differential equation has been written, the solution to it must be found. But even after the solution has been found, our work is not over, because the solution will not give us the position and velocity of the particle for all values of time. We must put in the INITIAL CONDITIONS of the particle's motion; that is, how did we start it? Did it start out with any velocity, and where did it start? A particle's motion will depend not only on the forces on it, but also on the position and velocity of the particle when the motion began. When the initial conditions have been properly included, the problem is solved.

A. Zero Net External Force on the Particle

Let's begin our discussion with the simplest case, where the sum of the external forces acting on our particle is zero. Newton's second law then becomes

$$\sum_i \mathbf{F}_i = 0 = m\frac{d\mathbf{v}}{dt} \qquad \text{or} \qquad \frac{d\mathbf{v}}{dt} = 0. \qquad (8.20)$$

Equation 8.20 is a differential equation, albeit a simple one. Our technique for solving it will be to *guess* the solution, plug the solution into the equation, perform the indicated operations, and see if it is indeed a solution.

You might object to this technique on two grounds. Your first objection might be: "How am I going to know what sort of solution to guess?" The answer is that after you've seen a few similar problems solved, you'll have acquired enough experience to judge rapidly what will work. You will learn to "read" the equation and express mathematically the physical solution you suspect is correct. "Fine," you say, "but suppose I find one solution that works, my roommate finds another solution that works, and the guy down the hall finds a third solution. How are we going to know which solution to choose? Are all these solutions correct?"

There's a sign in a famous Boston restaurant called Durgin Park which attests to the uniqueness of the restaurant; it reads: "There's no other place anywhere near this place that is just like this place, so this must be the place." Similarly, there's a theorem in mathematics called "the fundamental theorem of the calculus" (isn't that an impressive title?) which says that if you are given a function $a(t)$, and if you are able to guess a solution $v(t)$ to the equation $\frac{dv}{dt} = a(t)$, the most general solution of this differential equation is the solution you guessed plus a constant. This theorem says that you, your roommate, and the guy down the hall will only be able to find solutions that differ by a constant. Your solution is essentially *unique*. "There's no other solution that satisfies the differential equation that your solution satisfies, so your solution must be *the* solution—within an added constant."

To prove this theorem, suppose that we have *any* two solutions $v_1(t)$ and $v_2(t)$, both of which satisfy the same differential equation:

$$\frac{dv_1(t)}{dt} = a(t) \qquad \frac{dv_2(t)}{dt} = a(t).$$

Then $[v_1(t) - v_2(t)]$ will satisfy the following equation:

$$\frac{d}{dt}[v_1(t) - v_2(t)] = a(t) - a(t) = 0.$$

But the *only* function whose derivative is zero is a constant, so we deduce that $[v_1(t) - v_2(t)] = $ constant. In other words, if $v_2(t)$ is your solution, and $v_1(t)$ is any other solution, we've just proven that $v_1(t) = v_2(t) + C$.

Now that your mind is at ease about the propriety of guessing solutions to differential equations, let's return to Eq. 8.20. We pointed out in the last paragraph that the most general solution of the equation $\frac{dv}{dt} = 0$ is

$$v(t) = C \tag{8.21}$$

(where C is a constant, as yet undetermined), so our differential equation is solved. What value should the constant C be given? The differential equation describes the time evolution of the system of interest but does not tell you how the system started; you must insert this information into the mathematics. It is here that an initial condition of the particular problem comes in. Suppose that we are told the velocity of the particle at some instant of time—say that $v = v_0$ at time $t = 0$. Setting $t = 0$ on the right-hand side of Eq. 8.21 (since the right-hand side is independent of t in this case, the equation is unaffected), and setting the velocity on the left-hand side equal to v_0, we discover that $v_0 = C$, determining the constant C. Now that we know the value of the constant C, we can write our final result for the solution of the equation of motion 8.20 subject to the initial condition $v(t = 0) = v_0$:

$$v(t) = v_0. \tag{8.22}$$

We have just proven what you probably guessed from the beginning; that a particle that feels no net force moves with constant velocity.

Our next task is to find how the position of the particle changes with time. We can write Eq. 8.22 as

$$\frac{dx(t)}{dt} = v_0 \tag{8.23}$$

and once again we are faced with the problem of solving a differential equation. Equation 8.23 tells us that the derivative of $x(t)$ is a constant. Now try to think of a function whose first derivative with respect to time is a constant. It flashes into our minds that the time derivative of $C_1 t$ (C_1 is a constant) is a constant. Let's try out $x(t) = C_1 t$:*

$$\frac{d}{dt} C_1 t \overset{?}{=} v_0$$

or $$C_1 \overset{?}{=} v_0.$$

Yes, our inspiration did not deceive us; both the left and right sides are constant, so we have found a solution, but we see that $x(t) = C_1 t$ will not satisfy Eq. 8.23 unless C_1 is chosen equal to v_0.

We have found a solution to the differential Eq. 8.23, but it is *not* the most general solution. The most general solution is, according to the fundamental theorem,

$$x(t) = v_0 t + C_2 \tag{8.24}$$

(C_2 is a constant). You should differentiate Eq. 8.24 and check that it does satisfy Eq. 8.23. It only remains for us to determine the constant C_2. Again we refer to the initial condition for our problem. Suppose we are told that the particle is at position x_0 at time $t = 0$. If we set $t = 0$ on the right-hand side of Eq. 8.24, and $x(t = 0) = x_0$ on the left side of Eq. 8.24, we discover that $x_0 = C_2$.

In summary, a particle that feels no force, and that at time $t = 0$ is at position x_0 moving with velocity v_0, will be at position $x(t)$ at time t given by the expression

$$x(t) = x_0 + v_0 t. \tag{8.25}$$

This result can easily be checked by direct differentiation:

$$v(t) = \frac{dx}{dt} = \frac{d}{dt}(x_0 + v_0 t) = v_0$$

and $$a(t) = \frac{dv(t)}{dt} = \frac{d}{dt} v_0 = 0.$$

B. Sum of the Force Is Constant: $\sum_i \mathbf{F}_i = \mathbf{F}$

The next simplest case to examine is where the force is constant in time and space. An example is the force exerted by the earth's gravity on a small object near its surface, or the force exerted by a pair of electrically charged sheets of metal (parallel plates) on a charged particle inside. Before indulging in the mathematics of a problem, it is a good practice to anticipate the answer, based on physical reasoning. This is a good habit to get into. It is very easy to make a mathematical mistake, and an idea of how the particle ought to behave will

---

* The symbol $\overset{?}{=}$ means that we are asking the question: Is the left side of this equation equal to the right side, and if so, under what conditions?

help you spot an erroneous result. But the main reason to go through this exercise is that it will increase your physical insight.

It is very easy for us to guess how the position and velocity of a particle will change under a constant force because we have actually performed such experiments every time we've dropped something or thrown something into the air. For example, we expect that if **F** is negative, and the particle's initial velocity is negative, the particle will keep moving toward the negative direction, but more rapidly, for this is what happens when we throw objects down. We also expect that if the particle was moving initially with a positive velocity, it will continue to move, but will slow down until it stops, turns around, and moves in the opposite direction with greater and greater speed, for this is what happens when we throw something up into the air.

Now let's see if these predictions can be verified by applying Newton's laws:

$$\sum_i \mathbf{F}_i = \mathbf{F} = m\,\frac{d\mathbf{v}(t)}{dt}.$$

For one dimension (x-axis),

$$\frac{dv_x(t)}{dt} = \frac{F_x}{m}. \tag{8.26}$$

This differential equation is of the same form as Eq. 8.23 (derivative of something = a constant), so we could refer back to the discussion to obtain the mathematical form of the solution. Instead, let's take a slightly different approach that we can also use for other equations. Notice that the right-hand side of Eq. 8.26 is a polynomial in t (a very simple polynomial!). We know that the derivative of a polynomial in t is again a polynomial in t, of lower order. So without doing too much thinking, let's guess that the solution is a polynomial of the form

$$v = C_1 + C_2 t + C_3 t^2, \tag{8.27}$$

where C_1, C_2, and C_3 are all constants. Substituting Eq. 8.27 into Eq. 8.26 and performing the indicated differentiations, we have

$$\frac{d}{dt}(C_1 + C_2 t + C_3 t^2) = C_2 + 2C_3 t \stackrel{?}{=} \frac{F_x}{m}. \tag{8.28}$$

We see that our guess satisfies Eq. 8.26 provided that we set the constant C_2 equal to F_x/m and the constant C_3 equal to zero—there is no restriction on the constant C_1. How do we know this? Expression 8.28 *is not an equation;* it is our attempt to find a solution to the differential equation 8.26. For Eq. 8.27 to be the solution, it is necessary that the coefficients of various powers of t match *term by term* with the coefficients of the same power of t on the right side of the equation, after we perform the indicated operations of the differential equation on the trial solution. For this case

$$t^0 \text{ terms} \Rightarrow C_2 = \frac{F_x}{m}$$

$$t^1 \text{ terms} \Rightarrow 2C_3 t = 0 \qquad C_3 = 0.$$

Obviously, there was no need to include the $C_3 t^2$ term in our guess, because its derivative was bound to produce a term linear in t, a term we didn't need. But no harm was done—even if we didn't do too much thinking, the mathematics was thinking for us. Our solution to Eq. 8.26 is

$$v(t) = C_1 + \frac{F_x}{m} t. \qquad (8.29)$$

This happens to be the most general solution, because if we add a constant to the right-hand side of Eq. 8.29, we still obtain a solution of the same form: $\frac{F_x}{m} t + \text{constant}$. Now we determine the constant C_1 in terms of the initial condition. At time $t = 0$, the particle moves with velocity v_{0x}. Setting $t = 0$ on the right-hand side and $v(t = 0) = v_{0x}$ on the left-hand side of Eq. 8.29, we find that $v_0 = C_1$. Our complete result is then

$$v(t) = v_{0x} + \frac{F_x}{m} t. \qquad (8.30)$$

Let's check that Eq. 8.30 makes the predictions that we anticipated. If F_x is a negative number, we see that regardless of the value of v_{0x} for very large t the velocity v becomes more and more negative. Furthermore, if v_{0x} is positive, there is a time $t > 0$ when the particle is instantaneously at rest ($v = 0$ when $t = mv_{0x}/-F_x$), whereas this does not happen if v_{0x} is negative. So Eq. 8.30 agrees with our qualitative expectations. However, Eq. 8.30 says much more: it makes a *quantitative* prediction of the velocity *at every instant of time*!

Since we know $v(t)$, we can now find $x(t)$. Equation 8.30 can be written as a differential equation in $x(t)$ as

$$\frac{dx(t)}{dt} = v_{0x} + \frac{F_x}{m} t. \qquad (8.31)$$

The right-hand side of this differential equation is a polynomial in t. After a little thought we guess a solution of the form $x(t) = x_1 + C_2 t + C_3 t^2$ and try it out:

$$\frac{d}{dt} (C_1 + C_2 t + C_3 t^2) = C_2 + 2C_3 t \stackrel{?}{=} v_0 + \frac{F_x}{m} t.$$

Equating coefficients of the same power of t, we set $C_2 = v_{0x}$, and $2C_3 = F_x/m$. Our solution to Eq. 8.31 is

$$x(t) = C_1 + v_{0x} t + \frac{F_x}{2m} t^2$$

and it is already in the most general form. The constant C_1 is determined by the initial condition $x(t = 0) = x_0$, which requires that $x_0 = C_1$. Thus the position of a particle which is subjected to a constant force F, and which at time $t = 0$ is at position x_0 and has velocity v_0, is

$$x(t) = x_0 + v_{0x}t + \frac{F_x}{2m} t^2. \tag{8.32}$$

You should check right now if Eq. 8.32 agrees with our qualitative knowledge of how a ball thrown in the air behaves.

Equations 8.30 and 8.32 will be used quite often, and probably you'll memorize them. But should you ever forget them (or, for example, forget which letter comes where), don't panic. Just say to yourself: "I know that $x(t)$ has the form $x(t) = C_1 + C_2 t + C_3 t^2 + \cdots$" and take the derivative of $x(t)$ once to find the velocity, again to find the acceleration and determine C_1, C_2, C_3 from the initial conditions and Newton's law. It shouldn't take you more than a minute.

To summarize the *one-dimensional* kinematic equations of motion for a particle under the influence of a *constant force F_x*, we have

$$v_x(t) = v_{0x} + \frac{F_x}{m} t \tag{8.33a}$$

$$x(t) = x_0 + v_{0x}t + \frac{F_x}{2m} t^2 \tag{8.33b}$$

and two auxiliary equations which you can easily show by algebraic manipulations are consequences of the equations above:

$$v_x(t)^2 = v_{0x}^2 + 2 \frac{F_x}{m} [x(t) - x_0] \tag{8.33c}$$

$$x(t) = x_0 + \tfrac{1}{2}[v_{0x} + v_x(t)]t. \tag{8.33d}$$

Equation 8.33c is the only one that does not explicitly contain time. You can also write these equations in the form where a_x is substituted for $\frac{F_x}{m}$, as you will find in most elementary textbooks.

> *Caution:* *These four algebraic equations are valid **only** for particles moving under the influence of a constant resultant force.*

EXAMPLE 1 A rock is thrown vertically upward from 1.5 m above the ground. It leaves the thrower's fingers with a velocity of 12 m/s.

(a) What is the maximum height above the ground reached by the rock if we neglect all forces due to air resistance?

To use Newton's second law we must first determine all the forces on the rock. Since we are neglecting forces due to the air molecules moving past the rock, we are left with only the force of gravity, the earth's pull on the rock. In Chapter 2 we introduced the fundamental equation describing the gravitational force between two point objects of mass m_1 and m_2,

$$F_{12} = G \frac{m_1 m_2}{r_{12}^2}. \tag{2.1}$$

How do we use this expression to calculate the force between the earth and a small object near its surface, as depicted in Fig. 8.7?

The earth is obviously not a point object, but because it is nearly a perfect sphere, it can be shown that the gravitational force on any object that lies beyond its radius is the same as if all the mass of the earth were concentrated at its geometric center. That means that we can write the force on a small object of mass m due to the earth as

$$F_{\text{gravity}} = G \frac{M_e m}{r^2},$$

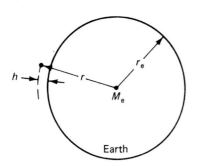

Fig. 8.7

where $r = r_e + h$ and h = height above the earth. For many practical problems the radius of the earth is much, much larger than the particle's height above the earth's surface. Thus we may approximate our expression for the gravitational force and replace r by r_e,

$$F_{\text{gravity}} = m \left(G \frac{M_e}{r_e^2} \right).$$

The constants inside the parentheses are

$$G = 6.673 \times 10^{-11} \text{ N-m}^2/\text{kg}^2$$

$$M_e = 5.97 \times 10^{24} \text{ kg}$$

$$r_e = 6.37 \times 10^6 \text{ m},$$

and substituting in the above, we obtain

$$F_{\text{gravity}} = m \left[6.673 \times 10^{-11} \text{ N-m}^2/\text{kg}^2 \frac{5.97 \times 10^{24} \text{ kg}}{(6.37 \times 10^6 \text{ m})^2} \right] = m(9.817) \text{ N/kg}$$

or

$$F_{\text{gravity}} = mg \text{ (directed toward the center of the earth)}$$

where $\quad g = 9.817$ N/kg or m/s^2

is called the ACCELERATION OF GRAVITY.
$$\tag{8.34}$$

Our result indicates that the force is independent of h, the height above the surface of the earth. Is that a reasonable approximation to make? How big an error did we make when we replaced r by r_e in Eq. 2.1? We can easily calculate this error by the following:

$$r^2 = (r_e + h)^2 = r_e^2\left(1 + \frac{h}{r_e}\right)^2 \approx r_e^2\left(1 + \frac{2h}{r_e}\right)$$

using the binomial expansion. The error then is represented by the term $\frac{2h}{r_e}$. If we consider objects that travel even as high as 10^3 m above sea level,

$$\frac{2h}{r_e} \approx \frac{2 \times 10^3}{6 \times 10^6} \quad \text{or about .03\% error.}$$

For our purposes, a .03% error may be neglected. We will assume that objects near the surface of the earth experience a constant force given by Eq. 8.34. Now back to our problem.

First order of business in these problems is to establish a coordinate system and to choose the positive and negative directions and the origin of the coordinates. Figure 8.8 shows our choice. The time origin is chosen to be the instant the rock leaves the thrower's hand. The net force on the rock is then

$$\mathbf{F}_{\text{rock}} = \mathbf{F}_{\text{gravity}} = -mg\mathbf{i}, \quad \textit{a constant force.}$$

This system, a particle that experiences a constant force, is one we have just solved. Equations 8.33 are the algebraic solutions.

How do we apply Eqs. 8.33 to this particular problem? When the rock reaches the maximum height it will have, at that instant, zero velocity. Thus we can substitute into Eq. 8.33a:

$$v_x(t) = v_{0x} + \frac{F_x}{m}t$$

$$0 = 12 \text{ m/s} + \left(-\frac{mg}{m}\right)t$$

or
$$t_{\text{max height}} = \frac{12 \text{ m/s}}{9.8 \text{ m/s}^2} = 1.22 \text{ s.}$$

Note that the gravitational force F_x acts in the negative x-direction, hence the minus sign. Substituting this time into Eq. 8.33b, we have

$$x(t) = x_0 + v_{0x}t + \frac{F_x}{2m}t^2$$

$$x(t) = 1.5 \text{ m} + (12 \text{ m/s})(1.22 \text{ s}) - \tfrac{1}{2}(9.8 \text{ m/s}^2)(1.22 \text{ s})^2$$

$$x_{\text{max height}} = 8.85 \text{ m.}$$

This is the maximum height of the rock. We could also have obtained this result in one step from Eq. 8.33c by direct substitution,

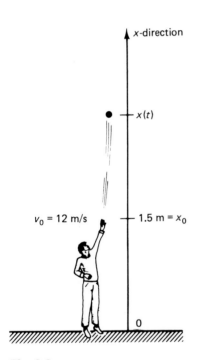

x-direction

$x(t)$

$v_0 = 12$ m/s

1.5 m = x_0

0

Fig. 8.8

noting again that the velocity of the particle at maximum height is zero:

$$v_x^2 = v_{0x}^2 + 2\left(\frac{F}{m}\right)(x - x_0)$$

$$0 = (12 \text{ m/s})^2 - 2(9.8 \text{ m/s}^2)(x - 1.5 \text{ m}).$$

Solving for x yields

$$x_{\text{max height}} = 8.85 \text{ m}.$$

(b) How long will it take for the rock to reach the ground after it leaves the hand of the thrower?

We can approach this problem in several ways. We already know how long it took to reach the top and how high it went, so the problem can be separated into two parts; the first part has already been completed. The second part can be stated: How long will it take a rock to fall 8.85 m if it starts with zero velocity? Using Eq. 8.33b, where $x = 0$ is the final position of the rock on the ground, we have

$$x(t) = x_0 + v_{0x}t + \frac{F_x}{2m}t^2$$

$$0 = 8.85 \text{ m} + 0 - \tfrac{1}{2}(9.8 \text{ m/s}^2)t^2$$

$$t^2 = 1.81.$$

Then

$$t = 1.34 \text{ s} \qquad \text{(to fall down)}$$

and

$$\text{total time (round trip)} = 1.22 \text{ s} + 1.34 \text{ s} = 2.56 \text{ s}.$$

The second method uses Eq. 8.33b also but starts the problem at $t = 0$, where $v_0 = 12$ m/s and $x_0 = 1.5$ m. Then

$$0 = 1.5 \text{ m} + (12 \text{ m/s})t - \tfrac{1}{2}(9.8 \text{ m/s}^2)t^2,$$

which is a quadratic equation in t. The solution can be found by recalling your high school algebra (see Appendix A.VIII):

$$t = \frac{-12 \pm [(12)^2 + (4)(1/2)(9.8)(1.5)]^{1/2}}{-9.8}$$

$$= \frac{-12 \pm 13.17}{-9.8} = 2.57 \text{ s}.$$

The negative time solution is not physical. Such solutions often occur from our mathematical analysis of a given problem and they must be dealt with using our knowledge of the particular system. In other words, the mathematics will not always give you the correct answer automatically. You must still think about the physics.

This example should make it clear that Eqs. 8.33 are tools that we can use to solve constant-acceleration problems, but that they do not provide automatic answers. You must establish a reference frame, find the constant forces on the body of interest, pick the correct equation, and think about the physics of the problem. In this example, the key to the solution is that when the rock reaches its maximum height it has zero velocity.

EXAMPLE 2   A driver moving along a straight road slams on his brakes, applying a uniform force in such a way as to reduce his speed from 50 km/h to 15 km/h in a distance of 75 m.

(a) What is the acceleration of the car?

Consider the direction of travel to be the positive x-direction, as shown in Fig. 8.9. Equation 8.33c is appropriate for this problem, but note that time is given in hours, velocity in km/h, and distance in meters. To use the constant force kinematic equation all the distances must be in the same units. In this example we will choose kilometers. Converting 75 m = .075 km gives

$$v_x^2 = v_{0x}^2 + 2a_x(x - x_0) \qquad \text{note that } a_x = \frac{F_x}{m}$$

$$(15 \text{ km/h})^2 = (50 \text{ km/h})^2 + 2a_x(0.75 \text{ km} - 0).$$

Solving for a_x, we obtain

$$a_x = -15{,}166 \text{ km/h}^2 = -1.17 \text{ m/s}^2.$$

The car is decelerating as indicated by the minus sign.

(b) If the car continues to decelerate at the same rate, how long will it travel before it comes to rest?

Equation 8.33a is appropriate since $v_x = 0$ when the car stops:

$$0 = 50 \text{ km/h} - (15{,}166 \text{ km/h}^2)t.$$

Solving for t yields

$$t = .0033 \text{ h} \quad \text{or} \quad 11.86 \text{ s}.$$

(c) How far will the car travel before it comes to rest?

Equation 8.33b, with $x_0 = 0$,

$$x = 0 + (50 \text{ km/h})(.0033 \text{ h}) - \tfrac{1}{2}(15{,}166 \text{ km/h}^2)(.0033 \text{ h})^2$$

$$= .0824 \text{ km} = 82.4 \text{ m}.$$

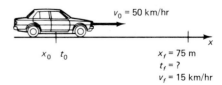

Fig. 8.9

THERE ARE many interesting problems where the net force on the particle is constant, but there are many more cases where the force depends on time, position, velocity, pressure, temperature, and a host of other variables. Although most of these systems are too complicated for us to analyze in this book, some of them can be studied now. We will need to have a mathematical function that describes how the force depends on the independent variable. For example, suppose that the force depends on time as

$$F_x(t) = \frac{C_1}{t^3} \quad \text{for } t > 0 \tag{8.35}$$

where C_1 is a constant and F_x is the force in the x-direction. Equation 8.35 gives us the value of the force for all times greater than zero. This expression can then be substituted into Newton's second law, and a solution to that differential equation for $v(t)$ and $x(t)$ obtained.

Is it always possible to solve such a differential equation if we have a mathematical function for the net force? The answer, unfortunately, is *no*, it is not always possible to find $v(t)$ and $x(t)$ given F. It is always possible to calculate a numerical solution on a computer to some arbitrary accuracy, given a large enough computer and the expression for the force, but analytical solutions for $x(t)$ and $v(t)$ cannot be guaranteed. However, I will make a deal with you. *All the problems in this book* (unless explicitly stated in the problem) do have analytical solutions and, what's more, all the solutions to these differential equations can be expressed as one of the following familiar functions or some combinations of these functions:

1. Polynomials of x, y, z, or t: for example, $(x^2 + y^2)$, $\frac{C}{t^4}$, etc.
2. Trigonometric functions, like $\cos(\omega t + \phi)$, $C\sin(\omega t + \phi)$
3. Exponential functions, as $e^{\alpha t}$, $Ce^{-\alpha t^2}$, $C(1 - e^{\alpha t})$
4. Combinations, $e^{-\alpha t}\cos(\omega t + \phi)$

It will help you a lot to become familiar with these functions and their derivatives. For example, everyone should have in mind that

$$\frac{d}{dt}[\cos(\omega t + \phi)] = -\omega \sin(\omega t + \phi)$$

and
$$\frac{d}{dt}e^{\alpha t} = \alpha e^{\alpha t}$$

as well as the derivatives of the polynomials.

To make this discussion concrete, let's do the example of a time-dependent force given in Eq. 8.35, where $C_1 = 15$ N-s^3. If the particle that experiences this force has a mass of .25 kg and a velocity of 35 m/s at $t = 1$ s, we can write the differential equation of motion of the system, using Newton's second law, as

$$F_x = m\,\frac{dv_x}{dt} \quad \text{or} \quad \frac{C_1}{t^3} = m\,\frac{dv_x}{dt} \quad \text{or} \quad \frac{dv_x(t)}{dt} = \frac{C_1}{m}\,\frac{1}{t^3}, \quad (8.36)$$

where we have kept the symbols C_1 and m so that we can substitute the numerical values after we have the complete solutions. Before trying to find a mathematical solution, let's "read" the equation and predict how this particle will move.

Equation 8.36 "says" that the *rate of change* of the *velocity* decreases as time increases (for $t > 0$). That does *not* mean the particle *slows down*, because, in fact, the force and the velocity are **in the same direction**. It *does* mean that the velocity of the particle *increases* but at a slower rate as time goes on. So we expect the velocity to increase from $v_1 = 35$ m/s, but the *rate of increase* decreases. Eventually, the force becomes so small that the particle's velocity will not change. A crude sketch of this prediction is given in **Fig. 8.10**.

Now it's time to do the mathematics. But remember my bargain. Because we know that

$$\frac{d}{dt}\left(\frac{1}{t^2}\right) = -\frac{2}{t^3},$$

a good guess for $v(t)$ might be

$$v(t) = \frac{C_2}{t^2} + C_3, \quad (8.37)$$

where C_2 and C_3 are constants to be determined. Let's try it by direct substitution into Eq. 8.36:

$$\frac{d}{dt}\left(\frac{C_2}{t^2} + C_3\right) = -\frac{2C_2}{t^3} + 0 \overset{?}{=} \frac{C_1}{m}\,\frac{1}{t^3}. \quad (8.38)$$

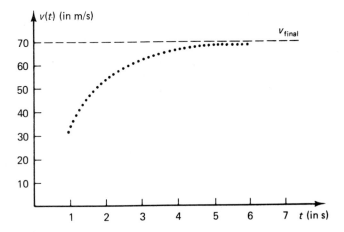

Fig. 8.10

For Eq. 8.37 to be a solution for all values of t, the coefficients of the same powers of t must be equal on each side of this "possible" solution (Eq. 8.38). That is a solution if

$$-2C_2 = \frac{C_1}{m} \quad \text{or} \quad C_2 = -\frac{C_1}{2m}.$$

Our solution is then

$$v(t) = -\frac{C_1}{2m}\frac{1}{t^2} + C_3.$$

The constant C_3 can be determined from the initial conditions, namely at $t = 1$ s, $v(1) = v_1$,

$$v_1 = -\frac{C_1}{2m} + C_3 \quad \text{or} \quad C_3 = v_1 + \frac{C_1}{2m}.$$

Thus we finally arrive at the solution for particular starting conditions,

$$v(t) = \left(v_1 + \frac{C_1}{2m}\right) - \frac{C_1}{2m}\frac{1}{t^2}, \tag{8.39a}$$

or, putting in the numerical values for the constants,

$$v(t) = 65 - \frac{30}{t^2} \quad \text{(in m/s)}. \tag{8.39b}$$

As $t \rightarrow \infty$, the velocity settles down to 65 m/s. A careful plot of Eq. 8.39b will demonstrate that our predictions were very good.

What if we had not guessed the correct mathematical function at the beginning? How would we know we were wrong?

Suppose we had assumed that

$$v(t) = C_2(1 - e^{-C_3 t}), \tag{8.40}$$

which is reasonable since the velocity does increase with time and level off at $t \rightarrow \infty$. If we substitute Eq. 8.40 into Eq. 8.36, we get

$$\frac{d}{dt}\left[C_2(1 - e^{-C_3 t})\right] = C_2 C_3 e^{-C_3 t} \stackrel{?}{=} \frac{C_1}{m}\frac{1}{t^3}.$$

What do you make of this expression? The left side of this expression cannot equal the right side for *all* values of t since an exponential function is on the left and the third power of t is on the right. True, you can find a *particular* value of t for which the two sides can be equal. But remember that this is supposed to be a solution to a differential equation; it must be true for *all* values of t, and there are no constants C_2 and C_3 that can make the two sides always equal.* It is *not* a solution. After a while, you will not have much difficulty guessing the correct solution on the first or second try.

---

* For values of time where the force is defined, which for this problem is $t > 0$.

A SOLID OBJECT moving through a fluid feels a force exerted by the fluid in a direction opposite to its instantaneous velocity. This force appears in the literature under several equivalent names: fluid drag, retarding force, viscous drag, or simply drag. All refer to the same phenomenon. Viscosity arises because as a body moves through a fluid medium it exerts forces on the atoms in the medium which, by Newton's third law, produce reaction forces on the moving body. Viscous drag is the net effect of a complex phenomenon, depending on speed, shape, surface texture, fluid compressibility, density, and viscosity.

Consider, for example, the drag force on a jet airplane in flight. Can you guess what kind of equation describes the drag on this object, where air is rushing over and under the wings and swirling around the fuselage? One would certainly not expect a simple relation such as $\mathbf{F} = m\mathbf{a}$ to describe the drag force. Suppose you know the equation that describes the drag. Would you want to call it a "law" in the same way that energy and momentum conservation are laws of nature? The answer is clearly no! Such an equation might only apply to a certain class and size of aircraft and would be the synthesis of experiments performed in a wind tunnel for a range of velocities, densities, air compositions, and so on. We call these "empirical" relationships or laws, and they are clearly lower in the hierarchy of "laws" of physics than the conservation principles. Nevertheless, these empirical relations are essential to engineers and physicists who design real systems that experience drag forces in their operation.

Sometimes nature is kind to us. Physicists have discovered that these extremely complicated phenomena can often be described reasonably well by simple mathematical relations. Two types of systems can be represented by such equations:

1. Small objects moving at low velocities
2. Large objects moving at high velocities

For class 1 systems, such as a grain of sand falling in a bucket of water, the drag force turns out to be proportional to the velocity. For the special case of a smooth, hard sphere of radius r, Stokes (in about 1850) derived a relationship for the drag force. Stokes's law is written

$$F_D = -6\pi\eta rv \quad \text{(class 1),} \tag{8.41}$$

where η is the fluid viscosity and v is the particle's velocity.* The use

---

* Viscosity is measured in kg/m-s or in g/cm-s. The latter has the name "poise."

of this law was essential for the first determination of the charge on the electron by R. A. Milikan (see Problem 50).

The resistance to motion that is experienced by an airplane, a skydiver, or a race car driver (class 2) usually turns out to be proportional to the *square* of the object's velocity expressed as

$$F_D = \tfrac{1}{2} C_D A \rho v^2 \qquad \text{(class 2)}, \tag{8.42}$$

where A is the cross-sectional area of the object, ρ is the mass density of the fluid, and C_D is a dimensionless quantity called the drag coefficient.

How small must the object be, and how slow must it move for the drag force to be proportional to v? Only an approximate answer can be given. A dimensionless quantity called the REYNOLDS NUMBER, defined as

$$R \equiv \frac{\rho v d}{\eta}, \tag{8.43}$$

where d is the object's diameter, provides a guideline as to which regime best describes the system of interest. Table 8.3 summarizes the empirical evidence.

TABLE 8.3

Reynolds Number	Law of Drag Force
0–10	Stokes's Law, Eq. 8.41
10–300	Transition region
300–3×10^5	Eq. 8.42
$>3 \times 10^5$	C_D is not a constant

EXAMPLE 1   Consider a particle that moves with a low Reynolds number, where the drag coefficient is proportional to the velocity. To simplify the notation, we write

$$\mathbf{F}_{\text{drag}} = -\alpha \mathbf{v}, \tag{8.44}$$

where α is a positive constant.

The system we will examine is shown in Fig. 8.11, where the only force on the small particle of mass m is the viscous drag, which acts in the opposite direction to the velocity. Newton's second law, as applied to this particle, is then

$$\sum_i \mathbf{F}_i = \mathbf{F}_{\text{drag}} = -\alpha \mathbf{v}(t) = m \frac{d\mathbf{v}(t)}{dt}.$$

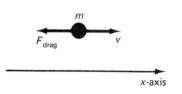

Fig. 8.11

For this one-dimensional example, the vector equation reduces to the scalar equation:

$$\frac{dv(t)}{dt} = -\frac{\alpha}{m} v(t). \tag{8.45}$$

The differential equation 8.45 is a type we have not seen before, because the right-hand side of the equation is not explicitly given as a function of time. The fundamental theorem of the calculus does not apply here, because both the right-hand and left-hand sides depend on v. That is, if we find a solution to Eq. 8.45, call it $v_1(t)$, then the solution $v_1(t) + C$ will not satisfy Eq. 8.45 for arbitrary C. This can be demonstrated by directly substituting both solutions into Eq. 8.45. Start with $v_1(t) + C$:

$$\frac{d}{dt}\left[v_1(t) + C\right] = -\frac{\alpha}{m}\left[v_1(t) + C\right]$$

or
$$\frac{dv_1(t)}{dt} = -\frac{\alpha}{m} v_1(t) - \frac{\alpha C}{m} \tag{a}$$

and the second solution $v_1(t)$:

$$\frac{dv_1(t)}{dt} = -\frac{\alpha}{m} v_1(t). \tag{b}$$

Since Eqs. (a) and (b) are obviously different differential equations, both expressions cannot be solutions to the origin differential equation, 8.45.

Nevertheless, if we are able to guess a solution to Eq. 8.45 or to any first-order differential equation* that *has an arbitrary constant somewhere within it*, that solution is the equation's most general solution. Once the constant is specified by the initial conditions (e.g., the initial velocity of the particle at time $t = 0$), the unique solution to the problem has been obtained. The mathematical proof of these statements will be left to your future math instructors (we have to leave something for them to teach).

Now what do you suppose would be a good guess for the solution to Eq. 8.45? If you "read" the equation, it "says": What function, when you differentiate it once with respect to time, gives back the function, times a negative constant? The only function that comes to mind with this characteristic is an exponential function of time. So let's try the solution

$$v(t) = C_1 e^{C_2 t}. \tag{8.46}$$

Substituting Eq. 8.46 into Eq. 8.45 gives

$$\frac{d}{dt}\left(C_1 e^{C_2 t}\right) = C_1 C_2 e^{C_2 t} \stackrel{?}{=} -\frac{\alpha}{m} C_1 e^{C_2 t}.$$

---

* A first-order differential equation is one where only first-order derivatives appear in the equation.

Dividing both sides by $C_1 e^{C_2 t}$, we see that Eq. 8.46 is a solution if

$$C_2 = -\frac{\alpha}{m}.$$

Since $-\frac{\alpha}{m}$ is a constant, we have obviously found the correct solution and determined the constant C_2. But what about C_1?

This is the arbitrary constant in the solution that must be determined by the initial condition on the physical system. If the mass m had a velocity v_0 at time $t = 0$, then

$$v(t) = C_1 e^{-(\alpha/m)t}$$
$$v(0) = v_0 = C_1 e^0 = C_1$$

or

$$C_1 = v_0.$$

Thus the unique equation that describes this motion of the mass m moving in a viscous medium starting with a velocity v_0 (and subjected to no other forces) is

$$v(t) = v_0 e^{-(\alpha/m)t}. \tag{8.47}$$

The equation makes physical sense. The particle's velocity is exponentially decreasing with time because of the viscous drag. The velocity-dependent force produces an exponentially decreasing velocity as shown in Fig. 8.12. The problem is solved.

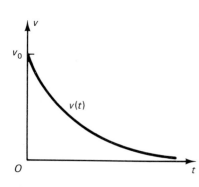

Fig. 8.12

EXAMPLE 2   The motion of a small spherical ball bearing falling in a bucket of water (Fig. 8.13) is a simple experiment that you can easily perform in the laboratory. The positive direction of motion will be taken to be downward in the direction of the gravitational force. Newton's second law can be written

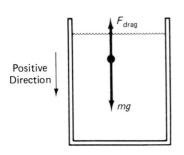

Fig. 8.13

$$\sum_i \mathbf{F}_i = \mathbf{F}_{\text{drag}} + \mathbf{F}_{\text{gravity}} = m\frac{d\mathbf{v}(t)}{dt}. \tag{8.48}$$

To solve Eq. 8.48 we must know what functional form is appropriate for this drag force. Is the drag force represented by Stokes's law (Eq. 8.41) or by Eq. 8.42, a v^2-dependent force? To answer this question we must estimate the Reynolds number for this system. We can calculate the Reynolds number if the characteristics of our system are specified:

1. Fluid density ρ (of water) $= 10^3$ kg/m^3
2. Velocity?
3. Particle radius $= 6 \times 10^{-4}$ m
4. Viscosity $= 1 \times 10^{-3}$ kg/m-s
5. Particle mass $= 2 \times 10^{-8}$ kg

But we still don't know the velocity of the particle. For the moment assume that Stokes's law applies to our small bearing, and then $\mathbf{F}_{\text{drag}} = -6\pi\eta r v$. The drag force always acts in the opposite direction to the velocity, so Eq. 8.48 becomes

$$m\frac{dv}{dt} = mg - 6\pi\eta rv. \qquad (8.49)$$

Suppose that the particle starts from rest. Then there is *no* viscous drag when the particle begins to fall, and very little drag for a while thereafter. The bearing acts initially as if it were in free fall. However, as the velocity increases, so does the viscous drag, until the viscous drag force equals the gravitational force. At that time the *net* force on the ball is *zero* and velocity stops changing: the bearing reaches its terminal or maximum velocity. This terminal velocity can easily be calculated from Eq. 8.49 since when the velocity no longer changes, $\dfrac{dv}{dt} = 0$ and

$$mg = 6\pi\eta rv_t \qquad \text{or} \qquad v_t = \frac{mg}{6\pi\eta r}. \qquad (8.50)$$

Putting in the number for our bearing in water,

$$v_t = \frac{2 \times 10^{-8}\ \text{kg} \times 9.8\ \text{m/s}^2}{6\pi \times 10^{-3}\ \text{kg/m-s} \times 6 \times 10^{-4}\ \text{m}} = 1.73 \times 10^{-1}\ \text{m/s},$$

which is the *largest* velocity the bearing can attain as it falls in the water. Now we are ready to see if our initial assumption was justified, namely using Stokes's law to describe the drag force. The Reynolds number can be calculated using the terminal velocity as the characteristic velocity of the motion (a conservative estimate since it is the largest velocity):

$$R = \rho \frac{vd}{\eta} = \frac{10^3\ \text{kg/m}^3 \times 1.73 \times 10^{-2}\ \text{m/s} \times 5 \times 10^{-4}\ \text{m}}{10^{-3}\ \text{kg/m-s}}$$

$$= 8.67.$$

Referring back to Table 8.3, we see that our assumption was justified, since systems with Reynolds numbers between zero and 10 can be well represented by Stokes's law. Now we have all the information necessary to solve for the acceleration, velocity, and the position of the particle as a function of time. To do this we must find the analytical solution to Eq. 8.49. The appearance of the equation can be simplified if we define the constant $\alpha \equiv 6\pi\eta r$ and Eq. 8.49 becomes

$$\frac{dv(t)}{dt} = g - \frac{\alpha}{m}v(t). \qquad (8.51)$$

What would be a good guess for the solution to Eq. 8.51, $v(t)$? We already solved the equation for the case where $g = 0$ and found that the solution was of the form $v(t) = C_1 e^{C_2 t}$, so we might guess that the solution for this equation might be of the form

$$v(t) = C_1 e^{C_2 t} + C_3. \qquad (8.52)$$

Let's try it. Substituting Eq. 8.52 into Eq. 8.51, we obtain

$$\frac{d}{dt}(C_1 e^{C_2 t} + C_3) = C_1 C_2 e^{C_2 t} \stackrel{?}{=} g - \frac{\alpha}{m}(C_1 e^{C_2 t} + C_3). \qquad (8.53)$$

Equation 8.53 can be satisfied for all values of time t provided that

$$C_1 C_2 = -\frac{\alpha}{m} C_1$$

and

$$0 = g - \frac{\alpha}{m} C_3.$$

These two equations tell us that $C_3 = \dfrac{mg}{\alpha}$, and that no matter what C_1 is, $C_2 = -\dfrac{\alpha}{m}$. Our solution to Eq. 8.52 is

$$v(t) = C_1 e^{-(\alpha/m)t} + \frac{mg}{\alpha}, \qquad (8.54)$$

which has an arbitrary constant C_1 in it and thus is the general solution. We determine the constant C_1 by the initial condition. The most general initial condition is that at time $t = 0$ the initial velocity is some value v_0 [i.e., $v(0) = v_0$].

Substituting these initial conditions into our general solution (Eq. 8.54) yields

$$v_0 = C_1 e^0 + \frac{mg}{\alpha} \qquad \text{or} \qquad C_1 = v_0 - \frac{mg}{\alpha}.$$

Then the unique solution to Eq. 8.51 is

$$v(t) = \left(v_0 - \frac{mg}{\alpha}\right) e^{-(\alpha/m)t} + \frac{mg}{\alpha}. \qquad (8.55)$$

In particular, if the particle begins its fall from rest, then $v_0 = 0$, and we have

$$v(t) = \frac{mg}{\alpha}(1 - e^{-(\alpha/m)t}) = v_t(1 - e^{-(\alpha/m)t}). \qquad (8.56)$$

A graph of Eq. 8.56 is given in Fig. 8.14. It shows what we expected from physical reasoning for a particle falling from rest and subjected to a velocity-dependent viscous drag. At very short times, the velocity is near zero and there is little viscous drag. Referring to the original differential eqution 8.51, when $(\alpha/m)\, v(t)$ is very small compared to g, the particle falls mainly under the influence of the gravitational force, and we know all about that: We expect its speed to increase linearly with time. The solution (Eq. 8.56) has this characteristic, which can easily be seen by expanding the exponential

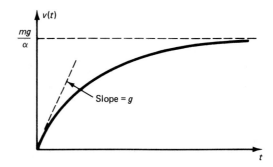

Fig. 8.14

term in a power series. Note that for small x, the exponential e^x can be approximated as

$$e^x = 1 + x + \text{(small higher-order terms for } x \ll 1\text{)}.$$

Thus

$$e^{-(\alpha/m)t} \approx 1 - \frac{\alpha}{m}\, t \qquad \left(\text{for } \frac{\alpha t}{m} < 1\right).$$

Using this expansion in Eq. 8.56 gives us

$$v(t) \approx \frac{mg}{\alpha}\left(1 - 1 + \frac{\alpha}{m}\, t\right) = gt,$$

which is the linear time dependence that we expected.

The quantity m/α has the dimensions of time. [$-\alpha v$ is a force, so it has the dimensions of

$$\text{force} = \text{mass}\,\frac{\text{length}}{(\text{time})^2}.$$

Then α has the dimensions of

$$\frac{\text{force}}{\text{velocity}} = \frac{\text{mass}}{\text{time}},$$

and m/α has the dimension of

$$\frac{\text{mass}}{\text{mass/time}} = \text{time}.$$

This quantity m/α is often called the *time constant*, and it is a measure of how rapidly the velocity approaches its terminal value. When $t = m/\alpha$ seconds, Eq. 8.56 tells us that

$$v(m/\alpha) = v_t(1 - e^{-1}) \approx v_t\left(1 - \frac{1}{2.72}\right) = .632 v_t$$

That is, after m/α seconds the velocity reaches about two-thirds of its terminal value. If m/α is large (large mass, or small viscosity, or both), it takes a long time for the particle's velocity to get close to terminal velocity, while if m/α is small (small mass or large viscosity, or both) the terminal velocity is achieved by the particle in a short time.

To find out how the position of the particle varies with time, we rewrite Eq. 8.56 as a differential equation in $x(t)$, as

$$\frac{dx}{dt} = \frac{mg}{\alpha}\left(1 - e^{-(\alpha/m)t}\right). \tag{8.57}$$

As before, let's guess what the solution will look like. We have noted that when t is small the particle falls mainly under the force of gravity (the viscous force is small because the particle is not moving rapidly). Therefore, for small t, we expect the position to be represented by the constant force expression $x = x_0 + v_0 t - \frac{1}{2}gt^2$. For *large* t, the particle falls with essentially the constant speed v_t, so we expect x to increase as $v_t t$.

Now we are ready to make our guess at the solution to Eq. 8.57. The right-hand side is an explicit function of t (so the fundamental theorem of the calculus *applies*). It is the sum of two terms, a constant and an exponential. A function whose derivative is a constant is $C_1 + C_2 t$, the function whose derivative is an exponential is $C_3 e^{C_4 t}$, so we are moved to try $x = C_1 + C_2 t + C_3 e^{C_4 t}$ as a solution:

$$\frac{d}{dt}\left(C_1 + C_2 t + C_3 e^{C_4 t}\right) = C_2 + C_3 C_4 e^{C_4 t} \stackrel{?}{=} \frac{mg}{\alpha} - \frac{mg}{\alpha}e^{-(\alpha/m)t}.$$

Our solution works if $C_2 = \dfrac{mg}{\alpha}$, $C_4 = -\dfrac{\alpha}{m}$, and $C_3 = \dfrac{m^2 g}{\alpha^2}$:

$$x(t) = C_1 + \frac{mg}{\alpha}t + \frac{m^2 g}{\alpha^2}e^{-(\alpha/m)t}. \tag{8.58}$$

The constant C_1 must be determined from the initial conditions.

8.11 Numerical Solutions

SUPPOSE WE had a machine that could do algebraic calculations very quickly but could not do any calculus. That is, our machine could add, subtract, multiply, divide, take exponents, and so on, but could not differentiate (or integrate). Could we devise a method to solve the problem of our particle slowly moving through a viscous medium? Are there numerical methods to solve a differential equation?

As you have probably guessed, the answer is yes. The basic strategy is to reduce the calculus operation of differentiation to an algebraic manipulation. This can only be done in an approximate form, but the approximation can be made arbitrarily accurate. Recall the definition of differentiation and using as an example the system described by Eq. 8.45,

$$\frac{dv(t)}{dt} \equiv \lim_{\tau \to 0} \frac{v(t + \tau) - v(\tau)}{\tau} = -\frac{\alpha}{m}v(t). \tag{8.59}$$

Suppose we don't take the mathematical limit as $\tau \to 0$, but simply make τ very small. By small, we mean a time interval over which the velocity $v(t)$ changes very little. For such a small τ the expression

$$\frac{v(t + \tau) - v(t)}{\tau} \tag{8.60}$$

is a very good approximation for the derivative dv/dt, even though it is not strictly an equality. Obviously, the smaller the time interval τ we pick, the better the approximation. A value of τ can always be chosen so that for a real system, the error in equating Eq. 8.60 to the derivative is less than the errors made in actually measuring $v(t + \tau)$, $v(t)$, t, and τ. So let's replace dv/dt in Eq. 9.59 by Eq. 8.60 to obtain the approximate relationship:

$$\frac{v(t + \tau) - v(t)}{\tau} \approx -\frac{(\alpha)v(t)}{m}. \tag{8.61}$$

Solving for $v(t + \tau)$ in terms of $v(t)$ by algebraic manipulation of Eq. 8.61, we obtain

$$v(t + \tau) \approx v(t)\left(1 - \frac{\alpha}{m}\tau\right). \tag{8.62}$$

If one knows the value of the velocity of the particle at some time t, $v(t)$, Eq. 8.62 can be used to estimate the particle's velocity at the time $t + \tau$ later, $v(t + \tau)$. Equation 8.62 will be an excellent approximation to the correct answer if the velocity does not change much over the time period τ.

Is this result consistent with the exact solution we obtain with calculus, Eq. 8.47? Consider the velocity of the particle a short time after $t = 0$. For short times (times where the quantity $\frac{\alpha}{m}t \ll 1$) the exponent can be approximated by the series expansion

$$e^{-(\alpha/m)t} \approx 1 - \frac{\alpha}{m}t + \text{higher-order terms (which we neglect)}.$$

Thus Eq. 8.47 becomes

$$v(t) \approx v_0\left(1 - \frac{\alpha}{m}t\right),$$

which agrees with Eq. 8.62 for $t = 0$.

What do we do if we want to calculate the velocity after a long time interval T? We cannot just plug the value of T into Eq. 8.62 since this equation is valid only for short time intervals. However, all is not lost. We can use a simple trick to solve the problem.

Any time interval can be divided into a sequence of n short time intervals. Mathematically, this can be written as

$$T = n\tau$$

where n is a large integer and τ is a short time interval. Now we will apply Eq. 8.62 sequentially, one short time interval after another, starting at time $t = t_0$. Then

$$v(t_0 + \tau) \approx v(t_0)\left(1 - \frac{\alpha}{m}\tau\right)$$

and the velocity at 2τ is then

$$v(t_0 + 2\tau) \approx v(t_0 + \tau)\left(1 - \frac{\alpha}{m}\tau\right)$$

$$= v(t_0)\left(1 - \frac{\alpha}{m}\tau\right)^2$$

and the velocity at 3τ is

$$v(t_0 + 3\tau) \approx v(t_0 + 2\tau)\left(1 - \frac{\alpha}{m}\tau\right)$$

$$= v(t_0)\left(1 - \frac{\alpha}{m}\tau\right)^3.$$

Obviously, then, the velocity at time $T = n\tau$ is

$$v(t_0 + T) \approx v(t_0)\left(1 - \frac{\alpha}{m}\tau\right)^n. \tag{8.63}$$

Equation 8.63 was derived without calculus, using only algebraic manipulations. It can also be derived from the exact solution using the series expansion of the exponential function:

$$v(T) = v(t_0)e^{-(\alpha/m)T} = v(t_0)e^{-(\alpha/m)n\tau}$$

$$= v(t_0)\left[e^{-(\alpha/m)\tau}\right]^n$$

$$\approx v(t_0)\left(1 - \frac{\alpha}{m}\tau\right)^n.$$

This method shows that once the right-hand side of a differential equation such as Eq. 8.45 is known at one instant of time, the solution at any later instant of time is fully determined—in fact, it can be found to any desired accuracy by using the step-by-step method illustrated above. So if you can find an exact solution to the differential equation that satisfies the initial conditions, you have found the fully determined solution. The step-by-step method provides an approximation to the exact solution. Approximate methods can also be used to program a computer to solve differential equations that have no analytical solutions.

8.12 Summary of Important Equations

$$v(t) = \lim_{\tau \to 0} \frac{x(t + \tau) - x(t)}{\tau} = \frac{dx}{dt}$$

$$a(t) = \lim_{\tau \to 0} \frac{v(t + \tau) - v(t)}{\tau} = \frac{dv}{dt}$$

$$\frac{dy}{dx} = \lim_{\varepsilon \to 0} \frac{y(x + \varepsilon) - y(x)}{\varepsilon}$$

$$\sum_i \mathbf{F}_i = \frac{d}{dt}\mathbf{p} \quad \text{or for Newtonian dynamics} \quad \sum_i \mathbf{F}_i = \frac{d}{dt}m\mathbf{v}$$

$$1 \text{ newton} = 1 \text{ kg-m/s}^2$$

Exponential functions:

$$e^x = 1 + x + \frac{x^2}{2!} + \frac{x^3}{3!} + \cdots \quad \text{where } e = 2.71828$$

$$\frac{d}{dx}e^x = e^x \qquad \frac{d}{dx}e^{ax} = ae^{ax} \qquad \frac{d}{dx}ae^x = ae^x$$

$$\text{if } y = e^x \quad \text{then} \quad \ln y = x$$

For constant-acceleration motion in one dimension:

$$v_x(t) = v_{0x} + \frac{F_x}{m}t$$

$$v_x(t)^2 = v_{0x}^2 + 2\frac{F_x}{m}[x(t) - x_0]$$

$$x(t) = x_0 + v_0 t + \frac{F_x}{2m}t^2$$

$$x(t) = x_0 + \tfrac{1}{2}[v_{0x} + v_x(t)]t$$

F_{gravity} (near the Earth's surface) $= mg$, where $g = 9.8$ m/s^2 directed toward the Earth's center.

Drag force:

$$F_D = \begin{cases} -6\pi\eta rv & 0 < R < 10 \\ \tfrac{1}{2}C_D A \rho v^2 & 300 < R < 3 \times 10^5 \end{cases}$$

Reynolds number:

$$R = \frac{\rho v d}{\eta}$$

Chapter 8 QUESTIONS

1. Does the speedometer on your automobile measure velocity as it is defined by physics?

2. Average velocity and average speed are generally different. Can you give some examples of motion where they are the same?

3. Can a particle have a changing velocity when its speed is constant? Is it possible for the velocity of a particle to increase while its acceleration decreases? If so, give examples.

4. Can a particle have zero velocity and nonzero acceleration? If so, give an example.

5. One of the great puzzles of the ancients concerning motion, called "Zeno's paradox," can be formulated as follows: The hare can never overtake the tortoise because to reach the tortoise, the hare must first cover the first half of the distance, then he must cover the next half $(\frac{1}{4})$, then the half of the remaining $(\frac{1}{8})$, and so on *ad infinitum,* never reaching his destiny. Explain what is wrong with this reasoning.

6. Can the instantaneous velocity of a particle ever be greater in magnitude than the average velocity? If your answer is yes, give an example.

7. My physics professor told me that a car traveling northward was accelerating in the southern direction. Does that make any sense? Explain.

8. Is it possible to have zero *average* velocity over a 5-s interval and still be accelerating? Is it possible to have zero *instantaneous* velocity over a 5-s second interval and still be accelerating? Explain your answers with examples.

9. Two children are playing tug-of-war. On separate diagrams for each child and the rope, draw the horizontal forces on each child and the rope. Which of these forces are equal in magnitude and opposite in direction as described by Newton's third law?

10. When two teams are engaged in a tug-of-war, Newton's third law tells us that the force exerted by the rope on one team is exactly equal to the force exerted by that rope on the other team, but in the opposite direction. How is it possible for one team to win and the other to lose if these forces are exactly equal in magnitude? Explain.

11. If you take air resistance into account, would you expect a ball thrown vertically up in the air to take longer to rise to its peak height than to fall back to the ground? Explain.

12. If a ball is thrown in the air, what is the acceleration of the ball when it reaches its maximum height and its acceleration just before it strikes the ground, neglecting air resistance?

13. Suppose that your car brakes provided constant deceleration no matter what your speed. How does doubling your initial speed affect the time necessary for you to come to a stop?

14. What are the dimensions of the Reynolds number?

15. The terminal speed of a baseball is about 95 mi/hr. Yet fast balls have been clocked at 100 mi/hr. How can this occur?

16. What are the microscopic processes that give rise to velocity-dependent drag force? Can you explain why this force should be proportional to velocity?

Chapter 8 PROBLEMS

1. Two sisters drove their separate cars to a dance at a country inn. Starting from their home, they traveled different routes, Jane going on a crowded expressway which headed directly north 40 miles to the inn, while Sarah took an empty, windy beach road. Both arrived at the same time with Sarah's odometer reading 48 miles.

  (a) Which sister moved with greater average speed?

  (b) Which sister traveled with greater average velocity?

  (c) If the trip took 45 minutes, what was the average velocity of each sister?

2. Starting with the definition of the derivatives, prove that

$$\frac{d}{dx}\,e^{ax} = ae^{ax} \quad \text{and that} \quad \frac{d}{dx}\,ae^{x} = ae^{x}$$

(where a is a constant).

3. (a) Prove that $\dfrac{d}{dx}\,ay(x) = a\,\dfrac{dy}{dx}(x)$ [where a is constant and $y(x)$ is any function of x].

(b) Prove that $\dfrac{d}{dx}\,xy(x) = x\,\dfrac{dy(x)}{dx} + y(x)$.

(c) Prove that $\dfrac{d}{dx}\,x^2y(x) = x^2\,\dfrac{dy(x)}{dx} + 2xy(x)$.

(d) Can you guess what $\dfrac{d}{dx}\,x^ny(x) = $?

4. The number of important science journals devoted to physics research seems to be increasing rapidly. In the early twentieth century (1910) there were only 11 significant physics journals, but by 1970 about 150 such journals existed. If we were to postulate that the number of these journals obeys the formula $N(t) = N_0 e^{t/\alpha}$, where t is the time in years past 1910, N_0 the number in 1910, and α the "time constant," we should be able to predict the number of journals that will exist in the future.

(a) Find the time constant α.

(b) How many significant physics journals should exist in 1990? Ask your librarian if your model is reasonable.

(c) How many journals will exist in the year 3000? Is this answer reasonable? Explain.

5. Using the fundamental definition of the derivative, show that

$$\frac{d}{dx}\cos ax = -a\sin ax.$$

6. A particle moves along the x-axis in such a way that it can reasonably be approximated by the graph shown in Fig. 8.15.

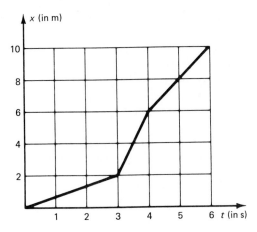

Fig. 8.15

(a) What are the velocities of the particle at 2, 3.5, and 6 s?

(b) Discuss the motion at 3 and 4 s.

(c) Does this particle accelerate? If so, when does it experience its greatest acceleration?

7. Consider a particle moving in one dimension as represented graphically by Fig. 8.16.

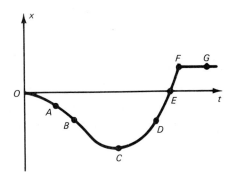

Fig. 8.16

(a) For every interval OA, AB, and so on, state if the velocity and acceleration is positive, negative, or zero.

(b) Plot velocity as a function of time, indicating the markers A, B, C, and so on, on your plot.

(c) Are there any intervals in which the acceleration is obviously not constant?

8. A woman runs a 14-s straight-line race with a velocity as shown in Fig. 8.17. How far will she have traveled when she has completed the race?

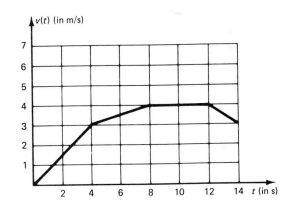

Fig. 8.17

9. The position of a particle along the x-axis is given by the equation $x = 1 + t^2/10$.

(a) What are the units of 10 and 1 if x is measured in meters and t in seconds?

(b) Draw a graph of x versus t and a "picture" of the motion, as in Fig. 8.3a, from $t = -3$ to $+3$ s.

(c) Determine $\langle v(0, 1) \rangle$ and $\langle v(0, 0) \rangle$ directly from the graph, by constructing the proper tangents and measuring the slope.

(d) Determine $\langle v(1, 1) \rangle$ and estimate $\langle v(1, 0) \rangle$ as in part (c) from the graph.

(e) Algebraically compute $v(t = 1)$ by using the limiting process using Eq. 8.6.

(f) Obtain an algebraic expression for the velocity at any time using Eq. 8.6.

(g) Compare your answer in part (f) with results graphically obtained in parts (c) and (d).

10. The position of a particle is given by the equation $x = t^3 + 2t - 3t^2$.

(a) Draw a graph of x versus t, labeling the times at which $x = 0$.

(b) The particle has zero velocity at two different times: once at a time between 0 and 1 s and again at a time between 1 and 2 s. Find these times by:

    (1) Calculating $v(t) = \lim\limits_{\tau \to 0} \dfrac{x(t + \tau) - x(t)}{\tau}$.

    (2) Solving the equation $v(t) = 0$ for the times when the velocity vanishes.

(c) Complete your labeling of the graph by indicating the times that the particle is at rest and the positions of the particle at these times.

(d) Estimate the velocity of the particle at time $t = 0$ from the graph, by construction.

(e) Estimate the velocity of the particle at time $t = 0$ by algebraically computing $\langle v(0, \tau) \rangle$ for a few small values of τ.

(f) Compare these two estimates with the actual value of $v(t)$ [found in part (b)] evaluated at $t = 0$.

(g) What are the units of 1, 2, and 3 in this equation if t is in seconds and x is in centimeters?

11. A particle's velocity is given by the equation $v(t) = \alpha t^3 - \beta t^2 + \gamma$, where α, β, and γ are all constant.

(a) What are the units of α, β, and γ if t is measured in seconds and v in meters/second?

(b) Find the equation for $x(t)$ that corresponds to this velocity.

(c) Will $x(t)$ + constant also correspond to this velocity? Explain your answer. What does this constant mean?

12. A particle's velocity is $v(t) = t^3 - 2t^2 + 6$. Find an expression for the position $x(t)$ that corresponds to this velocity. Note that $x(t)$ + any constant also corresponds to this velocity. Why?

13. A particle's position as a function of time is given by $x = 2 \sin 2\pi t$; the motion is called *simple harmonic motion*.

(a) Draw a graph of x versus t for -2 s $\le t \le 2$ s. Label the times at which $x = 0$ and the times for which x reaches its maximum and minimum.

(b) Find $v(t) = \dfrac{dx(t)}{dt}$. Draw a graph of $v(t)$ versus t. What are the maximum and minimum velocities of the particle? When is the particle instantaneously at rest? Label the graph appropriately.

(c) What is the period T of the motion, where T is the interval of time starting when the particle leaves a position with a certain velocity and ending when the particle returns to that position with the same velocity?

14. A particle moves along the x-axis with velocity $v(t) = A(e^{-bt} - 1)$, where t is measured in seconds, v is in meters/second, and A and b are positive. At $t = 0$, the particle is at the origin (i.e., $x = 0$).

(a) What are the units of A and b?

(b) Roughly sketch $v(t)$ for two values of b, b_1 and b_2, where $b_1 > b_2$. Draw both curves on the same graph, and label each curve with the appropriate value of b.

(c) Roughly sketch the displacement $x(t)$ versus t, for one value of b.

(d) What are approximate expressions for the acceleration for "very large" and "very small" values of t?

(e) Briefly describe a physical situation in which you would expect such a behavior to occur.

15. A car, starting from rest, moves along the x-axis according to the rule $x = bt^3$, where $b = 0.2$ m/s^3, until it reaches a maximum velocity of 60 m/s, after which time its velocity remains constant at this value.

(a) How long does it take for the car to reach its maximum velocity?

(b) Draw a graph of the car's velocity as a function of time.

(c) Draw a graph of the car's acceleration as a function of time.

(d) How far does the car go in the first 20 s of its motion?

16. An object moves along the x-axis and its velocity varies with time as shown in Fig. 8.18. Sketch this object's position $x(t)$ and acceleration as a function of time for values of t between 0 and 4 s. Assume that the particle started at $x = 0$ when $t = 0$.

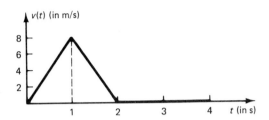

Fig. 8.18

17. A particle's position as a function of time is given by the expression $x = A \cos \omega t$, where A and ω are constants. Such motion is called *simple harmonic motion*.

  (a) What are the units of A and ω if x is in meters and t is in seconds?

  (b) Plot x versus t for $-4\pi/\omega \le t \le 4\pi/\omega$. Note your graph will be in terms of A and ω.

  (c) Calculate $v(t)$ and plot $v(t)$ versus t.

  (d) What are the maximum and minimum velocities of the particle? When is the particle at rest? Label these points on your graph.

  (e) This motion is obviously periodic. What is the period of the motion, that is, the time interval beginning at any arbitrary position with a certain velocity and returning to the same position with the *same velocity*: one full cycle of the motion?

18. Imagine a particle moving in one dimension trapped between two perfectly reflecting walls, as shown in Fig. 8.19. When the particle strikes the wall, it reverses direction and moves with constant velocity in the space between. Draw diagrams of the position versus time for the following cases.

Fig. 8.19

  (a) The walls are perfectly elastic; that is, the direction of the velocity changes, not the magnitude.

  (b) The velocity is reduced by the factor α after each collision, that is, $v_n = \alpha v_{n-1}$, where n is the number of the collision.

19. The amplitude of motion of a swinging pendulum can be described by the equation $x(t) = e^{-\alpha t} \cos \omega t$, where α and ω are positive constants.

  (a) What are the units of α and ω if t is measured in seconds?

  (b) Plot $x(t)$ versus t for three cycles of the cosine function for a given value of α.

  (c) Calculate the velocity and acceleration of the pendulum.

  (d) Sketch $v(t)$ and $a(t)$ as a function of time.

20. Consider a particle of mass m subjected to a constant force F in the negative x-direction. Assuming that this particle started with $x(0) = x_0$, $v(0) = v_0$,

  (a) Plot x versus t in terms of x_0, v_0, F, and m.

  (b) Find the two times when $x = x_0$.

  (c) Find the time when $x = 0$.

  (d) Find the maximum positive value of x that the particle reaches and the time at which this value of x is reached.

21. A particle moves with a constant force F_x acting on it. The initial conditions are that at time t_0 (t_0 is not necessarily equal to 0) the position and velocity of the particle are x_0 and v_0. Find the expressions for $v(t)$ and $x(t)$ in this case. Do they reduce to Eq. 8.30 when you set $t_0 = 0$?

22. A hunter shoots a bullet vertically upward from a cliff 500 m above a lake. The initial velocity of the bullet is such that it reaches a height of exactly 500 m above the starting point in 10 s. Neglecting air resistance:

  (a) What is its initial velocity?

  (b) How high will it go above the lake?

  (c) What is its velocity when it falls into the lake?

23. A show-off Texas cowboy decides to perform the ultimate gun shooting trick. He shoots one bullet in the air vertically and when it reaches its maximum height h, he shoots an identical bullet and makes the two collide in midair. In terms of g and h, find the following values. (Neglect air resistance in all your calculations.)

  (a) The time after the first bullet is shot that the collision occurs.

  (b) The height of the collision from the ground.

  (c) The velocity of the two bullets just before they collide.

24. Two drag racers A and B start down a straight track at the same time, moving at different constant accelerations. At the halfway marker A runs out of gas and has to coast to the finish at a constant velocity. B, having a smaller engine, keeps his constant acceleration to the finish. The result of the race is an exact tie; A and B finish together. Find the ratio of the accelerations of A and B.

25. A new space shuttle can accelerate in space with an acceleration equal to a free-falling body near Earth. It can do this for 100 days without running out of fuel.

(a) Will it have enough fuel to reach 15% of the speed of light if it starts from rest?

(b) How far will it travel if it does reach 15% of the speed of light?

26. Galileo was unable to openly explain his discoveries of astronomy and mechanics because they conflicted with the dogma of the Church. To circumvent the censors he published "Dialogues Concerning Two New Sciences," in which debate occurs among advocates of differing schools of science. The "old" school, represented by Sagredo discussing free-falling bodies, states: "Its speed increases in proportion to the space traversed, so that, for example, the speed acquired by a body in falling four cubits would be double that acquired in falling two cubits and this latter speed would be doubled that acquired in the first cubit." Write a mathematical expression that expresses Sagredo's discussion and show that it is not what is observed. (For Galileo's explanation one has to look for the voice of Salviati, under "Naturally Accelerated Motion, Dialogue of the Third Day.")

27. A medieval soldier standing on the castle ramparts wishes to throw boulders down upon his attacking enemy below with the largest possible velocity upon impact. Neglecting air resistance, should he throw them up or down, or doesn't it really matter? Prove your answer mathematically.

28. Juliet sees a bouquet of flowers go up to the top of her window and fall down past her window on the way back to Romeo. If the total time the flowers spent during the flight was 1.55 s and her window is 2 m tall, how far is she above Romeo? Should she jump into his arms? (Neglect all drag forces.)

29. Estimate the average force a person exerts on the bones of his body if he jumps onto firm ground from height of 1 m and stops his motion over a distance of a few centimeters. Do you understand why it is essential to bend your knees when you jump?

30. Two objects are thrown vertically up in the air, one with three times the initial velocity of the other. The one with the largest initial velocity has one-third the mass of the lower velocity object. Neglect air resistance in your calculations.

(a) What is the ratio of the times the objects spend in the air?

(b) What is the ratio of the maximum height the two objects will reach?

31. A large commercial jet aircraft fully loaded must reach speeds of 375 km/h to be able to lift off a runway. If we assume that the jet has constant acceleration from rest along a 1.9-km runway, what acceleration does it need for the takeoff?

32. Pauline is tied to the railroad tracks while her faithful dog Abercrombie stands guard over her, barking vociferously but helplessly. Suddenly (at $t = 0$) a train emerges out of the fog (traveling at a speed of 60 m/s toward Pauline) a scant 136 m away from her frail form. At that instant ($t = 0$) the engineer puts on the brakes, producing a constant deceleration of 15 m/s².

(a) At what time T does the train come to rest?

(b) How far from Pauline does the train come to rest?

At time $t = 0$, Abercrombie starts to race toward the train (starting from rest) at a constant acceleration. At the instant T that the train comes to rest, Abercrombie collides head-on with the train, but because the train is at rest and Abercrombie has a hard head, he is unhurt! Pauline is saved!

(c) What is Abercrombie's acceleration?

(d) What is Abercrombie's speed when he hits the train?

33. To bring a jetliner to rest after landing, the pilot must reverse the engines when it is on the ground. If the plane touches the ground moving at 275 km/h and comes to a halt 30 s later, what is the shortest landing field that can be used, assuming constant acceleration?

34. Two objects are dropped from the same place on a high cliff. If they are both released from rest but separated in time by an amount T, show that the objects will be separated by a vertical distance L at a time,

$$t_L = \frac{T}{2} + \frac{L}{Tg}.$$

35. Two boys jump off a 12-m-high cliff into a lake below. The first boy jumps vertically up with an initial velocity of 0.8 m/s and the second boy leaves the cliff with zero vertical velocity at the instant the first passes him.

(a) Find the velocity of both boys when they hit the water.

(b) Find the difference in time between the splashes, neglecting air resistance.

36. A particle of mass 1 kg moves under the influence of a time-dependent force $F_x = 1/t^3$ newtons, from time $t = 1$ to time $t = \infty$. (*Note:* 1 has the units of N-s^3.)

(a) Sketch how you think $v(t)$ and $x(t)$ will look if at time $t = 1$, $x(1)$ and $v(1)$ are positive. Give special attention to their behavior for large t.

(b) Find $v(t)$ if at time $t = 1$, $v(1) = v_0$.

(c) Find $x(t)$ if at time $t = 1$, $x(1) = x_0$.

(d) Draw a graph of $v(t)$ and $x(t)$ versus t. How fast does the particle move as t becomes very large?

37. A particle is subjected to a time-dependent force that produces a time-dependent deceleration given by the equation

$$a = \frac{a_0}{t_0}(t - t_0) \qquad (\text{for } 0 \le t \le t_0)$$

and $a_0 = 3$ m/s^2 and $t_0 = 10$ s. The particle initially moves at a velocity of 15 m/s.

(a) Find the expression for $v(t)$.

(b) Plot $v(t)$ and $a(t)$ on one graph with the same time scale.

(c) Show on your graph where $v(t) = 0$ and $a(t) = 0$.

(d) How far does the particle travel in the 10 s?

38. A tiny meteor fragment of mass 10^{-4} kg is moving in outer space with constant velocity 10^3 m/s, when at time $t = 0$ it encounters a cloud of hydrogen gas that exerts a viscous drag force $-10^{-10}v$ (newtons) on it.

(a) Find the expression for $v(t)$. How many days does it take before the velocity decreases to $1/e$ of its initial value?

(b) Assuming that the cloud is large enough, how many days does it take until the meteor fragment comes completely to rest?

(c) Find the expression for $x(t)$. How far through the cloud can the particle go, assuming that you wait as long as is necessary for the particle to come to rest? Does the fact that this distance is finite conflict with your answer to the question in part (b)?

(d) Now use the "step-by-step" method to compute v at times 1×10^4 s, 2×10^4 s, and 3×10^4 s. Compare these approximate results with the exact solution found in part (a).

39. A skydiver jumps out of an airplane and begins his fall in the gravitational field of the earth. Assuming that he starts at $x = 0$ with an initial velocity $v(0) = 0$ and is subjected to a drag force of the air whose magnitude is $(\frac{1}{2})C_D A \rho v^2$:

(a) Show that his terminal speed is given by $v_t = (2 \; mg/C_D A \rho)^{1/2}$.

(b) Show that his velocity is given by $v^2(t) = v_t^2(1 - e^{-x/x_c})$, where x_c is the characteristic distance, defined as $x_c = v_t^2/2g$.

40. Calculate the terminal speed of a ping-pong ball whose diameter is 4 cm, mass is 3 g, $C_D = .5$, and air density $\rho = 1.22 \times 10^{-3}$ g/cm^3, falling in air. Calculate the Reynolds number for this system taking the viscosity of air $\eta = 180 \times 10^{-6}$ poise.

41. A single particle of mass 3.5 kg starting from rest at the origin experiences a force in the x-direction which is described by the equation $F_x = 6e^{-\alpha t}$, where $\alpha = 5$ s^{-1} and 6 has the units of newtons.

(a) Write a brief description of the motion, explaining how the velocity, position, and acceleration depend on time.

(b) Sketch on a large graph the velocity and position of this particle as a function of time.

(c) Write the differential equation of motion of the particle. Does the fundamental theorem of calculus apply to the solution of this differential equation?

(d) Find the solution for the velocity. Plot velocity as a function of time. Does it agree with your original guess?

(e) Find $x(t)$ and plot it versus time and compare it with your original guess.

42. A particle of mass m starts from rest at the origin at time $t = 0$. It moves along the x-axis under the time-dependent force $F_1 = F_0 t^2/t_0^2$, where F_0 and t_0 are positive constants, whose dimensions are force and time, respectively.

(a) What is the particle's velocity at time $t = t_0$?

(b) Now at time $t = t_0$ the force F_1 is turned off and a new force which is constant, $F_2 = -F_0$, is turned on. The particle is then observed to slow down, stop, and reverse its previous directions of motion. At what instant of time does the particle come to rest?

43. A particle whose mass is 2 kg is instantaneously at rest at the origin at $t = 0$. For times greater than zero, the particle is subjected to the time-dependent force $F = 4 - 2t$ (newtons), directed along the x-axis.

  (a) What are the units of 4 and 2 in the expression for the force?

  (b) Sketch the velocity of this particle as a function of time for $t > 0$.

  (c) Find the expression for the velocity as a function of time. Find the time when the particle once again is instantaneously at rest.

  (d) Plot $v(t)$ versus t and compare this with your original sketch.

44. A 2-kg particle moves under the influence of a time-dependent force given by the equation

$$F = 6t$$

(6 has units of newton/seconds).

  (a) Sketch (before you mathematically solve the problem) how you think $v(t)$ and $x(t)$ will look at time $t = 0$. $v(0) = -1$ and $x(0) = 0$.

  (b) Find the expression for $v(t)$ and $x(t)$ under the same initial conditions as in part (a).

  (c) Plot the results of part (b) and compare them to your original sketches. How good was your physical intuition?

45. A small metal sphere whose mass is .0015 kg falls in a bucket of oil and reaches a terminal velocity .07 m/s. Find the coefficient of viscous damping, α, and the equation of motion of the particle, assuming that it started from rest.

46. In discussing the motion of small particles falling in fluid media, we have neglected one force on the particle. The ancient Greek scientist Archimedes discovered that a buoyant force, exactly equal to the weight of the fluid displaced, is exerted on a body immersed in any fluid. If the density of the fluid is ρ (kg/m^3) and the radius of the spherical particle is r:

  (a) What would be the "correct" differential equation of motion that describes the particle motion shown in Fig. 8.13?

  (b) What is the expression for the terminal velocity of the particle?

  (c) Find the solution for $v(t)$.

47. The force on a single particle moving along the x-axis is described by the equation $F_x = -kx$, where k is a positive constant and x is the displacement of the particle along the x-axis.

  (a) Discuss the motion of this particle if it is initially displaced a distance x_0 along the posi-

tive x-axis and released from rest. To aid you in your discussion, you should make three careful plots: (1) the acceleration as a function of position, (2) the velocity as a function of position, and (3) the position as a function of time.

  (b) Write the differential equation of motion for the system. Solve this equation for $v(t)$. Does this solution agree with your guess in part (a)?

  (c) Write the differential equation for $x(t)$ and solve it to find $x(t)$.

  (d) How good was your physical intuition in part (a)? Can you think of a real physical system that exhibits this kind of motion?

  (e) Describe how the particle would move if the constant k became negative.

48. Suppose that it was possible to exert a force $+\alpha v$ (α = positive constant) on a particle. Find the velocity and position of this particle if at time $t = 0$, its velocity and position are v_0 and x_0, respectively.

49. Suppose that you have a block of material in which at time $t = 0$ there are N (N is a large integer) radioactive nuclei. Each nucleus is independent of the other nuclei, so it "decides for itself" when it is going to emit an electron and an antineutrino, and thus become a nonradioactive nucleus. As time goes on, the number of radioactive nuclei decreases, and their law of decrease can be found as follows: If there are at time t, $n(t)$ radioactive nuclei still in the block, a certain fraction of these will radiate in the next ε seconds (ε is a small positive number), and that fraction is $\alpha \varepsilon$ (α is a positive number). Thus at time $t = \varepsilon$, there are no longer $n(t)$ radioactive nuclei left, for their number has been decreased by the $\alpha \varepsilon n(t)$ nuclei that have decayed. That leaves behind $n(t) - \alpha \varepsilon n(t)$, or

$$n(t + \varepsilon) = n(t) - \alpha \varepsilon n(t)$$

radioactive nuclei at time $(t + \varepsilon)$. We can rewrite this equation as

$$\frac{n(t + \varepsilon) - n(t)}{\varepsilon} = -\alpha n(t),$$

and if we let ε approach zero, we see that the number of radioactive nuclei obey the equation

$$\frac{dn(t)}{dt} = -\alpha n(t).$$

  (a) Solve this equation to find the number of radioactive nuclei at time t if there are N nuclei at time $t = 0$.

(b) Find an expression for the number of emitted electrons (or, if you prefer, antineutrinos) as a function of time.

50. In 1911, Robert Millikan, measured the terminal velocity of very small electrically charged oil drops in the presence of a gravitational and electrical force to determine the charge on the electron. Figure 8.20

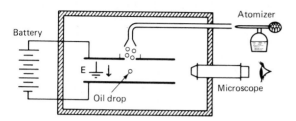

Fig. 8.20

shows a schematic diagram of his apparatus. It turns out that when oil is sprayed from an atomizer, the small droplets that are formed carry excess electrons ne, where n is an unknown integer and e is the charge on the electron. The determination of e, which was unknown at that time, was the goal of Millikan's experiments. Millikan knew that such a small charged drop would experience a drag force determined by Stokes's law and that the electrical force on the charged drop was $F_{\text{electrical}} = neE$, where E is the electric field (in volts/meter) and n is a positive integer representing the number of electrons on the drop. He was also able to obtain the following parameters of the apparatus:

$$\rho_0 = 8.83 \times 10^2 \text{ kg/m}^3 \quad \text{(density of oil)}$$

$$\rho_a = 1.29 \times 10^{-6} \text{ kg/m}^3 \quad \text{(density of air)}$$

$$\eta = 1.824 \times 10^{-5} \text{ N-s/m}^2 \quad \text{(viscosity of air)}.$$

(a) First Millikan measured the terminal velocity of the drop in free fall without an applied electric field in order to determine the size of the drop. Including the effect of air buoyancy (see Problem 46), write down the differential equation of motion for $v(t)$ of the drop moving under the gravitational force of the earth.

(b) Such a particle will quickly reach steady state, where it is no longer accelerating but moving with a constant terminal velocity. Derive the expression for the terminal velocity v_t.

(c) What is the radius of the drop whose terminal velocity is $v_t = 2.64 \times 10^{-5}$ m/s?

(d) Write down the differential equation for $v(t)$ when a vertical electric field E is turned on. What is the expression for the electric charge e in terms of the known quantities and the new terminal velocity in the presence of an electric field?

(e) An electric field $E = 10^5$ V/m is turned on, producing an electric force on the drop measured in part (c). We observe the following terminal velocities, where $+$ indicates the electric force is in the same direction as the gravitational force and $-$ the opposite direction

Trial No.	v_t (m/s)	Trial No.	v_t (m/s)
1	$+3.05 \times 10^{-4}$	3	$+5.84 \times 10^{-4}$
	-2.53×10^{-4}		-5.32×10^{-4}
2	$+4.91 \times 10^{-4}$	4	$+2.12 \times 10^{-4}$
	-4.39×10^{-4}		-1.60×10^{-4}

The numbers of electrons on the drop n has been changed for each trial by exposing the drop to a radioactive source. Using these data, find the charge on electron e.

51. One way to measure the acceleration of gravity is to photograph a small falling object (in a vacuum) with a stroboscopic light source. The camera records only the position of the object during the regular flashes of light. Assume that such an apparatus was assembled and the flashes started when the object was given an initial velocity of v_0 and repeated at time intervals Δt. Show that the distance traveled between the n and $(n + 1)$ flash is given by

$$\Delta x(n) = v_0 \, \Delta t + \frac{g}{2}(2n - 1)(\Delta t)^2.$$

9

DYNAMICS OF MOTION IN TWO DIMENSIONS

9.1 **Introduction**

IN Chapter 8 we carefully defined displacement, velocity, acceleration, force, and mass, yet we examined only single-particle motion restricted to one dimension using Newton's three laws. What happens when you connect objects together to make a simple machine? How does one use Newton's laws to describe such combined motion of several particles or several objects? How do we analyze objects that do not move in a straight line but are constrained to move in a circle? Are we now equipped to understand at a deeper level the motion of a charged particle in magnetic field? In this chapter we expand the application of Newton's laws of motion to idealized systems in two-dimensional space.

9.2 **Strategy**

WE BEGIN by examining different types of mechanical systems, simple machines consisting of small bodies connected to one another with cables, pulleys, and gears, that are guided along tracks or planes in a prescribed manner. Some examples are shown in Fig. 9.1. These systems may have many forces acting on each of their component parts and may have different forces acting on each part. It is hardly obvious how we are to apply Newton's three laws to such complex systems. We shall establish a "strategy of attack," a systematic procedure, whatever you wish to label it, to facilitate your analysis of the motion of these systems. The procedures may at first seem unnecessarily formal and rigid, particularly when you are analyzing a simple machine. But if I were your instructor, I would insist that each and every one of you follow the steps *exactly as prescribed.* If you do this now, you will develop the physical insight and analytical skills necessary to understand mechanical systems. Applying these

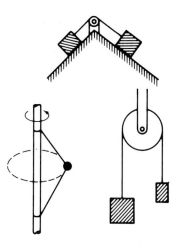

Fig. 9.1

"programmed" steps dissects a complex machine and simplifies solving for the motion of the system. I will outline the steps first and then explain each one with examples.

Newtonian Machines: A "Program"

1. Separate the system into its component parts, isolating each part.
2. Draw a diagram for *each* separated component body of the system. For each body, draw vectors representing the forces *on that body.* These are called "free-body" diagrams. (*Note:* There are only two kinds of forces, contact forces and forces acting-at-a-distance.)
3. Choose an appropriate coordinate system for each free-body diagram, labeling each axis. Make sure that the coordinate system is inertial.
4. Examine each free-body diagram carefully. Write Newton's second law, in component form, for *each* body (i.e., $\Sigma F_x = ma_x$, $\Sigma F_y = ma_y$, and $\Sigma F_z = ma_z$).
5. Write down *all* the constraint equations appropriate for this machine. Constraint equations may involve connecting the coordinate systems of one body to another coordinate system of a second body, or may describe the kinematic constraint placed on a particular body by the machine.
6. Using Newton's second law and the constraint equations, solve algebraically for the unknowns in the problems.
7. Check your answers to see if they make physical sense. Check their units and magnitudes.

A good way to help you to understand how to use this "program" is to study several examples I have provided for you. But don't be deceived. The only way you will really learn how to apply Newton's laws is to do many problems yourself. Follow the steps; they will help you.

EXAMPLE 1   Consider a single small block of mass m sliding without frictional resistance on a smooth, fixed, inclined plane, shown in Fig. 9.2. We are interested in finding the acceleration of the block. For this example, there is no need to isolate or separate the bodies since there is motion of only one part of the system. Our first task then is to draw the free-body diagram and to determine all the forces acting on the sliding block. Figure 9.3 shows these forces.

The earth's mass produces a gravitational force on the block, whose magnitude is mg, acting in the downward vertical direction.

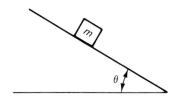

Fig. 9.2

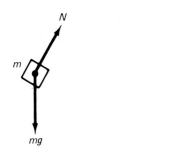

Fig. 9.3

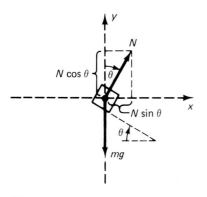

Fig. 9.4

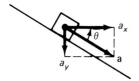

Fig. 9.5

This is an action-at-a-distance force. But what effect does the plane have on this block? It exerts a force perpendicular to the plane, a contact force, called in the jargon of the physicist, the NORMAL FORCE. There are no other forces in this problem, no ropes or springs attached, no wind blowing, no frictions between the plane and the block, no electrical forces.

Step 3 gives the student some choice, unlike steps 1 and 2. You may, in fact, choose any coordinate system you like, as long as it is at rest or moving with constant velocity with respect to the inclined plane. This is called an INERTIAL FRAME of reference.* Nothing in this problem suggests that a coordinate system moving with a constant velocity with respect to the plane benefits the analysis. Two coordinate systems, however, might occur to the reader. We will do the problem twice using the two systems.

First we will use the "traditional" coordinate, x-axis horizontal, y-axis vertical, as shown in Fig. 9.4. The origin of coordinates is taken through the center of the block, and all the forces are assumed to act through the center, since the block is treated like a point mass. If you remember your plane geometry, you can deduce that the normal force acts at the angle θ with respect to the y-axis, as shown.

The only constraint in this example is that the block must maintain contact with the plane, which means the acceleration is along the plane. For this coordinate system there are two components of the acceleration **a** (Fig. 9.5). This can be expressed mathematically as

$$\mathbf{a} = a \cos \theta \mathbf{i} - a \sin \theta \mathbf{j}$$

or
$$a_x = a \cos \theta \qquad a_y = -a \sin \theta \qquad \text{(a)}$$

Newton's second law is a *vector* relationship that must hold for *each orthogonal component x, y, and z independently!* Remember, two vectors can be equal only if their components are equal. Step 4 is

$$x\text{-comp.:} \quad \Sigma F_x = N \sin \theta + 0 = ma_x \qquad \text{(b)}$$

$$y\text{-comp.:} \quad \Sigma F_y = N \cos \theta - mg = ma_y. \qquad \text{(c)}$$

The rest is algebra. We are not concerned (in this problem) with the normal force, so we can eliminate it by solving the x-component [Eq. (b)] for N:

$$N = \frac{ma_x}{sin \ \theta},$$

---

* A more complete discussion of inertial frames of reference is given in Chapter 10.

and then substituting the expression for N into the y-component, Eq. (c):

$$\frac{(ma_x)(\cos\theta)}{\sin\theta} - mg = ma_y.$$

The constraint equations (a) can now be used to eliminate a_x and a_y and solve for the total acceleration a,

$$\frac{m(a\cos\theta)\cos\theta}{\sin\theta} - mg = -ma\sin\theta.$$

The mass m cancels out of this equation. If we now multiply by $\sin\theta$ and rearrange the equation, it reads

$$a(\cos^2\theta + \sin^2\theta) = g\sin\theta.$$

But the term in parentheses is equal to 1; thus we finally arrive at an expression for the total acceleration:

$$a = g\sin\theta,$$

a very simple, if not intuitively obvious result. Note for $\theta = \dfrac{\pi}{2}, a = g$, and for $\theta = 0, a = 0$. Both limits make physical sense, and the units of g are m/s^2; the $\sin\theta$ is dimensionless, so the units of the equation check.

That certainly was a long-winded explanation, but it attempted to address all the essential ingredients in the solution of these problems. Had we used what might be called the "natural coordinates," the solution would have been easier to obtain.

What is the natural coordinate system for this machine? The x' and y' Cartesian system shown in Fig. 9.6 is just that. For this coordinate system, one axis is *in the direction of the acceleration* of the block; in this case it is the x'-axis. The constraints can now be written simply as $a_{y'} = 0$, $a_{x'} = a$ (the only direction of acceleration). Newton's second law is

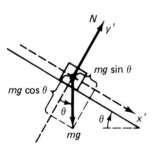

Fig. 9.6

x'-comp.: $\Sigma F_{x'} = mg\sin\theta = ma_{x'}$

y'-comp.: $\Sigma F_{y'} = N - mg\cos\theta = ma_{y'} = 0.$

Solving for the normal force from the y'-component, we have

$$N = mg\cos\theta,$$

and for the acceleration from the x'-component and the constraints,

$$a_{x'} = a = g\sin\theta.$$

We calculate the same acceleration, of course, but with the natural coordinates the answer was obtained in one line. I'm sure you have not missed the point that it is worth your time at the beginning to choose the natural coordinates for the particular machine you are analyzing. No simple scheme guarantees that you will always choose the most efficient coordinate system, but as you analyze many of these problems you will develop the skill and insight to pick the most efficient one quickly.

Two important ideas have emerged in our discussions of this very simple example, the *normal force* and the *inertial reference frame*. These are ideas worth elaborating. When we analyze a system containing an idealized plane, we need only consider the normal force produced by the plane. The normal force is one of the common constraining forces, that is, forces that restrict the freedom of motion of moving particles. Had it not been for the plane, the block would have traveled a different path. Close examination of the source of this force shows that it arises out of the repulsive interactions between the atoms in the block and the atoms in the plane. If it were not for this repulsion, the block would penetrate the plane. Actually, the atoms do penetrate each other very slightly and cause a small elastic deformation of the surface, but we will ignore such tiny effects in our discussions. Although we show the force acting at one point at the center of the block, the contact force actually acts uniformly over the entire bottom surface of the block.

Newton's second law is *valid only* in special reference frames. These are called *inertial reference frames*. We will carefully define and discuss these frames of reference in Chapter 10. For now imagine that one such reference frame is our laboratory and it is at rest.* Since you would have to make some very careful measurements on moving bodies to detect deviations from Newton's laws, this is an excellent approximation. All observers who are moving at a *constant velocity* with respect to our laboratory frame are also in inertial reference frames. These observers might be doing experiments in a train moving at constant speed in a straight line, for example. They would also find Newton's laws of motions obeyed.

EXAMPLE 2

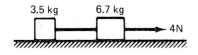

Fig. 9.7

Now let's examine a coupled system, two blocks of mass 3.5 and 6.7 kg connected by a fine wire and pulled by a constant force of 4 N acting in the horizontal direction (Fig. 9.7). The problem is to calculate the acceleration of the blocks and the tension in the connecting wire, neglecting any friction between the blocks and the plane over which they slide.

For this problem we shall require two free-body diagrams (Fig. 9.8) and two coordinate systems x, y and x', y', one for each block. First we must do a "force audit." There are only two kinds of forces to consider; (1) forces of contact, and (2) forces acting-at-a-distance. On the first mass m_1, there are two contact forces, the normal force

---

* Of course, it is not at rest. This is pure fiction. We inhabit a planet that spins on its axis, rotating around the sun and around our galaxy.

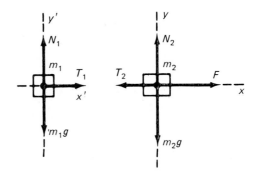

Fig. 9.8

produced by the plane N_1 and the tension exerted by the string $\mathbf{T}_1$. One force acts-at-a-distance, the force due to the gravitational field of the earth, $-m_1g\mathbf{j}'$. A similar force audit on the second mass m_2 shows that there are three contact forces $\mathbf{T}_2$ and $\mathbf{F}$ due to the wires and $\mathbf{N}_2$ due to the plane, and one action-at-a-distance, $-m_2g\mathbf{j}$.

The dynamic equations (Newton's second law) for *each* part of this machine are:

For m_1:

x'-comp.: $T_1 = m_1 a_{1x'}$

y'-comp.: $N_1 - m_1g = m_1 a_{1y'}$

For m_2:

x-comp.: $F - T_2 = m_2 a_{2x}$

y-comp.: $N_2 - m_2g = m_2 a_{2y}.$

The constraint equations for this system are:

$a_{1y'} = a_{2y} = 0$ (the blocks cannot accelerate in the vertical direction)

$a_{1x'} = a_{2x} = a$ (the wire does not stretch or break)

$|\mathbf{T}_1| = |\mathbf{T}_2| = T$ (magnitude of the tensions must be equal).

One way to understand that the tensions must be equal in magnitude (for this ideal massless wire) is to assume that tensions are not equal. In this case the vector sum of these two tensions is the net force acting on the wire, and by Newton's second law this net force will accelerate the wire by a_{wire} = net force/mass of wire. But we assumed that the wire was massless and thus *any net* force would produce an infinite acceleration of the wire. (You might see this more clearly if you isolate the connecting wire with its own free-body diagrams, then write down and solve Newton's second law.) Such infinite accelerations are not physical. Therefore, we conclude that there is no *net* force on the wire itself, and the tensions must be equal in magnitude and opposite in direction.

Substituting these constraints into the dynamic equations for the x and x' components yields

$$F - T_2 = F - T = F - m_1 a = m_2 a$$

or

$$a = \frac{F}{m_1 + m_2} = \frac{4 \text{ N}}{3.5 \text{ kg} + 6.7 \text{ kg}} = .392 \text{ m/s}^2.$$

To calculate the tension in the connecting wire T, we use the x'-component equation:

$$T_1 = m_1 a = \frac{m_1}{m_1 + m_2} F$$

or

$$T_1 = \left(\frac{3.5 \text{ kg}}{3.5 \text{ kg} + 6.7 \text{ kg}}\right) 5 \text{ N} = 1.37 \text{ N}.$$

These equations show that the tension in the connecting wire is *less* than the tension of the wire that pulls the system, by the ratio of the mass of the end block to the total mass. That result might surprise you, but upon reflection, it should be obvious. The middle wire accelerates only the end block, not both blocks as the front wire must, and thus it requires less tension.

You may wonder why I introduced the symbols m_1, m_2, and F and didn't work directly with the numbers given in the problem. There is a very important reason for this procedure. Because we used these symbols, we were able to derive two equations:

$$a = \frac{F}{m_1 + m_2} \quad \text{and} \quad T_1 = \frac{m_1}{m_1 + m_2} F.$$

which are true for *all* masses and forces on such a two-body configuration. I can now examine these expressions to see if they make physical sense, by imagining certain extreme limits for our machine. For example, as the mass m_1 gets smaller and smaller, one would also expect the tension to decrease toward zero, since if there were no mass m_1, no force would be required to accelerate it. Indeed, that is what the equation tells us. One would also expect the tension to go to zero if the external force becomes vanishingly small. That's what the mathematics tells us. Had we substituted the numerical values of the masses and the force at the early stages of the problem, it would not be possible to examine the solutions in these limits.

> *Problems should first be solved using algebraic symbols for the masses, forces, accelerations, and so on. At the very end, the numerical values for these parameters can easily be substituted into the final equations. The final equations should be tested in the appropriate limits to be sure that they make physical sense and that they have the correct units.*

These procedures will help assure you that you have made no mistakes in solving the problem.

EXAMPLE 3   Two masses connected by a rope over a pulley is called an Atwood's machine (Fig. 9.9). For our analysis we will consider both the pulley and the rope to have no mass and no friction existing anywhere in the system. The pulley only acts to "steer" the tension in the rope around itself and does not otherwise affect the motion of the system. There are two moving masses m_1 and m_2 in this machine, and thus two free-body diagrams, shown in Fig. 9.10. Two inertial reference frames are fixed to the axis of the pulley. The vertical up-direction has been chosen as positive for both the x and x' axes, even though I realize that the machine will accelerate in the minus x' and positive x directions, since $m_2 > m_1$. The mathematics will ultimately tell us the direction of acceleration, no matter what directions we arbitrarily pick as positive. However, we must be careful to be consistent when we write the constraint equations. The constraint equations are

$$a_{1x} = -a_{2x'}$$

(which is the mathematical way to state that for this machine if m_1 goes up, m_2 must go down), and

$$|\mathbf{T}_1| = |\mathbf{T}_2| = T.$$

Newton's second law:

x-comp.: $\qquad T_1 - m_1g = m_1a_{1x}$

x'-comp.: $\qquad T_2 - m_2g = m_2a_{2x'}.$

Substituting from the constraint equations yields

$$T = m_1g + m_1(-a_{2x'})$$
$$T = m_2g + m_2a_{2x'}.$$

Subtracting gives us

$$0 = m_1g - m_1a_{2x'} - m_2g - m_2a_{2x'}$$

$$a_{2x'} = \frac{m_1 - m_2}{m_1 + m_2}\,g.$$

If $m_2 > m_1$, the acceleration $a_{2x'}$ is negative, indicating that the machine increases its speed in the negative x'-direction or that it decreases its speed if it is moving initially in the positive x-direction, as suggested below. The analysis still holds if $m_1 > m_2$, in which case the machine accelerates in the opposite direction. It is incorrect to say that the machine moves in the x'-direction (if $m_2 > m_1$), since the machine could have been given an initial velocity at the start in the positive x'-direction. In that case it would move in the positive direction, slow up, stop, and then begin to move in the negative x'-direction. During all this time it is *always accelerating*

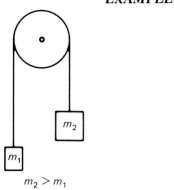

$m_2 > m_1$

Fig. 9.9

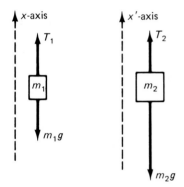

Fig. 9.10

in the negative x'-direction. (This is true even when it stopped. How can this be?)

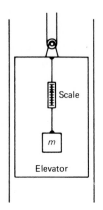

Fig. 9.11

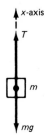

Fig. 9.12

EXAMPLE 4   A block of mass m is suspended inside an elevator by a spring scale that has been calibrated in newtons (Fig. 9.11). Find the reading on the spring scale for three cases: if the elevator is at rest, accelerating up (a_u), and when the elevator is accelerating down (a_d).

This problem has only one free-body diagram, Fig. 9.12 Only two forces act on the body, the tension exerted by the spring scale (the contact force), which is the "apparent" weight of the block, and the gravitational attraction, $-mg\mathbf{j}$ (the action-at-a-distance force). No other forces are exerted on the block. There are three constraints for the three different parts of the problem. It is important to realize that the reference frame *cannot* be fixed to the block or to the elevator since they are accelerating bodies. The reference frame must be inertial. In this case the frame is chosen to be fixed to the ground, at rest.

(a) Elevator at rest; $a = 0$. The Newton's second law gives

x-comp.: $T - mg = 0$ or $T = mg$

(b) Elevator accelerating up; $a = a_u$:

x-comp.: $T - mg = ma_u$ or $T = m(a_u + g)$

The apparent weight increased.

(c) Elevator accelerating down; $a = -a_d$:

x-comp.: $T - mg = m(-a_d)$ or $T = m(g - a_d)$

The apparent weight *decreased*. In fact, if the elevator accelerates in free fall ($a_d = g$), there is no tension on the scale—weightlessness!

9.3 **Special Effects**

A THOUGHTFUL student might suspect that our examples are from a Walt Disney cartoon script of *Alice in Wonderland* rather than a scientific description of natural phenomenon. Where in the world did we ever collect such a plethora of special effects and mythical props? Maybe this chapter should be titled "Mythical Physics," since we are studying massless strings, cables and pulleys, frictionless surfaces, perfect vacuums with no vicious drag, nondeforming rigid bodies, perfectly smooth surfaces, constant gravitational forces, exactly uniform electric fields, and so on. (The list will grow longer as you proceed.) None of these objects, none of the props in our play, exists in nature! You will never use a pulley that has no mass or friction in its bearings and between itself and the rope, and you will not find a vacuum that has no residual gas molecules in it or a perfectly uniform electric field. What a book—we are studying a collection of objects that don't exist! Is this science?

To make the analyses tractable for the beginning student, all the "props" we use are *idealizations* of the component parts of real machines. Even the primitive Atwood's machine becomes extremely complicated if we allow the pulley and rope to have mass and the rope to exert frictional forces on the pulley surface. Although it can be analyzed, for now the complications would only deflect your attention from the fundamentals. It is easy to construct a machine in the laboratory that closely approximates these idealizations. If the hanging masses are large compared to the mass of the real pulley and the pulley has a good bearing, we can measure machine accelerations that agree with our theory of idealized components to about 5%. This should convince you that the most important physical processes have been taken into account by the theory.

The theoretical techniques and principles you are now applying to these highly idealized model systems can later be used to study more realistic models of simple and complex machines. The physics is the same, but the analysis becomes increasingly complex as the models account more realistically for all the interactions in the real world. My advice is to enjoy yourself in this fantasy world—soon enough you may have to deal with the more complex problems of the "real" world.

9.4 **Sliding Friction of Rigid Bodies**

WITH THE exception of objects moving in outer space and elementary particles in a vacuum, frictional forces exist in almost every mechanical system. We have already considered the frictional drag on objects moving in fluids for two regions of the Reynolds number. There is a large number of mechanical devices where moving parts slide over one another. The relative motion of the two surfaces in contact results in a resistance to the motion. We will treat this resistance separately. It is called *sliding* friction and can be described reasonably well by *velocity-independent* forces. The equation representing sliding friction is simply

$$F_f = \mu N, \qquad (9.1)$$

where μ is called the coefficient of sliding friction and N is the normal force exerted on the object by the surface on which it is sliding. Notice that Eq. 9.1 is a scalar, *not* a vector equation. The frictional force does not act in the same direction as the normal force, even though the magnitude of this force is proportional to the magnitude of the normal force.

> *The sliding frictional force always acts in the direction oppo-
> site to the motion of the object relative to the surface over
> which it is sliding.*

Thus, without knowing the direction of motion of the body under consideration, it is not possible to write a mathematical expression that will always give the direction of the frictional force.

Although this equation appears as neat and simple as $\mathbf{F} = m\mathbf{a}$, the two equations represent vastly different understandings of the phenomenon they describe. True, both "laws" come from empirical observations, but the equation for sliding friction is really a "cover-up" for a collection of very complicated microscopic processes that are not simple. The more we examine the details of friction, the more complicated and specialized our equations must become to predict the observed motion accurately.

This is not the case with Newton's laws, at least within the limits we have established. To this day sliding friction is not very well understood. At one time it was believed that the frictional force arose from the motion of the surface irregularities. But this mechanism cannot account for the atomic and molecular vibrational processes that are observed, which generate heat and energy dissipation. It also cannot explain why this simple relationship does not hold for large velocities or for large normal forces. Many interesting and surprisingly straightforward experiments with sliding friction demonstrate that we are actually dealing with a very complex phenomenon. One of my favorites involves bringing to class a set of machinist's gauge blocks. These are precision machined, ground, and polished steel blocks of approximately 5- by 2-cm sides and different calibrated thickness. Machinists use them to measure fabricated parts accurately by comparison. I simply take the blocks out of their case one at a time, using gloves, clean them with solvent, and set them on top of one another until I have a stack about 15 cm tall. If the blocks have been carefully cleaned and handled with gloves, it is possible to pick up this entire stack by the bottom gauge block and *turn it upside down!* They all stick together. This experiment makes a mockery out of the equation $F_f = \mu N$.

There once were tables that purport to list the frictional coefficient for steel on steel, steel on brass, glass on steel, and so on. They were simply wrong. These tables disregard experiments such as those just described. In fact, the nature of the surface itself affects the frictional coefficient in an important but complicated way.

Some textbooks mention two kinds of sliding friction, static and dynamic. It is true that an experimenter can observe some differences in the frictional force between two objects when they are not moving relative to one another and when their surfaces are in relative motion. But it is also true that repeated measurements of these two "types" of friction do not generally yield reproducible results.

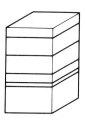

Fig. 9.13 Stacked gauge blocks

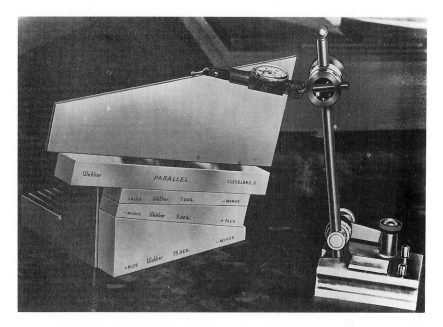

Gauge blocks and a dial indicator used for precision measurements

So skip this section and similar sections in other introductory texts. No, not quite so fast. There is a class of objects that do "obey" the simple empirical equation reasonably well. Since this is a messy subject and we are using only a crude mathematical approximation, we will represent friction in our analysis by a single coefficient μ. You may even perform experiments in the laboratory to bear this out. A description of the analysis of these traditional sliding friction experiments provides a good example of how to approach these problems.

EXAMPLE 1   Like the first example, Fig. 9.14 shows a block sliding down a surface inclined at an angle θ. But now we have an additional force, F_f, the frictional force that acts in the direction to oppose the motion of the block. Remember that in our previous analysis of the block sliding without friction, we calculated the normal force but did nothing with it. For this problem the normal force is of primary concern, since the frictional force is proportional to it.

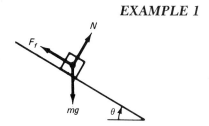

Fig. 9.14

Figure 9.15 is the free-body diagram, where we have chosen the natural coordinate system. The constraints are

$$a_y = 0 \qquad a_x = a$$

$$F_f = \mu N.$$

Newton's second law for the block in component form is

x-comp.: $\Sigma F_x = mg \sin \theta - F_f = ma_x$

y-comp.: $\Sigma F_y = N - mg \cos \theta = 0.$

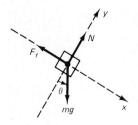

Fig. 9.15

Solving the y-component for N gives $N = mg \cos \theta$. Substituting this into the x-component equation (using Eq. 9.1) for the magnitude of the frictional force yields

$$mg \sin \theta - \mu mg \cos \theta = ma.$$

Again the mass cancels out of the problem and we obtain

$$a = g(\sin \theta - \mu \cos \theta). \tag{9.2}$$

Imagine performing an experiment with this apparatus. The angle of the incline of the plane is slowly decreased from a large angle to an angle where the block stops accelerating on the incline. At this critical angle, θ_c, the acceleration of the block is zero and Eq. 9.2 takes on a particularly simple form (since $a = 0$), namely

$$\mu = g\,\frac{\sin \theta_c}{\cos \theta_c} = g \tan \theta_c. \tag{9.3}$$

If the experiment is repeated many times, one measures reasonably consistent values of θ_c. This suggests a simple experimental way to measure μ, from a measurement of θ_c.

But what happens when you decrease the incline angle θ further, below θ_c? The block obviously remains at rest under these conditions. In fact, if $\theta = 0$, the block will not move. Equation 9.2 tells us that for $\theta = 0$, $\mu = g \sin (0) = 0$. But μ is supposed to be a constant, a number characteristic of the materials and the surfaces of the two sliding objects. Something is very wrong with our analysis.

Equation 9.2 does not represent the frictional force in all circumstances, only the maximum frictional force that exists when the body is moving. If the bodies are at rest, Eq. 9.1 is the maximum frictional force that can exist. These experiments suggest that the more accurate way to write the equation that describes the frictional force is

$$F_f \le \mu N, \tag{9.4}$$

an *inequality*, indicating that the *maximum* force of sliding friction that can oppose motion is μN, but the frictional force can indeed be less.

Sliding frictional force for rigid bodies exists under two conditions:

1. When there is a motion of one body relative to another on a common surface
2. When there is a tendency for relative motion to exist over a common surface

Equation 9.1 is appropriate to describe frictional forces in condition 1, where relative sliding motion actually is taking place. For condition 2, the frictional force is variable and Eq. 9.4 describes the physical situation. Consider, for example, a block at rest on a horizontal plane. Suppose that we apply a small horizontal force in the positive x-direction and observe that the block remains at rest. Newton's second law tells us that since the block did not accelerate in response to our initial nudge, the *total* force on the block must be zero. Obviously, friction between the block and the horizontal surface is the source of the other equal and opposite force on the block.

Suppose now that we apply a greater force but still the block remains at rest. This can happen only if the frictional force also increases, matching our applied force in magnitude but acting in the opposite direction. What happens if we apply the same force in the opposite direction, namely the negative x-direction? Once again the block remains at rest, but the frictional force has changed direction. In all these experiments the frictional force has responded to a tendency of one rigid body to move relative to another. On a microscopic level, the atomic and molecular bonds of the surface atoms are being stretched by the applied force up to some maximum value which, to a good approximation for many systems, is given by μN.

A good way to analyze any problem where condition 2 is applicable is first to assume that *no* frictional forces exist. In that case the system will move in some direction, some relative motion of one body over another. Since in the actual machine, the motion does *not* occur, it must be that the frictional force acts on the body in such a way as to oppose the relative motion. The frictional force will be just sufficient to make the total force on the body zero. With this in mind, consider a block at rest on a horizontal surface. Is there a frictional force acting on it?

The answer must be no, for several reasons. First, there is no tendency to motion in the absence of friction, therefore no frictional force. One might also ask: Which way would the frictional force act? If it acts in any horizontal direction, it would be the *only* force in this direction, and by Newton's second law this frictional force would cause the block to accelerate! That's pure nonsense. The student must be very careful in applying what appears to be simple mathematical relation (Eq. 9.4) to describe the effects of sliding friction of rigid bodies in mechanical systems.

Finally, as Leonardo da Vinci understood, the frictional force is independent of the surface area to a reasonably good approximation. You may have some idea why this is so. However, this is not the case for deformable solids (such as rubber), so it is still an advantage to purchase wider tires for your car so that it can stop more quickly on slippery surfaces.

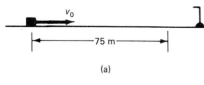

(a)

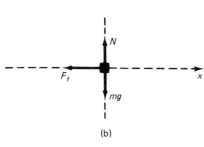

(b)

Fig. 9.16

EXAMPLE 2   A hockey puck is slapped by a defenseman all the way across the rink toward an open net. The defenseman got off only a weak shot, which gave the puck an initial velocity of 12 m/s (Fig. 9.16). If the open net was 75 m away and the puck had a coefficient of sliding friction of 0.1 with respect to the ice, find out if the puck will enter the net (and we win!) or if it stops short.

x-comp.: $-F_f = ma_x$

y-comp.: $N - mg = 0.$

This is case 1, so the frictional force is given by Eq. 9.1,

$$F_f = \mu N$$

Since $N = mg$, $-mg\mu = ma_x$ or $a_x = -g\mu,$

a constant-acceleration problem that we dealt with in Chapter 8. Using Eq. 8.30c, which is

$$v_{xf}^2 = v_{0x}^2 + 2\left(\frac{F}{m}\right)(x - x_0)$$

with $v_{0x} = 12$ m/s, and noting that $F/m = a_x$, $x_0 = 0$, and $v_{xf} = 0$ (the puck comes to rest at the maximum distance), we obtain

$$x = \frac{v_{0x}^2}{2a_x} = \frac{v_{0x}^2}{2\mu g} = \frac{(12\ \text{m/s})^2}{2(.1)(9.8\ \text{m/s}^2)} = 73.47\ \text{m}.$$

The puck never made it to the net—the game ends in a tie. Too bad!

EXAMPLE 3   An old man with a bad back wants to move a 43-kg block without exerting too much force. He decides to move it with a pulley arrangement, shown in Fig. 9.17, where he adds 1-kg weights to the lower end of the rope until the block begins to move. The coefficient of sliding friction between the block and the surface is 0.3.

(a) How many weights must be added before it begins to accelerate?

(b) What is the acceleration?

(c) What is the net force on the block when only two weights have been added? What is the frictional force on the block in this case?

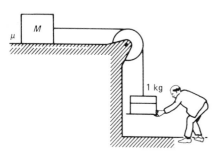

Fig. 9.17

There are two moving bodies in this problem, so we need two free-body diagrams (Fig. 9.18). Newton's second law:

x-comp.: $T_1 - F_f = Ma_x$

y-comp.: $N - Mg = Ma_y$

x'-comp.: $mg - T_2 = ma_{x'}.$

The constraints for this problem are

$|\mathbf{T}_1| = |\mathbf{T}_2| \equiv T$ (the rope steers the force around the pulley)

$$a_x = a_{x'} \equiv a \qquad a_y = 0 \qquad F_f = \mu N.$$

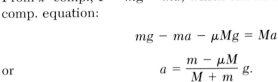

Fig. 9.18

Substituting in the constraints yields

$$N = Mg \quad \text{thus} \quad F_f = \mu Mg.$$

From x'-comp., $T = mg - ma$, which can be substituted into the x'-comp. equation:

$$mg - ma - \mu Mg = Ma$$

or

$$a = \frac{m - \mu M}{M + m} g.$$

If $\mu M < m$, the large block will not accelerate, since the acceleration in this problem cannot be in the negative x-direction. In this example

$$\mu M = .3 \times 43 \text{ kg} = 12.9 \text{ kg}.$$

He must add thirteen 1-kg weights before the system starts accelerating. Then the acceleration is

$$a = \frac{13 \text{ kg} - 12.9 \text{ kg}}{43 \text{ kg} + 13 \text{ kg}} \times 9.8 \text{ m/s}^2 = .0175 \text{ m/s}^2.$$

When only 2 kg has been attached, the net force on M is zero because the block does not accelerate. The frictional force is then exactly equal to the tension, which is the force of gravity on the lower weights, namely, $mg = 2 \text{ kg} \times 9.8 \text{ m/s}^2 = 19.6$ N.

9.5 Vector Differentiation

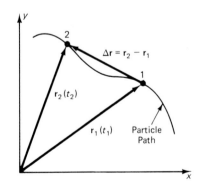

Fig. 9.19

WHAT DOES it mean to differentiate a vector quantity? Differentiation with respect to time yields an expression for the instantaneous time rate of change of the variable. Consider a particle moving along the two-dimensional path shown in Fig. 9.19, whose position with respect to the x–y coordinates is given by the vector $\mathbf{r}(t) = x(t)\mathbf{i} + y(t)\mathbf{j}$. At the times t_1 and t_2 the position vectors of the particle are $\mathbf{r}(t_1)$ and $\mathbf{r}(t_2)$. Vector addition shows that the displacement of the particle over the time interval $(t_2 - t_1)$ is:

$$\text{displacement} = \mathbf{r}_2 - \mathbf{r}_1 = \Delta\mathbf{r} = (x_2 - x_1)\mathbf{i} + (y_2 - y_1)\mathbf{j}.$$

Suppose that we wish to consider the general case where the time interval Δt has elapsed and the particle has changed its position. Now

$$\mathbf{r}_2(t + \Delta t) = \mathbf{r}_1(t) + \Delta\mathbf{r}$$

or

$$\Delta\mathbf{r} = \mathbf{r}(t + \Delta t) - \mathbf{r}(t),$$

dropping the subscripts 1 and 2, since the vectors are identified by their time argument. This single vector equation is equivalent to two scalar equations:

x-comp.: $\Delta x = x(t + \Delta t) - x(t)$

y-comp.: $\Delta y = y(t + \Delta t) - y(t).$

The instantaneous velocity of the particle is defined to be

$$\mathbf{v}(t) = \lim_{\Delta t \to 0} \frac{\Delta \mathbf{r}(t)}{\Delta t} \equiv \frac{d\mathbf{r}(t)}{dt},$$ (9.5)

which is equivalent to two definitions of the x and y components of velocity. The generalization to three dimensions should be obvious. What may not be immediately obvious is the interesting possibility that a particle may change its velocity but not its speed. It can do this simply by changing *direction*. We will soon return to this point and discuss it carefully.

Completing the formal definitions, using Cartesian unit vector notation for three dimensions, we obtain

$$\mathbf{r}(t) = x(t)\mathbf{i} + y(t)\mathbf{j} + z(t)\mathbf{k}$$

$$\mathbf{v}(t) = \frac{d\mathbf{r}(t)}{dt} = \frac{dx(t)}{dt}\mathbf{i} + \frac{dy(t)}{dt}\mathbf{j} + \frac{dz(t)}{dt}\mathbf{k}$$ (9.6)

$$\mathbf{a}(t) = \frac{d^2\mathbf{r}(t)}{dt^2} = \frac{d\mathbf{v}(t)}{dt} = \frac{d^2x(t)}{dt^2}\mathbf{i} + \frac{d^2y(t)}{dt^2}\mathbf{j} + \frac{d^2z(t)}{dt^2}\mathbf{k}.$$ (9.7)

9.6 **Circular Motion**

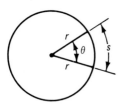

Fig. 9.20

BEFORE THE discussion focuses on circular motion, let's define the units of angular measurement which are almost always used in physics. These "natural" units are called RADIANS. Consider Fig. 9.20. The angle θ, measured in radians, is the ratio of the arc length S to the radius r, or

$$\theta \text{ (in radians)} \equiv \frac{S}{r}.$$ (9.8)

It is easy to remember how to convert radians to degrees. The arc length is equal to the circumference of the circle for θ equal to $360°$. So

$$360° = \frac{\text{circumference}}{\text{radius}} = \frac{2\pi r}{r} = 2\pi$$

or

$$2\pi \text{ radians} = 360° \quad \text{or} \quad 1 \text{ radian} = 57.3°. \tag{9.9}$$

Since the radian is defined as the ratio of two lengths, it is a dimensionless quantity.

Using the definition of radians and referring to Fig. 9.21, we see that for a small angle $\Delta\theta$ we have

$$\Delta S = r \, \Delta\theta. \tag{9.10}$$

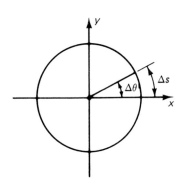

Fig. 9.21

A particle constrained to move along a circular arc sweeps out an angle $\Delta\theta$ and an arc ΔS in a time interval Δt. Forming the ratio

$$\frac{\Delta S}{\Delta t} = r \frac{\Delta\theta}{\Delta t}$$

and allowing the time interval Δt to get smaller and smaller in a limiting process as

$$\lim_{\Delta t \to 0} \frac{\Delta S}{\Delta t} = r \lim_{\Delta t \to 0} \frac{\Delta\theta}{\Delta t}$$

but

$$\lim_{\Delta t \to 0} \frac{\Delta S}{\Delta t} \equiv v \quad \text{(speed, a scalar)} \tag{9.11}$$

$$\lim_{\Delta t \to 0} \frac{\Delta\theta}{\Delta t} \equiv \omega \quad \text{(angular speed, a scalar).} \tag{9.12}$$

Thus

$$v = r\omega. \tag{9.13}$$

Equation 9.13 is a scalar equation, expressing a relationship between the magnitudes of the speed, radius, and angular speed for uniform circular motion. If the particle were to change the radius of the path during its rotation, Eq. 9.13 would not be valid.

Consider a particle traveling at constant speed v in a counterclockwise direction around a circle of radius r starting at $t = 0$ with $\theta = 0$. At some time t later, the *angular* displacement would be

$$\theta = \omega t \tag{9.14}$$

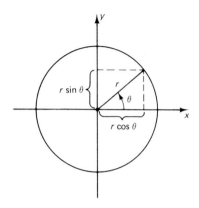

Fig. 9.22

and the position vector at the time t (Fig. 9.22) would be written in terms of Cartesian unit vectors **i** and **j** as

$$\mathbf{r}(t) = r \cos \omega t \, \mathbf{i} + r \sin \omega t \, \mathbf{j}. \qquad (9.15)$$

The particle's velocity can be calculated by differentiation of Eq. 9.15 as

$$\mathbf{v}(t) = \frac{d\mathbf{r}(t)}{dt} = -r\omega \sin \omega t \, \mathbf{i} + r\omega \cos \omega t \, \mathbf{j}$$

$$= r\omega[\sin \omega t \, (-\mathbf{i}) + \cos \omega t \, \mathbf{j}]. \qquad (9.16)$$

The magnitude of the velocity is

$$|\mathbf{v}(t)| = (r^2\omega^2 \sin^2 \omega t + r^2\omega^2 \cos^2 \omega t)^{1/2}$$

$$= r\omega \, (\sin^2 \omega t + \cos^2 \omega t)^{1/2} = r\omega,$$

the same result we obtained in Eq. 9.13. However, now that we have written the velocity in vector form, we can also determine its direction as well as its magnitude.

To find the direction, construct a tangent of length ωr to the circle at the point p, as shown in Fig. 9.23. The angle ωt is the included angle between the tangent and the constructed line parallel to the y-axis. The components of the constructed tangent vector are $-\omega r \sin \omega t$ (x-axis) and $\omega r \cos \omega t$ (y-axis). These are the x and y components of $\mathbf{v}(t)$ as given in Eq. 9.16.

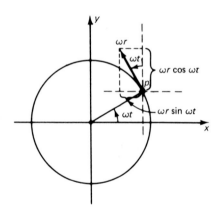

Fig. 9.23

An elegant way to show that the velocity and the position vectors are perpendicular utilizes the fact that the dot product of two perpendicular vectors is zero. Forming the dot product $\mathbf{v} \cdot \mathbf{r}$,

$$\mathbf{v} \cdot \mathbf{r} = [-r\omega \sin \omega t \, \mathbf{i} + r\omega \cos \omega t \, \mathbf{j}] \cdot [r \cos \omega t \, \mathbf{i} + r \sin \omega t \, \mathbf{j}]$$

$$\mathbf{v} \cdot \mathbf{r} = -r^2\omega \sin \omega t \cos \omega t + r^2\omega \cos \omega t \sin \omega t = 0.$$

Since the dot product is zero, the velocity must be perpendicular to the position vector. See the power of vectors!

The acceleration for uniform circular motion can now be calculated by differentiation of the velocity vector as

$$\frac{d\mathbf{v}(t)}{dt} = \mathbf{a}(t) = -r\omega^2 \cos \omega t \, \mathbf{i} - r\omega^2 \sin \omega t \, \mathbf{j}$$

$$= -\omega^2 [\underbrace{r \cos \omega t \, \mathbf{i} + r \sin \omega t \, \mathbf{j}}_{\mathbf{r}}].$$

Thus

$$\mathbf{a}(t) = -\omega^2 \mathbf{r} \qquad \text{(note the minus sign).} \qquad (9.17)$$

> *The acceleration of a particle moving with uniform (constant speed) circular motion is directed* **radially inward** *along the radius vector. It is called the* CENTRIPETAL *acceleration.*

The magnitude of this acceleration is

$$|\mathbf{a}_{\text{centripetal}}| = \omega^2 r = \frac{v^2}{r} \quad \text{(in m/s}^2\text{)} \tag{9.18}$$

since $\omega = v/r$ (Eq. 9.13).

In summary, a particle executing uniform circular motion has a time-varying velocity directed tangentially to the circle at the position of the particle, and an acceleration directed *inward* along the radial position vector.

Newton's second law tells us that a force must be acting on a particle if it is accelerating. In circular motion this is called the centripetal force. It acts in the same direction as the acceleration, inward along the radial position vector. The magnitude of this force can easily be calculated from Eq. 9.18 as

$$|\mathbf{F}_{\text{centripetal}}| = \frac{mv^2}{r} = m\omega^2 r. \tag{9.19}$$

EXAMPLE 1   A 0.5-kg ball, tethered by a 2-m string in outer space, is uniformly rotated once a second in a circular horizontal plane (Fig. 9.24). What is the tension on the string?

The angular speed ω is calculated by recalling that

$$\omega = \frac{\Delta\theta}{\Delta t} = \frac{\text{one angular revolution}}{\text{time period}}$$

$$= \frac{2\pi \text{ rad}}{1 \text{ s}} = 2\pi \text{ rad/s.}$$

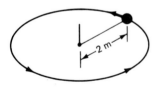

Fig. 9.24

But for this isolated ball in a gravitationally free environment, the string constrains its motion so that it moves in a circle. The only force then on the ball is the contact force on the string tension T. For the ball at the instant of time shown in Fig. 9.25, the x-component of Newton's second law is

$$x\text{-comp.:} \qquad -T = ma_x.$$

But the constraint imposed by the string (motion in a circle) is

$$\mathbf{a} = \mathbf{a}_{\text{cent}} = \omega^2 r(-\mathbf{i}).$$

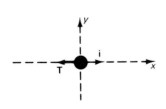

Fig. 9.25

Thus, substituting the constraint into Newton's second law yields

$$-T = -m\omega^2 r$$

or $$T = (.5 \text{ kg})(2\pi/\text{s})^2(2 \text{ m}) = 39.47 \text{ N (kg-m/s}^2).$$

EXAMPLE 2   A car of mass m with bald tires drives around a banked circular racetrack of a radius r (Fig. 9.26). Assuming that the track is covered with ice so that there is no frictional force on the bald tires, find the time T that the driver can go around the track without sliding.

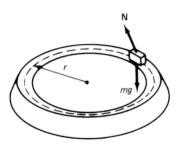

The car's uniform speed is equal to the circumference/time or $v = \dfrac{2\pi r}{T}$. The force that supplies the centripetal acceleration to the car comes from the component of the normal force acting in the direction of this acceleration (the negative x-direction at an instant represented by the free-body diagram, Fig. 9.27) or

$$\mathbf{F}_{\text{cent}} = -N \sin \theta \ \mathbf{i}.$$

Fig. 9.26

Since there is no frictional force on the tires in this example, this is the only force available in this example that can produce the centripetal acceleration. Thus

x-comp.: $-N \sin \theta = ma_{\text{cent}}$ (note the minus sign).

The car is constrained by the track to move in a circle, which means that it executes centripetal acceleration in the minus x-direction.

y-comp.: $N \cos \theta - mg = 0$ $N = \dfrac{mg}{\cos \theta}.$

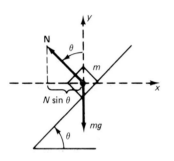

The car is constrained to move in a circle, so

$$\mathbf{a}_{\text{cent}} = -\frac{v^2}{r} \ \mathbf{i}$$

Fig. 9.27

Substitution into the x-comp. equation gives us

$$\frac{mg \sin \theta}{\cos \theta} = \frac{mv^2}{r}$$

or $$v = (rg \tan \theta)^{1/2},$$

but $$T = \frac{2\pi r}{v}.$$

Substituting for v gives us

$$T = \frac{2\pi r}{(rg \tan \theta)^{1/2}}.$$

(The mass does not enter the answer.) Without the banking of the road, the driver could not go around the ice-covered surface. Remember this when you drive!

EXAMPLE 3
The Conical Pendulum   A ball of mass m is suspended by a massless string of length L. It travels in a horizontal circle of radius r at a constant speed v, while the string makes an angle θ with respect to the vertical axis, as shown

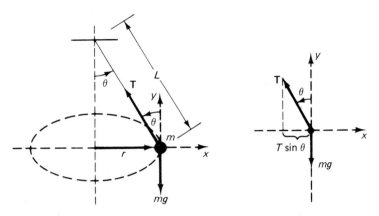

Fig. 9.28 The conical pendulum   **Fig. 9.29**

in Fig. 9.28. Find the speed of the ball, and the tension in the string, in terms of the other parameters.

From the free-body diagram (Fig. 9.29),

x-comp.: $\qquad -T \sin \theta = m a_{\mathrm{cent}}$

y-comp.: $\qquad T \cos \theta - mg = 0$ (no acceleration in y-direction)

The ball is constrained to move in a circle, so

$$\mathbf{a}_{\mathrm{cent}} = -\frac{v^2}{r}\,\mathbf{i}.$$

Substituting y-comp. into the x-comp. equation and using the constraint, we get

$$\frac{mg \sin \theta}{\cos \theta} = \frac{mv^2}{r} \qquad \text{or} \qquad \tan \theta = \frac{v^2}{rg}$$

$$v = (rg \tan \theta)^{1/2}.$$

To find the tension

$$T \sin \theta = \frac{mv^2}{r} \qquad \text{but} \qquad \sin \theta = \frac{r}{L}.$$

Thus $\qquad\qquad T = \frac{mv^2}{r}\frac{L}{r} = \frac{mv^2 L}{r^2} = m\omega^2 L.$

9.7 **Charged Particles in Magnetic Fields**

IN Chapter 3 a formula for the radius of curvature of a charged particle moving in a magnetic field was presented without proof. Let us now take up this important example, starting from first principles and developing the equations of motion for such a particle.

The magnetic force on a charged paticle is

$$\mathbf{F} \text{ (magnetic)} = q(\mathbf{v} \times \mathbf{B}), \qquad (9.20)$$

where q is the charge on the particle in coulombs, $\mathbf{v}$ is the velocity in meters/second, $\mathbf{B}$ is the magnetic field in tesla (1 tesla = 10^4 gauss), and the force is measured in newtons.* It is worth enumerating the unusual characteristics of the magnetic force, expressed mathematically in Eq. 9.20.

1. The force is proportional to the speed of the particle. This is an exact proportionality, not an approximation such as we encountered in the viscous drag force.
2. The force is always directed perpendicular to the direction of the velocity of the particle.
3. The force is always directed perpendicular to the direction of the magnetic field.
4. The force is proportional to the amount of charge.
5. The force is proportional to the magnitude of the magnetic field's strength.
6. The force reverses direction if any one of the following quantities is reversed: the sign of the electrical charge, the direction of the velocity, or the direction of the magnetic field.
7. The force is zero if the particle's velocity is parallel to the direction of the magnetic field, and varies from zero to its maximum value continuously as the sin θ, where θ is the angle between the velocity and magnetic field directions.

All that physics is contained in Eq. 9.20. (Mathematics is a powerful and terse language!)

First let's examine the special case where the charged particle's velocity is perpendicular to the magnetic field. The magnitude of the magnetic force is then simply

$$F_{\text{mag}} = qvB$$

and it is directed along a line perpendicular to both $\mathbf{B}$ and $\mathbf{v}$, as shown in Fig. 9.30. This force will continually change directions, as the velocity changes directions, producing a centripetal acceleration of the particle. By Newton's second law,

$$qvB = \frac{mv^2}{r}$$

or

$$r = \frac{mv}{qB} = \frac{\text{momentum}}{qB}, \qquad (9.21)$$

as we presented in Eq. 3.4.

---

* These are the standard SI units.

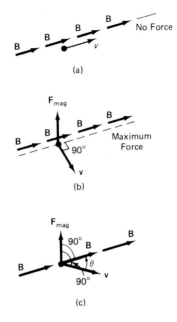

Fig. 9.30

The more general case, where the velocity can have any direction with respect to the magnetic field, is a thee-dimensional problem. The vector equation of motion is

$$\frac{d^2\mathbf{r}(t)}{dt^2} = \frac{d\mathbf{v}(t)}{dt} = \frac{q}{m}\,[\mathbf{v}(t) \times \mathbf{B}], \tag{9.22}$$

where we shall assume that $\mathbf{B}$ is constant, uniform, and directed along the z-axis:

$$\mathbf{B} = B_0\mathbf{k}, \tag{9.23}$$

and the velocity is

$$\mathbf{v}(t) = v_x(t)\mathbf{i} + v_y(t)\mathbf{j} + v_z(t)\mathbf{k}.$$

Equation 9.22 is the equivalent of three scalar differential equations, one for each Cartesian direction. From the rules for vector cross product, we have

$$(\mathbf{v} \times \mathbf{B})_{x\text{-comp.}} = v_y B_0 \qquad (\mathbf{v} \times \mathbf{B})_{y\text{-comp.}} = -v_x B_0$$

$$(\mathbf{v} \times \mathbf{B})_{z\text{-comp.}} = 0.$$

The three component equations of 9.22 are

x-comp.:
$$\frac{dv_x}{dt} = \frac{q}{m}\,v_y B_0 \qquad\qquad (9.24a)$$

y-comp.:
$$\frac{dv_y}{dt} = -\frac{q}{m}\,v_x B_0 \qquad\qquad (9.24b)$$

z-comp.:
$$\frac{dv_z}{dt} = 0. \qquad\qquad (9.24c)$$

Equation 9.24c is obviously the equation of zero acceleration in the z-direction. We have already examined its solution thoroughly.

$$v_z = v_{0z} = \text{const} \quad\text{and}\quad z(t) = v_{0z}t + z_0.$$

This is constant velocity motion in the z-direction. The motion in the x–y plane is more interesting. These differential equations involve time derivatives of one velocity component which are equal to a constant times the velocity of the other component. We have not seen this before. Let us look for solutions of the form

$$v_x(t) = v_0 \sin\omega t \qquad v_y(t) = v_0 \cos\omega t. \qquad (9.25)$$

Substituting Eq. 9.25 into Eqs. 9.24a and 9.24b gives us

$$v_0\omega \cos\omega t \overset{?}{=} \frac{qB}{m}\,v_0 \cos\omega t \qquad -\omega v_0 \sin\omega t \overset{?}{=} -\frac{qB}{m}\,v_0 \sin\omega t$$

Both equations are satisfied if

$$\omega = \frac{qB}{m}, \qquad\qquad (9.26)$$

where ω, the angular frequency, is called the cyclotron frequency. Equations 9.25 describe uniform circular motion, where v_0 is the constant speed of the particle tracing out a circle.

The time for one complete rotation T, called the period, is simply

$$\omega = \frac{2\pi}{T} \quad\text{or}\quad T = \frac{2\pi}{\omega} = \frac{2\pi m}{qB}, \qquad (9.27)$$

independent of the velocity and radius. The velocity and radius independence of the period of rotation of a charged particle in a magnetic field is critical to the design of the cyclotron. This three-dimensional motion is a combination of uniform circular motion in the plane perpendicular to the magnetic field and uniform translational motion parallel to the field. This is the spiral motion shown in Fig. 9.31.

Finally, let's calculate the rate of change of the magnitude of the velocity of the particle. To do this we calculate the time derivative of the square of the velocity ($\mathbf{v} \cdot \mathbf{v} = v^2$).

$$\frac{d}{dt}(\mathbf{v} \cdot \mathbf{v}) = \mathbf{v} \cdot \frac{d\mathbf{v}}{dt} + \frac{d\mathbf{v}}{dt} \cdot \mathbf{v}. \qquad (9.28)$$

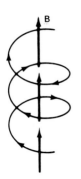

Fig. 9.31

But from Eq. 9.22,

$$\frac{d\mathbf{v}}{dt} = \frac{q}{m}(\mathbf{v} \times \mathbf{B}).$$

Thus

$$\frac{d}{dt}(\mathbf{v} \cdot \mathbf{v}) = \frac{q}{m}\mathbf{v} \cdot (\mathbf{v} \times \mathbf{B}) + \frac{q}{m}(\mathbf{v} \times \mathbf{B}) \cdot \mathbf{v} = 0;$$

this equation is equal to zero because each term is equal to zero. The vector $(\mathbf{v} \times \mathbf{B})$ is perpendicular to the vector $\mathbf{v}$ and the dot product of two orthogonal vectors is zero. The magnitude of the velocity, the speed, does not change with time; only the direction changes. This makes physical sense. The magnetic force, which always acts perpendicular to the velocity, cannot produce an acceleration parallel to the velocity. Thus the magnetic field cannot change the kinetic energy of the particle.*

9.8 Motion with Changing Speed

SO FAR we have considered only particles that move with constant speed in a circle. If the particle's speed changes as it traverses the circle, an additional acceleration must be considered. To analyze this generalized form of circular motion, cylindrical polar coordinates (Fig. 9.32) and their unit vectors are introduced.

Unlike the Cartesian unit vectors $\mathbf{i}$ and $\mathbf{j}$ that you are accustomed to, the polar coordinate unit vectors $\hat{\boldsymbol{\theta}}$ and $\hat{\mathbf{r}}$ *change their direction* with varying positions (i.e., they are a function of the polar angle θ).

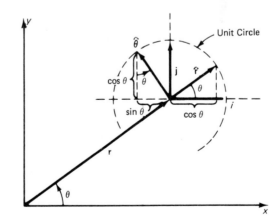

Fig. 9.32

---

* This is true only for time-independent, static, magnetic fields.

However, $\hat{\theta}$ and $\hat{r}$ still have unit length and are *always mutually perpendicular*. Any unit vector in the x–y plane can always be written as some linear combination of the Cartesian unit vectors **i** and **j**. From Fig. 9.32 one can show that

$$\hat{r} = \cos\theta\, \mathbf{i} + \sin\theta\, \mathbf{j}$$
$$\hat{\theta} = -\sin\theta\, \mathbf{i} + \cos\theta\, \mathbf{j}. \tag{9.29}$$

To show that $\hat{r}$ and $\hat{\theta}$ are perpendicular, we form the dot product:

$$\hat{r} \cdot \hat{\theta} = -\sin\theta\cos\theta + \sin\theta\cos\theta = 0.$$

For two dimensions, the position vector of the particle can be written in terms of polar unit vectors as

$$\mathbf{r}(t) = x(t)\mathbf{i} + y(t)\mathbf{j} = r(t)\hat{r}. \tag{9.30}$$

Equation 9.30 appears strange at first glance because the position vector is written as the product of the scalar $r(t)$ and only *one* unit vector $\hat{r}$. Doesn't a vector in two dimensions have to be written in terms of two variables? Yes, it does, but you must remember that $\hat{r}$ is a function of θ [$\hat{r} = f(\theta)$], as shown in Eq. 9.29, so everything is okay.

A particle's velocity may also be written in polar coordinates

$$\mathbf{v}(t) = \frac{d}{dt}\mathbf{r} = \frac{d}{dt}[x(t)\mathbf{i} + y(t)\mathbf{j}] = \frac{d}{dt}r\hat{r}$$

or

$$\mathbf{v}(t) = \hat{r}\frac{dr}{dt} + r\frac{d\hat{r}}{dt}, \tag{9.31}$$

since, in general, it is possible for *both* the unit vector $\hat{r}$ and the magnitude of the radial position $|\mathbf{r}|$ to change with time. We restrict our consideration here to circular motion where the magnitude $\mathbf{r}$, r, is time *independent* (motion with a fixed radius, circular motion). For fixed radius, $dr/dt = 0$ and Eq. 9.31 becomes

$$\mathbf{v} = r\frac{dr}{dt} = r\frac{d}{dt}(\cos\theta\,\mathbf{i} + \sin\theta\,\mathbf{j})$$

$$= r\left(-\sin\theta\frac{d\theta}{dt}\mathbf{i} + \cos\theta\frac{d\theta}{dt}\mathbf{j}\right) \tag{9.32}$$

$$= r\frac{d\theta}{dt}(-\sin\theta\,\mathbf{i} + \cos\theta\,\mathbf{j}) = r\omega\hat{\theta} = v\hat{\theta} \quad \text{(fixed radius)},$$

the same result that we obtained earlier in Eq. 9.16. However, we are now prepared to examine generalized particle acceleration using this formalism:

$$\frac{d\mathbf{v}}{dt} = \mathbf{a} = \frac{d}{dt}v\hat{\theta} = \hat{\theta}\frac{dv}{dt} + v\frac{d\hat{\theta}}{dt} \tag{9.33}$$

But

$$\frac{d\hat{\boldsymbol{\theta}}}{dt} = \frac{d}{dt}(-\sin\theta\,\mathbf{i} + \cos\theta\,\mathbf{j}) = -\cos\theta\,\frac{d\theta}{dt}\,\mathbf{i} - \sin\theta\,\frac{d\theta}{dt}\,\mathbf{j}$$

$$= -\frac{d\theta}{dt}(\cos\theta\,\mathbf{i} + \sin\theta\,\mathbf{j}) = -\omega\hat{\mathbf{r}}.$$

Substituting into Eq. 9.33 yields

$$\mathbf{a} = \frac{dv}{dt}\,\hat{\boldsymbol{\theta}} - \omega v\hat{\mathbf{r}} \qquad \text{or} \qquad \mathbf{a} = \mathbf{a}_t + \mathbf{a}_r. \tag{9.34}$$

Equation 9.34 shows that if a particle executes circular motion with a *variable speed*, the particle experiences *two* forms of acceleration;

$$
\begin{aligned}
&\textbf{1} \quad \text{Tangential acceleration:} \quad && \mathbf{a}_t \equiv \frac{dv}{dt}\,\hat{\boldsymbol{\theta}} \\[2mm]
&\textbf{2} \quad \text{Centripetal acceleration:} \quad && \mathbf{a}_r \equiv -\omega v\hat{\mathbf{r}}.
\end{aligned}
\tag{9.35}
$$

The magnitude of the total acceleration is

$$|\mathbf{a}| = (a_t^2 + a_r^2)^{1/2}.$$

A more generalized type of motion, such as elliptical motion, would involve terms originating from $\hat{\mathbf{r}}\,dr/dt$ in Eq. 9.31. Keep in mind that this entire discussion has assumed that no forces are acting in the z-direction.

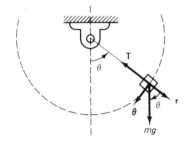

Fig. 9.33

EXAMPLE 1
The Pendulum

An example of a system that exhibits both radial and tangential acceleration is the pendulum, shown in Fig. 9.33. The tension in the string provides the radial acceleration. From Newton's second law,

$\hat{\mathbf{r}}$-comp.: $\qquad -T + mg\cos\theta = -ma_r = -\dfrac{mv^2}{r}.$

The tangential acceleration is provided by the component of the gravitational force in the $\hat{\boldsymbol{\theta}}$ direction,

$\hat{\boldsymbol{\theta}}$-comp.: $\qquad mg\sin\theta = ma_\theta \qquad \text{or} \qquad a_\theta = \dfrac{dv}{dt} = g\sin\theta.$

9.9 Projectile Motion

TO CONCLUDE our discussion of two-dimensional motion, we examine a particle moving in a plane with one constant force acting on it. If air resistance is neglected, particle motions near the surface of the earth covering distances small compared to the earth's circumference provide good examples of such motion. Electrons moving through the electric fields of a cathode ray tube's deflection plates also exhibit this kind of motion. This class of problems requires no new physics principles. To solve them, you just need some practice in learning the appropriate tricks.

The single most important idea to remember is this: The motion in the x-direction is independent of the motion in the y-direction. The force is traditionally aligned along the y-axis (gravitational attraction to Earth). This force has no effect on the motion along the x-axis. Since the forces are constant, and thus the acceleration is constant, we can generalize Eq. 8.29, making it a vector equation:

$$\mathbf{r}(t) = \mathbf{r}_0 + \mathbf{v}_0 t + \frac{\mathbf{F}}{2m}\, t^2 \qquad \text{where } \frac{\mathbf{F}}{m} = \mathbf{a}. \qquad (9.36)$$

Equation 9.36 is equivalent to three scalar equations, in the general case. For two-dimensional $(x-y)$ motion, with the *only* force acting in the y-direction, Eq. 9.36 reduces to

x-comp.: $x(t) = x_0 + v_{0x}t$

y-comp.: $y(t) = y_0 + v_{0y}y + \frac{1}{2}a_y t^2,$

and from Eq. 8.30, we have for the y-velocity,

$$v_y(t) = v_{0y} + a_y t,$$

and the x-component of the velocity is a constant of the motion. These are the special equations for projectile motion near Earth's surface, neglecting frictional drag.

EXAMPLE 1   A cannon shoots a ball with a muzzle velocity $\mathbf{v}_0$ at an angle of θ degrees (Fig. 9.34). How far will the ball travel, what is its maximum height, how long will it be in the air, and what is the shape of its trajectory, neglecting air resistance?

Consider the x-axis motion first. The particle never changes its velocity in the x-direction during the flight, since there are no forces acting on it in this direction. (Remember that we are neglecting air

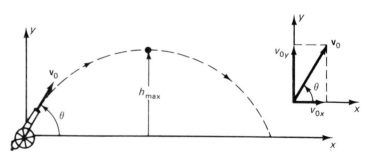

Fig. 9.34

resistance.) The initial velocity in the x-direction is

$$v_{0x} = v_0 \cos \theta.$$

The x-coordinate of the projectile is then

$$x = x_0 + v_{0x}t = (v_0 \cos \theta)t \qquad \text{since } x_0 = 0. \qquad (9.37)$$

The gravitational force acts in the y-direction producing an acceleration in the minus y-direction. The initial velocity in the y-direction is $v_{0y} = v_0 \sin \theta$. The projectile's position in the y-direction is thus

$$y = y_0 + v_{0y}t + \tfrac{1}{2}a_y t^2 = (v_0 \sin \theta)t - \tfrac{1}{2}gt^2 \qquad (9.38)$$

since $y_0 = 0$.

From the x-component

$$t = \frac{x}{v_0 \cos \theta}.$$

Substituting into Eq. 9.38 yields

$$y = (\tan \theta)x - \frac{g}{2v_0^2 \cos^2 \theta} x^2, \qquad (9.39)$$

which is an equation of the form $y = Ax - Bx^2$, the equation of a parabola.

When the ball hits the ground at the end of its flight ($y = 0$), we have from Eq. 9.38

$$0 = (v_0 \sin \theta)t - \tfrac{1}{2}gt^2$$

or

$$t = \frac{2v_0 \sin \theta}{g} \qquad \text{time of flight.} \qquad (9.40)$$

The total distance traveled comes from the x-axis (Eq. 9.37) and the time of flight:

$$x = (v_0 \cos \theta)t = \frac{2v_0^2 \sin \theta \cos \theta}{g}. \qquad (9.41)$$

The maximum height can be calculated by recalling that at the peak height the ball has zero velocity in the y-direction, and using this in the equation

$$v_y = v_{0y} + a_y t \qquad \text{or} \qquad 0 = v_0 \sin \theta - gt_{\text{max height}}$$

$$t_{\text{max height}} = \frac{v_0 \sin \theta}{g}$$

$$y_{\text{max}} = v_0 \sin \theta \frac{v_0 \sin \theta}{g} - \frac{1}{2} g \left(\frac{v_0^2 \sin \theta^2}{g^2} \right)$$

$$y_{\text{max}} = \frac{v_0^2 \sin \theta}{2g}. \qquad (9.42)$$

EXAMPLE 2

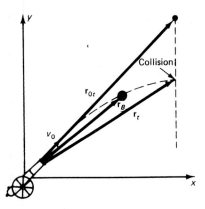

Fig. 9.35

A well-known lecture demonstration involves shooting a ball at a falling target when the target has been released at the same instant that the ball leaves the gun. The apparatus is set up so that the gun is aimed at the initial position of the target as shown in Fig. 9.35. The target is electrically released when the gun goes off and it falls directly down in the minus y-direction. To the student's surprise, the ball always hits the target, no matter what initial muzzle speed is imparted to the ball. The only thing that changes is the point of impact!*

To understand what is going on in this demonstration, consider the mythical case where gravitation is "turned off." The target remains at rest and the ball travels in a straight line and hits it. If we "turn on" the gravitational force, both the ball and the target will fall with exactly the same acceleration (gravitational acceleration being independent of mass). Thus the two particles must collide.

Instead of solving the problem by components, an elegant proof of this can be carried out using vector equations. First consider the position vectors of the two colliding objects: $\mathbf{r}_B$, for the ball, and $\mathbf{r}_t$ for that of the target. From Eq. 9.36,

$$\mathbf{r}_B = \mathbf{v}_0 t - \tfrac{1}{2}gt^2\mathbf{j}$$

$$\mathbf{r}_t = \mathbf{r}_{0t} - \tfrac{1}{2}gt^2\mathbf{j}.$$

For a collision to occur, $\mathbf{r}_t = \mathbf{r}_B$ or

$$\mathbf{v}_0 t - \tfrac{1}{2}gt^2\mathbf{j} = \mathbf{r}_{0t} - \tfrac{1}{2}gt^2\mathbf{j}$$

or $$\mathbf{v}_0 t = \mathbf{r}_{0t},$$

which says that $\mathbf{v}_0$ must be in the same direction as $\mathbf{r}_{0t}$ (since t is a scalar). But such is the case since the gun is aimed at the taget. The time of the collision is then

$$t = \frac{|\mathbf{r}_{0t}|}{|\mathbf{v}_0|}.$$

See the power of vectors!

Harold Edgerton, the acknowledged master of stroboscopic photography, demonstrated this effect beautifully in Fig. 9.36. This photograph shows two golf balls launched simultaneously from the top left corner, one allowed to fall freely from rest and the other given an initial velocity in the horizontal direction. The photograph is taken by opening the camera's shutter in a dark room and illuminating the balls with a bright flashing strobe light at equal time intervals. The photograph shows the location of both balls at the same instant, spaced at equal time intervals.

It is clear that the vertical coordinate of each ball is the same. The race to hit the ground is a tie. You can see quite clearly that both

---

* The muzzle velocity of the ball must be sufficient for the ball to reach the target before it hits the ground.

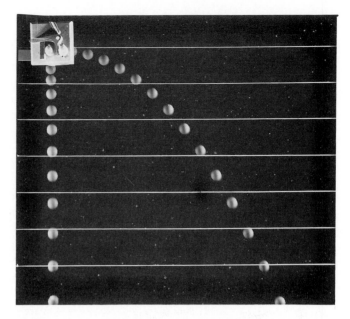

Fig. 9.36 A stroboscopic picture of two golf balls. One ball was released from rest at the instant the other was launched with a horizontal velocity. The simultaneous vertical coordinates of each ball are the same.

balls are accelerating in the vertical direction, since the vertical spacing between the images is not equal but increases with time. If you draw a vertical grid on the paper, you can also see that the ball going to the right is moving with a constant velocity.

9.10 **Discovering New Forces**

IN THIS chapter we have focused on the important problem of predicting the details of the motion of objects when they are subjected to known forces. This deductive process is clearly one of the triumphs of physics. Physicists, however, get excited by the prospect of discovering new laws and new forces in nature. Not only do they observe systems before and after their interactions to search for new conservation principles, but they also observe the motion of these systems (when possible) during the interaction, to discover unknown interactions and forces. Rutherford observed the scattering of α-particles off atoms of gold, noticing an unexpectedly large number of α-particles which were scattered backward. This led him to predict the existence of the massive charged nuclear core of atoms. The motion of electrons in semiconductors tells us how these electrons interact with the crystal lattice of atoms through which they move. Such motion must be understood in terms of quantum mechanical models of solids. The inductive process is certainly as important as the deductive exercises we have focused on, but it requires the mastery of the deductive methods of reasoning first.

I hope you do not think that all the forces in nature have already been discovered and that nothing is left for you to do but to work out problems using known theories. As recently as 1986, a group of physicists at Purdue University reanalyzed the seminal experiment done 80 years ago by Baron Roland von Eötvös.* They suggested that gravity doesn't behave as either Newton or Einstein had predicted, but that a so-called "fifth force" is present in nature. That really stirred things up! Now there are experimental groups all over the world measuring gravitational effects, even repeating Galileo's experiment by dropping two objects to see if different masses have exactly the same free-fall acceleration in Earth's gravitational field. Of course, they do these experiments in a vacuum with very fancy and precise electronics, hoping to confirm or reject this proposal. What are their conclusions? As of this writing, the problem is unresolved.†

9.11 **Summary of Important Equations**

$$F_f \leq \mu N \qquad \text{for sliding friction}$$

$$\theta \equiv \frac{S}{r} \qquad \theta \text{ in radians}$$

$$1 \text{ radian} = 57.3° \qquad 2\pi \text{ radians} = 360°$$

For uniform circular motion:

$$v = r\omega$$

$$\theta = \omega t$$

$$\mathbf{a} = -\omega^2 \mathbf{r} \qquad \text{centripetal acceleration}$$

$$\mathbf{r}(t) = r \cos \omega t \mathbf{i} + r \sin \omega t \mathbf{j}$$

$$|\mathbf{a}|_{\text{centripetal}} = \omega^2 r = \frac{v^2}{r}$$

Polar coordinate unit vectors:

$$\hat{\mathbf{r}} = \cos \theta \, \mathbf{i} + \sin \theta \, \mathbf{j}$$

$$\hat{\boldsymbol{\theta}} = -\sin \theta \, \mathbf{i} + \cos \theta \, \mathbf{j}$$

* E. Fischback, D. Sudansky, A. Szafer, C. Talmage, and S. Aronson, *Physical Review Letters*, **56**, 3 (1986).

† A brief review of the state of the search in 1988 appears in the July 1988 issue of *Physics Today*, p. 21.

Position and velocity vectors in terms of polar unit vector:

$$\mathbf{r} = r\hat{\mathbf{r}}$$

$$\mathbf{v} = r\omega\hat{\boldsymbol{\theta}}$$

Circular motion with changing speed:

$$\mathbf{a}_{\text{tangential}} = \frac{dv}{dt}\,\hat{\boldsymbol{\theta}}$$

$$\mathbf{a}_{\text{centripetal}} = -\omega v\hat{\mathbf{r}}$$

Chapter 9   QUESTIONS

1. Is it possible for an object to move in a direction opposite to the direction of the net force that it experiences? Explain.

2. The vector sum of all the forces acting on a particular body is zero. What can you conclude about the velocity of that body?

3. Most of you have heard someone say, "I was thrown forward when the driver suddenly stepped on the brakes." What's wrong with this expression? How would you say it correctly?

4. "You gain weight when you ride in an elevator." Comment on this crazy statement. How can you lose weight? Do you recommend this for weight-watchers? Shouldn't Weight-Watchers' name be changed to "Mass-Watchers"?

5. Is it possible for μ to be greater than unity? What does it mean?

6. A block placed on a flat horizontal surface experiences a frictional force between the surface of the block and the horizontal surface. Since the frictional force will act in the horizontal direction, why doesn't the block accelerate in the direction of the frictional force? This friction, by Newton's law, will cause the block to accelerate along the horizontal surface. What's wrong with all this reasoning? Explain.

7. You are asked by your kid sister for the best way to move a sled with your baby brother aboard. Should

Fig. 9.37

she push or pull the sled? Does it matter? Does one require less force than the other if she does it as shown in Fig. 9.37? Explain.

8. Two blocks are stacked on top of one another. A force is exerted on the lower block, as shown in Fig. 9.38. If there is no friction anywhere, describe what happens to the system.

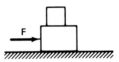

Fig. 9.38

9. Friction is often considered an annoying problem. Give examples where it plays an important, even essential role in the operation of some machines.

10. Consider the Atwood's machine shown in Fig. 9.9. Describe qualitatively the motion of the machine if the following "realities" are taken into consideration (one at a time).

   (a) The pulley has mass.

   (b) The rope has mass.

   (c) The bearing has friction.

11. Explain why race tracks have banked curves. Why must old racetracks be rebuilt?

12. Old-fashioned stunt aircraft were sometimes built with open cockpits and no seat belts. Why doesn't the pilot fall out when these planes fly upside down?

13. When an airplane takes off from rest, where does the sitting passenger experience an extra force? If the passenger were casually standing in the aisle, what might happen to that person?

14. When a roller coaster rapidly comes to the top of a steep slope, the passengers often leave their seats for a moment. How is that possible? What happened to gravity? Draw a diagram and explain.

15. When you are in a car that turns a sharp corner, you get the impression that a force is trying to push you toward the outside of the curve. Is there such a force? What is going on? Explain.

16. A young girl swings a ball tethered on a light string over her head as shown in Fig. 9.39. Describe the motion of the ball if the string breaks, starting the instant the string breaks.

Fig. 9.39

17. Is it possible for an object with no net force on it to move in a curved path? Explain.

18. A well-known demonstration of the laws of dynamics is depicted in Fig. 9.40. The professor exerts a force slowly on the lower string T_1. The upper string T_2 breaks while the lower one remains intact. The experiment is repeated, but this time the force is exerted quickly and the lower string T_1 breaks while the upper one remains intact. Explain what is going on.

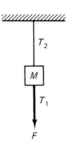

Fig. 9.40

19. Large space platforms are being planned. Designers will use the rotation of the platform to produce "gravity" artificially. Are they really creating a force? Explain.

20. Does the rotation of the Earth affect your weight? What do you mean by weight? Do I lose weight if I go to the moon? What happens to my weight if I go to the north pole?

21. In the expression $\mathbf{F} = q\mathbf{v} \times \mathbf{B}$, some of the vectors are perpendicular to others, while others can be at any angle to one another. Which are perpendicular? What is the magnitude of $\mathbf{F}$?

22. A physics professor pedals a bicycle around a large oval track, gradually and uniformly slowing as he tires. Draw arrows on Fig. 9.41 at the points indicated to show the acceleration on him as he travels.

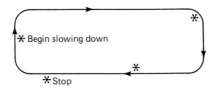

Fig. 9.41

Chapter 9   PROBLEMS

1. A 25-kg mass hangs by a supporting system of massless ropes shown in Fig. 9.42. Find the tension in all the ropes T_1, T_2, T_3.

2. Two masses are suspended in equilibrium by the complex of ropes and pulley shown in Fig. 9.43. Find the angle θ and the tension in all the ropes.

3. The simple "machine" shown in Fig. 9.44 can be used to measure charge. It consists of two small light spheres suspended by fine insulating thread of length L. When the two spheres are each given a charge q, they repel each other and reach a static equilibrium angular separation of θ radians. Show that the charge on each sphere is given by

$$q = 2L \sin \frac{\theta}{2} \left(4\pi\varepsilon_0 mg \tan \frac{\theta}{2}\right)^{1/2}$$

(*Hint:* See Eq. 2.2 for the force between two stationary charges.)

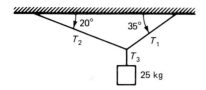

Fig. 9.42

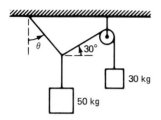

Fig. 9.43

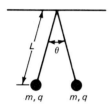

Fig. 9.44

4. A constant force **F** acts on the system of two blocks sliding over a frictionless plane shown in Fig. 9.45.

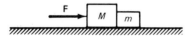

Fig. 9.45

   (a) Find the acceleration of the system.
   (b) Find the contact force between the two blocks.
   (c) If the smaller mass m is placed to the left of the larger mass M so that F acts directly on m, what is the acceleration of the system and the contact force between the two blocks?

5. A force of 26 N pulls on a system of three masses as shown in Fig. 9.46. Find the acceleration of the system and tension on the connecting strings T_a and T_b, neglecting sliding friction.

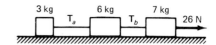

Fig. 9.46

6. Two masses shown in Fig. 9.47 sliding on a frictionless surface are connected by a massless string through a frictionless eyelet. Find the acceleration of the system and the tension in the string in terms of θ, M, m, and g.

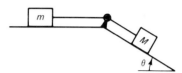

Fig. 9.47

7. Two blocks are connected through a massless pulley arrangement shown in Fig. 9.48. They both slide on a frictionless surface. Find the acceleration of each block M and m and the tensions in the rope T_1 and T_2 when a constant force **F** is applied to the block m.

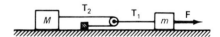

Fig. 9.48

8. Two blocks are connected through a massless pulley arrangement shown in Fig. 9.49. Neglecting all friction and pulley masses, find the two tensions T_1 and T_2 and the acceleration of the m_1 and m_2.

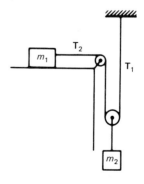

Fig. 9.49

9. A constant force of 10 N is applied to the massless pulley shown in Fig. 9.50. Find the acceleration of the 7-kg mass and the tension in the rope supporting it.

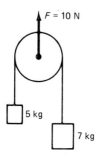

Fig. 9.50

10. The apparatus in Fig. 9.51 is sometimes called a double Atwood's machine. Neglecting all friction and the mass of all pulleys and connecting cables and considering that $m_1 > (m_2 + m_3)$ and $m_3 > m_2$, find the acceleration of m_1 and m_2 and the tensions T_1 and T_2 of this crazy system.

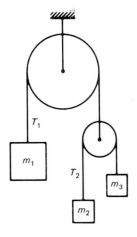

Fig. 9.51

11. A window washer of mass M wants to raise himself up on a platform of mass m as shown in Fig. 9.52. If the washer is able to exert a constant downward force **F** on the rope, find the acceleration on the platform and the tension in the rope that holds the massless pulley to the top of the building.

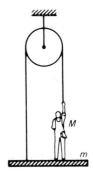

Fig. 9.52

12. Three masses are connected by a massless string over a massless pulley shown in Fig. 9.53. The two masses have a coefficient of sliding friction $\mu = .15$. Find the acceleration of the system and the tension in both strings.

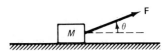

Fig. 9.53

13. A block of mass M slides with friction over a horizontal surface shown in Fig. 9.54. The frictional coefficient is μ and a constant force **F** is applied at an angle θ by a string.

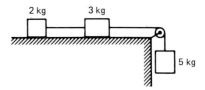

Fig. 9.54

(a) Find the acceleration of the system.
(b) If the magnitude of the force remains constant, find the angle θ that will produce the maximum acceleration of the system.

14. Three masses are connected by strings over pulleys in the configuration shown in Fig. 9.55. Mass m_2 slides with frictional coefficient μ over the horizontal surface. The masses are such that the system does accelerate. Find the acceleration and the tension in the two strings T_1 and T_2 if $m_3 < m_1$.

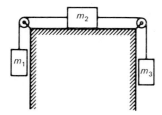

Fig. 9.55

15. Consider two blocks connected by a massless string sliding over a surface whose frictional coefficient is .03, as shown in Fig. 9.56. Neglect any effects of the pulley.

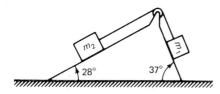

Fig. 9.56

    (a) Find the acceleration of the m_2 if $m_2 = 12.5$ kg and $m_1 = 3.5$ kg.

    (b) If the system starts from rest, how long will it take m_2 to move 2 m along the inclined plane?

16. Three frictionless pulleys are attached to the vertices of a triangular fixture of polished metal. One small block of mass m is placed on the 30° incline of the slab. Another small block, also of mass m, is placed alongside the vertical face. Light string is run over the pulleys and tied to each block, as Fig. 9.57 shows. Assume that all surfaces are frictionless. Find the acceleration of either block. Which block rises?

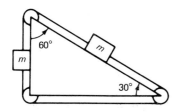

Fig. 9.57

17. Consider the mechanical system shown in Fig. 9.58, where m_1 slides with frictional coefficient μ over the horizontal surface and m_2 slides without friction

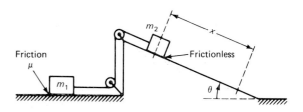

Fig. 9.58

over the inclined plane. If the system is initially at rest, show that the final velocity of m_2 after it has traveled a distance x is

$$v_f = \left[\frac{2xg(m_2 \sin \theta - \mu m_1)}{m_1 + m_2} \right]^{1/2}$$

18. Find the magnitude of the horizontal force **F** necessary to keep the block in Fig. 9.59 from sliding down the wall if the block's mass is 26.5 kg and the coefficient of friction with the wall is $\mu = 0.4$.

Fig. 9.59

19. Two blocks of equal mass m, connected by a massless string, are sliding on a plane inclined at an angle θ as shown in Fig. 9.60. They both slide with friction but with different coefficients of sliding friction μ_1 and μ_2. If $\mu_2 > \mu_1$:

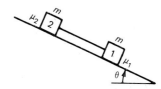

Fig. 9.60

    (a) Find the acceleration of the system.

    (b) Find the tension in the string.

    (c) Find the angle θ_c at which the blocks will slide with a constant velocity.

20. A constant force **F** is applied to the two-block system shown in Fig. 9.61. Block m_2 slides with negligible friction, but m_1 has a coefficient of sliding friction μ with the horizontal surface. Find the acceleration of the system and the contact force between the two blocks.

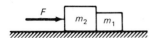

Fig. 9.61

21. One way to construct a simple apparatus to measure the acceleration of your automobile is to hang a small ball by a string from the inside of your car's roof and measure the angle the string makes with the vertical during acceleration (and with the windows closed). Derive the expression for the acceleration in terms of the angle and any other relevant parameters. Does the ball go forward or backward during forward acceleration of the car? What happens when you put on the brakes? What happens when you travel at constant velocity?

22. A 15-kg block with a 5-kg block on top is accelerated by a horizontal constant force **F** (Fig. 9.62). The 15-kg block slides without friction, but the 5-kg block has a coefficient of sliding friction $\mu = .3$ with respect to the surface of the 15-kg mass. What is the maximum force **F** that can be applied before the 5-kg block begins to slide over the 15-kg mass below it? What is the acceleration of the 15-kg block with this force applied?

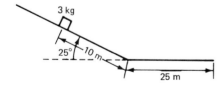

Fig. 9.62

23. A 3-kg block slides down a 10-m slope inclined at 25° and then continues on a horizontal surface for 25 m before it comes to rest (Fig. 9.63). Assuming the

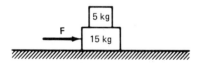

Fig. 9.63

coefficient of friction is the same for both the inclined and horizontal surfaces, find the coefficient of friction μ.

24. How far up an inclined plane will a block of mass m travel if it is given an initial velocity v_0 along the plane, has a coefficient of sliding friction with the plane of μ, and the plane is inclined at an angle θ? (See Fig. 9.64.)

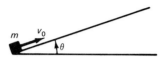

Fig. 9.64

25. What force **F**, acting on the inclined plane of mass M in Fig. 9.65, is required to keep the small block m from sliding with respect to the inclined plane? Neglect all friction in the problem.

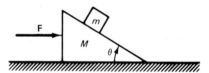

Fig. 9.65

26. A 2-kg block sits on top of a 30° slope inside an elevator (Fig. 9.66). What acceleration of the elevator is necessary to keep the 2-kg block from sliding with respect to the 30° inclined plane? Ignore all frictional forces.

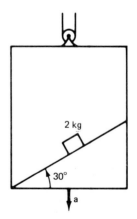

Fig. 9.66

27. A fly of mass m rests comfortably on a stationary turntable at a radius r from the center. How long can the fly hold its position if the turntable, starting from rest, is turning with a constant angular acceleration α, given that the fly's coefficient of friction is μ?

28. Three masses rotate at a constant speed around in a horizontal circle on a frictionless surface shown in Fig. 9.67. If the masses make one rotation every minute, find the tensions T_1, T_2, and T_3 in the connecting rope.

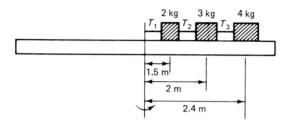

Fig. 9.67

29. If you define your weight as the reading you observe on a spring scale, calculate how much weight you would apparently "lose" if you weighed yourself at the geographic north pole and compared it to your weight at the geographic equator. Assume that Earth is a perfect sphere.

30. A small cube of mass m rides inside a conical-shaped enclosure at a distance d above the vertex (Fig. 9.68). If the coefficient of friction between the cube and the cone's surface is μ, find the shortest and longest periods of rotation possible so that the cube will not slide on the conical surface, in terms of m, d, θ, μ, and g.

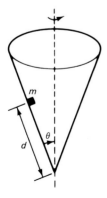

Fig. 9.68

31. A physics student of mass m rides a Ferris wheel of radius R that rotates in a clockwise direction at a rate of 1 revolution every τ seconds (Fig. 9.69). Calculate the force *exerted by* the student on the bottom of his seat (in terms of m, τ, R, and g) when the student is at positions A, B, C, and D.

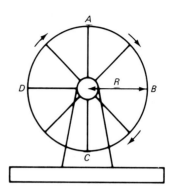

Fig. 9.69

32. A speeding car reaches the top of a hill which has a circular profile of radius 300 m, as shown in Fig. 9.70. The car's wheels leave the surface of the road for an instant. What is the speed of the driver?

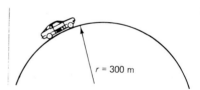

Fig. 9.70

33. A mass m is rotated in the horizontal plane at an angular frequency ω as shown in Fig. 9.71. The mass is suspended by two equal-length strings L. Find the tension in each string.

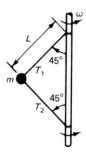

Fig. 9.71

34. Prove that two conical pendulums with different lengths will have the same period of rotation if the pendulums are started in such a way that the vertical distance from the pivot to the plane defined by the circular orbit are the same.

35. A spinning vertical drum can be rotated with sufficient speed that people will "stick" to the wall and the floor can be removed (Fig. 9.72). If the coefficient of friction between a person and the wall of the cylinder is .4 and the cylinder has a 5-m radius, what is the slowest angular velocity ω that the drum can rotate without the people slipping?

Fig. 9.72

36. Consider a particle moving in a straight line in the x-direction with constant velocity $\mathbf{v} = v_{0x}\mathbf{i}$. If the particle begins at the position $y(0) = y_0$ and $x(0) = 0$ at $t = 0$, find the expression for this particle's velocity in terms of the cylindrical polar coordinates $\hat{\boldsymbol{\theta}}$ and $\hat{\mathbf{r}}$ and time.

37. A bead of mass m slides on a vertical wire hoop of radius r which is rotated at an angular frequency ω. Neglecting all friction, find the angle θ that the bead positions itself on the hoop in terms of m, r, ω, and g (Fig. 9.73).

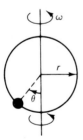

Fig. 9.73

38. A velocity selector or velocity filter is an important element in some types of gaseous electronic instrumentation. A collimated beam of charged ions enters a region of uniform electric and magnetic fields. Only ions that have a certain velocity are not deflected in this region. These are the ions that pass through the output collimation port of the velocity selector.

    (a) What are the directions of the electric and magnetic fields relative to the velocity of the incoming ion beam?

    (b) Calculate the velocity of the undeflected beam in terms of E, B, m, mass of ion, and q, the charge of the ion.

39. A proton (mass m, charge e) is moving with velocity v_0 along the x-axis.

    (a) As it passes the origin, it encounters a uniform magnetic field in the square region which is shaded in Fig. 9.74. The proton's trajectory is such that it emerges from the square at the point $(x = L, y = L)$, moving with speed v_0 in the y-direction. What are the magnitude and direction of the uniform magnetic field that caused this motion?

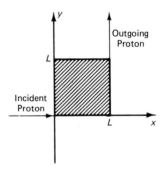

Fig. 9.74

    (b) Now suppose that the square region contains only a uniform electric field (*no* magnetic field) but that the proton emerges from the square at the upper right-hand corner with velocity $v_0\mathbf{j}$ as before. What are the magnitude and direction of the uniform electric field that caused this motion?

40. You are designing a space "platform" for people to live on for long periods of time. Since there is no gravitational force in space, you decide to "create" artificial gravity by rotating the ship as shown in Fig. 9.75. If the platform has a 100-m radius, what angular velocity do you need to give each passenger the same "weight" as they have on Earth?

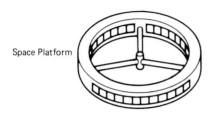

Fig. 9.75

41. Two cannons, at rest on Earth, are fired simultaneously (Fig. 9.76). Cannon 1 makes a 45° angle with the horizontal. Cannon 2 is fired vertically, with a muzzle speed of v_0. The y-component of the muzzle velocity (v_{1y}) for cannon 1 is also v_0. For what value (or values) of v_0 will the two projectiles collide? Express your answer in terms of L (the separation between cannons) and g.

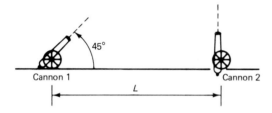

Fig. 9.76

42. Prove, as Galileo understood, that the range of a projectile is the same for angles of projection greater than 45° or less than 45° by the same amount. That is, for example, the range is the same for 40° as it is for 50°, given the same muzzle speed.

43. A cannon ball is shot from a 100-m-high cliff with a muzzle speed of 175 m/s at an angle of 86° with respect to the horizon.

(a) Calculate the distance the ball travels.

(b) What is the *velocity* (magnitude and direction) of the projectile at impact?

(c) How long was the projectile in flight?

44. A battleship is cruising 5000 m away from enemy shores. An alert seaman notices a puff of smoke onshore and 10 s later a shell falls 1000 m short of the ship.

(a) What is the muzzle *velocity* of the shell as it leaves the gun on shore?

(b) How much farther away from shore must the ship move to avoid being hit by shells from this gun? You may assume that all the shells are the same; only the gun's angle can be changed by the gunner on shore.

45. An electron enters a 1-cm-long electric field deflecting plate with a velocity of 5×10^5 m/s in the x-direction (Fig. 9.77). If the deflecting plate has an electric field that exerts a force of 1.6×10^{-17} N on the electron, find the velocity and position of the electron beam when it strikes the screen 30 cm away.

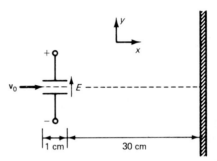

Fig. 9.77

46. An Ultimate Frisbee player throws a Frisbee at an angle of 35° with an initial speed of 20 m/s at a distance of 1.5 m off the ground. A second player 60 m away starts to run toward the Frisbee the instant it is thrown. How fast must he run to catch the Frisbee before it hits the ground? Neglect all air resistance (which is nonsense for a Frisbee!).

47. Mass spectrometers are important research tools for many fields of science. One early design is shown in Fig. 9.78. Positively charged ions enter a region of uniform electric field E, are accelerated a distance y_0, then enter a region of uniform magnetic field B out of the paper (zero electric field), and are deflected in a semicircular path to a detecting film on a plane. If all the ions have the same charge q but have two differ-

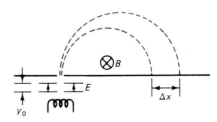

Fig. 9.78

ent masses m and $m + \Delta m$, find their physical separation Δx when they strike the detecting surface.

48. Consider the mechanical system shown in Fig. 9.79, where the mass m_1 is such that the system moves in the direction shown by the arrows. If the mass m_1 has frictional coefficient μ with its surface, and the small sphere moves in the liquid with a drag described by Strokes's law, find the expression for the velocity of the mass m_1 as a function of time if it starts with a velocity v_0 at time $t = 0$. The mass m_2 has

radius r and moves in a liquid with viscosity η, and density ρ.

49. A solid steel cylinder of radius a and length L is dropped into a bucket of water (Fig. 9.80). Find the expression for the acceleration of the cylinder as a function of position. In your analysis, you may neglect the frictional drag of the fluid, but you must account for the buoyant force of the water. Remember Archimedes: "The buoyant force is equal to the weight of the fluid displaced."

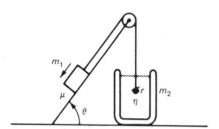

Fig. 9.79

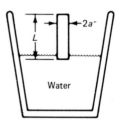

Fig. 9.80

ANOTHER POINT OF VIEW: COORDINATE TRANSFORMATIONS

10.1 **Introduction**

THE CHOICE of reference frame, the vantage point from which a system is observed and measured, can sometimes greatly simplify the analysis of both complex motion and collisions. This analytical technique is called COORDINATE TRANSFORMATION. We have already applied this technique (without explicitly stating it) to the important example of particle creation in a high-energy two-body collision (Section 6.6). There we calculated the minimum projectile kinetic energy necessary to create an antiproton in the collision

$$p + p \rightarrow p + \bar{p} + p + p.$$

The reasoning was as follows:

1. View the collision of a projectile proton with a stationary target proton from a special moving reference frame. In this new frame the total momentum of the initial system of two particles was zero (the projectile and the target had equal but opposite momentum.)
2. Recognize that *in such a reference frame* all four product particles *could* remain at rest after the collision.
3. Recognize that *should* the product particles remain at rest after the collision, this would be the most "efficient" of all possible collision processes, since *all* the kinetic energy (as measured in this special zero-momentum frame) would have been transformed into rest energy of the newly created particle.
4. Transform back into the original laboratory frame, and calculate the initial kinetic energy of the projectile particle, as measured in the laboratory frame.

There are many other examples of the application of coordinate transformations where the system, when studied in the new reference frame, becomes more tractable to physical and mathematical analysis. Transformations to both linearly translating and rotating frames of reference are explored in this chapter.

Before we examine the mathematical methods used to execute these various coordinate transformations, we must address a crucial physics question: Do the same laws of physics apply in *all* the various coordinate systems that one might dream up to observe and analyze a particular system? For example, in the proton–proton collision, is it still true that the conservation of total linear momentum, total energy, baryon number, strangeness, and so on, are valid laws for the observer who makes measurements of this collision from the vantage point of a rocketship moving at a constant velocity? In Chapter 6 we did not raise the question; we simply assumed that it was true and went ahead with the analysis. That was a cheat! Now we must address this crucial question.

10.2 **Principle of Relativity**

> *All the laws of physics are the same in every inertial reference frame.*

That deceptively simple statement is called the "principle of relativity." Newton was the first to enunciate this principle, although he applied it only to his laws of mechanics and gravitation. It was Einstein who generalized the principle to include all the laws of nature, particularly the laws of electrodynamics. The principle of relativity was taken by Einstein as one of the two cornerstones of his special theory of relativity.*

To understand and apply this principle, we must first explain what is meant by an *inertial reference frame.* It is a frame of reference that has the following properties:

1. If a small test particle is observed in this reference frame, it obeys Newton's first law of motion. That is, if the test particle was initially at rest, it will remain at rest, and if it was moving with a velocity **v**, it will continue to move in a straight line with the same velocity throughout that entire region of space.

2. The inertial frame is operationally defined in a *local* region of space (or more precisely, space–time).†

---

* Einstein's first paper on the theory of special relativity [*Annalen der Physik*, **17**, 891–921 (1905)] did not contain the word "relativity" in the title. It was called "On the Electrodynamics of Moving Bodies."

† An excellent discussion of the concept of inertial reference frame can be found in *Space–Time Physics*, by J. A. Wheeler and E. F. Taylor (San Francisco: W. H. Freeman, 1963), Chap. 1, p. 5.

This still leaves one question unanswered: How small is "small" for the test particle? From the abstract viewpoint, the test particle can be arbitrarily small, but from the viewpoint of real experimental science, a better statement is: "The particle must have so little mass that, to within some specific accuracy, its presence will not affect the motion of the particles nearby."

What is the significance of the principle of relativity to our discussion? First, it means that the intrinsic properties of all particles, such as rest mass, charge, family membership, the velocity of the photon in a vacuum, strangeness, and isotopic spin, are the same in all inertial reference frames. It also means that the conservation laws of charge, energy, and linear and angular momentum are obeyed in all inertial reference frames, as are the laws of gravity, electromagnetism, strong, and weak forces.

It does *not* mean that the kinetic energy of a particular particle has the same value in all inertial reference frames. It does *not* mean that the linear momentum of a particle, or even a system of particles, is the same in all inertial frames, or that two events that occur simultaneously in one frame occur simultaneously in another. It does *not* even mean that clocks keep the same time in all inertial frames. The fact that a particle's momentum may have different values in two inertial reference frames in *no* way implies that the *total* momentum of an isolated system changes as it interacts, regardless of the inertial reference frame from which the interaction is viewed. The principle of relativity requires that the *laws* of physics are the same in all inertial reference frames, not that all the dynamical parameters of particles be the same in all inertial reference frames.

Consider, for example, the conservation of linear momentum. The vector sum of the linear momentum of a collection of isolated particles is a constant of the motion; it does not change, no matter how the system interacts or decays. Viewed from different inertial reference frames, the total linear momentum of a system of particles is generally not the same, but the initial total momentum of the system as measured in one inertial frame equals the final momentum of the system when it is measured in the same frame.

Suppose that you and I observe the linear momentum of a system of n particles interacting and transforming themselves into a new collection of $a \rightarrow z$ particles. You remain in the lab to watch the show while I climb aboard a rocket ship and measure the particles' momentum while traveling at a constant velocity with respect to your laboratory. You measure

$$\mathbf{P}_i^{\text{lab}}(1), \mathbf{P}_i^{\text{lab}}(2), \ldots, \mathbf{P}_f^{\text{lab}}(a), \ldots, \mathbf{P}_f^{\text{lab}}(z),$$

while I determine

$$\mathbf{P}_i^{\text{shp}}(1), \mathbf{P}_i^{\text{shp}}(2), \ldots, \mathbf{P}_f^{\text{shp}}(a), \ldots, \mathbf{P}_f^{\text{shp}}(z).^*$$

---

* The nomenclature is getting complicated. We need to identify three parameters for each value of the momentum; $\mathbf{P}_i^{\text{lab}}(a)$. The superscript "lab" or "shp" indicates the reference frame, the subscript i stands for the initial state (f for the final state), and the quantity in parentheses designates the particle of interest.

The particles are referred to by numbers in the initial state 1, 2, 3, . . . , n, and by letters $a, b, c, . . . , z$ in the final state to remind you that the particles may not be the same after the interaction. In general, $\mathbf{P}_i^{lab}(3) \neq \mathbf{P}_i^{shp}(3)$, for example, and $\mathbf{P}_f^{lab}(b) \neq \mathbf{P}_f^{shp}(b)$ and even the vector sum of the momenta ($\Sigma \, \mathbf{P}_i^{lab} \neq \Sigma \, \mathbf{P}_i^{shp}$) are not usually equal. The conservation of linear momentum holds only for all measurements made on the isolated system in the *same* inertial reference frame. This is shown in Table 10.1.

TABLE 10.1

Initial Linear Momentum of a System of Particles	Final Linear Momentum of the System of Particles

LABORATORY INERTIAL REFERENCE FRAME

$$\{\mathbf{P}_i^{lab}(1) + \mathbf{P}_i^{lab}(2) + \cdots + \mathbf{P}_i^{lab}(n)\} \; = \; \{\mathbf{P}_f^{lab}(a) + \mathbf{P}_f^{lab}(b) + \cdots + \mathbf{P}_f^{lab}(z)\}$$

⇑ ⇑

generally not equal   generally not equal

⇓ ⇓

$$\{\mathbf{P}_i^{shp}(1) + \mathbf{P}_i^{shp}(2) + \cdots + \mathbf{P}_i^{shp}(n)\} \; = \; \{\mathbf{P}_f^{shp}(a) + \mathbf{P}_f^{shp}(b) + \cdots + \mathbf{P}_f^{shp}(z)\}$$

ROCKETSHIP INERTIAL REFERENCE FRAME

The principle of relativity is much more than a theoretical conjecture; it is a hypothesis that has been carefully verified in high-precision experiments over many years. Although many clever experimenters have searched for even the smallest exception, none has been found to date. It stands as one of the cornerstones of modern physics. Why should it be that nature "chose" inertial reference frames to be so special—to play such an important role in describing all the physical laws? Why is it that all the laws of nature are the same in all reference frames, where "when a test particle at rest remains at rest and when it is moving with a velocity $\mathbf{v}$, it continues to move with the same velocity $\mathbf{v}$?" These questions remain unanswered.

Given a particular inertial reference frame, how can we find another one without going through the same experiments with our test particle? The answer is quite simple.

> *All reference frames in the same local region that move at a constant velocity with respect to the original inertial frame are themselves inertial reference frames.*

Consider four observers in four different reference frames measuring the fall of a small ball enclosed within a vacuum chamber near the surface of the earth. The observers are shown in Fig. 10.1, all

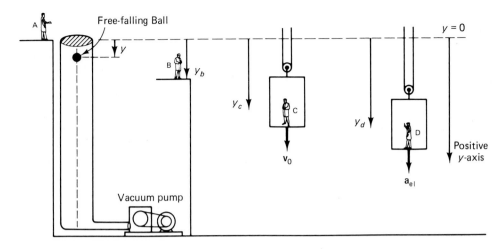

Fig. 10.1

watching the same falling ball, which starts from rest at the origin $y = 0$.

We discussed the motion of such a free-falling object as measured in the laboratory coordinates (observer A) in Chapter 8 (Eq. 8.33b). The ball's y-coordinate varies with time according to the equation

$$y^A = \tfrac{1}{2}gt^2. \tag{10.1}$$

Observer B, who is also at rest with respect to A, but displaced a distance y_b from the origin $y = 0$, measures

$$y^B = y - y_b = \tfrac{1}{2}gt^2 - y_b. \tag{10.2}$$

Observer C, in an elevator that is falling at a constant velocity v_0, records

$$y^C = y - y_c = \tfrac{1}{2}gt^2 - v_0 t. \tag{10.3}$$

Finally, observer D in an elevator that is accelerating in the same direction as the ball, with an acceleration a_{el}, observes that

$$y^D = y - y_d \quad \text{but} \quad y_d = \tfrac{1}{2}a_{\text{el}}t^2,$$

so
$$y^D = \tfrac{1}{2}(g - a_{\text{el}})t^2. \tag{10.4}$$

Four different "views" of the same phenomenon: the free fall of a small object near the surface of the Earth. The velocity of the ball at any time t can be determined by differentiation of the four equations with respect to time.

For A:

$$v^A = \frac{dy^A}{dt} = \frac{d}{dt}\left(\frac{1}{2}gt^2\right) = gt$$

For B:

$$v^B = \frac{dy^B}{dt} = \frac{d}{dt}\left(\frac{1}{2}gt^2 - y_b\right) = gt$$

For C: (10.5)

$$v^C = \frac{dy^C}{dt} = \frac{d}{dt}\left(\frac{1}{2}gt^2 - v_0 t\right) = gt - v_0$$

For D:

$$v^D = \frac{dy^D}{dt} = \frac{d}{dt}\left[\frac{1}{2}\left(g - (a_{el})t^2\right)\right] = (g - a_{el})t$$

Finally, the acceleration of the ball as measured by the four observers can be obtained by a second differentiation:

For A:

$$a^A = \frac{dv^A}{dt} = g$$

For B:

$$a^B = \frac{dv^B}{dt} = g$$

For C: (10.6)

$$a^C = \frac{dv^C}{dt} = g$$

For D:

$$a^D = \frac{dv^D}{dt} = g - a_{el}$$

Observers A, B, and C all measure the same acceleration of the ball. Each of them would agree, if they conferred and shared their observational data, that Newton's laws of motion for a particle in a uniform gravitational field are correct. They were all in inertial reference frames. But D was not. She witnessed the event from the vantage point of an accelerated coordinate system and measured a smaller acceleration of the falling ball, $g - a_{el}$, than one would calculate from Newton's laws for a particle that experiences Earth's gravitational force near its surface. The accelerating elevator is obviously *not* an inertial reference frame.

10.3 Galilean Transformation

IN THE previous example we made several important implicit assumptions which must now be made explicit. We used what is known as the GALILEAN TRANSFORMATION to calculate A, B, C, and D's observations. Imagine two Cartesian coordinate systems S and S' shown in Fig. 10.2, where S' is moving with a constant velocity $(u\mathbf{i})$

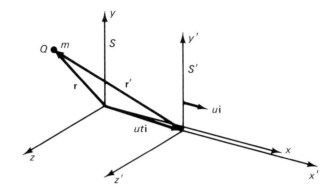

Fig. 10.2

along the x-axis. (S' is displaced slightly in the z-direction in the drawing for artistic clarity.)

Two observers, Primed and Unprimed, will be measuring the position, velocity, and acceleration of a particle from the vantage point of these different reference frames. Each observer has his own measuring apparatus, metersticks and clocks, and independently determines the parameters of the moving particle. Each sees the particle from his own "point of view." We should like to develop a set of mathematical equations that will relate the measurements made by one observer to the measurements made by the other. These equations are called "coordinate transformations."

To simplify the mathematics without degrading the usefulness of these transformation equations in any essential way, we will assume that the origins of both reference frames O and O' are coincident at zero time on both clocks, $t = t' = 0$. At a time t later, Unprimed notes that Prime's reference frame has moved away from him a distance ut down the positive x-axis, as shown in Fig. 10.2. At this time Prime observes the particle m at the position $\mathbf{r}'$ and Unprimed determines the same particle to be located at $\mathbf{r}$ with respect to their respective origins O' and O. If we construct the vector displacement from O to O', namely $ut\mathbf{i}$, then straightforward vector addition shows that

$$\mathbf{r} = \mathbf{r}' + ut\mathbf{i}, \tag{10.7}$$

or in component form,

$$
\begin{aligned}
x &= x' + ut' \\
y &= y' \qquad z = z' \\
t &= t',
\end{aligned}
\tag{10.8}
$$

where x, y, z, t and x', y', z', t' are the coordinates of a particle as measured in the unprimed and primed coordinates, respectively. To find the velocities of a particle as measured in the two frames, we take differentials of Eq. 10.7 (remembering u is a constant) thus:

$$dr = dr' + udt\mathbf{i}. \qquad (10.9)$$

Dividing Eq. 10.9 by the scalar quantity dt yields

$$\frac{d\mathbf{r}}{dt} = \frac{d\mathbf{r}'}{dt} + u\mathbf{i}. \qquad (10.10)$$

If $dt = dt'$ (the same time kept in both coordinate systems), we have

$$\frac{d\mathbf{r}}{dt} = \frac{d\mathbf{r}'}{dt'} + u\mathbf{i} \qquad \text{or} \qquad \mathbf{v} = \mathbf{v}' + u\mathbf{i}. \qquad (10.11)$$

Or, in component form,

$$v_x = v_{x'} + u \qquad v_y = v_{y'} \qquad v_z = v_{z'}. \qquad (10.12)$$

Finally, in a similar way, one can easily show that the equation for the transformation of the acceleration gives

$$a_x = a_{x'} \qquad a_y = a_{y'} \qquad a_z = a_{z'}. \qquad (10.13)$$

The acceleration is the *same* for both observers S and S'.

The last equation of the group Eqs. 10.8 expresses the Newtonian concept of absolute time. It assumes that time flows continuously and is *invariant* in all inertial reference frames.* This concept seems sensible and logical and was in fact pretty much taken for granted by the physics community before 1900. But as Einstein remarked, common sense consists of all the prejudices one learns before the age of 18. What experimental evidence was there to support this assumption? For the moment let us accept this important assumption and apply Newton's ideas.

Fig. 10.3

EXAMPLE 1   Curling is a popular Canadian sport played on ice. A player launches a heavy, smooth stone toward a target circle at the far end of the rink and the stone is guided by a second player who sweeps the ice in front of the moving stone. A 12-kg stone is sent down the ice with an initial velocity of 2.4 m/s directly toward an identical stone at rest. The sweeper starts at midpoint on the rink and sweeps toward the stationary stone at a constant velocity of 1.8 m/s.

---

* An invariant physical quantity means that it is identical in all coordinate systems in which it is appropriately measured.

(a) Find the linear momentum of the projectile and target stone in the frame of reference of the sweeper.

For this problem the sweeper is the primed reference frame. Transposing Eq. 10.11 gives

$$\mathbf{v}_1' = \mathbf{v}_1 - u\mathbf{i}$$

$$= 2.4 \text{ m/s} - 1.8 \text{ m/s} = .6 \text{ m/s} \quad \text{(along the } x'\text{-axis)}$$

$$\mathbf{P}_1' = m_1\mathbf{v}_1' = 12 \text{ kg}(.6 \text{ m/s}) = 7.2\mathbf{i}' \text{ kg m/s} \quad \text{(projectile)}$$

The target:

$$\mathbf{v}_2' = \mathbf{v}_2 - u\mathbf{i}$$

$$= 0 - 1.8 \text{ m/s} = -1.8 \, \mathbf{i}' \text{ m/s}$$

$$\mathbf{P}_2' = m_2\mathbf{v}_2' = 12 \text{ kg}(-1.8 \text{ m/s}) = -21.6\mathbf{i}' \text{ kg m/s} \quad \text{(target)}$$

(b) What is the velocity of a moving reference frame where the total momentum of both stones is zero?

Let's call this new reference frame double primed. In that frame

$$\mathbf{P}_1'' + \mathbf{P}_2'' = 0 \qquad \text{or} \qquad m_1\mathbf{v}_1'' = -m_2\mathbf{v}_2''.$$

Since the two masses are equal,

$$\mathbf{v}_1'' = -\mathbf{v}_2''$$

or, from Eq. 10.11,

$$\mathbf{v}_1 - u\mathbf{i} = \mathbf{v}_2 + u\mathbf{i}.$$

For our one-dimensional example where $\mathbf{v}_2 = 0$,

$$u = \frac{v_1}{2}$$

in the positive x-direction.

Using the Galilean transformation, it is possible to show that the *change* in the kinetic energy of an arbitrary two-body nonrelativistic collision is an invariant quantity in all inertial reference frames. That is, if you calculate the total initial kinetic energy of two colliding particles (KE_i) and the final kinetic energy after they collided (KE_f) in the laboratory frame of reference, then take the difference $\Delta\text{KE} = \text{KE}_f - \text{KE}_i$, this difference will be the same in all inertial reference frames. This statement can be shown to be true regardless of the nature of the collision, be it elastic, explosive, or completely inelastic. It can be expressed as

$$\Delta\text{KE} = \Delta\text{KE}' = \Delta\text{KE}'', \tag{10.14}$$

where the prime refer to the various inertial reference frames (see Problem 5).

FOR ANY nonzero-rest-mass particle one can always find an inertial reference frame which is moving with the same velocity as the particle, and thus, from the point of view of this frame, the particle has zero linear momentum. Even for a collection of particles, it is still always possible to find an inertial reference frame where the *total* linear momentum of the *system* of particles (the vector sum) is identically zero. This zero-momentum frame is also called the center-of-mass (CM) coordinate system. Examining collisions in the zero-momentum frame of reference often makes the analysis much easier. We begin by considering a nonrelativistic collision of two unequal masses shown in Fig. 10.4. The total linear momentum of the two particles as measured in the laboratory frame is

$$\mathbf{P}_{total} = m_1 \mathbf{v}^{lab}(1) + m_2 \mathbf{v}^{lab}(2).$$

In the second inertial frame, the center-of-mass frame, which travels at a constant velocity $\mathbf{u}_{CM}$ and has its origin coincident with 0 at $t = t' = 0$, the total linear momentum is

$$\mathbf{P}^{CM}_{total} = m_1 \mathbf{v}^{CM}(1) + m_2 \mathbf{v}^{CM}(2). \tag{10.15}$$

Using the velocity transformation Eq. 10.8, we obtain

$$\mathbf{P}^{CM}_{total} = m_1(\mathbf{v}^{lab}(1) - \mathbf{u}_{CM}) + m_2(\mathbf{v}^{lab}(2) - \mathbf{u}_{CM}).$$

Since we wish the center-of-mass frame to be the zero-momentum reference frame, we set $\mathbf{P}^{CM}_{total} = 0$ and solve for $\mathbf{u}_{CM}$:

$$\mathbf{u}_{CM} = \frac{m_1 \mathbf{v}^{lab}(1) + m_2 \mathbf{v}^{lab}(2)}{m_1 + m_2}. \tag{10.16}$$

The position of the origin of the center-of-mass coordinates is still arbitrary; only the velocity has been fixed by Eq. 10.16. To find

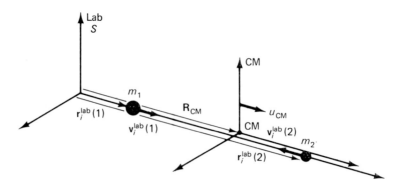

Fig. 10.4

the location of the center of mass, we must integrate Eq. 10.16. Begin by recalling that

$$u_{CM} = \frac{d\mathbf{R}_{CM}}{dt} \qquad \mathbf{v}^{lab}(1) = \frac{d\mathbf{r}(1)}{dt} \qquad \mathbf{v}^{lab}(2) = \frac{d\mathbf{r}(2)}{dt}. \qquad (10.17)$$

Substituting into Eq. 10.16 yields

$$\frac{d\mathbf{R}_{CM}}{dt} = \frac{1}{m_1 + m_2}\left[m_1\frac{d\mathbf{r}(1)}{dt} + m_2\frac{d\mathbf{r}(2)}{dt}\right]. \qquad (10.18)$$

Multiplying Eq. 10.18 by the differential dt and integrating term by term, we have

$$\int \frac{d\mathbf{R}_{CM}}{dt}\,dt = \frac{1}{m_1 + m_2}\left[m_1\int\frac{d\mathbf{r}(1)}{dt}\,dt + m_2\int\frac{d\mathbf{r}(2)}{dt}\,dt\right]. \qquad (10.19)$$

And arrive at

$$\mathbf{R}_{CM} = \frac{m_1\mathbf{r}(1) + m_2\mathbf{r}(2)}{m_1 + m_2} + \text{constant of integration.} \qquad (10.20)$$

By convention, the constant of integration is chosen equal to zero, leaving

$$\boxed{\mathbf{R}_{CM} = \frac{m_1\mathbf{r}(1) + m_2\mathbf{r}(2)}{M} \qquad M = m_1 + m_2.} \qquad (10.21)$$

It should be noted that $\mathbf{R}_{CM}$, $\mathbf{r}(1)$, and $\mathbf{r}(2)$ are in general time-dependent position vectors.

The center-of-mass position becomes a particularly useful concept when we discuss extended objects that cannot be treated as point masses. How are these concepts useful for analyzing collisions? Consider Eq. 10.16 again and rewrite it in a more compact form

$$\mathbf{u}_{CM} = \frac{\mathbf{P}^{lab}_{total}}{M}. \qquad (10.22)$$

But since $\mathbf{P}^{lab}_{total}$ is a constant in any collision—elastic, inelastic, or explosive—the velocity of the center of mass must also be constant. (Of course, we knew this all along since the CM frame was constructed as an inertial reference frame.) This means that

$$\frac{d\mathbf{u}_{CM}}{dt} = \frac{d^2\mathbf{R}_{CM}}{dt^2} = 0. \qquad (10.23)$$

There is no acceleration of the center of mass. This is consistent with the absence of a net external force acting on the system. All these derivations can easily be generalized to systems with more than two particles. Equations 10.22 and 10.23 also hold for many-particle nonrelativistic systems.

Now examine the kinetic energy of a system as measured in the CM coordinate system. Once again we consider first a two-particle collision to simplify the algebra and focus our attention on the physics. In the laboratory frame,

$$\text{KE}^{\text{lab}} = \tfrac{1}{2}m_1 \mathbf{v}^{\text{lab}}(1) \cdot \mathbf{v}^{\text{lab}}(1) + \tfrac{1}{2}m_2 \mathbf{v}^{\text{lab}}(2) \cdot \mathbf{v}^{\text{lab}}(2). \qquad (10.24)$$

Writing the kinetic energy in the lab frame in terms of the velocities in the center of mass frame and the relative velocity of the two frames (using Eq. 10.12) yields

$$\mathbf{v}^{\text{lab}}(1) = \mathbf{v}^{\text{CM}} + \mathbf{u}_{\text{CM}}. \qquad (10.25)$$

Substituting Eq. 10.25 into Eq. 10.24 and carrying out the indicated scalar product, we obtain

$$\text{KE}^{\text{lab}} = \tfrac{1}{2}m_1[v^{\text{CM}}(1)]^2 + \tfrac{1}{2}m_2[v^{\text{CM}}(2)]^2$$
$$+ \mathbf{u}_{\text{CM}} \cdot [m_1 \mathbf{v}^{\text{CM}}(1) + m_2 \mathbf{v}^{\text{CM}}(2)]$$
$$+ \tfrac{1}{2}(m_1 + m_2)\mathbf{u}_{\text{CM}} \cdot \mathbf{u}_{\text{CM}}. \qquad (10.26)$$

However, since $\mathbf{P}^{\text{CM}}_{\text{total}} = 0$, the term on the second line of Eq. 10.26 is equal to zero and Eq. 10.26 reduces to a simpler form. The first line is easily recognized as the kinetic energy of the two particles as measured in the center-of-mass system. The last line is new. It is mathematically identical to the kinetic energy, in the lab frame, of a *nonexistent particle* of mass $M = m_1 + m_2$, which is moving at the same velocity as the center of mass.

It appears that we have decomposed the total kinetic energy in the lab frame into two parts:

1. The kinetic energy of the individual parts of the system relative to the center of mass.
2. The kinetic energy of a fictitious particle of mass $m_1 + m_2$ moving at the velocity of the center of mass.

It is important to stress that such a decomposition can be done *only* for the zero-momentum inertial reference frame. It is *not* true for all reference frames.

For n particles this theorem can be generalized and written

$$\sum_{j=1}^{n} \tfrac{1}{2}m_j(v_j^{\text{lab}})^2 = \sum_{j=1}^{n} \tfrac{1}{2}m_j(v_j^{\text{CM}})^2 + \tfrac{1}{2}u_{\text{CM}}^2 \sum_{j=1}^{n} m_j \qquad (10.27)$$

or

$$\sum_j \text{KE}_j^{\text{lab}} = \sum_j \text{KE}_j^{\text{CM}} + \text{KE of CM}.$$

We have already shown that the velocity of the center of mass is a constant of the motion; that is, it does not change during or after the collision. This implies that the last term in Eq. 10.27, the kinetic energy of the CM, is also a constant that cannot change regardless of the nature of the collision. The kinetic energy of the CM is therefore *not "available"* for deformation, internal energy, elastic energy, light output, or particle creation. It is, so to speak, "locked up" in the center-of-mass motion. This type of analysis becomes particularly important in high-energy physics experiments where one is trying to use the kinetic energy of the projectile to create new particles in the collision. The kinetic energy of the center of mass motion *cannot* be converted to mass energy of a new particle because this kinetic energy must still be the center of mass after the collision. Notice that the mathematical expression for high-velocity particles is *not* the same as Eq. 10.27, since both the velocity transformation equations and the expression for the kinetic energy are different. Yet one can still think of the kinetic energy of a system of relativistic particles as decomposed into the kinetic energy as measured in the center-of-mass frame plus the kinetic energy of the center-of-mass frame.

10.5 **Two-Dimensional Collisions, Revisited**

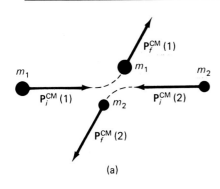

(a)

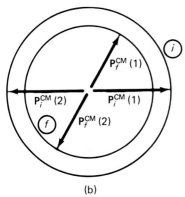

(b)

Fig. 10.5

LET US again examine two-dimensional collisions, this time to see what can be learned about them in the zero-momentum frame. We are not yet ready to study high-velocity collisions, for we are familiar only with the nonrelativistic transformations of velocities. But many of the ideas that will be developed are also applicable at velocities near the speed of light.

Consider first a Newtonian collision of two particles in the x–y plane. In the center-of-mass frame the two particles m_1 and m_2 have initial momentum of equal magnitude but opposite direction, as shown in Fig. 10.5a. But the total *final* momentum must also be zero. Thus

$$\mathbf{P}_f^{CM}(1) + \mathbf{P}_f^{CM}(2) = 0 \qquad \text{or} \qquad \mathbf{P}_f^{CM}(1) = -\mathbf{P}_f^{CM}(2),$$

which means that the magnitudes of the final momentum of the two particles must be equal. This is shown graphically in Fig. 10.5b. The details of the particular collisions determine the relative scattering angle of $\mathbf{P}_f^{CM}(2)$ to $\mathbf{P}_i^{CM}(2)$ (for hard spheres this corresponds to how directly the two spheres hit), and the magnitude of the final momentum depends on the nature of the collision (elastic, inelastic, or possibly some kind of explosion). For an ideal elastic collision the magnitude of the initial and final momentum of each particle remains the same, so the two construction circles (which are the locus of all final momentum vectors) have the same radius. If the collision is perfectly inelastic, the magnitude of each final momentum (in the CM frame) is zero, and circle ⓕ has zero radius. If the two balls

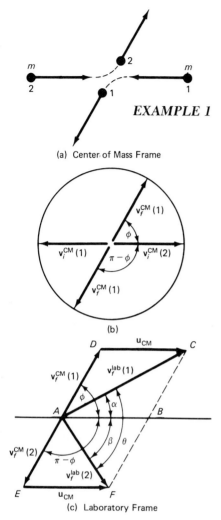

(a) Center of Mass Frame

(b)

(c) Laboratory Frame

Fig. 10.6

carry some explosive charge, which is set off at the impact, it is possible that circle (f) is actually larger than the initial momentum circle (i). If you know the final momentum of one particle in the CM frame, you can immediately construct the momentum vector of the second particle for a two-body collision.

EXAMPLE 1   Show that for a perfectly elastic Newtonian glancing collision of two *equal masses*, the angle between the two emerging particles after the collision is always 90°, as measured in the reference frame where the target particle was initially at rest (the laboratory frame).*

We shall analyze this collision in the CM frame. One immediate simplification that the CM frame offers is that there is only one magnitude of velocity in the problem, $|\mathbf{v}^{CM}| = v^{CM}$, since both particles have the same speed before and after the collision (as shown in Fig. 10.6b). If particle 1 scatters through an angle ϕ, it follows be that particle 2 scatters $\pi - \phi$ (as viewed in the CM frame). But what does this collision look like in the laboratory? The final velocities in the lab can be found by using our old friend Eq. 10.11:

$$\mathbf{v}^{lab} = \mathbf{v}^{CM} + u_{CM}\mathbf{i}, \qquad (10.11)$$

which "says" that we must add the vector $u_{CM}\mathbf{i}$ to the velocities in the center-of-mass frame to obtain the particle's velocity in the laboratory frame. This is shown graphically in Fig. 10.6c.

Now the trick is to show that $\theta = 90°$. It is all geometry. The vector $\mathbf{v}_f^{lab}(1)$ is the diagonal bisector of the parallelogram $ABCD$ and thus

$$\alpha = \frac{\phi}{2} \quad \text{and} \quad \beta = \frac{\pi - \phi}{2}.$$

Adding these two gives us

$$\alpha + \beta = \theta = \frac{\pi}{2}.$$

EXAMPLE 2   In our laboratory we observe the elastic collision of two particles of different mass m_1 and m_2, where the more massive particle m_2 is the target at rest. We want to analyze this collision in detail, focusing on the transfer of kinetic energy from one particle to the other. Figure 10.7a shows the collision process from the point of view of an observer in the laboratory.

To begin, we change our perspective by transforming to the center-of-mass frame of reference. The velocity of the CM can readily be calculated from Eq. 10.16:

$$\mathbf{u}_{CM} = \frac{m_1\mathbf{v}^{lab}(1) + m_2\mathbf{v}^{lab}(2)}{m_1 + m_2} = \frac{m_1}{m_1 + m_2} v_i^{lab}(1)\mathbf{i}. \qquad (10.28)$$

_____

* Head-on collisions are the exception to the rule. They are one-dimensional events.

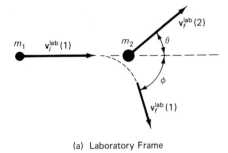

(a) Laboratory Frame

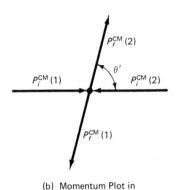

(b) Momentum Plot in
     Center-of-Mass Frame

Fig. 10.7

In the CM frame, the magnitude of the momentum vectors for the particles both before and after the collision are the same, and the collision appears only to cause a rotation of the coordinates (Fig. 10.7b). However, the velocity vectors of the two particles do not have the same magnitude since their masses are different. Nevertheless, the magnitude of the velocity of a given particle is not changed by the collision, as viewed in the CM frame of reference. To calculate the magnitude of each particle's initial velocity in this frame, we again use Eq. 10.11 as:

For the target m_2:

$$|\mathbf{v}_i^{CM}(2)| = |\mathbf{v}_f^{CM}(2)| = -u_{CM}.$$

For the projectile m_1:

$$|\mathbf{v}_i^{CM}(1)| = |\mathbf{v}_f^{CM}(1)| = |\mathbf{v}_i^{lab}(1) - u_{CM}\mathbf{i}|.$$

The *velocity* vector diagram of the collision in the CM frame is shown in Fig. 10.8. It is now a straightforward task of adding velocity vectors to determine the final velocities in the laboratory frame. Equation 10.11 tells us that can be accomplished by adding on the velocity of the center of mass $u_{CM}\mathbf{i}$. The dashed lines in Fig. 10.8 are the final velocities of each particle in the laboratory frame. Note that the scattering angles in the CM frame are *not* the same as we observed in laboratory.

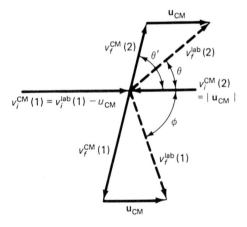

Velocity Diagram

Fig. 10.8

Let's now examine how the kinetic energy of the projectile particle changes due to an *elastic* collision with a stationary target particle. If the projectile misses the target (scatters through an angle 0°), then obviously the final kinetic energy is

$$\text{KE}_f(\phi = 0) = \tfrac{1}{2}m_1 v_i^{lab}(1)^2. \qquad (10.29)$$

If the projectile bounces back, scattering through a 180° angle, the velocity in the lab frame is easily calculated (see Fig. 10.6c):

$$\text{KE}_f(\phi = \pi) = \tfrac{1}{2}m_1[-(v_i^{\text{lab}}(1) - u_{\text{CM}}) + u_{\text{CM}}]^2.$$

Substituting for u_{CM} from Eq. 10.28 gives us

$$\text{KE}_f(\phi = \pi) = \tfrac{1}{2}m_1\left[v_i^{\text{lab}}(1) + \frac{2m_1}{m_1 + m_2}v_i^{\text{lab}}(1)\right]^2,$$

which reduces to

$$\text{KE}_f(\phi = \pi) = \tfrac{1}{2}m_1 v_i^{\text{lab}}(1)^2\left(\frac{m_1 - m_2}{m_1 + m_2}\right)^2. \tag{10.30}$$

The largest change in the kinetic energy of the projectile particle occurs when it backscatters. In that case the projectile loses the largest possible amount of kinetic energy, giving it to the target. In fact, if $m_1 = m_2$, Eq. 10.30 predicts that the projectile particle will have no kinetic energy left (in the lab frame) after impact. This should not surprise any pool or shuffleboard player. If you propel a nonrotating cue ball so that it strikes another ball head on, the cue ball will stop and the target ball will move off with the same velocity that the cue ball had at the moment of impact.

I have nothing against a good game of pool, but physicists have a more important use for this kind of scattering. Elastic scattering is used to "moderate" or slow down free neutrons which are generated in nuclear reactors. It turns out that very slow neutrons are surprisingly more effective in causing certain types of nuclear fission (spontaneous splitting of the nuclei of some heavy element, particularly some isotopes of uranium) than are the fast neutrons generated in the reactor. Low-energy neutrons are also used to study atomic and molecular vibrational motion in various crystalline solids and liquids.

The most efficient moderator, as we can see from Eq. 10.30, would be a collection of free neutrons. But one cannot confine them with high density since they decay in about 15 minutes. The next closest practical material would be hydrogen, since the mass of its nucleus (the proton) is only slightly less than the mass of the free neutron. However, protons have the nasty habit of capturing the neutrons (by strong forces) and becoming deuterium, which is not what we want. We nomally use very light elements, such as carbon (in paraffin), beryllium, and deuterium (in heavy water) because their nuclear mass is closer to the neutron's nuclear mass and they don't easily form bound nuclear states with it.

These examples indicate a general approach for examining two-body collisions using center-of-mass coordinate transformations:

1. Find the velocity of the center of mass $\mathbf{u}_{\text{CM}}$.
2. Subtract $\mathbf{u}_{\text{CM}}$ from the initial lab velocities of the two particles to obtain their velocities in the CM frame.

3. Calculate the momenta in the CM. These momenta, $P^{CM}(1)$ and $P^{CM}(2)$, are now equal and opposite.

4. The effect of the collision is to rotate the line of relative motion and to increase, decrease, or leave unchanged the magnitude of $\mathbf{P}_f^{CM}(1)$ and $\mathbf{P}_f^{CM}(2)$, depending on whether energy is released, absorbed, or unchanged during the collision.

5. Add $\mathbf{u}_{CM}$ on to the final CM velocities of the two particles to obtain the final velocities in the laboratory frame.

10.6 **Lorentz Transformation**

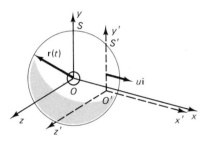

Fig. 10.9

THE GALILEAN transformations eventually get us into trouble. Consider a pulse of light that is sent off in all directions by a flashbulb and is detected and recorded by our two observers, S (at rest in the flashbulb's laboratory) and S' moving with a constant speed in the x-direction as shown in Fig. 10.9. The pulse propagates from the source at the origin of S with a spherical wavefront traveling at a speed c. To our laboratory observer (S) the radius of this wavefront increases with time according to the relation

$$|\mathbf{r}(t)| = ct.$$

For the x-direction alone,

$$x = ct. \tag{10.31}$$

The moving observer, S', traveling at a constant velocity $u\mathbf{i}$, observes the same wavefront. To predict her observations for the wavefront motion in the x-direction, we invoke the Galilean transformation,

$$x' = x - ut = (c - u)t. \tag{10.32}$$

The velocity of the wavefront in the primed reference frame can be found by direct differentiation of Eq. 10.32 (or by inspection):

$$\frac{dx'}{dt'} = v_{x'} = c - u, \tag{10.33}$$

which appears reasonable enough but is contrary to *all* experimental evidence. Every time experimentalists measure the speed of a light pulse or a wavefront from either a moving source or a moving observer (and there have been thousands of such careful experiments), they always get the same value (2.997924×10^8 m/s). Einstein had made this the second cornerstone of the special theory of relativity *before* almost all of these measurements had been carried out. He based this theoretical premise *not* on experimental evidence, but on his own "scientific faith" that this was one of the fundamental laws of nature. He turned out to be correct!

If the speed of light is an invariant quantity in all inertial reference frames, the Galilean transformation must be wrong! Where did we go astray?

To begin with, we assumed time to be an invariant quantity: $t = t'$. Einstein proposed that time was *not* an absolute continuously flowing quantity but rather, since it is measured by the periodic behavior of a physical system (the revolution of the earth, the electrical oscillation of an atom, the periodicity of a pendulum's motion), the properties of time must be connected to the laws that govern the behavior of the clock. Newton, on the other hand, had stated in the *Principia*: "Absolute, true, and mathematical time, of itself and by its own true nature, flows uniformly on, without regard to anything external."

By adopting the Galilean transformation equations we implicitly accepted the Newtonian view of time, which is *not* correct. We require a set of transformation equations that will preserve the two essential physical principles:

1. The principle of relativity
2. The invariant nature of the speed of light

We do not have the time to derive and explain all the interesting consequences of the correct transformation equations, but we will write them down and use them in several important examples. They are called the LORENTZ TRANSFORMATIONS.* By convention, the origins of the two coordinate systems O and O' are coincident at $t = t' = 0$ and S' moves along the x-axis with a constant velocity $u\mathbf{i}$. Under these conditions, the coordinates of any event (like the detection of a light pulse or the position of a particle) is given by Eq. 10.34.

The Lorentz Transformations

(a)	**(b)**

$$x' = \frac{x - ut}{\left(1 - \dfrac{u^2}{c^2}\right)^{1/2}} \quad \text{or} \quad x = \frac{x' + ut'}{\left(1 - \dfrac{u^2}{c^2}\right)^{1/2}}$$

$$= \gamma(x - ut) \qquad\qquad = \gamma(x' + ut')$$

$$y' = y \quad z' = z \qquad\qquad y = y' \quad z = z' \tag{10.34}$$

$$t' = \frac{t - \dfrac{ux}{c^2}}{\left(1 - \dfrac{u^2}{c^2}\right)^{1/2}} \qquad\qquad t = \frac{t' + \dfrac{ux'}{c^2}}{\left(1 - \dfrac{u^2}{c^2}\right)^{1/2}}$$

$$= \gamma\!\left(t - \frac{ux}{c^2}\right) \qquad\qquad = \gamma\!\left(t + \frac{ux'}{c^2}\right)$$

---

* Many excellent textbooks give a full derivation of the Lorentz transformations. I recommend *Space–Time Physics* by E. F. Taylor and J. A. Wheeler (San Francisco: W. H. Freeman, 1966).

where x, y, z, t and x', y', z', t' are the coordinates of the *same* event as measured in the unprimed (lab) and primed ("moving") coordinate systems. One immediately sees that time is no longer an invariant quantity, for it depends on the relative speed and location in the moving frame. However, it is also apparent that if the velocity of the moving frame is small compared to the speed of light $u \ll c$, the Lorentz and Galilean transformations become indistinguishable. Thus all the problems we worked out at the beginning of this chapter are still applicable, although in a fundamental way they are incorrect.

Do the Lorentz transformations demonstrate the invariance of the speed of light? The equation describing the spherical propagation of a light wavefront from a point source as observed in the laboratory frame is

$$x^2 + y^2 + z^2 = c^2 t^2, \tag{10.35}$$

where x, y, z are the positions of the wavefront at the time t (as measured by the clocks in the laboratory Fig. 10.9). Using the Lorentz transformation to calculate the coordinates of the *same wavefront* as measured by an observer in the moving (primed) frame (substituting Eq. 10.34 into Eq. 10.35) gives

$$\frac{(x' + ut')^2}{1 - \dfrac{u^2}{c^2}} + y'^2 + z'^2 = c^2 \frac{\left(t' + \dfrac{ux'}{c^2}\right)^2}{1 - \dfrac{u^2}{c^2}}.$$

Squaring, canceling $\left(1 - \dfrac{u^2}{c^2}\right)$ on both sides, and regrouping yields

$$(x'^2 + y'^2 + z'^2) - \frac{u^2}{c^2}(x'^2 + y'^2 + z'^2) = t'^2(c^2 - u^2)$$

or

$$\left(1 - \frac{u^2}{c^2}\right)(x'^2 + y'^2 + x'^2) = c^2 t'^2 \left(1 - \frac{u^2}{c^2}\right)$$

or

$$x'^2 + y'^2 + z'^2 = c^2 t'^2, \tag{10.36}$$

which is the equation of an expanding spherical wavefront, as measured in the *primed* coordinates, whose radius is increasing with a velocity c. Both observers see the same wavefront expanding with the *same speed*—in agreement with the principle of special relativity.

For our application we are more interested in the velocity transformation equations. To obtain them from Galilean transformations, we merely differentiated the coordinate equations with respect to "any old time," but now we must be more careful. Time is *not* an invariant. When observer S' measures velocity, she means the rate of change of the position of the particle (as she measures it) with respect to *her* clock's time t' or dx'/dt', *not* dx'/dt. This is an important distinction. We can obtain the correct velocity transformation equations by taking the differentials of Eq. 10.34, such as

$$dx' = \frac{dx - u\,dt}{\left(1 - \dfrac{u^2}{c^2}\right)^{1/2}} \tag{10.37}$$

$$dt' = \frac{dt - \dfrac{u}{c^2}\,dx}{\left(1 - \dfrac{u^2}{c^2}\right)^{1/2}} \tag{10.38}$$

$$dy' = dy \quad \text{and} \quad dz' = dz. \tag{10.39}$$

Dividing Eq. 10.37 by Eq. 10.38, we obtain

$$\frac{dx'}{dt'} = \frac{dx - u\,dt}{dt - \dfrac{u}{c^2}\,dx} = \frac{\dfrac{dx}{dt} - u}{1 - \dfrac{u}{c^2}\dfrac{dx}{dt}}$$

or

$$v_{x'} = \frac{v_x - u}{1 - \dfrac{uv_x}{c^2}} \quad \text{or} \quad v_x = \frac{v_{x'} + u}{1 - \dfrac{uv_{x'}}{c^2}}. \tag{10.40a}$$

In a similar manner (see Problem 13) one can readily show that

$$v_y = \left(1 - \frac{u^2}{c^2}\right)^{1/2} \frac{v_y}{1 - \dfrac{uv_x}{c^2}} \tag{10.40b}$$

$$v_{z'} = \left(1 - \frac{u^2}{c^2}\right)^{1/2} \frac{v_z}{1 - \dfrac{uv_x}{c^2}}. \tag{10.40c}$$

Equations 10.40 are the correct velocity transformation equations, relating the velocities of a particle v_x (measured in the lab frame) to the velocity of the *same* particle as measured in the moving frame S', $v_{x'}$. Again we notice that when the relative velocities of the two reference frames u is small compared to c, these equations reduce to the Galilean transformations, as they must. Let's look at a few examples.

EXAMPLE 1   Show that the speed of the photon is the same in all inertial reference frames.

We will choose our coordinates so that the x-direction is the direction of the velocity of the photon, $v_{\text{photon}} = c\mathbf{i}$. From Eq. 10.40a,

$$v_{x'} = \frac{v_x - u}{1 - \dfrac{uv_x}{c^2}} = \frac{c - u}{1 - \dfrac{uc}{c^2}} = \frac{c - u}{\dfrac{c - u}{c}} = c.$$

It works!

EXAMPLE 2   Two rocket ships are traveling toward each other as shown in Fig. 10.10. The observer on the ground measures the speed of *each* ship to be $.95c$. What is the speed of rocket ship B as measured by the observer in rocket ship A?

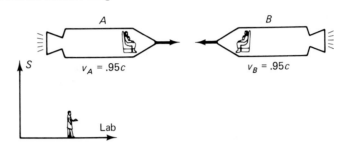

Fig. 10.10

Your quick answer might be $V_{\text{rel}} = .95c + .95c = 1.9c$, which is wrong! Nothing has ever been observed that travels faster than the speed of light. Nothing!

To answer this question correctly, we must carefully define the coordinate system. Let A be the observer in the primed reference frame S'. Then $u = .95c$. The speed of B that A would measure is

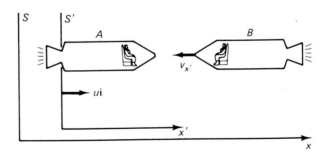

Fig. 10.11

then $v_{x'}(B)$. The speed of B as measured in the lab S is $v_x(B) = -.95c$. Using Eq. 10.44a gives

$$v_{x'}(B) = \frac{-.95c - .95c}{1 - \dfrac{.95c}{c^2}(-.95c)} = -\frac{1.9c}{1.9025c} = -.9987c.$$

Relativity gives unexpected results—but they are the correct observed ones!

EXAMPLE 3   Consider the so-called "uphill" decay. The problem is as follows: Can a particle of rest mass m_0 and an arbitrary amount of kinetic energy decay into a particle (or particles) of mass m, where $m > m_0$? That is, can one convert the kinetic energy of a particle into rest energy *in a free decay*?

We already answered this question by a straightforward application of conservation principles, but let us now approach the problem using an inertial coordinate transformation. Climb aboard your rocket ship and chase after the initial particle m_0 until you are traveling at the same velocity as the particle. Now look out the window. What you see is a particle of rest mass m_0 *at rest*. If the particle were suddenly to decay into one or more particles whose rest mass adds up to more than m_0, it would obviously violate energy conservation. Thus, from the perspective of the rocket ship, the event cannot occur. Since all inertial reference frames are equivalent, it cannot occur in any inertial frame, including the original lab frame.

EXAMPLE 4   Why can't a photon spontaneously decay into one or more nonzero-rest-mass particles?

Consider first that a single photon did decay into a material particle, $\gamma \rightarrow m$, in free space. One could always find a rocket ship that would travel at the speed of the particle m, and in this frame the particle could have zero momentum. But, of course, in this frame the photon is *still* moving at the velocity c and has momentum E/c. From this frame of reference, it would appear that a particle with initial *nonzero* momentum (the photon) transforms itself into a particle of rest mass m but no linear momentum. Even if the photon transforms itself into two or more nonzero-rest-mass particles, one can still find an inertial reference frame where the total linear momentum *after* the decay is zero. But the initial momentum of the photon can *never* be zero. Thus such decays cannot and have not been observed to take place.*

10.7 Relativistic Zero-Momentum Coordinate Frame

NOW THAT we have the relativistic velocity transformation equations, we can examine high-energy collisions in the center-of-mass frame of reference. The procedure is fundamentally the same as before. We look for the reference frame where the total linear momentum of the system of particles is identically zero. As you recall, the expression for relativistic linear momentum is

$$\mathbf{p} = \frac{m\mathbf{v}}{\left(1 - \dfrac{v^2}{c^2}\right)^{1/2}} = \gamma m\mathbf{v} \tag{5.13}$$

---

* It is worth noting that particle accelerators and the concept of elementary particles did not exist when Einstein proposed his special theory of relativity.

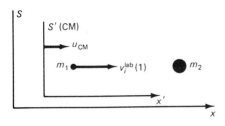

Fig. 10.12

Consider a one-dimensional example (Fig. 10.12) where a projectile of mass m_1 and velocity $\mathbf{v}_i(1)$ (as measured in the lab frame) strikes a stationary target of mass m_2. This problem can be transformed to the center-of-mass coordinate system by finding a moving frame S', where

$$\mathbf{P}_i^{CM}(1) + \mathbf{P}_i^{CM}(2) = 0.$$

But

$$\mathbf{P}_i^{CM}(1) = \frac{m_1 \mathbf{v}_i^{CM}(1)}{\left[1 - \dfrac{v_i^{CM}(1)^2}{c^2} \right]^{1/2}} \quad \text{and} \quad \mathbf{P}_i^{CM}(2) = \frac{m_2 \mathbf{v}_i^{CM}(2)}{\left[1 - \dfrac{v_i^{CM}(2)^2}{c^2} \right]^{1/2}}, \quad (10.41)$$

and to find $\mathbf{v}_i^{CM}(1)$ and $\mathbf{v}_i^{CM}(2)$ we need to use the velocity transformation

$$v_{x'} = \frac{v_x - u_{CM}}{1 - \dfrac{u_{CM} v_x}{c^2}}.$$

Relativistic transformations require more algebraic manipulations than the Galilean transformation, but for some problems it is worth the effort: for example, the class of problems concerning particle creation in a two-body collision with a stationary target. In the CM frame, all the kinetic energy of the two colliding particles can be

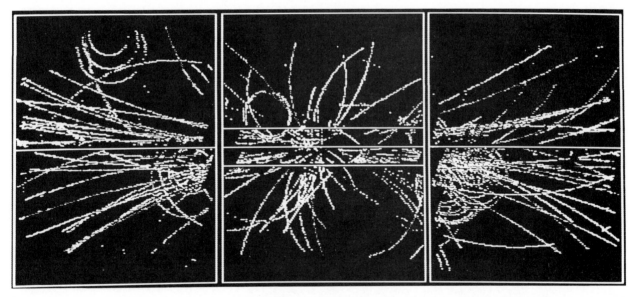

A proton–antiproton collision as observed in the center-of-mass coordinate system at CERN in the UA1 detector. In this case the CM frame and the lab frame are the same since this detector is placed at the intersection point of two colliding beams. The myriad of tracks all come from just two initial projectile particles colliding at the center of the detector.

transformed into rest mass of a new particle. All this kinetic energy is "available." But this is true *only* for the CM frame. In the lab frame all the initial kinetic energy is *not* available; some of it is locked up in the momentum of the system.* For this class of problems the CM coordinate transformation can help to find the solution. Since the initial and final magnitude of the momentum remains constant in the CM, we can use the same geometric construction techniques that we applied to the Newtonian collisions. However, you must remember to use the Lorentz equations when transforming back to the laboratory frame. Because the two transformation equations are different, elastic scattering of identical particles at high energies does not occur with a 90° relative scattering angle as it does in Newtonian collisions. In fact, it turns out to be less than 90°, with the angle decreasing as the kinetic energy of the particles increases.

10.8 Local Conservation

NOW THAT we have considered the idea of viewing collision processes from a variety of inertial reference frames, another question comes to mind. Imagine that the conservation principles means that a certain quantity, for example the electrical charge, will always be constant in a collision but may disappear from one place and reappear *at the same time* at another place. That is, the total charge in some region would indeed remain constant, but it would somehow move instantaneously from one place to another.

One might incorrectly guess that such nonlocal conservation would be possible because, after all, isn't the total charge conserved at all instants of time in the system? The reason that this is not "kosher" can be understood by considering further implications of the principle of relativity. That principle states that the laws of physics, including of course the conservation of electrical charge, must be true for *all* observers in *all* inertial reference frames. This means that two observers in two different reference frames must *both* agree that in any interaction charge is always conserved.

Consider two observers, A and B, as shown in Fig. 10.13. Observer A is in a fast rocket ship moving at a constant velocity **v** with respect to two events that occur at positions x_1 and x_2 which are separated by a distance L. Assume that at $t = 0$ a charge q disappears from position x_1 and at that same instant reappears at x_2. Suppose we have a device that emits a light signal when a charge either appears or disappears. Observer B, who is stationed exactly equidistant from x_1 and x_2, observes light signals at the same instant of time from x_1 and x_2, and confirms that although q has disappeared from x_1 it has simultaneously reappeared at x_2. Charge, he asserts, is conserved.

---

* The one exception is when the lab frame and the CM frame are the same.

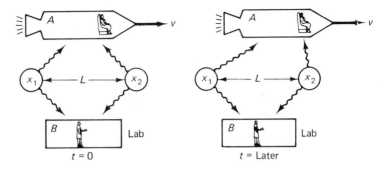

Fig. 10.13

Observer A strongly disagrees. She states quite emphatically that the charge was created at x_2 before it disappeared at x_1, since the light signal from x_2 reaches her first. Thus she says that for some period of time, charge is *not* conserved. The two observers cannot agree that charge is a conserved quantity. This, of course, violates the principle of relativity since *both* observers are in inertial reference frames. Thus *nonlocal* conservation laws and the principle of relativity are incompatible.

If the principle of relativity is indeed a law of nature, and all experimental evidence so far indicates that it is, *all* conservation principles *must be local*. That is, the quantity conserved must be conserved at the same point in space–time. This is the only way that *all* observers, in *all* inertial reference frames, can agree that the laws of nature hold in their reference frame. A careful examination of the Lorentz transformations reveals that two events separated in space will not appear simultaneous to observers in both S' and S.

10.9 Noninertial Reference Frames

SO FAR we have limited our discussion to inertial reference frames. They occupy a unique place in physics since they are all equivalent, and they are the frames in which the laws of physics are expressed. Another way of stating this is as follows:

> *One cannot perform any physics experiment within the confines of a given inertial reference frame and from that experiment determine the velocity of that frame with respect to any other inertial frame.*

But what about reference frames that are accelerating with respect to an inertial reference frame? We will examine two types of accelerating frames: (1) linearly accelerating and (2) rotating reference frames. The simpler of the two, linearly accelerating coordinate systems, is a good place to begin.

We already touched on this topic at the beginning of this chapter when we considered the four coordinate systems used to measure the free-falling ball near the surface of the earth. Equation 10.6d,

$$\frac{dv^D}{dt} = g - a_{\text{el}},$$

showed that the ball appears to the observer in the accelerating elevator to be falling with a reduced acceleration $(g - a_{\text{el}})$. Consider another example. The acceleration of an automobile (or any horizontal accelerating object) can be determined by measuring the angle subtended by a small mass suspended by a thread inside a closed vehicle (Fig. 10.14). (This was given as Problem 21 in Chapter 9; maybe you have already solved it.) How is θ related to the horizontal acceleration?

First we analyze the problem from the point of view of an inertial reference frame where Newton's laws are valid (this is *not* a relativistic car!). The free-body diagram, Fig. 10.15, shows the only two forces on the mass: a contact force **T**, the tension in the string, and the action-at-a-distance force $-mg\mathbf{j}$, the gravitational attraction of the earth. The inertial reference frame *x,y fixed to the ground* has its origin at the mass m at the instant the free-body diagram is "constructed." We will assume that the system has been accelerating for a long time, so that the ball is in equilibrium (no swinging), the mass m is accelerating at the same rate as the car. Newton's second law gives:

x-comp.: $T \sin \theta = ma_{\text{car}}$

y-comp.: $T \cos \theta - mg = 0$ or $T = \dfrac{mg}{\cos \theta}.$

Substitution in the above yields

$$\tan \theta = \frac{a_{\text{car}}}{g}. \tag{10.42}$$

There is another way to solve the same problem, looking at the system "from another point of view." But before we do that, let us consider the more general case of measuring a nonrelativistic system from a linearly *accelerating* coordinate reference frame. If S is an inertial frame where Newton's laws hold, and S' is a second frame that is accelerating with respect to S (Fig. 10.16), the position of the particle m in both frames is related by the vector equation

$$\mathbf{r}(t) = \mathbf{R}(t) + \mathbf{r}'(t). \tag{10.43}$$

By straightforward differentiation, for this nonrelativistic system,

$$\frac{d^2\mathbf{r}}{dt^2} = \frac{d^2\mathbf{R}}{dt^2} + \frac{d^2\mathbf{r}'}{dt^2}. \tag{10.44}$$

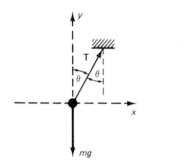

Fig. 10.14

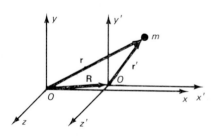

Fig. 10.15

Fig. 10.16

Newton's second law tells us that if a mass is subject to a net external force, it will accelerate:

$$\sum \mathbf{F} = m\,\frac{d^2\mathbf{r}}{dt^2} = m\left(\frac{d^2\mathbf{R}}{dt^2} + \frac{d^2\mathbf{r'}}{dt^2}\right)$$

or

$$\sum \mathbf{F} = m\mathbf{a}_{\text{ref}} + m\mathbf{a'}, \qquad (10.45)$$

where $\mathbf{a}_{\text{ref}}$ is the accelerating of the reference frame S' with respect to S and $\mathbf{a'}$ is the acceleration of m as measured by S'. Rearranging Eq. 10.45, we obtain

$$\sum \mathbf{F} - m\mathbf{a}_{\text{ref}} = m\mathbf{a'}. \qquad (10.46)$$

In this form one can think of the term $m\mathbf{a}_{\text{ref}}$ as a kind of *force*, a "pseudoforce," that exists by virtue of the acceleration of the reference frame in which we are observing the motion. Is it a force? The answer to that question depends on your philosophical point of view, and physicists don't agree. I do not consider it to be a force because it does *not* depend on contact or action-at-a-distance with any other body. It depends *only* on the reference frame. Yet if you are riding in an accelerating car and are holding a massive object, it may feel like a force is pushing on the object.

Some physicists believe that the introduction of pseudoforces simplifies the problems. Maybe so, but I warn you that it is easy to make mistakes using them. You should keep in mind the following:

> *The use of pseudoforce is* never *necessary to solve any problem in dynamics. All such problems can be solved using Newton's (or Einstein's) laws in inertial reference frames.*

Having said my piece about pseudoforces, we will go on to use them, because they are commonly used in textbooks, and for certain systems (mostly rotating coordinates) they can simplify the analyses.

Let us return to the accelerating car. If we draw a free-body diagram (Fig. 10.17) for a coordinate system *fixed on the car*, it must include an extra force, the pseudoforce $-ma_{\text{car}}$ and the equations of "motion" become

$$x'\text{-comp.:} \qquad T \sin\theta - ma_{\text{car}} = 0$$

$$y'\text{-comp.:} \qquad T \cos\theta - mg = 0.$$

Dividing these equations yields

$$\tan\theta = \frac{a_{\text{car}}}{g}.$$

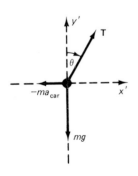

Fig. 10.17

Not terribly exciting! But there are important differences here. We are no longer analyzing the problem from an inertial reference frame where Newton's laws hold. Because of this, we are required to add a pseudoforce to the system. We have transformed this particular system so that it has become a *statics* problem, where the sum of the "forces" is equal to zero. In *this* accelerating frame of reference there is no motion of any kind.

The more interesting category of problems involves rotation. The rotating coordinate system is fundamentally an accelerating frame, even if the angular velocity ω is constant.* Let's jump on a merry-go-round and perform a few simple dynamics experiments. We might even imagine that the laboratory is enclosed, large, and rotating rather slowly, so we are not normally aware that we are on a rotating platform.

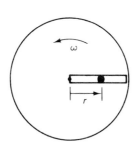

Fig. 10.18

For our first experiment we place a puck inside a smooth frictionless slot on the platform and let it go (Fig. 10.18). It might surprise the observer on the platform to see the puck accelerate away from him *as if* a force was acting on it. In fact, if the observer on the platform were to attach a spring scale to the puck (Fig. 10.19) and hold the other end stationary, he would observe a deflection of the spring, recording a force that depended linearly on the distance of the puck from the center of the platform. If the experimenter changed the angular velocity of the platform, he would discover that this "force" (called the "centrifugal force") depended quadratically on the angular velocity following the formula

$$F_{centrifugal} = m\omega^2 r. \qquad (10.47)$$

An observer, looking at this experiment from the inertial frame of the ground, analyzes the experiment differently. To this observer, the spring scale is applying the *only* horizontal force to the puck. This force causes the puck to be accelerated *inward* toward the center. She describes the experiment by the force:

$$F_{spring} = -m\omega^2 r.$$

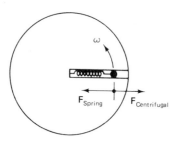

Fig. 10.19

The observer in the *noninertial* rotating frame says that there are two forces; $\mathbf{F}_{spring}$, acting radially inward, and $\mathbf{F}_{centrifugal}$, acting radially outward. The sum of the "forces" is zero and thus the puck remains at rest on the platform. Again the pseudoforce has reduced this dynamics problem to a *statics* problem in the noninertial frame.

We may now reexamine the conical pendulum (Example 3 in Section 9.6). Viewed from a coordinate system that rotates with the same angular velocity as the mass, the free-body diagram appears as Fig. 10.20 and the statics equations for this noninertial reference frame are:

$$x'\text{-comp.:} \qquad -T \sin\theta + m\omega^2 r = 0$$

$$y'\text{-comp.:} \qquad T \cos\theta - mg = 0.$$

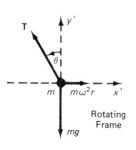

Fig. 10.20

---

* We will restrict our discussion to rotating references frames with constant angular velocities. The case of angularly accelerating coordinates is much more complicated.

The concept of "centrifugal force" is often incorrectly used. Consider a description of an artificial satellite orbiting Earth. An *incorrect* description of the motion of such an object states that the satellite does not fall to Earth because there is no net force on it: the centrifugal force and the gravitational force balance each other. Newton's first law tells us that if there is no net force on an object, the object moves in a straight line. But the satellite clearly does not do that; it orbits Earth. Only *in the rotating frame* can you invoke the pseudoforce. Indeed, in the frame of reference rotating with the satellite, the satellite is stationary. But viewed from an inertial frame, the satellite is executing circular or elliptical motion. Such common mistakes, and many others like it, are the reasons I have little stomach for this topic.

One last item for rotating coordinate systems is the Coriolis "force." This is another pseudoforce which is very useful in describing large-scale phenomena near the Earth's surface. Imagine yourself as a long-range artillery gunner in World War II stationed near the geographic north pole (Fig. 10.21). If you fired shells a few hundred miles over the north pole, you would discover that the trajectories of the shells act *as if* a side force, perpendicular to the velocity, deflected them from their original projected paths. To an observer watching from the north star, the shells move in straight lines, but the earth rotated underneath during their flight. An estimate of the magnitude of this deflection, and its functional dependence, can be obtained from an analysis of a simple special case.

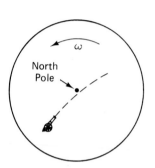

Fig. 10.21

Consider our flat rotating platform again (Fig. 10.22). A man stands at the center, draws his pistol and aims it at his potential victim a distance r away at the perimeter (high drama!). He pulls the trigger and the bullet travels in a straight line (as observed in an inertial frame) toward his victim, *but* the victim is rotating at an angular velocity ω and moves out of the way of the bullet. The rate of change of the angle θ is

$$\frac{d\theta}{dt} = \omega \qquad \text{but} \qquad S = r\theta.$$

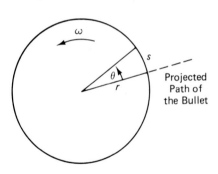

Projected
Path of
the Bullet

Fig. 10.22

The bullet, traveling with a velocity of v, reaches a distance r in a time t given by

$$t = \frac{r}{v} \qquad \text{or} \qquad r = tv.$$

In that time t, the target has rotated a distance S given by

$$S = r\theta = r\omega t = \omega v t^2,$$

which is mathematically the same as

$$S = \tfrac{1}{2}(2\omega v)t^2 = \tfrac{1}{2}a_{\text{cor}}t^2,$$

where

$$a_{\text{cor}} \equiv 2\omega v \qquad\qquad (10.48)$$

is called the Coriolis acceleration.

To the intended victim (who is glad that this effect exists), the bullet appears to be deflected by a side force, for he sees a curved trajectory of the traveling bullet. The fact that the deflection from the straight-line path obeys a $\frac{1}{2}at^2$ law indicates that this "force" is constant. The Coriolis "force" is obviously

$$F_{\text{Cor}} = ma_{\text{Cor}} = 2m\omega v. \qquad (10.49)$$

For the general case, the correct nonrelativistic expression for the Coriolis force is

$$\mathbf{F}_{\text{Cor}} = -2m(\boldsymbol{\omega} \times \mathbf{v}'), \qquad (10.50)$$

where $\mathbf{v}'$ is the velocity vector measured in the rotating reference frame and $\boldsymbol{\omega}$ is the angular velocity *vector*.*

The use of the Coriolis force is helpful in understanding complicated motion in rotating systems. On a spherical rotating surface, one must untangle both centrifugal and Coriolis effects. The

This satellite photograph of Hurricane Allen shows detail of the cloud structure as this hurricane filled much of the Gulf of Mexico. This photograph was taken on August 8, 1980, and has 1/2 mile visual resolution.

---

* The vector properties of the angular velocity are discussed in detail in Chapter 14.

weather patterns on the surface of the earth are good examples of this kind of motion. Hurricanes or cyclones can be understood by invoking these pseudoforces. A low-pressure center, usually located over a large expanse of water, causes the surrounding airmass to be driven radially inward toward the low-pressure center as shown in Fig. 10.23. The Coriolis force is responsible for the side deflection. It causes the storm in the northern hemisphere to rotate counterclockwise (as viewed from above) and to rotate clockwise in the southern hemisphere.

Fig. 10.23

You may have noticed one unusual feature of the pseudoforces; they are *all* proportional to the *mass* of the object on which they act. This is a universal characteristic of the pseudoforce. Can you think of any other force that is also proportional to mass? You probably have guessed correctly: gravitational force has the same property. Is it not possible that gravity is simply due to the fact that we do not have the "correct" reference frame? Einstein believed that there is *no* way an observer can distinguish between an accelerated coordinate system in distant outer space and a coordinate system at rest near a massive object like the Earth. That is, accelerated coordinate systems and gravitational fields *are equivalent*. This is called the principle of equivalence, and it is one of the cornerstones of the general theory of relativity. It is this theory that has so accurately predicted the bending of starlight as it passes near the sun, the advance of the perihelion of the planet Mercury, and the change in clock rate in varying gravitational fields. This is another fascinating subject you might some day wish to explore.

10.10 Where Are the Inertial Frames?

IF WE wish to test the theories of mechanics and relativity, we need an inertial reference frame. Where do we find one? You will all do experiments in the laboratory, but as you know, the lab is attached to spaceship Earth and is rotating! In fact, the centripetal acceleration of Earth due to its rotation about its geometric axis is about 3.4×10^{-2} m/s^2 at the equator. That is not all. Earth is also orbiting the sun, and that orbital motion is responsible for an acceleration of about 6×10^{-3} m/s^2 toward the sun. The sun is rotating about our galactic center (the Milky Way), producing an acceleration of 10^{-10} m/s^2 and there is good evidence that our galaxy is part of a cluster of galaxies (called the "local group") which are rotating about each other, and so on. We have not discovered an unaccelerated reference frame to which we can attach our inertial frame. We must then rely on our operational definition of an inertial frame—a local region of space–time where Newton's first law is obeyed by our test particles—as the only way to fix an inertial frame.

There is still considerable controversy and speculation over the existence of inertial properties of matter. Possibly the most original thinking on this subject came from the German scientist-philosopher Ernst Mach, who profoundly influenced Einstein. In 1872 he wrote:

> For me, only relative motions exist . . . and I can see, in this regard, no distinction between rotation and translation. Obviously it does not matter if we think of the earth as turning round on its axis, or at rest while the fixed stars revolve around it. . . . But if we think of the earth at rest and the fixed stars revolving around it, there is no flattening of the earth, no Foucault's experiment and so on—at least according to our usual conception of the law of inertia. Now one can solve the difficulty in two ways. Either all motion is absolute, or our law of inertia is wrongly expressed. . . . I prefer the second way. The law of inertia must be so conceived that exactly the same thing results from the second supposition as from the first. By this it will be evident that in its expression, regard must be paid to the masses of the universe.*

The idea that the inertial properties of all matter depend on the existence and distribution of other matter in the universe is called Mach's principle. Verification of this principle and observation of its consequences in modern cosmology remain an important but unresolved problem of modern physics.

10.11   Summary of Important Equations

Galilean transformations; nonrelativistic particles:

$$\mathbf{r} = \mathbf{r}' + ut\mathbf{i}$$

$$\mathbf{v} = \mathbf{v}' + u\mathbf{i}$$

$$\mathbf{a} = \mathbf{a}'$$

$$\mathbf{R}_{CM} = \frac{m_1\mathbf{r}^{lab}(1) + m_2\mathbf{r}^{lab}(2)}{m_1 + m_2} \quad \text{(two particles)}$$

$$\mathbf{u}_{CM} = \frac{m_1\mathbf{v}^{lab}(1) + m_2\mathbf{v}^{lab}(2)}{m_1 + m_2} \quad \text{(two particles)}$$

$$\mathbf{u}_{CM} = \frac{\mathbf{P}_T}{M}$$

$$\frac{1}{2}\sum_{j=1}^{m} m_j(v_j^{lab})^2 = \frac{1}{2}\sum_{j=1}^{m} m_j(v_j^{CM})^2 + \frac{1}{2}(u^{CM})^2\sum_{j=1}^{m} m_j$$

---

* E. Mach, *History and Root of the Principle of the Conservation of Energy*, 2nd ed. (Leipzig: Barth, 1909).

Lorentz transformations:

$$x' = \frac{x - ut}{\left(1 - \dfrac{u^2}{c^2}\right)^{1/2}} = \gamma(x - ut)$$

$$y' = y \qquad z = z'$$

$$t' = \frac{t - \dfrac{ux}{c^2}}{\left(1 - \dfrac{u^2}{c^2}\right)^{1/2}} = \gamma\left(t - \frac{ux}{c^2}\right)$$

$$v_{x'} = \frac{v_x - u}{1 - \dfrac{uv_x}{c^2}} \qquad v_x = \frac{v_{x'} + u}{1 + \dfrac{uv_{x'}}{c^2}}$$

Pseudoforces:

$$\sum \mathbf{F} - ma_{\text{ref}} = ma' \qquad F_{\text{centrifugal}} = m\omega^2 r \qquad F_{\text{Coriolis}} = 2m\omega v$$

10.12 References on Special Relativity

1. *Spacetime Physics*, E. F. Taylor and J. A. Wheeler, W. H. Freeman, San Francisco (1966).
2. *Special Relativity*, A. P. French, W. W. Norton, New York (1966).
3. *Einstein's Theory of Relativity*, M. Born, Dover, New York (1962).
4. *Relativity: The Special and the General Theory*, A. Einstein, Crown, New York (1961).

Chapter 10   QUESTIONS

1. Why doesn't the magnitude of the velocity change in a Newtonian elastic collision of two particles, as viewed in the CM frame?

2. How is it possible for the relativistic momentum of a particle in the x-direction to depend on the velocity of that same particle in the y-direction?

3. Give a simple example of how the total momentum of a system of particles is *not* an invariant quantity in all inertial reference frames.

4. Explain why the center-of-mass reference frame simplifies the analyses of a two-body collision.

5. Design an experiment to measure a violation of the "principle of relativity." What would you measure, and how would you determine if relativity had indeed been violated?

6. What would light look like if you could observe it from a rocketship moving at the speed of light? This is a question that young Einstein asked himself. Comment on this question.

7. Explain how you could use a simple pendulum and a stopwatch to get a good estimate of how far a plane traveled before it left the ground, assuming that you were a passenger in the plane and had a hand calculator.

Chapter 10 PROBLEMS

1. When a biker is 250 m from a cliff traveling at a velocity of 3.8 m/s, someone drops a ball directly down in the biker's path (Fig. 10.24).

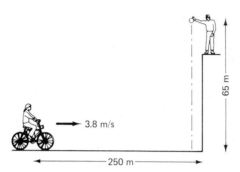

Fig. 10.24

(a) Find the expression for the position of the ball as a function of time as measured in the reference frame of the biker.

(b) Find the expression for velocity as a function of time as measured by the biker.

(c) What is the velocity of the ball (as measured by the biker) when the ball has fallen halfway down the cliff?

(d) Find the acceleration as measured by the biker.

2. Two particles collide and stick together after an inelastic collision. Particle A had a mass of 5.0 kg and moved with a velocity of $\mathbf{v}_a = 4\mathbf{i} + \mathbf{j} - 3\mathbf{k}$ m/s and particle B had a mass of 2.0 kg and a velocity $\mathbf{v}_b = 3i - 2\mathbf{j} + 5\mathbf{k}$ m/s.

(a) Find the velocity of the composite particle after they stick together.

(b) Find the kinetic energy of the particles in the CM frame before this collision.

3. A 1-kg puck is moving in the positive x-direction at a speed of 6 m/s when it collides with a 2-kg puck at rest. The 1-kg puck is observed moving off at 45° in the x–y plane at a speed of 2.828 m/s after the collision.

(a) Find the velocity of the 2-kg puck after the collision.

(b) In the CM frame, find the fraction of kinetic energy that was lost in the collision.

(c) In the CM frame, find the angle of deflection of the 1-kg puck.

4. Show that the conservation of linear momentum is true in all inertial reference frames using the Galilean transformation.

5. Using the Galilean transformation equations, show that the *change* in kinetic energy of a two-body collision is an invariant quantity in all inertial reference frames, regardless of the degree of inelasticity of the collision. That is, the ($KE_{final} - KE_{initial}$) of the system is the same quantity no matter what inertial reference frame the collision is viewed.

6. Two large clumps of intergallactic mud make an inelastic collision and stick together. Clump one has a mass of 489 kg and was moving with a velocity of $8.9 \times 10^3\mathbf{i} - 2.1 \times 10^3\mathbf{j}$ m/s before colliding with the second clump whose mass was 96 kg and velocity was $-3.0 \times 10^4\mathbf{i} + 1.6 \times 10^4\mathbf{j}$ m/s as measured in the laboratory frame.

(a) What is the velocity of the center of mass of the system?

(b) What is the final momentum of the combined masses in the laboratory frame?

(c) What is the final momentum in the center-of-mass frame?

(d) What fraction of the total kinetic energy is "lost" in the collision? Where did it go?

7. Two particles move in the x–y plane with velocities $\mathbf{v}_1 = 3.8\mathbf{i}$ m/s and $\mathbf{v}_2 = 4.5\mathbf{i} - 6.7\mathbf{j}$ m/s. The particles have masses $m_1 = 5.4$ kg and $m_2 = 1.6$ kg, respectively.

(a) Find the velocity of the center of mass.

(b) Find the total linear momentum of the two particles.

(c) Find the velocities of each particle from the point of view of an observer in the CM frame.

8. Show that the "available" kinetic energy in a two-body Newtonian collision is given by the expression

$$KE(available) = \tfrac{1}{2}\mu v_{rel}^2$$

where

$$\mu = \frac{m_1 m_2}{m_1 + m_2}$$

is called the reduced mass and

$$v_{Rel} \equiv v_2 - v_1$$

is the relative velocity of the particles.

9. Calculate the "available energy" in each of the following collisions.

 (a) A 20-MeV neutron strikes a stationary proton.

 (b) Two 20-MeV protons collide head-on.

 (c) A 1-MeV neutron hits a stationary carbon nucleus. (Assume the mass of the carbon nucleus to be 12 times the mass of the neutron.)

10. A puck of radius a and mass m slides over a frictionless surface with a velocity **v** and makes a glancing elastic collision with an identical puck that has been pinned rigidly to the horizontal surface.

 (a) Find the relationship between the scattering angle θ and the impact parameter b, as defined in Fig. 10.25. (*Hint:* You can greatly

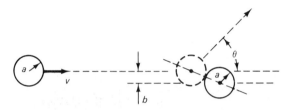

Fig. 10.25

simplify the problem by a simple coordinate transformation. Note that the only force on the projectile puck comes from the contact with the stationary target and that force acts along a line connecting the two pucks' centers. A stationary coordinate system that is oriented parallel to and perpendicular to the line connecting the centers is the natural system to analyze the collision. The initial momentum of the projectile puck then has two components. Remember, this is an elastic collision.)

 (b) What is the magnitude of the momentum change of the projectile puck as a function of impact parameter b?

11. Reanalyze Problem 10 for the case where the stationary target puck is free to slide along the frictionless surface. First calculate the velocity of the center of mass. Then solve the problem in the center-of-mass coordinate system and finally transform back into the laboratory frame to calculate the scattering angle as a function of impact parameters.

12. Prove the inverse velocity transformation equation,

$$v_x = \frac{v_{x'} + u}{1 + \dfrac{uv_x}{c^2}}.$$

13. Derive the velocity transformation for $v_{y'}$ and $v_{z'}$, in Eqs. 10.44b and 10.44c.

14. Two young physicists decide to check the concept of local conservation by experimenting with the idea of simultaneity. They arrange two flash bulbs on a very fast moving train (the world's first relativistic train: built in Japan, of course) at positions equidistant from the center at o (Fig. 10.26). One of the sci-

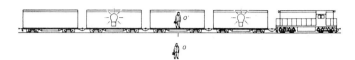

Fig. 10.26

entists boards the train and connects each bulb to an electronic trigger, so that each bulb will go off at exactly the same instant. He plans to stay on the train and verify that this has occurred by detecting the time difference (if any) between the two light signals he observes from the two flashes. His collaborator agrees to observe these two flashes as well, but decides to measure this time difference from a position right next to the train but on the platform, at rest, as the train goes by. The experiment is conducted, and the flashes go off in such a way that O and O' are next to each other at the time they both receive the flashes. Both of them observe *zero* time difference. Since they are near each other when the light arrives, when O measures no time *difference*, O' will also measure no time difference. They conclude from this experiment that both observers in different inertial reference frames *do* measure two separated events as simultaneous, and that local conservation is *not* required by special relativity. Their measurements are correct, but their conclusion is *wrong!* Reanalyze this experiment and show that if O and O' think about it carefully, they will conclude that the two events did *not* occur simultaneously.

15. A particle of mass m has linear momentum **p** and kinetic energy K in a reference frame S. Find the expressions for the linear momentum and kinetic energy of this particle in an inertial frame S$'$ whose velocity is $u\mathbf{i}$ with respect to the frame S. Do it for the general case where **p** has three components for a low velocity particle.

16. Two particles moving with the same *very high* speed approach each other on a collision course. The particles have different masses, as shown in Fig. 10.27.

Fig. 10.27

(a) Find the velocity of the center of mass of this system.

(b) From the point of view of an observer in this center-of-mass frame, find the velocities of each mass.

17. Consider the same two particles described in Problem 16 moving at nonrelativistic velocities. For this low-velocity system find the center-of-mass velocity as well as the velocities of each particle as measured from the center-of-mass frame.

18. Two identical particles of rest mass m are approaching each other with equal and opposite velocities v, as measured in the laboratory frame of reference (Fig. 10.28). Show that the *total* relativistic energy of particle 2, as measured by an observer at rest with respect to particle 1, is

$$E_T = \frac{c^2 + v^2}{c^2 - v^2}\, mc^2.$$

Fig. 10.28

19. Prove that for a single particle the quantity $(E_T^2 - p^2 c^2)$ is invariant in all inertial reference frames. That is,

$(E_T^2 - p^2 c^2)_{\text{lab}} = (E_T'^2 - p'^2 c^2)$ any other inertial frame,

where E_T is the kinetic plus rest energy of the particle and p is the linear momentum of the particle.

20. Calculate the threshold kinetic energy per particle (proton, antiproton) for the creation of the antineutron if the two colliding baryons make a head-on collision in a storage ring. The reaction is

$$p + \bar{p} \rightarrow p + \bar{p} + n + \bar{n}.$$

21. Consider the experimental problem of creation of a pion in a proton-proton collision using a single particle accelerator and a stationary target. The reaction is

$$p + p \rightarrow n + p + \pi^+.$$

(a) Calculate the minimum velocity of the proton in both the lab and center-of-mass frames of reference necessary for this reaction to go.

(*Hint:* To simplify the algebra, use 940 meV for the mass of both the proton and neutron. Solve the problem first in the center-of-mass frame, and use the velocity transformation to calculate the laboratory velocity of the projectile.)

(b) Calculate the minimum kinetic energy for the projectile proton necessary for this reaction to go.

22. An incident γ-ray strikes a proton at rest and creates a π^0 in the following reaction:

$$\gamma + p \rightarrow p + \pi^0.$$

Using the technique of transforming to the CM coordinate system, show that the threshold energy of the photon in the center-of-mass frame for creation of the π^0 is

$$E_\gamma^{\text{CM}} = m_{\pi^0} c^2\, \frac{m_p + \frac{1}{2} m_\pi}{m_p + m_\pi}.$$

23. An elevator accelerating upward at a rate of $g/4$ contains an ideal Atwood's machine with a 3-kg and a 2-kg weight (Fig. 10.29).

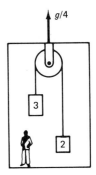

Fig. 10.29

(a) Calculate the acceleration of the 2-kg weight relative to an observer in the elevator.

(b) Calculate the acceleration of the 2-kg weight with respect to an inertial frame at rest on the ground.

(c) Calculate the force on the pulley exerted by the support that attaches it to the top of the elevator.

24. The frames of reference shown in Fig. 10.30 have their origins coincident at $t = 0$. (The displacement of the frames in the diagram is for artistic clarity only.) Frame S is the laboratory inertial frame and S' and S'' are noninertial frames accelerating in the x-direction, $a\mathbf{i}$. Frame S'' has an initial velocity of $v_0\mathbf{i}$ while S' starts from rest.

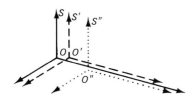

Fig. 10.30

(a) Write the expression that describes the position of O' and O'' with respect to O as a function of time.

(b) Write down the transformation equations that relates $\mathbf{r}$ to $\mathbf{r}'$, $\mathbf{r}$ to $\mathbf{r}''$, and $\mathbf{r}'$ to $\mathbf{r}''$.

(c) A single particle experiences a force $\mathbf{F}$ in the laboratory frame. What are the equations of motion of the particle as measured by an observer in the S' and S'' frames?

(d) Do any fictitious forces appear in any of the equations derived in part (c)?

(e) A particle moves with a velocity $\mathbf{v}'$ in S'. What is the velocity of this particle as measured by an observer in S''?

25. The driver of a long flatbed truck applies his brakes suddenly when he is moving at a speed of 25 m/s down a straight road. A 12-kg box at the rear of the flatbed begins to slide with a coefficient of friction of .12 along the bed (Fig. 10.31). If the truck is able to

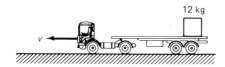

Fig. 10.31

stop in 125 m, and assuming that the deceleration of the truck is uniform:

(a) Find the acceleration of the box with respect to the truck bed during the period of time the truck is decelerating.

(b) Find the acceleration of the box with respect to the ground during the time the truck is decelerating. Will the box continue to slide after the truck has stopped? Explain.

(c) How far on the bed will the box slide?

(d) Find the velocity of the box with respect to the ground at the time the truck stops.

(e) What information given is unnecessary?

26. A man standing on a uniform linear accelerating horizontal platform wishes to throw a ball into the air in such a way that he can catch it without moving his feet. Neglecting all air resistance, find the angle at which the ball must be thrown so that this can be accomplished. Do the problem first using pseudo-forces and then again using inertial reference frames.

27. The ultracentrifuge is an important scientific tool used to separate objects of different densities in fluid media. Consider the problem of separation of a virus particle suspended in water. In principle the denser virus particle will drift to the bottom of the container under the influence of the gravitational field. This process can take an impractically long time, but it can be greatly sped up using "artificial gravity" created by spinning the sample of virus and water. The ultracentrifuge has the following characteristics:

Radius of rotation: 5 cm
Rotation speed: 1.6×10^3 revolutions per second

The virus has a mass of 7×10^{-19} kg and a diameter of 1.1×10^{-7} m.

(a) What is the effective g from the point of view of an observer rotating with the centrifuge? What is the net centrifugal force, including the effects of gravity?

(b) The motion of the virus is complicated by several factors. The virus experiences a viscous drag force, described in Section 8.7, Example 2. Calculate the terminal velocity of the virus with respect to its container inside the centrifuge when it is rotating at its maximum rate.

28. A man swings a bucket of water in a circle in the x–z plane with uniform angular velocity ω on two different accelerating platforms: (1) on an elevator with upward vertical acceleration $a_{el}\mathbf{j}$, and (2) on a horizontally accelerating platform $a_b\mathbf{i}$. Find the angular velocity necessary in *each* case to swing the bucket (at a radius r) so that no water will fall out.

11

WORK AND POTENTIAL ENERGY

11.1 Introduction

AT THE very beginning of this book I introduced the concept of energy as a fundamental "organizing principle," necessary to unravel and predict complicated particle collisions, interactions, and decays. You were given the mathematical expression for the kinetic and mass energy of a particle at that time, and by now you are all "experts" at using these expressions, with the appropriate conservation laws, to solve many kinds of problems. Later we introduced the basic principles of particle dynamics, the rules that describe the time evolution of position, velocity, and acceleration of particles subjected to known forces. There *appear* to be two distinct and separate ways to describe the motion of mechanical systems. Are they related in any way?

One would certainly hope so! In this chapter as we explore the connection, we will "discover" another form of energy, potential energy, and develop new methods of describing particle motion. These new methods are particularly useful if one is *not* concerned about knowing the time variations of the dynamical variables, such as position or velocity, but is satisfied with predicting, for example, the velocity of a particle at a certain position, whenever it gets there. We deal principally with Newtonian systems in this chapter, but in Section 11.6 the expression for the kinetic energy of high-velocity particles will be derived from the relativistic equations of motion. But enough of these generalities. Let's begin.

11.2 Infinitesimal Work

PHYSICISTS have the bad habit of using common words, with well-known meanings in ordinary speech to express quite different meanings in the physics lexicon. We certainly witnessed this in the discussion of particle physics, when the names color, charm, strangeness, and so on, appeared. The quantity "work" has a precise meaning in physics, which is related to, but by no means the same as, the meaning of the word in ordinary language. Please keep that in mind as you apply this idea in mechanics.

An infinitesimal amount of WORK dW done by an applied force **F** to a rigid body is defined by the expression

$$dW \equiv \mathbf{F} \cdot d\mathbf{s}, \qquad (11.1)$$

where $d\mathbf{s}$ is the infinitesimal displacement of the rigid body. No work is done on a body if the body is not displaced, regardless of the magnitude of the force. Since work is defined as a dot product of two vectors, force and displacement, it is a scalar quantity that depends on the cosine of the angle between the force and the displacement. If more than one force acts on the body, one can still calculate the work done by each force separately, or by the resultant of all the forces.

EXAMPLE 1

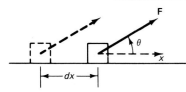

Fig. 11.1

A constant force **F** pulls on a rigid block and displaces it along the x-axis an infinitesimal distance dx as shown in Fig. 11.1. Since the force acts at an angle θ with respect to the direction of the displacement dx, the work done during this displacement is

$$dW = \mathbf{F} \cdot d\mathbf{s} = F(\cos \theta) \, dx. \qquad (11.2)$$

If **F** is measured in Newtons and dx in meters, work is measured in the SI units of joules or

$$1 \text{ joule} = 1 \text{ newton-meter} = 1 \text{ kg-m}^2/\text{s}^2. \qquad (11.3)$$

EXAMPLE 2

A simple pendulum consists of a mass m suspended on a rigid massless rod of length l, which is pivoted by a frictionless bearing at the top end. How much work is done by the gravitational force $-mg\mathbf{j}$ as the pendulum moves an infinitesimal distance $d\mathbf{s}$ along the arc?

In this example the body in question moves in a two-dimensional plane. The angle between the force of gravity $-mg\mathbf{j}$ and the displacement $d\mathbf{s}$ is $(\pi/2 - \theta)$, as shown in Fig. 11.2. Thus from our definition, Eq. 11.1, the infinitesimal work done by gravity on the pendulum is

$$dW = mg \cos \left(\frac{\pi}{2} - \theta \right) ds = mg \sin \theta \, ds. \qquad (11.4)$$

However, the magnitude of the displacement $d\mathbf{s}$ is related to the infinitesimal y-displacement dy (Fig. 11.3) by

$$-dy = \sin \theta \, ds, \qquad (11.5)$$

and thus

$$dW = -mg \, dy. \qquad (11.6)$$

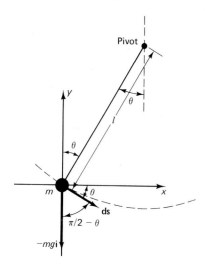

Fig. 11.2

This result can also be derived by decomposing the displacement ds into its x and y components:

$$d\mathbf{s} = ds \cos \theta \, \mathbf{i} - ds \sin \theta \, \mathbf{j} \qquad (11.7)$$

and substituting into the definition, Eq. 11.1, gives us

$$dW = -mg\mathbf{j} \cdot (ds \cos \theta \, \mathbf{i} - ds \sin \theta \mathbf{j})$$
$$= mg \sin \theta \, ds = -mg \, dy.$$

The work done by the gravitational force is positive, since the force acts in the $(-y)$-direction, and the displacement $(-dy)$ is also in the negative y-direction.

> *The work done is positive when the force and the displacement are in the same direction and the work done is negative when the force and displacement are in the opposite direction.*

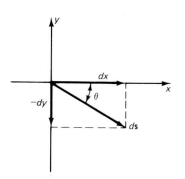

Fig. 11.3

If a force $\mathbf{F}$ acts on a body at right angles to its displacement $d\mathbf{s}$, no work is done on that body by that force. This is apparent from the definition of work. For example, examine carefully the forces on the tethered ball, which forms a conical pendulum (Fig. 11.4; see also Fig. 9.28). The tension exerted by the string $\mathbf{T}$ *always* acts perpendicular to the ball's displacement, as it revolves in the horizontal plane. Thus the tension does no work on the ball. What about the gravitational force mg? It acts perpendicular to the displacement $d\mathbf{s}$, so it also does no work on the ball. In this example, all the external forces act perpendicular to the displacement and do no work.

If a block slides down an inclined plane, as shown in Fig. 11.5, the plane exerts a normal force $\mathbf{N}$ on the block. At all points on the plane this normal force is perpendicular to the block's displacement and therefore does no work on the block. The tethered ball and the sliding block are just two examples of constrained motion. In each case the forces that are responsible for constraining the motion to a predetermined path act perpendicular to the displacement and do no work on the body. These two examples are not special cases. They represent the following general principle:

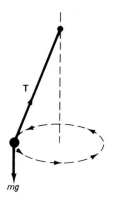

Fig. 11.4

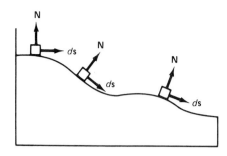

Fig. 11.5

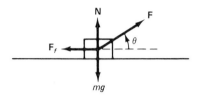

Fig. 11.6

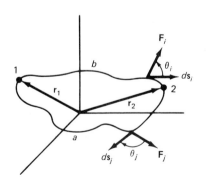

Fig. 11.7

> *No work is done on a body by those forces that constrain the body to move in a predetermined path.*

These examples should make it clear that although a force may act on a body, it does not necessarily do work on that body. It should also be apparent from the definition that work can be both positive and negative. Consider the "exciting" example of a block pulled by a force **F** over a rough horizontal plane (Fig. 11.6). In this case friction acts to oppose the displacement. The work done by the frictional force alone is

$$dW \text{ (friction)} = -F_f \, dx. \qquad (11.8)$$

The total work done by all forces on this block is

$$dW \text{ (total)} = (F \cos \theta - F_f) \, dx. \qquad (11.9)$$

Because the gravitational force and the normal (constraint) force act perpendicular to the motion, they do no work on the system. If, in this particular example, the block were to move with a constant velocity, the resultant force must be equal to zero. For the x-component of forces, that requires

$$F \cos \theta - F_f = 0, \qquad (11.10)$$

which implies that $dW = 0$. There is zero *net* work done on the block.

11.3 **Finite Work and Path Integrals**

TO OBTAIN the total work done when a body is moved from point 1 to point 2 (Fig. 11.7), all the infinitesimal contributions dW must be added up. That is,

$$W_{1 \to 2} = \sum_i dW_i = \sum_i \mathbf{F}_i \cdot d\mathbf{s}_i. \qquad (11.11)$$

This is a kind of mathematical summation which you probably have not seen before. Formally, Eq. 11.11 requires that one add up the dot product of the force and the infinitesimal displacement at *each* position along the path, starting at point 1 and going to end point 2. In general, this summation will depend on what path is chosen, since the displacements and force will generally be different along different paths. For example, the summation carried out for the particle moving from 1 to 2 along path (a) will be different from the summation of the infinitesimal work between the same two end points along path (b). In the limit, where the finite sum becomes an infinite sum, Eq. 11.11 becomes the integral,

$$W_{12} = \oint_c \mathbf{F} \cdot d\mathbf{s},$$

(11.12)

where this integral $\int$, with the inserted c (for contour), is called the line integral.*

One way to get a handle on this kind of integral is to represent it graphically for a one-dimensional case where the force depends on position. If we plot the spatially varying force $F(x)$ as a function of x (Fig. 11.8) and divide the interval between x_1 and x_2 into n small intervals, the approximate total work done is given by

$$W_{1\to2} \approx \sum_{n=1}^{N} F(x_n) \, \Delta x_n,$$

(11.13)

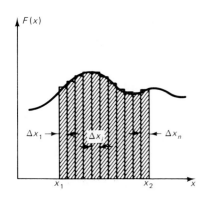

Fig. 11.8

which is the shaded area between x_1 and x_2. If we make the displacements Δx_n smaller and smaller, the force will approach a constant value over the interval, and in the limit of $N \to \infty$ and $\Delta x \to 0$, the sum becomes a definite integral. The relationship is no longer approximate, and we write

$$W_{1\to2} = \lim_{n\to\infty} \sum_{n=1}^{N} F(x_n) \, \Delta x_n = \oint_{x_1}^{x_2} F(x) \, dx,$$

(11.14)

which is the one-dimensional case of Eq. 11.12. The equivalence of the infinite sum and the definite integral is called "the fundamental theorem of calculus," discovered independently by Isaac Newton and Gottfried Leibniz (a German mathematician).

EXAMPLE 1   As is our custom, we begin to explain this new concept with a simple one-dimensional example. The block sliding across a horizontal plane, with a constant force acting on it (Fig. 11.1) is a good example. If the force is 5 N at an angle of 30°, and the block slides from the origin to a distance $x = 2$ m, how much work is done on the block?

$$W = \oint_{x_1}^{x_2} F \cos\theta \, dx = F \cos\theta \int_{x_1}^{x_2} dx = F \cos\theta \, (x_2 - x_1).$$

(11.15)

Substituting in the particular values for this problem, we obtain

$$W = 5 \text{ N}(.866)(2 \text{ m} - 0) = 8.66 \text{ J}.$$

———————————

* In some books it is referred to as the path integral.

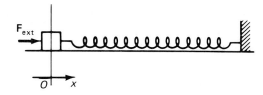

Fig. 11.9

EXAMPLE 2   Calculate the work done by the external force $\mathbf{F}_{ext}$ in compressing the spring from its equilibrium position $x = 0$ to some arbitrary position x (Fig. 11.9).

To solve this problem you must first know the behavior of the spring. An excellent empirical relationship exists between the compression (or the elongation) from the equilibrium position of a "good" spring, and the force exerted by the spring. As long as the spring does not exceed its elastic limits (i.e., stretched or compressed so far that it becomes permanently distorted), the spring force is given by the expression

$$F_{spring} = -kx, \tag{11.16}$$

where k is a constant, known as the spring constant or stiffness coefficient, and x is the extension or compression *from the equilibrium position*. This expression is known as Hooke's law. The minus sign is needed since the spring force *always* acts in the opposite direction to its displacement from equilibrium.

The total work done by the *external force* in compressing a spring a distance x is calculated using Eq. 11.12 as

$$W_{0 \to x} = \oint_0^x \mathbf{F}_{ext} \cdot d\mathbf{s} = \int_0^x kx\, dx = k \int_0^x x\, dx \tag{11.17}$$

$$W_{0 \to x} = \tfrac{1}{2}kx^2. \tag{11.18}$$

The external force is equal to, but opposite the spring force; $F_{ext} = -F_{spring} = kx$. The work we calculated is a positive quantity because the force and the displacement are in the same direction.

This is an example of a spatially varying force in a one-dimensional system. For this example we can easily evaluate the integral graphically. A plot of the external force versus x yields a straight line (Fig. 11.10). The shaded triangular area is

$$\text{area} = \text{work} = \tfrac{1}{2}(kx)(x) = \tfrac{1}{2}kx^2,$$

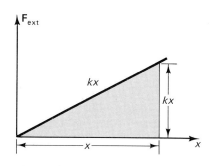

Fig. 11.10

exactly as we had calculated using integration. You should keep in mind that we have calculated the work done *by* the external force *on* the spring. The work done by the spring on the person pushing it is the negative of this.

EXAMPLE 3   Calculate the work done on a 5000-kg rocket by the gravitational force if it is launched with a sufficiently large initial velocity in the vertical direction, so that it will ultimately escape that gravitational pull of the Earth and drift into the galaxy (Fig. 11.11).

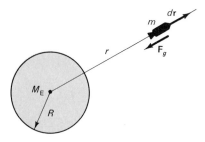

Fig. 11.11

This is another important example of a variable force in a one-dimensional problem. The gravitational force between the rocket of mass m and the Earth M_E is given by Newton's universal gravitational law,

$$\mathbf{F}_g = -G\,\frac{M_E m}{r^2}\,\hat{\mathbf{r}}. \qquad (11.19)$$

The total work *on* the rocket as it moves from the Earth's surface out into the galaxy is

$$W_{R\to\infty} = \int_R^\infty \mathbf{F}_g \cdot d\mathbf{r} = -GM_E m \int_R^\infty \frac{dr}{r^2} \qquad (11.20)$$

$$= GM_E m \left[\frac{1}{r}\right]_R^\infty = -\frac{GM_E m}{R}. \qquad (11.21)$$

Rather than looking up the mass of the Earth, we note that near the Earth's surface the force of gravity is mg, so

$$F_g = G\,\frac{M_E m}{R^2} = mg \quad \text{or} \quad g = \frac{GM_E}{R^2} \quad \text{or} \quad \frac{GM_E}{R} = Rg. \quad (11.22)$$

Substituting into Eq. 11.21 yields

$$W_{R\to\infty} = -gRm. \qquad (11.23)$$

The constants in Eq. 11.23 are $g = 9.8$ m/s^2, $R = 6.4 \times 10^6$ m, and $m = 5 \times 10^3$ kg. Substituting them in yields

$$W_{R\to\infty} = -3.13 \times 10^{11} \text{ J}.$$

The minus sign is correct, since the force of gravity and the displacement of the rocket are in opposite directions.

11.4 Work–Energy Theorem

OF WHAT use is this new quantity, work? Its usefulness can be best understood by studying a particular example of one-dimensional motion: a body with a constant *net* force acting on it. Consider the glider on a frictionless air track with a constant force Fi acting on it, as shown in Fig. 11.12. The gravitational force and the normal force

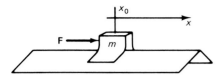

Fig. 11.12

on the glider are equal but opposite and thus Fi is the net external force on the glider. Since the net force acts in the direction of the motion, then

$$dW = \mathbf{F}_{net} \cdot d\mathbf{s} = F\,dx$$

and if the force acts on the glider as it travels from point x_0 to x, the total work done by this force is

$$W_{x_0 \to x} = \int_{x_0}^{x} F\,dx = F \int_{x_0}^{x} dx = F(x - x_0). \qquad \text{Trivial!}$$

But we studied the dynamics of a body that is subject to a constant net force back in Chapter 8. In fact, Eq. 8.33c is a relationship between the velocity of a particle of mass m and the distance traveled for a given constant force:

$$v_x^2 = v_{0x}^2 + 2\left(\frac{F_x}{m}\right)(x - x_0). \qquad (8.33c)$$

Rearranging this equation by multiplying by m, dividing by 2, and transposing one term yields

$$\tfrac{1}{2}mv_x^2 - \tfrac{1}{2}mv_{x_0}^2 = F(x - x_0). \qquad (11.24)$$

Now look at Eq. 11.24. The left-hand side is the Newtonian expression for the change in kinetic energy (i.e., the kinetic energy at x minus the initial kinetic energy at x_0), and the right-hand side is the work done by the external force in moving the glider from x_0 to x. We can write Eq. 11.24 in symbolic form:

$$\Delta KE = W. \qquad (11.25)$$

What happens to the glider if the external force $\mathbf{F}$ is removed at the point x? Newton's first law tells us that the glider will continue to move down the air track in the positive x-direction with the same velocity that it had at point x. As long as no net external force acts on the glider, its velocity will remain constant, as will its kinetic energy. With no net external force on the glider, no work is done on it, and once again

$$\Delta KE = 0 = W,$$

so Eq. 11.25 still holds.

After the glider has drifted for some time, imagine that the original force is again applied. This time, however, its direction is reversed ($-\mathbf{F}_{ext}$), and the glider begins a deceleration (Fig. 11.23). Equation 11.24 is still appropriate to describe the motion, as long as we reverse the sign of $\mathbf{F}$. It tells us that the work done *on* the glider is negative, but *so is the change in kinetic energy*. Once again the change in kinetic energy of the glider is equal to the work done on it

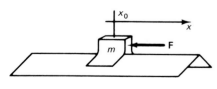

Fig. 11.13

by the net external force. As you suspect by now, this is a general relationship. So far we have only demonstrated that it holds for a one-dimensional constant-force system.

Equation 11.25, the so-called WORK–ENERGY theorem, can be derived by integrating Newton's second law of motion. For the class of problems that can be treated as one-dimensional systems, we write Newton's second law as

$$F(x) = m \frac{dv(x)}{dt}, \tag{11.26}$$

where the net external force is explicitly written as a function of the position and the mass is assumed to be constant. Formally, we integrate Eq. (11.26) along a one-dimensional path from x_1 to x,

$$\oint_{x_1}^{x} F(x)\, dx = m \oint_{x_1}^{x} \frac{dv(x)}{dt}\, dx, \tag{11.27}$$

where the left side of Eq. 11.27 is work done on the mass m by the net force $F(x)$ in moving from x_1 to x. Evaluation of this integral requires knowledge of both the functional dependence of the force with distance and the path of the particle's motion. The right side of Eq. 11.27 can be integrated by a mathematical trick that converts this integral, which includes position, velocity, and time variables, to a single variable, velocity. To change variables we note that

$$dx = \frac{dx}{dt}\, dt = v(x)\, dt. \tag{11.28}$$

Substituting Eq. 11.28 into Eq. 11.27 gives

$$m \oint_{x_1}^{x} \frac{dv(x)}{dt}\, dx = m \oint_{x_1}^{x} \frac{dv(x)}{dt}\, v(x)\, dt$$

$$= m \oint_{v_1}^{v} v(x)\, d(v(x))$$

$$= m \oint_{v_1}^{v} \tfrac{1}{2} d(v(x)^2) = \tfrac{1}{2} m v^2 \Big|_{v_1}^{v}$$

$$= \tfrac{1}{2} m v^2 - \tfrac{1}{2} m v_1^2. \tag{11.29}$$

Finally,

$$\oint_{x_1}^{x} F(x)\, dx = \tfrac{1}{2} m v(x)^2 - \tfrac{1}{2} m v_1(x)^2. \tag{11.30}$$

Equation 11.30 proves that the work–energy theorem holds for one-dimensional systems *even if the external force depends on position.*

It is absolutely essential to remember that $F(x)$ is the *net force* on the particle of interest.

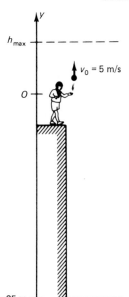

Fig. 11.14

EXAMPLE 1   A boy stands on a cliff 25 m above the ground. He tosses a ball vertically in the air with an initial velocity of 5 m/s. Neglecting air resistance:

    (a) How high will it go?
    (b) What is its velocity when it passes the boy on the way down?
    (c) What is its velocity when it hits the ground?

    Obviously, this problem could be solved using the kinematic equations for one-dimensional constant force motion, developed in Chapter 8. But now I want to demonstrate the applicability of the work–energy theorem. Since all the action takes place near the surface of the Earth, the force on the ball can be considered to be a constant mg in the negative y-direction. To answer part (a), we first evaluate the work done on the ball by gravity in going from its initial position (here defined as $y = 0$) to its maximum height h_{max}. This is

$$W_{0 \to h} = \int_0^{h_{max}} \mathbf{F} \cdot ds = \int_0^{h_{max}} -mg \, dy = -mg \int_0^{h_{max}} dy = -mgh_{max}.$$

By the work–energy theorem, this work must equal the change in the kinetic energy of the ball,

$$-mgh_{max} = \tfrac{1}{2}mv_f^2 - \tfrac{1}{2}mv_i^2,$$

where v_f is the final speed of the ball. But when the ball reaches its maximum height, its speed is zero, so this expression reduces to

$$-mgh_{max} = -\tfrac{1}{2}mv_i^2.$$

The mass cancels out. Solving for the maximum height yields

$$h_{max} = \frac{v_i^2}{2g} = \frac{(5 \text{ m/s})^2}{2 \times 9.8 \text{ m/s}^2} = 1.28 \text{ m}.$$

When the ball passes the boy on the way down, the net work on the ball by the gravitational force is zero. Gravity does $-mgh_{max}$ on the way up and $+mgh_{max}$ on the way down. If the net work on the ball is zero, there is no change in the kinetic energy and no change in the speed of the ball. Only the direction of the velocity changes.

    Finally, we can easily calculate the velocity of the ball when it strikes the ground. The work done on the ball in moving from $y = 0$ to $y = -25$ meters is

$$W_{0-25} = \int_0^{-25} -mg \, dv$$

$$= 25 \, mg \quad \text{(positive because the force and displacement are in the same direction)}.$$

This work must be equal to the change in kinetic energy, or

$$25mg = \tfrac{1}{2}mv_f^2 - \tfrac{1}{2}m(5)^2$$

$$v_f = [2(25 \text{ m} \times 9.8 \text{ m/s}^2 + 25 \text{ m}^2/\text{s}^2)]^{1/2} = 23.23 \text{ m/s}.$$

The mass of the ball does not appear in the answer because of the nature of the gravitational force.

EXAMPLE 2   Calculate the initial velocity needed to propel a 5000-kg rocket into outer space, free from the gravitational attractive force of the Earth. In Example 3 in Section 11.3 we calculated the work done by the gravitational force of the Earth on this rocket and found it equal to $-gR_Em$ (Eq. 11.23). Now we must equate this amount of work to the change in kinetic energy of the rocket. If we assume that the rocket has zero (or very small) speed at extremely large distances, and that it was launched with an initial velocity v_i, the work–energy relationship gives

$$-gR_Em = 0 - \tfrac{1}{2}mv_i^2$$

$$v_i = (2gR_E)^{1/2} \quad \text{(independent of mass)}$$

$$= (2 \times .98 \text{ m/s} \times 6.4 \times 10^6 \text{ m})^{1/2} = 11.2 \times 10^3 \text{ m/s}.$$

This is called the "escape velocity." This velocity, given to any object near the surface of the Earth, will be sufficient to allow the object to escape the Earth's gravitational field and travel into the solar system (neglecting air resistance). The escape velocities for various planets and the moon are given in Table 11.1.

TABLE 11.1

	Radius (m)	Local Gravitational Acceleration (m/s²)	Escape Velocity (m/s)
Moon	1.74×10^6	1.62	2.3×10^3
Mercury	2.44×10^6	3.76	4.3×10^3
Venus	6.05×10^6	8.88	10.3×10^3
Mars	3.39×10^6	3.73	5.0×10^3
Jupiter	69.5×10^6	26.2	60×10^3
Saturn	58.1×10^6	11.2	36×10^3
Uranus	24.5×10^6	9.75	22×10^3
Neptune	24.6×10^6	11.34	24×10^3

EXAMPLE 3   For Example 2 of Section 11.3, find the final velocity of the block of mass m after the spring has been compressed a distance x_f from its equilibrium position.

This is almost a trick question. Your first impulse might be to equate the work done ($W = \tfrac{1}{2}kx_f^2$) by the external force $\mathbf{F}_{\text{ext}}$ in moving the block a distance x_f to the final kinetic energy of the block. This is wrong! You should recall that the work–energy theorem can only be applied when you calculate the work done on a body by the net

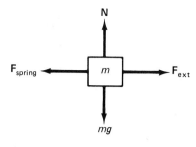

N

F_{spring} ← $\boxed{m}$ → F_{ext}

mg

Fig. 11.15

external force. Examine the free-body diagram, Fig. 11.15. The spring exerts a force equal in magnitude but opposite in direction to the external force $\mathbf{F}_{ext}$. The normal force is equal in magnitude to the gravitational force, so the net force on the block is zero! No net work is done on the block and there is no change in its kinetic energy. The block remains at rest. Nonsense!

The block obviously moves when the spring is compressed, so our analysis has a flaw. But the flaw is not as serious as you might suspect. The block does momentarily accelerate from rest and attain some small velocity. The external force does a small amount of work for a short time, but then the block continues to move at the same small velocity as the spring compresses. Our calculation is only slightly incorrect. (*Warning:* You must be careful when you use this theorem!)

If after the spring has been compressed a distance x_f and the block brought to rest, you release it by removing F_{ext}, the block will accelerate in the minus x-direction. The work–energy theorem can be used to calculate the velocity of the block when it reaches the origin:

$$\tfrac{1}{2}kx_f^2 = \tfrac{1}{2}mv_f^2$$

or

$$v_f = \left(\frac{kx_f^2}{m}\right)^{1/2}.$$

In this case the spring force *is the net force* on the block. The work done by the spring force over a distance x_f is the same as the work done by the external force in compressing the spring.

EXAMPLE 4

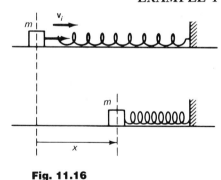

Fig. 11.16

A block of mass m and velocity $\mathbf{v}_i$ slides along a frictionless surface (Fig. 11.16). If it collides with an ideal massless spring of spring constant k, how far will it compress the spring?

The maximum compression will occur when the block has come to rest and converted all its kinetic energy into work on the spring. From the work–energy theorem,

$$\oint \mathbf{F} \cdot d\mathbf{s} = \tfrac{1}{2}kx^2 = \tfrac{1}{2}mv_i^2, \tag{11.31}$$

where x is the distance the spring is compressed. Solving for x, we obtain

$$x = v_i \left(\frac{m}{k}\right)^{1/2}.$$

Throughout our discussion we have assumed, without explicitly stating it, that the masses are in fact ideal rigid bodies with no internal motion. Remember that the work–energy theorem was derived from a straightforward integration of Newton's second law. This law, as written, is valid for objects that can be treated as point masses with *no internal motions* such as rotations or vibrations. We have not included in our analysis the possibility that the object can have internal energy. Thus we are once again dealing with "special effects," since real objects clearly do have internal energy.

11.5 Generalization to Three Dimensions

SO FAR we have only derived the work–energy theorem for a one-dimensional world. As you might guess, the theorem has a broader applicability. To prove this, we begin with Newton's second law for constant mass m:

$$\mathbf{F}(\mathbf{r}) = m \frac{d\mathbf{v}(\mathbf{r})}{dt} \tag{11.32}$$

and integrate it along the path the particle travels.

$$\oint_{r_1}^{r_2} \mathbf{F}(\mathbf{r}) \cdot d\mathbf{r} = m \oint_{r_1}^{r_2} \frac{d}{dt} \mathbf{v}(\mathbf{r}) \cdot d\mathbf{r}. \tag{11.33}$$

The left side of Eq. 11.33 is the work done by the net force on the particle as it traverses the path. This integral should be thought of as the summation of the dot products of the force at each point $\mathbf{r}_i$ with the displacement of $d\mathbf{r}_i$ along the path (Fig. 11.17), or

$$\oint_{r_1}^{r_2} \mathbf{F}(r) \cdot d\mathbf{r} = \sum_i \mathbf{F}_i(\mathbf{r}_i) \cdot d\mathbf{r}_i = \sum_i F_i(r_i) \, dr_i \cos \theta_i. \tag{11.34}$$

This is not a difficult mathematical operation to conceptualize, but to carry out some line integrals requires sophisticated techniques.

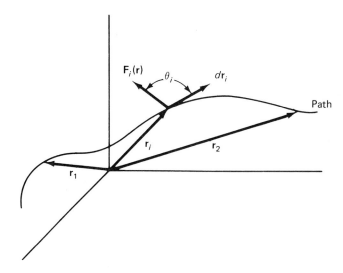

Fig. 11.17

What about the right side of Eq. 11.33? We can use the same mathematical trick we used before for one dimension, but now we must use three-dimensional vectors to execute the variable change. Consider the vector differential $d\mathbf{r}$, which can be written

$$d\mathbf{r} = \frac{d\mathbf{r}}{dt}\, dt = \mathbf{v}(\mathbf{r})\, dt. \qquad (11.35)$$

Substituting Eq. 11.35 into the right side of Eq. 11.33 yields

$$m \oint_{r_1}^{r_2} \frac{d\mathbf{v}(\mathbf{r})}{dt} \cdot d\mathbf{r} = m \oint_{r_1}^{r_2} \frac{d\mathbf{v}(\mathbf{r})}{dt} \cdot \mathbf{v}(\mathbf{r})\, dt. \qquad (11.36)$$

But for any vector $\mathbf{A}$,

$$\frac{d}{dt}(\mathbf{A} \cdot \mathbf{A}) = \frac{d\mathbf{A}}{dt} \cdot \mathbf{A} + \mathbf{A} \cdot \frac{d\mathbf{A}}{dt} = 2\mathbf{A} \cdot \frac{d\mathbf{A}}{dt}. \qquad (11.37)$$

Using this vector identity in Eq. 11.36 gives us

$$m \oint_{r_1}^{r_2} \frac{d\mathbf{v}(\mathbf{r})}{dt} \cdot d\mathbf{r} = m \oint_{r_1}^{r_2} \frac{1}{2}\frac{d}{dt}(\mathbf{v} \cdot \mathbf{v})\, dt$$

$$= m \oint_{v_1}^{v_2} \frac{1}{2}\, d(\mathbf{v} \cdot \mathbf{v})$$

$$= \frac{1}{2}\, mv(\mathbf{r}_2)^2 - \frac{1}{2}\, mv(\mathbf{r}_1)^2. \qquad (11.38)$$

Sure enough (and to no one's surprise) the work–energy theorem holds for three dimensions. We have found a way to express the vector quantities of motion, force, acceleration, velocity, and position in terms of scalar quantities, work, and kinetic energy. For some circumstances this approach has distinct advantages. The price we pay is the complete loss of time information. That is, although you may be able to calculate the final velocity of a mass after it has been catapulted by a spring, you have no knowledge of *when* the mass attains this velocity. You can easily calculate the velocity of a small ball tossed into the air when it strikes the ground, but you cannot use this theorem directly to find the time when this occurs. Nevertheless, for some problems, time is not relevant, and the work–energy theorem greatly simplifies the calculation.

EXAMPLE 1 Find the speed a simple pendulum of mass .50 kg and length 1.5 m when it reaches its lowest point in the arc if it starts from rest at an angle of 35° (Fig. 11.18).

We started to examine this problem at the beginning of the chapter. Equation 11.7 gives the relationship between the differential of work dW done by the gravitational force and the motion along the constrained arc. As we showed,

$$dW = -mg\, dy.$$

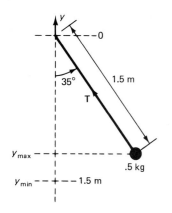

Fig. 11.18

Please note that although the mass moves in both the x and y directions (along its circular arc), the gravitational force is the only force that does work on the mass. The tension in the rod acts as a constraint force doing no work ($\mathbf{T} \cdot d\mathbf{s} = 0$, everywhere). Since the gravitational force acts in the negative y-direction, the only work on the mass occurs for its motion in that direction. The total work done as the mass falls to its lowest point is

$$\oint \mathbf{F} \cdot d\mathbf{r} = \oint_{y_{\max}}^{y_{\min}} (-mg)(dy) = -mg(y_{\min} - y_{\max}). \qquad (11.39)$$

From the geometry of the pendulum

$$y_{\min} = -1.5 \text{ m}$$

$$y_{\max} = -1.5 \cos 35° = -1.23 \text{ m}$$

$$\text{work} = -(.5 \text{ kg})(9.8 \text{ m/s}^2)(-1.5 + 1.23)\text{m} = 1.33 \text{ J}.$$

By the work–energy theorem, this must be equal to the change in the kinetic energy

$$1.33 \text{ J} = \tfrac{1}{2}(.5 \text{ kg})v_f^2 - 0$$

$$v_f = 2.31 \text{ m/s}.$$

If you examine this calculation, it is easy to see that mass is an irrelevant parameter, since velocity at the minimum (or at any other point) is independent of mass. The velocity depends only on how far the mass falls in the y-direction. In fact, the mass acquires exactly the same kinetic energy as *if* it fell the same distance in a *free* fall. Show this!

EXAMPLE 2

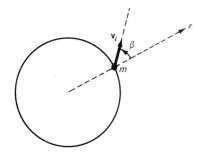

Fig. 11.19

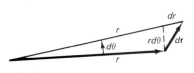

Fig. 11.20

Let's return to the problem of calculating the escape velocity of an object from the Earth's surface, but this time consider the possibility that the rocket takes off at an arbitrary angle β with respect to the radial (vertical) direction (Fig 11.19). The problem can best be solved by using two-dimensional polar coordinates, with $\hat{\mathbf{r}}$ and $\hat{\boldsymbol{\theta}}$ unit vectors. We begin by calculating the work done by the gravitational force and noting that the element of integration along the path $d\mathbf{r}$ is given by (Fig. 11.20)

$$d\mathbf{r} = dr\,\hat{\mathbf{r}} + r\,d\theta\,\hat{\boldsymbol{\theta}}. \qquad (11.40)$$

Then

$$\mathbf{F} \cdot d\mathbf{r} = -\frac{GM_E m}{r^2}\,\hat{\mathbf{r}} \cdot (dr\,\hat{\mathbf{r}} + r\,d\theta\,\hat{\boldsymbol{\theta}})$$

$$= -\frac{GM_E m}{r^2}\,dr \qquad \text{since } \hat{\mathbf{r}} \cdot \hat{\boldsymbol{\theta}} = 0. \qquad (11.41)$$

The total work done by the gravitational force on the rocket while it moved from the Earth's surface out into the galaxy is found by integrating Eq. 11.41:

$$\text{work} = \oint_{R_E}^{\infty} - \frac{GM_E m}{r^2} \, dr = -GM_E m \left(-\frac{1}{r} \right)_{R_E}^{\infty} = \frac{GM_E m}{R_E}. \quad (11.42)$$

This must equal the change in kinetic energy of the rocket. If we again assume that the rocket has little or no kinetic energy at infinity, the change in kinetic energy is just $-\frac{1}{2}mv_i^2$ and thus

$$0 - \frac{1}{2} \, mv_i^2 = - \frac{GM_E m}{R_E}$$

or

$$v_i = \left(\frac{2GM_E}{R_E} \right)^{1/2}.$$

But

$$g = \frac{GM_E}{R_E^2}. \quad (11.43)$$

Finally,

$$v_i = (2gR_E)^{1/2}, \quad (11.44)$$

the same result as we calculated for the one-dimensional problem. Neither the mass of the rocket nor the angle with respect to the vertical (β) enters into the expression for the escape velocity. Of course, this is true only because we have neglected air resistance. A longer flight path through the atmosphere (β not equal to 90°) would require a larger initial velocity for the rocket to "escape" because a real rocket would experience the drag force due to the Earth's atmosphere.

11.6 **Relativistic Work–Energy Theorem**

IS THERE A work–energy theorem for high-energy particles? How do we treat particles moving at relativistic speeds? The answer comes from a straightforward application of the same mathematical techniques to the relativistic expressions for force and linear momentum. Newton's second law, written in the form

$$\mathbf{F} = \frac{d}{dt} \, \mathbf{p},$$

is correct, even for high-energy particles, provided that the relativistic expression for linear momentum is used, namely,

$$\mathbf{F} = \frac{d}{dt} \, \mathbf{p} = \frac{d}{dt} \, (\gamma m \mathbf{v}). \quad (11.45)$$

If we form the path integral in the same way that we wrote the nonrelativistic expression, Eq. 11.45 becomes

$$\oint \mathbf{F} \cdot d\mathbf{s} = \oint \frac{d}{dt}(\gamma m v) \cdot d\mathbf{s}. \qquad (11.46)$$

To simplify the algebra we shall consider the one-dimensional case where $ds = dx$. Changing variables, noting that

$$dx = \frac{dx}{dt} dt = v\, dt,$$

Eq. 11.46 becomes

$$\text{work} = \oint_0^v \frac{d(\gamma m v)}{dt} v\, dt = \oint_0^v v\, d(\gamma m v), \qquad (11.47)$$

where we integrate over a path that starts with the particle having zero velocity. But

$$d(\gamma m v) = m\gamma\, dv + m v\, d\gamma \qquad \text{(for constant mass)} \qquad (11.48)$$

and

$$d\gamma = d\, \frac{1}{\left(1 - \dfrac{v^2}{c^2}\right)^{1/2}} = \frac{v\, dv}{c^2\left(1 - \dfrac{v^2}{c^2}\right)^{3/2}}.$$

Thus

$$d(\gamma m v) = m\left[\frac{dv}{\left(1 - \dfrac{v^2}{c^2}\right)^{1/2}} + \frac{v^2\, dv}{c^2\left(1 - \dfrac{v^2}{c^2}\right)^{3/2}} \right]$$

$$= \frac{m\, dv}{\left(1 - \dfrac{v^2}{c^2}\right)^{3/2}}. \qquad (11.49)$$

Substituting Eq. 11.49 into Eq. 11.47, we arrive at

$$\text{work} = m\int_0^v \frac{v\, dv}{\left(1 - \dfrac{v^2}{c^2}\right)^{3/2}} = mc^2\left[\frac{1}{\left(1 - \dfrac{v^2}{c^2}\right)^{1/2}} - 1 \right]$$

or

$$\boxed{\text{work} = \gamma mc^2 - mc^2 = (\gamma - 1)mc^2.} \qquad (11.50)$$

You should all recognize the right side of this equation. It is the relativistic expression for the kinetic energy of a particle of mass m moving with a velocity v (Eq. 5.1). We have established a relationship between the work done on mass m and the change in the relativistic kinetic energy of the particle. In most books you will find this proof given as the derivation for the relativistic expression for kinetic energy.* Here we are using this derivation to justify our earlier ad

---

* R. Resnick, *Introduction to Special Relativity* (New York: Wiley, 1968), Chap. 3; and P. Tipler, *Foundation of Modern Physics* (New York: Worth, 1969), Chap. 1, for example.

hoc assumptions of the forms of the relativistic kinetic energy and linear momentum.

Now we can see that the work–energy theorem is a general relationship that applies to all particles regardless of their velocity. It is a fundamental concept in physics.

11.7 **Frictional Forces**

FRICTIONAL forces have thus far been left out of the discussion. Does the work–energy theorem apply if there are frictional forces in the system? Let's go back to our example of the glider on the horizontal air track moving at a constant velocity (zero net external force) and imagine that the compressed air source for the track breaks down. The cushion of air that provides the nearly perfect friction-free surface is gone, and a metal-to-metal contact is established between the glider and the air track. If the glider was moving in the positive x-direction before the breakdown, it will continue in that direction now experiencing a frictional force μmg in the negative x-direction. This force will decelerate the glider and eventually bring it to rest. Since the net force and displacement are in opposite directions, the frictional force does negative work on the glider. The work–energy theorem tells us that the *change* in kinetic energy must also be negative; that is, the final kinetic energy must be less than the initial kinetic energy. That is obviously the case here, since the final kinetic energy is zero. So it appears, and you can readily prove, that the work–energy principle also applies when frictional forces are present.

But, you ask, what happened to the kinetic energy of the glider? Where did it go? When we applied a force to the friction-free glider, we did work on the glider and witnessed the products of our "labors," in that the kinetic energy of the glider increased. When the moving block collided with the relaxed spring, the kinetic energy of the block compressed the spring and stored the energy in the spring. But what happened to the work done by the frictional force in stopping the glider?

The key to the explanation of the "missing energy" lies in a standard fiction that we have incorporated into our analysis, which must now be exposed. The fiction is that the glider (and the track) are rigid bodies. They are not. After the glider comes to rest on the track, all the initial kinetic energy has *not* disappeared; only the "organized" kinetic energy is gone. While the glider as a whole has come to rest, if you could examine the glider with a super electron microscope, you would see that the atoms had increased their random vibrational motion. The organized, collective motion of the glider has been transformed into random vibrational motion of the atoms in this "rigid" body. If you don't have a million dollar microscope, try a good $50 thermometer, because if you do the experiment carefully, you will observe a slight but perceptible rise in the temperature of both the glider and the track. None of the energy is lost. It has simply taken a new form, internal thermal energy of the system. So if you do a complete energy audit, you must include the internal motions of the atoms in these so-called "rigid" bodies. You will study this form of energy in detail when you come face to face with thermodynamics.

11.8 **Power**

WE HAVE NOT yet considered time in our discussion of work and energy. You obviously do the same amount of work no matter how slowly or quickly you raise a granite block from the floor and place it on a table. However, the RATE at which work is done can vary. The rate at which work is done is defined as POWER; mathematically, it is written

$$\text{POWER} = P \equiv \frac{dW}{dt}, \tag{11.51}$$

a scalar quantity with the SI units of JOULES/SECOND = WATTS. The English units for power are ft-lb/sec and

$$550 \text{ ft-lb/sec} = 746 \text{ watts} = 1 \text{ horsepower.} \tag{11.52}$$

Some of these archaic units are still in common usage in American mechanical equipment.

If the power used is constant in time, one can integrate Eq. 11.51:

$$\int P \, dt = \int dw \quad \text{or} \quad \boxed{W = Pt \quad \text{(constant power).}} \tag{11.53}$$

Finally, recalling the definition of infinitesimal amount of work $dW = \mathbf{F} \cdot d\mathbf{s}$, and if the force acts over an infinitesimal amount of time dt, the rate of doing work is given by

$$\frac{dW}{dt} = \mathbf{F} \cdot \frac{d\mathbf{s}}{dt} = \mathbf{F} \cdot \mathbf{v} = P. \tag{11.54}$$

EXAMPLE 1   A newly designed electric automobile uses its own electric motor for braking, by converting the motor to a generator when the driver steps on the brake. If the car's mass is 900 kg and it is traveling 50 mi/hr when the brakes are activated, causing a deceleration of 1.5 m/s², calculate the *maximum* rate at which energy can be returned to the car's storage battery. Assume 100% efficiency of the generator:

$$P_{max} = \mathbf{F} \cdot \mathbf{v}_{max}.$$

But what is $\mathbf{F}$? It is the braking force, but you are not told the force. Instead, you are given the acceleration. Therefore, the maximum force is given by Newton's second law:

$$F_{max} = ma = 900 \text{ kg} \times 1.5 \text{ m/s}^2 = 1350 \text{ N}$$

$$P_{max} = -1350 \text{ N} \times 50 \text{ mi/hr} \times .447 \text{ (m/s)(mi/hr)}^{-1}$$

$$= -3.02 \times 10^4 \text{ W}.$$

The negative sign is there because the force and velocity were in opposite directions, indicating that work was being done *on* the car.

EXAMPLE 2   The car in Example 1 requires 60 hp to travel at a constant speed of 50 mi/hr. What is the forward force on the car exerted by the electric engine?

To get everything in SI units (since there are no metric horses)

$$60 \text{ hp} \times 746 \text{ W/hp} = 44{,}760 \text{ W}$$

and

$$50 \text{ mi/hr} \times .447 \frac{\text{m/s}}{\text{mi/hr}} = 22.35 \text{ m/s}.$$

Since

$$P = Fv,$$

$$F = \frac{P}{v} = \frac{44{,}760 \text{ W}}{2{,}235 \text{ m/s}} = 2003 \text{ N}.$$

This is not the net force on the car! If it were, the car would accelerate. The problem stated that the car is traveling at a constant speed, which means that the net force on the car is zero. Thus the 2003-N force is balanced by the friction, wind resistance, and other drag forces the car also experiences. It takes about 2000 N just to keep the car moving at 50 mi/hr!

11.9 Path-Independent, Path-Dependent Work

WE HAVE defined total work through Eq. 11.12, which indicates that the infinitesimal quantities dW must be summed along the path that the particle travels. In general, the total work done depends on the path chosen. However, we can divide forces into two categories for our discussion of mechanics: (1) those forces for which the total work done is *independent* of the path followed and depends only on the initial and final positions (the end points); and (2) those forces for which the total work done *does* depends on the path as well as the end points. The first type of force is called a conservative force and is defined by the following statement:

> *The work done by a* CONSERVATIVE FORCE *is independent of the path, depending only on the end points.*

This definition of a conservative force may appear restrictive, but most of the common forces we will study belong to this category.

EXAMPLE 1   Consider a uniform and constant force **F** such as the gravitational force near the Earth's surface or the electric force on a charged particle inside parallel conducting plates. These forces do *not* depend on the position, velocity, or acceleration of the particle or on time. Such a force is depicted by the arrows in Fig. 11.21. It acts on a point mass as the particle moves from position 1 to position 2.

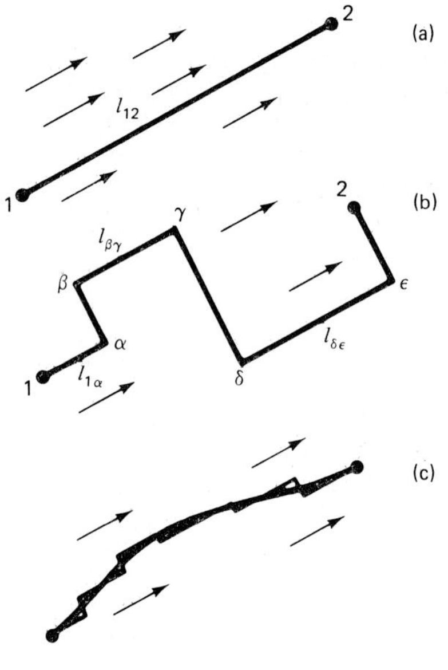

For the path in Fig. 11.21a, the particle moves in a straight line parallel to the direction of the force **F**. Calculating the work done by this force is trivial; it is

$$W_{1 \to 2}(\text{a}) = \oint \mathbf{F} \cdot d\mathbf{s} = Fl_{12}. \tag{11.55}$$

Case (b) is a bit more complicated, since the path indicated is rather tortuous. We will calculate the work done along each straight segment of path (b) and then add them up to get the total work done. That is,

$$W_{1 \to 2}(\text{b}) = \oint_1^\alpha \mathbf{F} \cdot d\mathbf{s} + \oint_\alpha^\beta \mathbf{F} \cdot d\mathbf{s} + \oint_\beta^\gamma \mathbf{F} \cdot d\mathbf{s} + \oint_\gamma^\delta \mathbf{F} \cdot d\mathbf{s}$$
$$+ \oint_\delta^\varepsilon \mathbf{F} \cdot d\mathbf{s} + \oint_\varepsilon^2 \mathbf{F} \cdot d\mathbf{s}. \tag{11.56}$$

Even though this equation looks like a mess, it is actually quite easy to carry out this sum. There are really only two types of segments in path (b): those parallel to **F**, $1 \to \alpha$, $\beta \to \gamma$, and $\delta \to \varepsilon$; and those perpendicular to **F**, $\alpha \to \beta$, $\gamma \to \delta$, and $\varepsilon \to 2$. No work is done along the path segments that are perpendicular to the displacements, so Eq. 11.56 reduces to

$$W_{1 \to 2}(\text{b}) = \oint_1^\alpha \mathbf{F} \cdot d\mathbf{s} + \oint_\beta^\gamma \mathbf{F} \cdot d\mathbf{s} + \oint_\delta^\varepsilon \mathbf{F} \cdot d\mathbf{s}, \tag{11.57}$$

and since **F** and $d\mathbf{s}$ are parallel and **F** is constant,

$$W_{1-2}(\text{b}) = Fl_{1\alpha} + Fl_{\beta\gamma} + Fl_{\delta\varepsilon}.$$

But

$$l_{1\alpha} + l_{\beta\gamma} + l_{\delta\varepsilon} = l_{12},$$

(a)

(b)

(c)

Fig. 11.21

so $\qquad\qquad\qquad W_{1-2}(b) = Fl_{12} \qquad$ [path (b)], $\qquad\qquad$ (11.58)

which is the same result that we got for path (a).

Finally, imagine a general path, as shown in (c). We can approximate this path by a sequence of short, straight segments Δl which are either parallel or perpendicular to the uniform force **F**. No work will be done along the perpendicular segments, and if one adds up the parallel segments, $\Sigma\,\Delta l = l_{12}$, the total parallel length of the straight-line segment in path (a). If you will accept the idea that an arbitrary path can always be made up of an infinite number of infinitesimal parallel and perpendicular segments, it appears that the work done in this uniform force field is path independent. Thus we can say that constant and uniform forces are conservative forces.

EXAMPLE 2 — A central force is a force that acts only in the radial direction from, and has a magnitude that depends only on, the distance to the origin. The source of such a force is at the origin (Fig. 11.22). Mathematically, all forces in this category can be written in the form

$$\mathbf{F}_{\text{radial}} = f(r)\hat{\mathbf{r}} \qquad\qquad (11.59)$$

using polar coordinates.

For simplicity, we shall consider the two-dimensional case where the displacement $d\mathbf{r}$ can be written in terms of the polar coordinates' unit vector $\hat{\mathbf{r}}$ and $\hat{\boldsymbol{\theta}}$, shown in Fig. 11.23 as

$$d\mathbf{r} = dr\,\hat{\mathbf{r}} + r\,d\theta\,\hat{\boldsymbol{\theta}}. \qquad\qquad (11.60)$$

The work done by a central force in moving a body along an arbitrary path from a to b is (Fig. 11.22) is

$$W_{a\to b} = \oint_a^b \mathbf{F}\cdot d\mathbf{s} = \oint_a^b f(r)\hat{\mathbf{r}}\cdot(dr\,\hat{\mathbf{r}} + r\,d\theta\,\hat{\boldsymbol{\theta}})$$

$$= \oint_a^b f(r)\,dr, \qquad\qquad (11.61)$$

since $\hat{\mathbf{r}}\cdot\hat{\boldsymbol{\theta}} = 0$. The work done does not depend on θ.* The evaluation of Eq. 11.61, for any function $f(r)$, gives the work done. One can see that it depends *only* on the end points $\mathbf{r}_a$ and $\mathbf{r}_b$ and *not* on the path. The path has dropped out of the problem. Thus we have shown that *any* central force must be conservative.

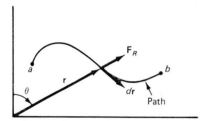

Fig. 11.22

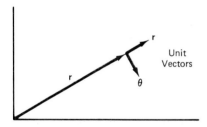

Fig. 11.23

A. Work Done around Closed Loops (Conservative Forces)

Now let's examine the amount of work done by a conservative force on a particle as it traverses a closed path and returns to its starting

---

*If you examine the three-dimensional case, it becomes apparent that the work done by a radial force does not depend on $\boldsymbol{\phi}$, the azimuthal angle. For three dimensions, the general displacement $d\mathbf{r} = \hat{\mathbf{r}}\,dr + \hat{\boldsymbol{\theta}}r\,d\theta + \hat{\boldsymbol{\phi}}r\sin\theta\,d\phi$.

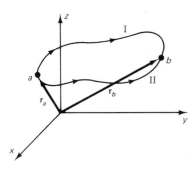

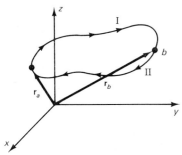

Fig. 11.24

point. The work done by a conservative force is path independent, depending only on the end points of the path. In Fig. 11.24 we have constructed two paths from $\mathbf{r}_a$ to $\mathbf{r}_b$, which are labeled I and II. Since the force or forces on the particle are assumed to be conservative,

$$W_{\mathbf{r}_a\to\mathbf{r}_b}(\mathrm{I}) = W_{\mathbf{r}_a\to\mathbf{r}_b}(\mathrm{II}), \tag{11.62}$$

where $W_{\mathbf{r}_a\to\mathbf{r}_b}(\mathrm{I})$ is the work done on the particle in going from $\mathbf{r}_a$ to $\mathbf{r}_b$ along path *I* and $W_{\mathbf{r}_a\to\mathbf{r}_b}(\mathrm{II})$ is the work done along path II also going from $\mathbf{r}_a$ to $\mathbf{r}_b$. If we now reverse the direction of motion of the particle, going from $\mathbf{r}_b \to \mathbf{r}_a$ along the same path II, everything remains the same *except* the direction $d\mathbf{r}$, which is reversed. Thus

$$W_{\mathbf{r}_a\to\mathbf{r}_b}(\mathrm{II}) = -W_{\mathbf{r}_b\to\mathbf{r}_a}(\mathrm{II}). \tag{11.63}$$

The work done in going from $\mathbf{r}_a$ to $\mathbf{r}_b$ is the negative of the work done in going from $\mathbf{r}_b$ to $\mathbf{r}_a$ along the same path. Substituting Eq. 11.63 into Eq. 11.62, we have

$$W_{\mathbf{r}_a\to\mathbf{r}_b}(\mathrm{I}) = -W_{\mathbf{r}_b\to\mathbf{r}_a}(\mathrm{II})$$

or

$$W_{\mathbf{r}_a\to\mathbf{r}_b}(\mathrm{I}) + W_{\mathbf{r}_b\to\mathbf{r}_a}(\mathrm{II}) = 0. \tag{11.64}$$

Look at what Eq. 11.64 is telling us! The work done around a closed loop, where the particle returns to the starting point, is zero. Since paths I and II, as well as end points $\mathbf{r}_a$ and $\mathbf{r}_b$, were arbitrary, we have proven a general theorem:

> *The work done by a conservative force on a particle around a closed loop, returning to the starting point, is always zero.*

B. Nonconservative Forces

Not all forces are conservative. The first nonconservative path-*dependent* force that comes to mind is friction. Since the sliding frictional force acts to oppose motion, it will do negative work on the moving body. Consider the last example. Equation 11.63 would *not* hold for frictional forces, since reversing the direction on the path not only reverses the differential of displacement $d\mathbf{r}$, but also reverses the direction of the force. Thus the total work by frictional forces around the closed loop is *never* zero.

Other types of forces are also nonconservative, even forces that depend only on distance. An example of such a force is

$$\mathbf{F} = Cy\mathbf{i}, \tag{11.65}$$

a force that always acts in the x-direction, but whose magnitude and sign depends only on the y-coordinate (C is just a constant). In Fig. 11.25 we have indicated the "appearance" of this force field by drawing arrows designating the magnitude and direction of the force for selected points on a two-dimensional grid.

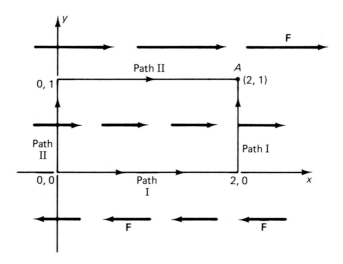

Fig. 11.25

We will now calculate the work done by this force on a test particle that moves along two separate paths, starting at the origin (0, 0) and both ending at point A (2, 1). First, calculate the total work done by the force $\mathbf{F}$ when the particle follows path I in Fig. 11.25. This can be done by adding two path integrals (0, 0) to (2, 0) and (2, 0) to (2, 1) as

$$W_{\mathrm{I}} = \int_{\substack{x=0 \\ y=0}}^{\substack{x=2 \\ y=0}} f(y = 0)\, dx + \int_{\substack{x=2 \\ y=0}}^{\substack{x=2 \\ y=1}} F(x = 2)\, dy = 0. \quad (11.66)$$

The first term in Eq. 11.66 is zero because the force is zero when $y = 0$. The second term is also zero, because the force is perpendicular to the displacement at every point along this segment of the path.

Consider now the work done along path II, which can also be broken up into two integrals.

$$W_{\mathrm{II}} = \int_{\substack{x=0 \\ y=0}}^{\substack{x=0 \\ y=1}} F(x = 0)\, dy + \int_{\substack{x=0 \\ y=1}}^{\substack{x=2 \\ y=1}} F(y = 1)\, dx. \quad (11.67)$$

The first term is zero, because the force is perpendicular to the displacement everywhere along this segment of the path. But the second term is *not* zero. Since the force is parallel to dx and constant along this segment,

$$W_{\mathrm{II}} = \int_{\substack{x=0 \\ y=1}}^{\substack{x=2 \\ y=1}} Cy\, dx = C \int_{x=0}^{x=2} dx = 2C. \quad (11.68)$$

This particular position-dependent force is *not* conservative. It produces path-*dependent* work when it acts on a particle.

CONSERVATIVE forces have another very important property. Remember that the work done by a conservative force acting on a body as it moves along some path depends *only* on the end points of the path. That is, the value of the path integral is independent of the path chosen, and depends only on the initial and final positions on the path. Mathematically, this is expressed as

$$\oint_{\mathbf{r}_a}^{\mathbf{r}_b} \mathbf{F} \cdot d\mathbf{r} \equiv - [U(\mathbf{r}_b) - U(\mathbf{r}_a)], \qquad (11.69)$$

where $U(\mathbf{r})$ is a mathematical function called the POTENTIAL ENERGY. The minus sign in Eq. 11.69 is a matter of convention, but its rationale will shortly become apparent. The potential energy functions are *scalar* quantities, even though in general they depend on the vector quantity $\mathbf{r}$.

Potential energy (PE) has one noteworthy and unusual property which may at first appear quite perplexing. The work done by the net conservative force is equal to the *difference* between two quantities, $U(\mathbf{r}_b)$ and $U(\mathbf{r}_a)$. If, for fun (or spite), I decided to add a constant scalar quantity Q to *both* $U(\mathbf{r}_a)$ and $U(\mathbf{r}_b)$, Eq. 11.69 would remain valid. When we speak of the "potential at a point," what we really mean is the potential energy at that point *relative to a reference point*. It makes no sense (but people do it all the time) to state that at the point $\mathbf{r}_b$ the potential is 4 J. Since the potential energy in principle is arbitrary to a constant, the potential at $\mathbf{r}_p$ can be any scalar number. What we *mean* when we say the potential energy at $\mathbf{r}_p$ is 4 J is that the potential energy at $\mathbf{r}_p$ is 4 J *with respect to the potential energy at the reference point* $\mathbf{r}_{\text{REF}}$. Once the reference point has been chosen, and the value of the PE at that point has been established, the PE at *all* other points in space is also *fixed*. You are no longer free to play around with the value of the PE after you fix the location of the reference point and the value of the PE at that point. To summarize:

> *The potential energy at a point in space relative to the potential energy at the reference point is defined as the negative of the work done by a conservative force on an object as it moves from the reference point to the point under consideration. The location and the potential energy of the reference point are arbitrary, but by convention are usually chosen where the force on the object is zero. The value of the reference potential energy is usually taken to be zero.*

Our definition of the potential energy (Eq. 11.69), combined with the work–energy theorem, yields

$$W_{\mathbf{r}_a \to \mathbf{r}_b} = -U(\mathbf{r}_b) + U(\mathbf{r}_a) = KE_{\mathbf{r}_a} - KE_{\mathbf{r}_b}. \tag{11.70}$$

Rearranging terms gives us

$$U(\mathbf{r}_a) + KE_{\mathbf{r}_a} = KE_{\mathbf{r}_b} + U(\mathbf{r}_b). \tag{11.71}$$

This is an unusual equation. The left-hand side is a function of $\mathbf{r}_a$ only and the right-hand side is a function of $\mathbf{r}_b$ only. But $\mathbf{r}_a$ and $\mathbf{r}_b$ are arbitrary vectors that can take on any reasonable values that we wish, yet Eq. 11.71 must *always* be satisfied. How can this be true?

The only way that Eq. 11.71 can be satisfied for all values of $\mathbf{r}_a$ and $\mathbf{r}_b$ is for both sides of the equation to equal the same constant.* If this constant is written E, then

$$U(\mathbf{r}_a) + KE_{\mathbf{r}_a} = U(\mathbf{r}_b) + KE_{\mathbf{r}_b} = E. \tag{11.72}$$

Equation 11.72 is called the conservation of mechanical energy, and E is the total MECHANICAL ENERGY. This is *not* the same as the conservation of energy, which is a fundamental law of nature. It is a special case of the universal conservation principle, which is applicable for low-velocity systems that are unable to convert mass to energy, and which involve *only conservative forces*. It "says" that under these restrictive conditions, a mechanical system will retain the sum of its potential and kinetic energy at all times as the system progresses along any arbitrary path. It defines for us another form of energy, potential energy, to add to our list, which up to now had included only kinetic and mass energy.

We have already calculated the potential energy for two important systems, but I kept it a secret. Go back and look at our discussions of the work done by the gravitational force on a pendulum (Eq. 11.39). Here we calculated the work done as the pendulum swings from its maximum height y_{max} to its lowest point y_{min}:

$$\text{WORK} = -\int_{y_{max}}^{y_{min}} mg \, dy = -mg(y_{min} - y_{max}), \tag{11.73}$$

but by Eq. 11.69 this is equal to the negative of the change in the potential energy, or

$$-U(y_{max}) + U(y_{min}) = -mgy_{min} + mgy_{max}. \tag{11.74}$$

It is clear from Eq. 11.73 that the potential energy of this pendulum has the functional form

$$U \text{ (gravitational near the Earth)} = mgy. \tag{11.75}$$

---

* This is the same mathematical argument that we used in Chapter 8 to verify assumed solutions to differential equations.

You can easily show that any point mass near the Earth's surface has the same form for its gravitational potential energy mgy, where y is the vertical distance from the Earth's surface.

The reference point for the PE is arbitrary as long as one keeps away from points where the function blows up. For example, we will soon show that the gravitational potential energy function for any object far from the Earth is given by $U(r) = K/r$, where K is a constant and r is the distance of the center of Earth from the object. One would not choose $r = 0$ as the reference point here, because the potential function is not defined at that point. According to convention, the reference point is taken at very large distances, where the gravitational force tends toward zero. For gravity this is at $r = \infty$, where $U(r) = 0$. However, you should understand that this is a convention, not a requirement, and it is quite legitimate to choose $r = 1$ meter and set $U(r = 1) = 4.658$ J as a reference. I cannot think of one good reason to do it, however.

This still leaves a puzzling point. If I am allowed to choose any arbitrary reference point and to define the potential energy as any value I wish at that point, then by Eq. 11.72 I can make E any value I like. Isn't E the total energy? How is it possible to make the total energy anything I want? Isn't that getting something for nothing?

No, E is *not* the total energy; it is the total *mechanical energy*, a very different bird. It is the sum of the kinetic and potential energies of a system that is subjected only to conservative forces. It is true that one can define the reference of the PE to be as small or as large as you please, but it is only the *changes in potential energy* that matter. It is these *changes* that are converted to kinetic energy of motion, not the absolute value of the PE. The words *potential energy* are well chosen, for it is the energy that has the potential to be converted to energy of motion.

EXAMPLE 1   Find the potential energy of a rocket as a function of its distance from Earth.

This is a typical incomplete question in textbooks and or journal articles. The correct complete statement of the question would ask you to find the potential energy of the rocket with respect to a reference PE chosen at infinity where the gravitational force on the rocket is zero. This can easily be done by calculating the negative of the work done in moving a rocket from infinity to some arbitrary distance from Earth, or

$$\text{PE} = -\int_{\infty}^{r} \mathbf{F}_g \cdot d\mathbf{r}. \qquad (11.76)$$

But we have already calculated this integral earlier in the chapter when we studied rockets leaving the Earth (Eqs. 11.20–11.21). Since the force on the rocket is zero at infinity and since the force and the displacement $d\mathbf{r}$ are in the same direction, the integral in Eq. 11.76 is intrinsically positive. The minus sign is there because the

PE is the negative of the work done on the body. The potential energy is then

$$PE = U(r) = -\frac{GM_E m}{r}.$$

(11.77)

EXAMPLE 2

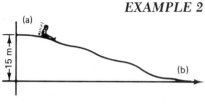

Fig. 11.26

A 25-kg boy slides down a frictionless water slide starting from the top, which is 15 m above the ground (Fig. 11.26). Find the velocity of the boy when he reaches the bottom of the slide.

Since the slide is frictionless, the only forces on the boy are the varying normal force of the slide and the gravitational force. The normal force (the constraint force) always acts perpendicular to the motion, so it does no work on the boy. We may use Eq. 11.71, conservation of mechanical energy, to solve this problem. The initial mechanical energy is all potential energy while the final mechanical energy is all kinetic, if we choose the bottom of the slide as our zero of PE. Thus

$$U(a) + KE(a) = U(b) + KE(b)$$

$$mgh + 0 = 0 + \tfrac{1}{2} mv_F^2$$

or

$$v_f = \sqrt{2gh} = (2 \times 9.8 \text{ m/s} \times 15 \text{ m})^{1/2} = 17.2 \text{ m/s}.$$

The mass of the boy was irrelevant to the calculation.

Darien Lake's "Viper" rollercoaster gives the rider a real life experience with the conservation of mechanical energy.

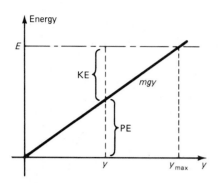

Fig. 11.27

ANALYSIS OF potential energy diagram provides us with insights into important features of one-dimensional motion, such as the one shown in Fig. 11.27. Here we have plotted *two* quantities on the vertical axis, total mechanical energy E and potential energy $U(y)$. On the horizontal axis is the relevant spatial parameter, which might be x, y, z, or r. The energy diagram can be used to understand particle motion, but it is a more abstract physical picture than the usual position versus time diagram. This is because one is plotting two quantities on the same vertical axis and time does not appear in the diagram. A single diagram can represent more than one example of motion. Modern physics uses these diagrams to understand systems, particularly in atomic and nuclear studies, more than force representations. Similarly, advanced topics in chemistry, molecular biology, nuclear physics, solid-state physics, and engineering analysis utilize this type of graphical representation.

The motion of a point mass moving in the vertical direction near the Earth's surface (neglecting air resistance) provides a good first example. We showed that the potential energy function for the object is of the form

$$U(y) = mgy. \tag{11.78}$$

If the ground position is defined as $y = 0$, we automatically set the reference potential $U(y = 0) = 0$. In Fig. 11.27, $U(y) = mgy$ is plotted as a straight line whose slope is mg. This single diagram can now be used to describe several examples of free-fall motion in one dimension: for example, a ball of mass m that has been raised to a height y_{max} above the ground and then allowed to free fall to the ground. As the ball falls, simultaneously the kinetic energy increases and the potential energy decreases, but the total mechanical energy remains constant. This is represented in the energy diagram. The constant mechanical energy is the horizontal straight line, independent of y. The kinetic energy can easily be seen because

$$E = KE + PE \quad \text{or} \quad KE = E - U(y). \tag{11.79}$$

When $y = 0$, the potential energy is zero, but that is the point where the kinetic energy has its maximum value.

$$KE(y = 0) = mgy_{max} = E = \tfrac{1}{2}mv_{max}^2$$

$$v_{max} = (2gy_{max})^{1/2}. \tag{11.80}$$

The basic diagram (Fig. 11.28) represents the problem of the same ball at a height y_0 thrown up in the air with an initial velocity v_i. Here the total mechanical energy is the sum of its initial kinetic and potential energies

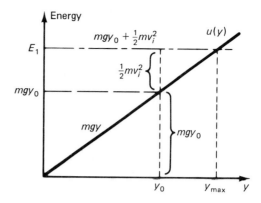

Fig. 11.28

$$E_1 = mgy_0 + \tfrac{1}{2}mv_i^2. \tag{11.81}$$

The ball travels upward, gaining potential energy and losing kinetic energy until the kinetic energy is zero at y_{max}. It cannot travel past the "intersection" of E_1 and $U(y)$, because to do so would make its kinetic energy *negative* (Eq. 11.79). For all classical particles (both relativistic and Newtonian), the kinetic energy is an intrinsically *positive* quantity. It makes no physical sense to have a negative kinetic energy.* When the ball reaches the maximum height y_{max}, it will stop, reverse direction, and begin to fall toward the ground, gaining kinetic energy as it loses potential energy. During the entire flight the total mechanical energy remains constant. It should be obvious from the energy diagram that the ball's velocity when it strikes the ground is the same in this example as it was in case 1 (assuming that the total mechanical energy is the same in both examples).

Another illustration of the conservation of mechanical energy and the use of potential energy diagrams is the harmonic oscillating system. An example of this kind of motion is a block, attached to an idealized spring, sliding across a frictionless surface (Fig. 11.29).† Equation 11.18 is the expression for the work done by an *external force* in compressing a Hooke's law ($F = -kx$) spring. But to calculate the potential energy, we use the force exerted *on the mass by the spring*, or

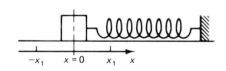

Fig. 11.29

$$U(x) - U(0) = -\int_0^x -kx \, dx = \tfrac{1}{2}kx^2, \tag{11.82}$$

---

* Quantum mechanics, in its usual perverse way, violates this "logic." Particles can have negative kinetic energy according to the Heisenberg uncertainty principle, but that's the subject of another course.

† This looks like one of our "special effects," but it turns out that nature has many real examples of systems that behave like this fictitious one.

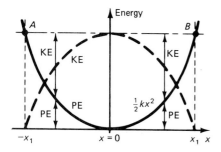

Fig. 11.30

which is the same result as the work done by the external force on the spring.* The zero of potential energy is chosen at $x = 0$, the equilibrium position [i.e., $U(0) = 0$]. A plot of this parabolic potential energy is shown in Fig. 11.30.

The total mechanical energy of the system is the sum of the kinetic energy and potential energies, or

$$E = \tfrac{1}{2}mv^2 + \tfrac{1}{2}kx^2. \tag{11.83}$$

This is a system where both the potential and kinetic energies are even functions of x, so both are always positive. Imagine starting our system at rest with the spring compressed so that the mass is located at x_1. All the energy is initially potential: $E = \tfrac{1}{2}kx_1^2$. The mass is released, the potential energy decreases, and the kinetic energy increases, until the mass reaches $x = 0$. Here the system's energy is all kinetic energy of motion of the mass:

$$E = \tfrac{1}{2}mv_{\text{max}}^2 = \tfrac{1}{2}kx_1^2$$

or

$$v_{\text{max}} = \left(\frac{kx_1^2}{m}\right)^{1/2}.$$

Of course, the mass continues on past $x = 0$, losing kinetic energy and elongating the spring. This increases the potential energy until it reaches $-x_1$ (point A), where it stops and reverses velocity. The cycle repeats itself indefinitely. If the system is truly frictionless (including internal friction of the spring), the mass will continue to execute this oscillating motion indefinitely. In fact, we can readily obtain the differential equation of motion of our system from the conservation of mechanical energy (Eq. 11.81). This can be rewritten

$$E = \tfrac{1}{2}m\left(\frac{dx}{dt}\right)^2 + \tfrac{1}{2}kx^2, \tag{11.84}$$

which is a first-order differential equation.† Newton's laws for the same system yielded a second-order differential equation

$$m\frac{d^2x}{dt^2} = -kx. \tag{11.85}$$

As an exercise, you should prove to yourself that the solution to Eq. 11.84 is of the form

$$x = A \cos \omega t,$$

---

* In these expressions for potential energy and work, it is the minus signs that drive everyone crazy, including me. But there is an easy way to figure out whether you have done the problem correctly. If *you* do work *on* a system, it must increase its energy. If you lift a stone, compress a spring, or put charge on a capacitor, all must increase their energy. In Eq. 11.81, if we compress or elongate the spring, we must increase its potential energy.

† A first-order differential equation is one that contains only first derivatives of the independent variable.

where A is the amplitude of the motion and ω is the angular frequency of the periodic motion. In Chapter 15 we discuss this very important system more fully.

Finally, we can use a variation of this diagram to describe an ideal bouncing ball. To carry out this analysis we must think carefully about what happens when the ball strikes the ground. Figure 11.31 is a reasonable approximation of the "correct" energy diagram for a bouncing ball system. When the ball hits, it deforms both the ground surface (hence the potential in the negative y-direction) and itself. During contact, all the kinetic energy of the ball is rapidly converted to potential energy of elastic deformation of both the ball and the ground surface. Although we really have no knowledge of the details of this deformation process, it is clear that $U(y)$ will be some rapidly increasing function of y, when y becomes negative. This is displayed as the steeply ascending line to the left of the origin. Now there are two places where E intersects $U(y)$: points A and B. These are aptly called the turning points, where the direction of the ball's motion changes and where its velocity is instantaneously zero. If the ball went past these turning points, it would acquire a negative kinetic energy. In this idealized model, the ball continues to travel back and forth between the turning points, forever.

Real balls don't do that! A good superball will bounce many times, but it will never return to its starting height, and eventually it will stop bouncing altogether. After some time it will lie quietly on the ground. What's wrong with our theory? It doesn't predict this. The answer, of course, is that we have considered only mechanical energy, and have neglected both air resistance as well as the energy transferred to the internal motions of the ground surface and the ball itself. These involve the so-called nonconservative forces, whose effects are not included in the mechanical energy. One might think of them as a kind of "loss of organization of energy." By this I mean that the kinetic energy of the ball is gradually converted into kinetic energy of the random motion of molecules and atoms in the air, the ground surface, and the ball. If you were to measure carefully the temperature of the air, the ground, and the ball after it had been allowed to bounce repeatedly until it stopped, you would discover that their temperatures had increased slightly from their initial values. This slight rise in temperature is a measure of the increase in the random kinetic energy of the parts of the system we did not include in our idealized model.

There are examples of systems where we *can* calculate the amount of energy that has been transformed to random internal motions (internal energy). For such systems one can write the energy conservation principle as follows:

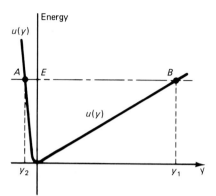

Fig. 11.31

$$
\boxed{
\begin{aligned}
&\textit{Initial mechanical energy = final mechanical energy} \\
&\qquad\qquad\qquad\quad \textit{+ work done by nonconservative forces}
\end{aligned}
}
\qquad (11.86)
$$

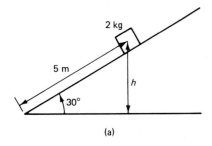

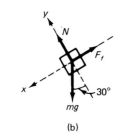

Fig. 11.32

EXAMPLE 1 Consider the 2-kg block sliding down a 30° slope with a frictional coefficient $\mu = .15$ (Fig. 11.32a). If the plane is 5 m long and the block started from rest, how fast will the block be moving when it reaches the bottom of the plane?

Figure 11.32b is the free-body diagram of the block. From it we can determine the normal force as:

$$y\text{-comp.:} \qquad N - mg \cos \theta = 0 \qquad \text{or} \qquad N = mg \cos \theta.$$

Then the frictional force is

$$F_f = \mu mg \cos \theta = .15 \times 2 \text{ kg} \times 9.8 \text{ m/s}^2 \times \cos 30° = 2.5 \text{ N}.$$

The work done by the frictional force as the block slides 5 m,

$$W_{\text{friction}} = \oint \mathbf{F}_f \cdot d\mathbf{x} = F_f x = 5 \times 2.54 = 12.7 \text{ J}.$$

The block starts from rest with only an initial potential energy

$$U_i = mgh,$$

where h is the height above the ground (taking the zero of potential at ground level). But $h = 5 \text{ m} \times \sin 30° = 2.5 \text{ m}$. Thus

$$U_i = 2 \text{ kg} \times 9.8 \text{ m/s}^2 \times 2.5 \text{ m} = 49 \text{ J}.$$

Conservation of energy requires that

$$U_i = \text{KE}_f + \text{work done by nonconservative forces}$$

$$49 \text{ J} = \tfrac{1}{2}mv_f^2 + 12.7 \text{ J}$$

or

$$v_f = \left(\frac{2 \times 36.3 \text{ J}}{2 \text{ kg}}\right)^{1/2} = 6.02 \text{ m/s}.$$

Even in this case, where mechanical energy is *not* conserved, we still applied the conservation principle, because the energy "lost" by the system due to nonconservative forces was calculable. Actually, no energy was lost: *No energy is ever lost; it just changes form.* For all forces and interaction, one can, in principle, always account for all the energy of a system.

11.12 Bounded and Unbounded Motion

THE MASS on a spring is an example of bounded motion. If the mechanical energy is increased (possibly by giving the mass a larger initial velocity or a greater spring compression), the motion of the mass remains bounded; only the turnaround points move farther from the equilibrium position. This is true only for our idealized machine, whose potential energy keeps increasing with increasing x.

Real potentials, of course, do not behave that way. A potential energy function that reasonably describes some types of interaction

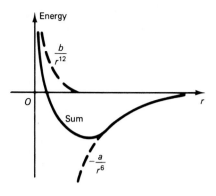

Fig. 11.33

between two atoms is shown in Fig. 11.33. At large separation there is a weak attractive force, which is represented by the potential function $-a/r^6$. At small distances between the atoms there is a sharply increasing repulsive interaction of the form b/r^{12}, where r is the internuclear separation and a and b are constants. The total interaction potential for this two-atom system is then the sum of these two terms:

$$U(r) = \frac{b}{r^{12}} - \frac{a}{r^6}. \qquad (11.87)$$

This is called the Lennard-Jones potential. Diatomic gas molecules such as H_2 and HCl can be described by this potential energy function. The zero of PE is taken at infinite separations, where there is no force between the two atoms.

Two classes of motion can be analyzed using this interaction potential (Fig. 11.34). First, consider the case where the total mechanical energy of the two atoms, E_1, is greater than zero ($E_1 > 0$). This is an example of unbounded motion. If the two atoms approach each other, their potential energy decreases (becomes more negative), but their kinetic energy increases until they reach a separation of r_0. Past r_0, the two atoms experience a repulsive interaction; their kinetic energy decreases, while their potential energy sharply rises. The repulsive force prevents them from getting any closer than r_{min}. At r_{min}, they reverse direction and begin to fly apart. If these were the only two atoms in the world, the repulsive interaction would push them away from each other, and they would never return. In effect, they escape each other's attractive force. If these atoms exist in a confined bottle of gas, they may approach each other again by

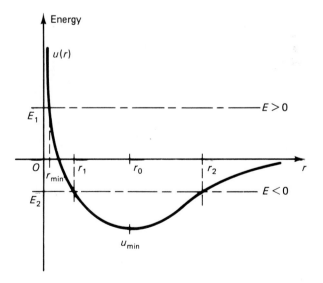

Fig. 11.34

chance during their random motion with the same results (assuming that they retain the same total mechanical energy). This will happen with all pairs of atoms whose *total* mechanical energy is greater than zero.

The case of two atoms whose total mechanical energy is negative ($E < 0$) is quite different. They are represented in Fig. 11.34 as having mechanical energy E_2. If two such atoms get close together, they will form a BOUND system: a molecule whose separation will oscillate between r_1 and r_2. In fact, it is possible to detect this vibrational oscillation using infrared optical techniques. From such optical measurements one can obtain information about the nature of the interaction potential between the atoms.

The Lennard–Jones interaction potential energy function illustrates a noteworthy feature of all bound systems:

> *The potential energy function must be at a minimum for the system to be in stable equilibrium.*

If there is no minimum in this function, it is not possible to have a stable bound system. Figure 11.35 shows a magnified view of a minimum region of an arbitrary potential energy function for a two-particle system. The PE function $U(r)$ has its minimum value at r_0. If the system has a total energy E_0, the two-particle separation will remain r_0, but if the total energy is raised by a small perturbation to E_1, the two particles will oscillate with their separation going from r_a to r_b.

Let's carefully examine this oscillatory behavior brought on by the small disturbance. You are all aware from your calculus course that a well-behaved function can be represented, in a limited region, by an infinite series called a Taylor series, where each term in the series contains derivatives of the function itself. We wish to represent a general potential energy function $U(r)$ in the region about the minimum (r_0) by such a Taylor series as

$$U(r) = U(r_0) + (r - r_0)\frac{dU}{dr}\bigg|_{r=r_0} + \frac{1}{2}(r - r_0)^2 \frac{d^2U}{dr^2}\bigg|_{r=r_0} + \cdot \cdot \cdot \quad (11.88)$$

This *infinite* series will accurately reproduce the actual function for small values of $(r - r_0)$. However, to a very good approximation, only a few terms in the series will describe small excursions about the potential minimum at r_0. The fact that we are expanding about a minimum value of $U(r)$ means that the first derivative of the function, evaluated at the minimum separation r_0, is zero:*

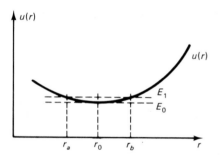

Fig. 11.35

---

* The fact that $dU/dr = 0$ at the minimum tells us that the force at that point is also zero (see Eq. 11.99).

$$\frac{dU(r)}{dr}\bigg|_{r=r_0} = 0 \tag{11.89}$$

and thus

$$U(r) \approx U(r_0) + \frac{1}{2}(r - r_0)^2 \frac{d^2U}{dr^2}\bigg|_{r=r_0}, \tag{11.90}$$

neglecting terms in $(r - r_0)^3$ and other higher-order terms in the series.

The absolute value of the potential at any point is arbitrary, as we have discussed previously. We choose to set the reference potential at r_0 so that

$$U(r_0) = 0 \tag{11.91}$$

and the second derivative of the PE function evaluated at the minimum r_0 is just a number that we shall call k, or

$$k = \frac{d^2U}{dr^2}\bigg|_{r=r_0}. \tag{11.92}$$

Finally, if we define x as the separation of the two bodies from their equilibrium position, or

$$x \equiv r - r_0. \tag{11.93}$$

then Eq. 11.90 reduces to

$$U(r) \approx \tfrac{1}{2}kx^2. \tag{11.94}$$

But this approximate expression is the same as Eq. 11.82, the potential energy function for a harmonic oscillator (a mass connected to an ideal spring). To the degree that our approximation holds (small excursions), we can identify the *curvature* of the PE function at the minimum as the spring constant or spring stiffness. In Chapter 15 we show that this spring constant k is linearly related to the square of the frequency of oscillation. Thus a flat PE curve indicates low-frequency oscillation, while a steep, large curvature function represents a system that will oscillate at a high frequency. One can experimentally determine the shape of the PE function by measuring the frequency of absorption of infrared light in diatomic gases (KCl, KBr, HCl) and thereby obtain basic information on the binding mechanism of these molecules.

11.13 **Force from Potential Energy**

IF WE CAN integrate a force times a distance along a prescribed path to obtain a potential, we might suspect that the mathematical process could be reversed. That is, a potential energy could be differentiated with respect to a spatial coordinate to obtain a force. This hunch is correct. To demonstrate this formally, we consider only the one-dimensional case. Recall that the potential energy at point x was defined through the equation

$$U(x) - U(a) = - \oint_{x_a}^{x} F(x) \, dx, \qquad (11.95)$$

where $U(a)$ is the reference potential at the point $x = a$, and x is any position along the x-axis that we choose to determine the potential. Consider the change in the potential energy as the particle moves from x to $x + \Delta x$, over a sufficiently small distance where the force can be considered constant. Then Eq. 11.95 becomes

$$U(x + \Delta x) - U(x) = - \int_{x}^{x+\Delta x} F(x) \, dx \approx -F(x) \int_{x}^{x+\Delta x} dx$$

$$\approx F(x)[(x + \Delta x) - x] \approx -\Delta x \, F(x). \quad (11.96)$$

But we define

$$U(x + \Delta x) - U(x) \equiv \Delta U(x), \qquad (11.97)$$

so
$$F(x) \approx - \frac{\Delta U(x)}{\Delta x}. \qquad (11.98)$$

Equation 11.98 is not a true equality, because we have taken $F(x)$ out of the integral equation 11.96, since it was assumed constant over distance Δx. This is true only in the limit of $\Delta x \to 0$, where Eq. 11.98 becomes a derivative

$$F(x) = - \frac{dU(x)}{dx}. \qquad (11.99)$$

This equation is just what we might have guessed. The force on an object is equal to the negative derivative of its potential energy with respect to a spatial coordinate. For three dimensions, one has three components of force F_x, F_y, and F_z. Each of them can be found by taking this spatial derivation with respect to the three corresponding spatial coordinates:

$$\boxed{F_x = - \frac{\partial U(x, y, z)}{\partial x} \quad F_y = - \frac{\partial U(x, y, z)}{\partial y} \quad F_z = - \frac{\partial U(x, y, z)}{\partial z}.} \quad (11.100)$$

Here we have introduced the operation of partial derivative, which is probably new to you. We will not use these equations extensively, but you will probably see partial differentiation in the future. The idea of such an operation is to take a derivative with respect to one variable at a time, while assuming that all the other variables are constant. For example, if the potential energy is given by the expression

$$U(x, y, z) = xy^2 + yz^3,$$

then

$$F_x = -\frac{\partial U}{\partial x} = -y^2 \quad (y, z \text{ assumed constant})$$

$$F_y = -\frac{\partial U}{\partial y} = -2xy - z^3 \quad (x, z \text{ assumed constant})$$

$$F_z = -\frac{\partial U}{\partial z} = -3yz^2 \quad (y, x \text{ assumed constant}).$$

That was just for fun. Let's get back to our one-dimensional world, where a force can easily be calculated from the slope of the potential energy diagram. Reexamine Fig. 11.34. At r_{min} the slope of the potential energy curve has its maximum negative value. This indicates that at r_{min} the largest positive force exists on the atoms. A positive force, in this case, means one that increases r (the separations). This, then, is a repulsive force. For the bound state E_2, when the atoms are separated a distance r_1, the slope is negative. The force is then positive or repulsive. When the atoms are at the separation r_2, the slope is positive and the force is negative, which is an attractive interaction.

If we know the potential energy curve at all points in space, we can calculate the force at all points. This entire discussion applies *only* to conservative forces, which are the only forces that can be characterized by a potential.

11.14 Stability

POTENTIAL ENERGY curves can also be used to determine the mechanical STABILITY of an object. By stability we mean that if the body is perturbed by a small external force from its equilibrium position, it will experience a restoring force, which will tend to return the body to its equilibrium position. This will happen regardless of the direction in which the perturbing force is applied. Think of an object the shape of an ice cream cone. Place it on a table with the point of the cone down and you have a classic case of an *unstable* body. At the slightest jarring of the table the cone will topple over. If you place it "ice cream side down" (with the round side on the table), the object would sustain a small motion of the table or other perturbation and return to its upright and stable position.

The potential energy curve directly reveals the stability of the system it represents. For example, consider a simple pendulum, a mass m supported by a rigid rod of length l (Fig. 11.36). The poten-

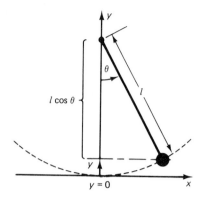

Fig. 11.36

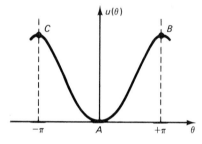

Fig. 11.37

tial energy of the pendulum is just

$$U(y) = mgy. \tag{11.101}$$

But from the geometry of the pendulum

$$y = l - l \cos \theta. \tag{11.102}$$

Thus $$U(\theta) = mgl(1 - \cos \theta). \tag{11.103}$$

Figure 11.37 is a plot of the potential energy of the pendulum as a function of θ. For three values of θ, A, B, and C, it is obvious that there is no net force on the pendulum. These are the places where the first derivative of the potential $dU/d\theta$ is zero. Consider B and C first. They are, in fact, the same position; the pendulum is upside down. They are also unstable; if the pendulum is slightly perturbed from exactly $\theta = \pm\pi$, it begins to fall and to execute periodic motion from $+\pi$ to $-\pi$ radians. On the other hand, if the pendulum is placed at position A, a small perturbing force would lead only to small oscillation around $\theta = 0$. A is a stable equilibrium position. Can we see this from Fig. 11.37?

If the pendulum is slightly displaced from $\theta = 0$ to $\theta = 0 + \Delta\theta$, the potential curve shows that the pendulum experiences a force in the minus θ-direction (the slope is positive, but the force is $-dU/d\theta$). If the pendulum is displaced $\theta = 0 - \Delta\theta$, the pendulum experiences a force in the plus θ-direction. No matter which way the perturbing force acts, the system responds with a restoring force that accelerates the system back toward its equilibrium position.

The situation is quite different at $\theta = \pm\pi$, points C or B. Although these are equilibrium positions, the net force is zero, yet they are *not* stable equilibrium points. If the pendulum is displaced ($\theta = \pi + \Delta\theta$) from either C or B, the slope of the potential curve becomes negative and the force on the pendulum is positive. The force accelerates the pendulum in the positive θ-direction away from these equilibrium points.

A straightforward mathematical test can be applied to determine the nature of the equilibrium positions. It is:

$$
\begin{array}{ll}
\text{EQUILIBRIUM POSITION} & \dfrac{dU}{dx} = 0 \\[2em]
\text{STABLE EQUILIBRIUM} & \dfrac{d^2U}{dx^2} > 0 \\[2em]
\text{UNSTABLE EQUILIBRIUM} & \dfrac{d^2U}{dx^2} < 0.
\end{array}
\tag{11.104}
$$

The second derivative of the potential energy curve at a potential minimum (equilibrium position) reveals the nature of the stability of this equilibrium.

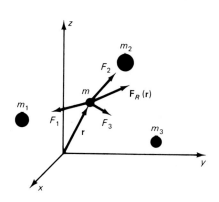

Fig. 11.38

THE CONCEPT of a field is one of the most pervasive ideas in modern physics. It is particularly important in the study of electricity, magnetism, and particle physics, but it is also a useful concept in mechanics. Suppose that we have three large masses m_1, m_2, and m_3 fixed somehow in space, as shown in Fig. 11.38. A small test mass m is placed at an arbitrary position $\mathbf{r}$. Newton's universal law of gravity predicts that each of the large masses will produce an attractive force on the test mass m. The vector sum of these forces is another vector force, the resultant force on the test mass $\mathbf{F}_R(\mathbf{r})$. In fact, at every position $\mathbf{r}$ in space, we can calculate the resultant force on this test particle, $\mathbf{F}_R(\mathbf{r})$. Since the gravitational force on the object is proportional to this mass, we write

$$\mathbf{F}_R(\mathbf{r}) = m\mathbf{Q}(\mathbf{r}), \qquad (11.105)$$

where $\mathbf{Q}(\mathbf{r})$ is the vector force field. We say that the three objects m_1, m_2, and m_3 generate the GRAVITATIONAL FORCE FIELD $\mathbf{Q}(\mathbf{r})$ in space. That is, these three masses *create a condition* for every point in space (at a fixed time).

But if the gravitational force can be represented by a field, the gravitational potential energy also has a field associated with it. In fact, all our plots of potential energy curves in one dimension could be thought of as plots of the gravitational potential *fields*. That is, for every point in space, there is a *scalar* quantity, call it $\psi(x, y, z)$, which, when multiplied by the mass of the test particles, gives the gravitational potential energy of the particle. So what?

What have we accomplished with all this discussion of fields? To begin with, fields have a life of their own! If the three large masses were suddenly to disappear, at that instant, some distance away, the fields would *not* change. The information that the masses have disappeared takes a finite time to travel to where the test mass resides. Consider our starry sky. We are probably observing electromagnetic fields (which we call light) from some stars far out in the universe that long ago burned out and are no longer giving off light. Yet the fields continue to exist.

Scalar potential fields are useful in solving some types of problems, because they add *algebraically*. That is, the field caused by m_1 at point $\mathbf{r}$, $\psi_1(\mathbf{r})$, the field from m_2 at point r, $\psi_2(\mathbf{r})$, and from m_3 at the same point, $\psi_3(\mathbf{r})$, can be added together to get the total potential field due to all three masses at r:

$$\psi_T(\mathbf{r}) = \psi_1(\mathbf{r}) + \psi_2(\mathbf{r}) + \psi_3(\mathbf{r}). \qquad (11.106)$$

This is an algebraic addition of three scalar quantities. Once the scalar potential field $\psi_T(\mathbf{r})$ is known at all points in space, Eq. 11.100

can be used to obtain the vector force field at all points also. It is often much easier to calculate a scalar field (by algebraic addition) than a vector field, where you must deal with the three components of the force at each point. If you think about the meaning of Eq. 11.100, you will see that the force field is proportional to the spatial rate of change of potential energy. If the potential is constant over some region of space, the force is zero in that region. That is consistent with our discussion, since no work is needed to move a body through a region where the potential is constant.

A few words of caution. This entire discussion applied to conservative force. One cannot define a potential for a nonconservative force. If you have only the value of the potential at *one* point, you cannot calculate the force at that point. It is *not* possible to calculate a rate of change from a knowledge of one point; the potential in the region around the point of interest is needed. In general, we must know the potential field in an extended region of space to calculate the force field in the same region.

11.16 Binding Energy: Relativity

HOW DOES relativistic mechanics treat the concept of potential energy and work? We have already derived the work–kinetic energy theorem for relativistic particles in Section 11.6, and we have alluded to the way that relativity deals with potential energy when we discussed the energy–mass equation $\Delta E = \Delta mc^2$. Recall that the mass of the proton plus the mass of the neutron is larger than the mass of the two particles bound together as the nucleus of the neutron. Where has the mass gone? We said it went into the binding energy of the compound nucleus. That is, the two particles separated have a larger combined energy than when they are bound together. Another way to understand this phenomenon is to be aware that energy must be put into the system of two bound particles to separate them.

Relativity is truly democratic, for it treats all forms of energy equally. If you put energy into a system its mass will change. A glass of hot water has a greater mass than an equivalent glass of cold water. A compressed spring has a greater mass than one in equilibrium. Classical physics treats each form of energy, such as mechanical energy, elastic energy, heat energy, and so on, as a separate entity. Relativity puts all forms of energy on an equal footing.

$$dW = \mathbf{F} \cdot d\mathbf{s}$$

$$W_{1-2} = \oint_{1}^{2} \mathbf{F} \cdot d\mathbf{s}$$

$$W = \Delta KE \qquad \text{(work done by } net \text{ force)}$$

$$F_{\text{spring}} = -kx \qquad \text{(Hooke's law for ideal spring)}$$

$$\text{power} = \frac{dW}{dt}$$

$$W = Pt \qquad \text{(constant power)}$$

$$PE = -\int_{r_{\text{ref}}}^{r} \mathbf{F} \cdot d\mathbf{r} = -W$$

$$U(\text{gravitational, near Earth}) = mgy$$

$$U(\text{gravitational, far from Earth}) = -\frac{GM_E m}{r}$$

$$d\mathbf{r} = dr\hat{\mathbf{r}} + r\, d\theta\, \hat{\boldsymbol{\theta}} \qquad \text{(two dimensions, polar coordinates)}$$

$$F(x) = -\frac{dU(x)}{dx}$$

$$1 \text{ joule} = 1 \text{ newton-meter} = 1 \text{ kg-m}^2/\text{s}^2$$

$$550 \text{ ft-lb/s} = 746 \text{ watts} = 1 \text{ horsepower}$$

Chapter 11 QUESTIONS

1. The planets move around the sun. Does the sun do work on the planets? Explain.

2. In his experiments Rutherford directed α-particles at a gold foil and measured the scattering of the α-particles. He observed some of these particles deflecting through an angle of 180°, coming back toward the source. What work does the nucleus of gold do *on* the α-particle on its way in? What work is done on the α-particle during a round trip?

3. A cyclist pedals down a hill into a stiff headwind in such a way that her velocity remains constant. Describe the various forces that are acting on the system. Are these forces conservative or nonconservative in nature? What remains constant in the system: the potential energy, the kinetic energy, the mechanical energy, the work done by the conservative forces, the work done by all the forces, or the work done by the nonconservative forces?

4. Consider a particle moving in a conservative force field described by a potential energy diagram Fig. 11.39. Consider two particles moving in the same po-

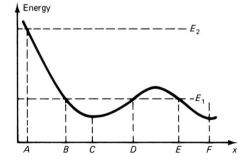

Fig. 11.39

tential, but with different starting conditions and thus different total mechanical energy E_1 and E_2. Discuss the motion of both particles, particularly at the points A, B, C, D, E, and F.

5. A spring is kept in a compressed state by a string with its ends tied together. If the spring is placed in an acid bath and the spring dissolves, what happens to the potential energy that was stored in the spring?

6. How would you represent a satellite in a circular orbit on a potential energy diagram? How would you represent the same satellite in an elliptical orbit?

7. A ball is thrown straight up in the air and subsequently caught at the point from which it was originally thrown.

    (a) Does the gravitational force do net work on the ball?

    (b) Suppose that air friction was significant and the speed of the ball was reduced when it was caught. Does gravity do net work on the ball? Explain.

    (c) In part (b), does the air do positive or negative work on the ball? Explain.

Chapter 11 PROBLEMS

1. A boy pulls on a 25-kg block with a rope inclined at an angle of $27°$ with respect to the ground (Fig. 11.40).

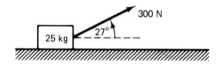

Fig. 11.40

    (a) If he exerts a force of 300 N and the block has a coefficient of sliding friction of .17, how much work does the boy do in moving the block 15 m along the ground?

    (b) Calculate the net work done on the block by the resultant force.

2. A 75-kg block of ice slides down an inclined plane with a man exerting a force on it parallel to the plane to keep it moving at a constant velocity. If the plane is 6 m long and 4 m high and the coefficient of friction of the ice on the plane is a constant $\mu = .2$, calculate the following:

    (a) The force exerted by the man.

    (b) The work done by gravity.

    (c) The work done by the man.

    (d) The work done by friction.

    (e) The work done by the resultant force.

    (f) The change in kinetic energy of the block of ice.

3. A body of mass 15 kg experiences a horizontal force of 150 N as it moves up a $15°$ inclined plane (Fig. 11.41). If the coefficient of sliding friction $\mu = .2$, and the incline is 15 m long:

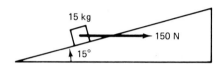

Fig. 11.41

    (a) Find the total work done by the force $F = 150$ N.

    (b) Find the work done by friction.

    (c) Find the work done by gravity.

    (d) If the body starts at the bottom with a speed of 2 m/s, find the speed at the top of the plane.

4. A bullet traveling at a speed of 600 m/s penetrates 15 cm into a wooden block. If the mass of the bullet is 45 g, what is the *average* force that the bullet exerts on the block? Assume that the wooden block exerts constant force on the bullet.

5. A block of mass 25 kg is subjected to a spatially varying force F whose values are plotted in Fig. 11.42. If the block starts out with an extremely small velocity at $x = 0$, find its speed at $x = 10$ m and 20 m.

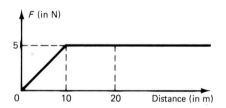

Fig. 11.42

6. Consider the mechanical system shown in Fig. 11.43. Mass m_1 moves along a surface with a frictional coefficient μ, and m_2 slides down a frictionless plane a distance x. Find the velocity of m_1 after m_2 has moved a distance x, assuming that the system was initially at rest. Solve the problem two ways: (a) using the work–energy principles, and (b) using forces.

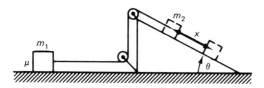

Fig. 11.43

7. Find the velocity of the mass m_1 after the mass m_2 has fallen a distance l in Fig. 11.44. Assume no friction in the problem and that the string and pulley have negligible mass.

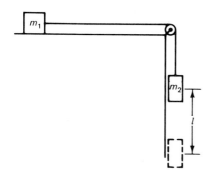

Fig. 11.44

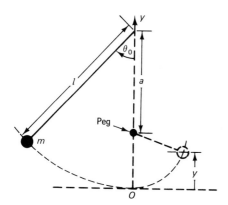

Fig. 11.45

8. A simple pendulum consists of a mass m suspended by a string of length l (Fig. 11.45). It starts from rest at an angle θ_0. The mass swings down, and when $\theta = 0$, the string hits a rigid peg at $(y = l - a)$. Find the height y that m will travel swinging with the new arc of radius $(l - a)$. Galileo performed this experiment.

9. A small block of putty of mass m falls a distance h, landing and sticking to a platform of mass M that is supported by a spring (Fig. 11.46). If the spring constant is k, find the maximum excursion Δy of the platform after the collision. Assume that $M = 7.0$ kg, $m = .045$ kg, $k = 9.3$ N/m, and $h = 6.0$ m.

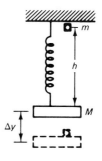

Fig. 11.46

10. A small block of mass m is tethered by a light string of length L on a friction-free inclined plane as shown in Fig. 11.47. If the tension in the string at the lowest point in the orbit is T:

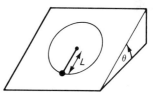

Fig. 11.47

    (a) Find the speed of the block at the lowest point.

    (b) Find the tension in the string and the speed at the highest point in terms of the given parameters.

11. A block of ice with mass M has a circular arc of radius R carved in it (Fig. 11.48). A small steel block of mass m slides down the circular arc and leaves the larger block. If the entire system is initially at rest, if there is no friction anywhere (even the large block slides without friction along the horizontal plane), and if the little block starts from rest at the top of the arc, find the velocity of the small mass as it leaves the larger block for

    (a) $M = m$, equal masses; and for **(b)** $M > m$.

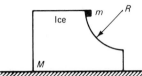

Fig. 11.48

12. A ballistic pendulum (Fig. 11.49) is a simple device that can be used to determine the velocity of a bullet fired from a gun. A large, usually wooden block is suspended by four strings, as shown in the diagram. The bullet is fired into the wooden block, where it quickly comes to rest, and the block swings up through an angle θ. Prove that the initial velocity of the bullet is related to the angle of swing by the equation

$$v_0^2 = \frac{m_1 + m_2}{m_1^2}[2g(1 - \cos\theta)],$$

where m_1 is the mass of the bullet and m_2 the mass of the wooden block.

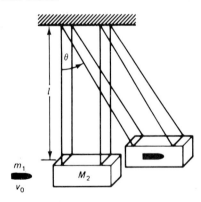

Fig. 11.49

13. A block of mass m slides across a frictionless table traveling at a speed of v_0 and collides with a massless spring (Fig. 11.50). This spring does *not* obey Hooke's law but produces a restoring force given by the expression

$$F(x) = -\alpha x - \beta x^3.$$

Fig. 11.50

How far will the mass travel *after* it collides with the spring, before it comes to a complete stop?

14. An ideal massless spring can be compressed .7 m by a 75-N force. This spring is placed at the bottom of a 30° inclined plane. A mass of 8 kg is released from rest at the top of this plane and is brought instantaneously to rest after compressing the spring 2.5 m (Fig. 11.51). What is the total distance along the plane that the mass slides before coming to momentary rest, assuming no friction?

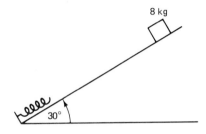

Fig. 11.51

15. A small block of mass m starting from rest slides on a frictionless spherical surface of radius r (Fig. 11.52).

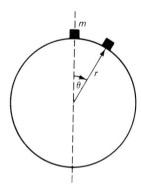

Fig. 11.52

    (a) Write the expression for the potential energy as a function of θ if the zero of potential is taken at the *top* of the sphere.

(b) Find the expression for the radial and tangential acceleration of the block as a function of θ, assuming that the block is in contact with the sphere.

(c) Find the angle θ at which the block leaves the surface of the sphere.

(d) Imagine that the surface *has* friction. Does the block remain in contact with the sphere for a longer, shorter, or the same distance? Explain.

16. Can a horse deliver 1 hp? Consider a single horse pulling a barge with a long rope. If the tension in the rope is 385 N at an angle of 45° with respect to the direction of the barge, and the horse walks at a speed of 5.5 km/h, is the horse delivering 1 hp?

17. Only a fraction of the water in the Niagara River is diverted to the hydroelectric generators. About 6000 m³/s of water flows over the falls and drops 50 m.

(a) How many watts of gravitational power is unused?

(b) How much energy is unused every year?

18. An energy-conscious student is given an electric toothbrush that uses 85 W of power. He estimates that he will brush his teeth 12 minutes a day, every day. To avoid consuming more energy than he did before he received this device, he decides to reduce the use of his car. If his car delivers 90 hp at 100% efficiency, how much driving will he have to give up in one year to keep his energy consumption constant?

19. How fast can a 30,000-kg truck with a 400-hp engine travel along a road that has a 7° incline? (neglect air resistance.)

20. An automobile travels at a constant speed of 30.0 m/s along a level road. Its motor is producing energy at a rate of 300 kW. What is the equivalent total resistive force on the automobile?

21. An automobile weighs 1000 kg and is moving 36 km/h.

(a) Find its kinetic energy (in joules).

(b) Its bumper is mounted on springs that allow 10 cm of motion before they are completely compressed. If the bumper just succeeds in preventing damage in a 36-km/h collision with a solid wall, what is the spring constant (magnitude and units)?

(c) At the instant during the collision when the kinetic energy is half its initial value, find the power being transferred from car motion to spring compression.

22. A world-class weight lifter raises a 195-kg mass a vertical distance of 2 m in 3 s. What is the average power output of his body during this time?

23. How much mass is converted into energy in one year by a nuclear reactor electric power generating station that produces 100 MW of electric power and waste heat at a rate of 150 MW.

24. Use energy considerations to show that the maximum height of a bullet fired from a gun is

$$y_{max} = \frac{v_0^2 \sin^2 \theta}{2g},$$

where v_0 is the initial speed of the bullet and θ is the angle of the initial trajectory above the horizontal (Fig. 11.53). Use energy considerations to calculate the ratio of the speed at the top of the trajectory to the initial speed.

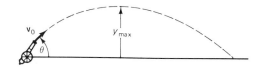

Fig. 11.53

25. A small block of mass m slides on a frictionless track in Fig. 11.54. If the block starts from rest:

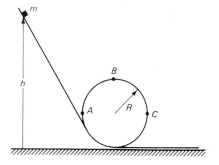

Fig. 11.54

(a) How high above the ground does the mass have to start so that it will make the entire loop without losing contact with the track?

(b) What is the velocity of the block at points A, B, and C?

(c) What is the velocity of the block when it exits this loop?

(d) Assuming that the block is raised *above* the minimum height necessary to keep it on the track, find the expression for the normal force at point *B* that the block exerts on the track.

26. A ball is whirled in a vertical circle by a massless string so that its total *mechanical energy* is always *constant*. Show that under these conditions, the tension in the string when the ball is at the bottom of the circle is greater than the tension at the top of the circle by six times the weight of the ball.

27. Consider the system of a bead of mass *m* attached to a spring (of spring constant *K*) that slides over a vertical hoop of radius *R* without friction (Fig. 11.55). If the equilibrium length of the spring is

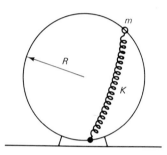

Fig. 11.55

taken to be *zero* (to make the problem simple), show that the velocity of the bead at the bottom of the hoop is given by

$$v_f = 2 \left(gR + \frac{KR^2}{m} \right)^{1/2}$$

if the bead started at the top of the hoop.

28. A .025-kg block is pressed with a force of 25 N against a spring that has been compressed to half its equilibrium length of .05 m (Fig. 11.56). The block is then released and propelled by the spring, traveling 3.6 m after it leaves the spring. Find the coefficient of sliding friction between the block and the horizontal surface.

Fig. 11.56

29. A 2.6-kg block slides with friction on a 26° slope as shown in Fig. 11.57. It is attached to an ideal spring whose constant $k = 26$ N/m. If the block starts off at a position where the spring is unstretched and slides .32 m before stopping, find the coefficient of friction of the block with the plane.

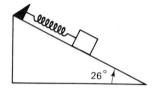

Fig. 11.57

30. The Demon Drop at Cedar Point amusement park in Sandusky, Ohio, is a popular ride. The "victims" are caged in an enclosed car and elevated *h* m above the ground, where they are released on a metal track (Fig. 11.58). The car drops in free fall on the track, then turns and rises on a *θ* slope for *l* m before coming to rest. If the only friction between the car and the track occur on the slope, find this coefficient of friction.

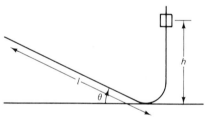

Fig. 11.58

31. A 2500-kg elevator is falling downward at 2 m/s, 20 m above a safety spring at the bottom when its cable breaks (Fig. 11.59). An automatic safety device instantaneously clamps the elevator to the running track, producing a constant frictional force of 3000 N on the elevator. This clamp slows down the elevator's descent. If the safety spring has a force constant $k = 80,000$ N/m:

(a) Find the speed of the elevator just before it hits the spring.

(b) Calculate the distance the spring is compressed.

(c) Find the height that the elevator will bounce back.

(d) Use the conservation principles to find the approximate *total* distance the elevator will move before coming to rest.

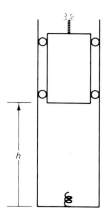

Fig. 11.59

32. A spherically symmetric force acts on a particle according to the formula

$$F(r) = \frac{C}{r^3}\,\hat{\mathbf{r}},$$

where r is the radial distance. Find the expression for the potential energy as a function of r if $U(r) = 0$ for $r = \infty$.

33. Obtain the expression for the potential energy of an electron in a uniform electric field E. Where is the zero of potential energy?

34. One early theoretical model of nuclear forces predicted a functional form for the potential energy associated with the interaction between two nucleons to be

$$U(r) = -U_0\,\frac{b}{r}\,e^{-r/b},$$

where U_0 and b are constants. This is called the Yukawa potential.

(a) Plot $U(r)$ versus r, using $U_0 = 50$ MeV and $b = 1.5 \times 10^{-15}$ m.

(b) Find the expression for the force between the particles. Is it attractive or repulsive?

(c) Compute the ratio of the forces at $2b$, $4b$, and $6b$ to the force at b.

35. Consider a force field described by the function

$$\mathbf{F} = y^2\mathbf{i} + 2xy\mathbf{j}.$$

Compute the line integral from the origin $(0, 0)$ to some point on the x–y plane (x_0, y_0) along a path made up of two straight sections: from $(0, 0)$ to $(x_0, 0)$ and $(x_0, 0)$ to (x_0, y_0); and do it again for another path from $(0, 0)$ to $(0, y_0)$ and $(0, y_0)$ to (x_0, y_0). Compare these two path integrals. Is this force conservative?

36. A particle moves in a one-dimensional world subject to a force described by the expression $F(x) = 3x^3 - x^2$ (in newtons).

(a) Find the work done on a particle by this force if it moves from the original to a distance 1.8 m.

(b) Find the total work done by this force if it moves from the origin to 1.8 m and returns to the origin.

37. Show that if you lived in a one-dimensional world and all the forces depended only on position, all these forces would necessarily be conservative.

38. Show that a force described by the expression

$$F(r) = \frac{C}{r}\,\hat{\boldsymbol{\theta}},$$

which acts in the $\hat{\boldsymbol{\theta}}$ direction but depends *only* on the radial distance from the origin r, appears to be a conservative force if the particle's path does not *enclose* the origin. Is the force conservative? How much work is done if the path does enclose the origin?

39. You have taken a space trip to a far-off galaxy and made an amazing discovery. You discovered planet Z, which is composed of a new form of matter that obeys a different law of gravitational attraction. This law can be written

$$\mathbf{F}_{12} = G'\,\frac{m_1 m_2}{r_{12}^3}\,\hat{\mathbf{r}},$$

where F_{12} is the force between mass m_1 and m_2 and r_{12} is the distance between the two objects (Fig. 11.60). If planet Z has the same radius and the same mass as Earth:

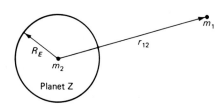

Fig. 11.60

(a) Find the expression for the potential energy as a function of r for a particle of mass m.

(b) Calculate how much initial velocity you need on your rocketship of mass m_R to leave planet Z and escape its attractive force forever.

(c) Draw a neat energy diagram to represent the motion of your rocketship and another rocketship that did not have sufficient fuel to escape and is caught forever on planet Z.

40. Consider a classical model of the hydrogen atom where the electron is orbiting the proton at a distance of 1.5×10^{-8} cm and the proton is at rest.

(a) Find the velocity of the electron by equating the centripital and electrostatic forces.

(b) What is the kinetic and potential energy of the system? Give your answer in both joules and electron-volts.

(c) How much energy is needed to ionize this atom; that is, how much energy is necessary to remove the electron to infinity with no final kinetic energy?

41. A two-dimensional force field is represented by a potential function

$$U(x, y) = \alpha(x^2 + y^2).$$

(a) Find the expression for the force components F_x and F_y.

(b) Find the expression for the force components F_r and F_θ in polar coordinates.

(c) What real physical object might be described by such a force?

42. The potential energy of a particle is given by the expression

$$U(r) = -\frac{\alpha}{r},$$

where α is a constant, and r is the length of the radius vector in two dimensions. Derive the x and y components of force on the particle.

43. The Lennard-Jones expression for the energy between two atoms in a diatomic molecule is

$$U(r) = -\frac{A}{r^6} + \frac{B}{r^{12}},$$

where r is the separation between atoms and A and B are numerical constants.

(a) Find the equation for the force between the atoms as a function of r.

(b) Find the equilibrium position of the two atoms. Is this a stable equilibrium point?

(c) What energy is necessary to separate the atoms far apart from their equilibrium configuration? This is called the dissociation energy.

44. A particle of mass m moves in a one-dimensional conservative force field described by a potential $U(x)$ with a total mechanical energy E_0.

(a) Show that the speed of the particle is given by

$$\left[\frac{2(E_0 - U(x))}{m} \right]^{1/2}.$$

(b) Prove that the distance dx traveled in a time dt is

$$\frac{dx}{[E - U(x)]^{1/2}} = \left(\frac{2}{m} \right)^{1/2} dt.$$

(c) If the potential is $U(x) = \alpha x^2$ and the total energy is $E_0 = \alpha A^2$ (where A is the maximum displacement), integrate the expression above to find the total time the mass takes to move from $x = -A$ to $+A$.

45. The force of a nonlinear (not a Hooke's law) spring can be written

$$F(x) = -kx(1 - \alpha x^2),$$

where x is the displacement (compression or elongation) from the equilibrium position. Find the potential energy function for this spring.

46. A particle experiences a conservative force given by the expression

$$\mathbf{F} = a(y\mathbf{i} + x\mathbf{j}),$$

where a is a constant whose units are newton/meter. Find the total work done on the particle in going from the origin to the position $x = 3$ m, $y = 5$ m in the x–y plane.

47. The cyclotron, invented by E. O. Lawrence and M. S. Livingston in 1932, is a device for accelerating charged particles to high speed. The trajectory of a proton in a proton cyclotron is indicated schematically in Fig. 11.61. A slowly moving proton enters a region of a magnetic field B that is perpendicular to its direction of motion. The magnetic field exerts a force that causes the proton to move in a circular orbit. When the proton orbits 180°, it enters a small gap between the metal dees, where it experiences a strong electric field (magnitude E) which accelerates

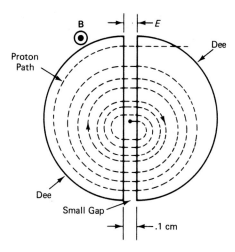

Fig. 11.61

it. Upon leaving the gap, the magnetic field once more bends it in a circular orbit (this time of larger radius, since the proton is moving faster) until it once more enters the gap. However, in this portion of the gap it finds itself in an electric field whose direction has been reversed, so that once again the proton is accelerated. In this manner the proton spirals out-

ward from the center of the cyclotron with ever-increasing speed until it reaches the rim of the device.

(a) If the electric field in the gap is $E = 15 \times 10^6$ V/m and the gap is .1 cm wide, what is the change in velocity of the proton as it moves through the region of the gap?

(b) If we assume that the proton enters the gap with zero velocity and that $B = 16$ tesla, what is the speed of the proton as it leaves the gap for the first time, and what is the radius of its first semicircular orbit?

(c) Find the speed of the proton after it passes through the gap for the nth time.

(d) What is the radius of the nth semicircular orbit?

(e) What is the final speed of the proton after 10^3 passes through the gap? Should we use relativistic rather than Newtonian equations to describe this motion? (Assume that the proton starts at rest.)

(f) What is the necessary minimum diameter of the cyclotron?

(g) How much time does the proton spend in the cyclotron? (Neglect the amount of time it spends traversing the gap.)

12

RIGID BODIES

12.1 Introduction

OUR DEVELOPMENT of mechanics has come a long way, but throughout this discussion we have treated each body as if it were a small particle, with the entire mass concentrated at its geometric center and with the applied forces acting through that mathematical point. Our discussion of the force laws and conservation principles has been in terms of these "point" masses. Since every macroscopic object can be thought of as made up of many pointlike particles, we could assume that our discussion is now complete. In reality, the motion of an extended object is the combined motion of its many point masses. If we know the forces on each point mass and we know the laws of motion that govern each point mass's response to these forces, we can, *in principle*, calculate the trajectories of every point mass in the extended body and solve the problem. Of course, if we adopt this strategy, we would have to solve about 10^{25} simultaneous equations to describe an ordinary ball rolling down an inclined plane. Surely there must be an easier way!

In this chapter we study the motion of RIGID BODIES; that is, objects whose distances between *any* pair of its point masses remain constant at all times. In fact, real extended objects always deform to some extent when they are subjected to external forces, and their constituent parts vibrate about their equilibrium positions. But we must retain this "special effect" to keep the discussion conceptually and mathematically manageable. Henceforth, unless explicitly noted, all our extended objects will be treated as if they were ideal rigid bodies. We will explore Newton's laws as they apply to extended bodies, introducing the concepts of center of mass and static equilibrium. As a bonus, we will follow Einstein's reasoning in the celebrated "gedanken experiment," which led him to discover the relationship between energy and mass.* All that and more in the next pages.

12.2 Center of Mass of Extended Bodies

HAVE YOU ever watched a baton twirler throw her baton in the air and catch it while marching to the beat of the band? You might wonder if the equations of motion you have been learning in physics describe the apparently complicated motion of the baton. You might also wonder how she ever catches it! In simple translational motion of an extended object, each part of the body experiences the same displacement, so that one may treat this motion the same as the motion of a single point mass. However, if the body is rotating or vibrating as well as translating, it is clear that all parts of the body do not move with the same velocity and acceleration at all times. Complicated as the motion of this rotating, spinning, and flying projectile appears to be, there is one *point* on (or near) the object that moves just as an imaginary point particle would move if this fictitious particle were subjected to the same external forces. This point on or near the body is called the CENTER OF MASS of the body. Figure 12.1 shows the parabolic trajectory of the center of mass of an asymmetric baton in flight near the Earth's surface. The center of mass acts just like a point particle, moving in space in a rather uncomplicated parabolic path that one can calculate using the equation in Section 9.9. In fact, Newtonian mechanics has been kind to us, for it permits the "dissection" of the motion of an extended body into two parts: the motion *of* the center of mass and the motion *about* the center of mass. We will shortly demonstrate that the motion *of* the center of mass is particularly straightforward.

The location of the center of mass of a many-particle system such as the one shown in Fig. 12.2 can be inferred from our definition of the center of mass for a two-particle system (Eq. 10.21). It is

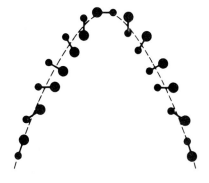

Fig. 12.1

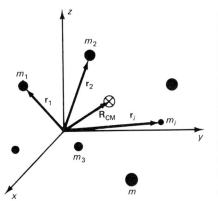

Fig. 12.2

$$\mathbf{R}_{CM} = \frac{1}{M} (m_1\mathbf{r}_1 + m_2\mathbf{r}_2 + m_3\mathbf{r}_3 + \cdots + m_N\mathbf{r}_N)$$

$$= \frac{1}{M} \sum_{i=1}^{N} m_i\mathbf{r}_i, \qquad (12.1)$$

where $M = \sum_{i=1}^{N} m_i$ (the total mass of the system).

---

* A gedanken experiment is a thought experiment; you never carry it out but instead, think through the process and predict the results.

The center of mass is a mass-weighted average of the individual positions of each particle that composes the extended object. It is important to recognize that the center of mass may be located at a position in space where there are *no* particles. For example, the center of mass of a donut (not the jelly kind) lies at the geometric center of the hole. The center of mass is a mathematical abstraction useful for our study of dynamics; *it is not a physical feature of the extended body.*

Equation 12.1 is a compact definition of the location of the center of mass. You may be more comfortable in calculating the center of mass of a system of mass using Eq. 12.1 in its component form. Constructing a Cartesian coordinate system for the system, it is clear the Eq. 12.1 can be written as three equations:

$$x_{CM} = \frac{1}{M}(m_1 x_1 + m_2 x_2 + \cdots + m_n x_n)$$

$$y_{CM} = \frac{1}{M}(m_1 y_1 + m_2 y_2 + \cdots + m_n y_n) \qquad (12.2)$$

$$z_{CM} = \frac{1}{M}(m_1 z_1 + m_2 z_2 + \cdots + m_n z_n).$$

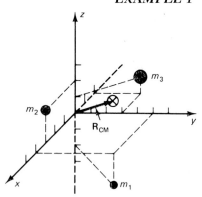

Fig. 12.3

EXAMPLE 1   Find the center of mass of the discrete three-particle system shown in Fig. 12.3, if $m_1 = m$, $m_2 = 2m$, and $m_3 = 5m$, and the position vectors of the three masses are

$$\mathbf{r}_1 = 4\mathbf{i} + 5\mathbf{j} - 2\mathbf{k}$$

$$\mathbf{r}_2 = 3\mathbf{i} + 2\mathbf{k}$$

$$\mathbf{r}_3 = -2\mathbf{i} + 3\mathbf{j} + \mathbf{k}.$$

From the definition of the location of the center of mass (Eq. 12.1),

$$\mathbf{R}_{CM} = \frac{1}{m + 2m + 5m}[(4m + 3 \times 2m - 2 \times 5m)\mathbf{i}$$

$$+ (5m + 3 \times 5m)\mathbf{j}$$

$$+ (-2m + 2 \times 2m + 1 \times 5m)\mathbf{k}]$$

$$= \frac{1}{8m}(0\mathbf{i} + 20m\mathbf{j} + 7m\mathbf{k})$$

$$= 2.5\mathbf{j} + .875\mathbf{k}.$$

Note that the masses have all canceled out, as they should, since $\mathbf{R}_{CM}$ is a position vector whose SI units are meters. Here, as in many other examples, the center of mass is located at a point in space where there is no particle. We can get the same result using the three-component equations 12.2.

How do we find the center of mass of an extended continuous body such as a meterstick, a solid rod, or a pyramid? The number of atoms in a macroscopic object is so large, and the spacing between atoms so small, that we may treat such objects as though they have a continuous distribution of mass. In principle we can then divide the object into N small elements each with mass Δm_i, located at the position $\mathbf{r}_i = x_i\mathbf{i} + y_i\mathbf{j} + z_i\mathbf{k}$ (Fig. 12.4). Forming the sums of the N mass elements and the N position coordinates, we have from our definition of center of mass in component form:

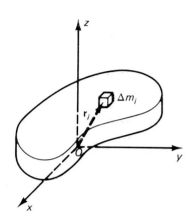

Fig. 12.4

$$x_{CM} = \frac{\sum\limits_{i=1}^{N} \Delta m_i x_i}{\sum\limits_{i=1}^{N} \Delta m_i} \qquad y_{CM} = \frac{\sum\limits_{i=1}^{N} \Delta m_i y_i}{\sum\limits_{i=1}^{N} \Delta m_i} \qquad z_{CM} = \frac{\sum\limits_{i=1}^{N} \Delta m_i z_i}{\sum\limits_{i=1}^{N} \Delta m_i}. \qquad (12.3)$$

If we now further subdivide the mass elements into smaller elements and continue this division as N tends toward infinity, the position of each element becomes more precise, and the summations become integration (by the fundamental theorem of calculus).

$$x_{CM} = \lim_{\Delta m_i \to 0} \frac{\sum\limits_{i=1}^{N} \Delta m_i x_i}{\sum\limits_{i=1}^{N} \Delta m_i} = \frac{\int x\,dm}{\int dm} = \frac{1}{M}\int x\,dm$$

$$y_{CM} = \lim_{\Delta m_i \to 0} \frac{\sum\limits_{i=1}^{N} \Delta m_i y_i}{\sum\limits_{i=1}^{N} \Delta m_i} = \frac{\int y\,dm}{\int dm} = \frac{1}{M}\int y\,dm \qquad (12.4)$$

$$z_{CM} = \frac{1}{M}\int z\,dm.$$

Equations 12.4 can all be compressed into one vector equation:

$$\mathbf{R}_{CM} = \frac{1}{M}\int \mathbf{r}\,dm. \qquad (12.5)$$

If the density of a body varies with position, it is more useful to write Eq. 12.5 in terms of the density $\rho(x, y, z)$ and the volume of integration dV. The smallest element of mass dm may be written

$$dm = \rho(x, y, z)\,dV \qquad (12.6)$$

$$\mathbf{R}_{CM} = \frac{\int \mathbf{r}\rho(x, y, z)\, dV}{\int \rho(x, y, z)\, dV} \tag{12.7}$$

Equation 12.7 is correct and complete, but sometimes it can be a formidable integral to evaluate. In its most general form, the volume element in Cartesian coordinates is $dV = dx\, dy\, dz$, which is an infinitesimal parallelpiped with sides dx, dy, and dz. Calculation of $\mathbf{R}_{CM}$ may require performing three different integrations, the limits of which may be interdependent. In our discussion we consider only examples that can be reduced to integrals over a single variable.

EXAMPLE 2
Center of Mass of a
Semicircular Hoop

Figure 12.5 shows a solid rod of mass M and a cross-sectional radius small compared to R bent in the form of a semicircular hoop. Find its center of mass.

The coordinate system chosen to solve this problem reflects the symmetry of the physical system. For example, it should be apparent that the center of mass must lie on the y-axis, since for every element at $-x$ there is an equal mass element at $+x$. The problem then reduces to finding the y-component of $\mathbf{R}_{CM}$:

$$y_{CM} = \frac{1}{M} \int y\, dm.$$

If we do the integration in polar coordinates, all mass elements are located at the same radius,* R. The infinitesimal element of mass dm is the mass per unit length times the infinitesimal length of arc $ds = R\, d\theta$, so

$$dm = \alpha R\, d\theta,$$

where α = the mass per unit length = $M/\pi R$ and $y = R \sin \theta$. Substituting into the equation for y_{CM} gives

$$y_{CM} = \frac{1}{M} \int_0^\pi R \sin \theta\, \alpha R\, d\theta = \frac{R^2 \alpha}{M} \int_0^\pi \sin \theta\, d\theta = \frac{R^2 \alpha}{M} \left[-\cos \theta \right]_0^\pi$$

$$= \frac{2R^2 \alpha}{M} = \frac{2R^2 M}{M \pi R} = \frac{2R}{\pi}.$$

That was easy. Now let's examine the answer. The center of mass is about two-thirds of the way up the y-axis, *not* in the middle at $R/2$. Why is that? The mass is symmetrically distributed with respect to the x-axis, equal mass elements on either side of $x = 0$, but the mass is not equally distributed about the midpoint $R/2$ on the y-axis. There is more mass above $y = R/2$ than below it, because of the

_____

* We are making an approximation here, which is justified by the fact that the rod diameter is very much smaller than R.

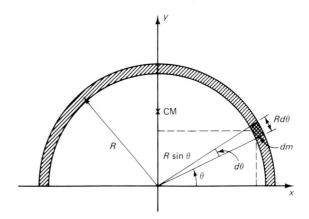

Fig. 12.5

curvature of the hoop. The hoop is nearly horizontal on top and vertical near the bottom. The calculus does the infinite sum along the y-axis for you and gives the exact answer, but your eye should tell you approximately what the result should be.

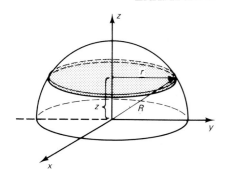

Fig. 12.6

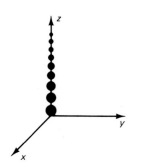

Fig. 12.7

EXAMPLE 3   Find the center of mass of a solid hemisphere of radius R shown in Fig. 12.6. It can easily be shown from symmetry arguments alone that the center of mass must lie along the z-axis of the hemisphere. The problem is then reduced to finding the z-component of $\mathbf{R}_{\mathrm{CM}}$, or

$$z_{\mathrm{CM}} = \frac{1}{M} \int z \, dm = \frac{1}{M} \int z\rho \, dV. \qquad (12.8)$$

The hemisphere is a real three-dimensional object and thus $dV = dx \, dy \, dz$ and Eq. 12.8 is really a triple integral. If, as I suspect, most of you have not yet studied triple integrals, we can use a trick to reduce the problem to a single integral over one variable, z. The hemisphere can be sliced up into an infinite number of infinitesimal thin slices, each slice a circular disk with a different radius (something like the end of a Jewish salami). The mass of each disk will vary as the radius varies, for the density of the hemisphere is uniform. It should be apparent from symmetry arguments that the center of mass of each slice is at its geometric center—along the z-axis. The problem is then reduced to finding the center of mass of a string of point masses (one from each slice) along the z-axis, as indicated in Fig. 12.7.

$$\text{Volume of a slice} = \pi r^2 \, dz = dV,$$

but

$$r^2 = R^2 - z^2.$$

Thus

$$dV = \pi(R^2 - z^2) \, dz.$$

The mass of a slice is $dm = \rho dV$, but

$$\rho = \frac{\text{mass}}{\text{volume}} = \frac{M}{\frac{\frac{4}{3}\pi R^3}{2}} = \frac{3M}{2\pi R^3}$$

or

$$dm = \frac{3}{2}\frac{M}{\pi R^3}\pi(R^2 - z^2)\,dz. \tag{12.9}$$

Now we can substitute Eq. 12.9 into Eq. 12.8:

$$z_{\text{CM}} = \frac{1}{M}\int_0^R \frac{3}{2}\frac{M}{\pi R^3}z\pi(R^2 - z^2)\,dz$$

$$= \frac{3}{2R^3}\left(\frac{R^2 z^2}{2} - \frac{z^4}{4}\right)\Big|_0^R = \frac{3}{2R^3}\left(\frac{R^4}{2} - \frac{R^4}{4}\right) = \frac{3}{8}R$$

$$x_{\text{CM}} = y_{\text{CM}} = 0$$

Fig. 12.8

One last comment on the center of mass of continuous bodies. Suppose that we know the center of mass of several individual macroscopic objects, such as those drawn in Fig. 12.8. If these objects are rigidly connected together, one can easily show that the center of mass of the composite structure is at the same point as the center of mass would be of three imaginary *point* masses m_1, m_2, and m_3 located at $\mathbf{R}_1^{\text{CM}}$, $\mathbf{R}_2^{\text{CM}}$, and $\mathbf{R}_3^{\text{CM}}$. In other words, each of the parts of the rigid object can be replaced by their entire mass concentrated at their center of mass, for calculations of the center of mass of the entire object. As long as the composite body remains rigid, this can be written as a general theorem:

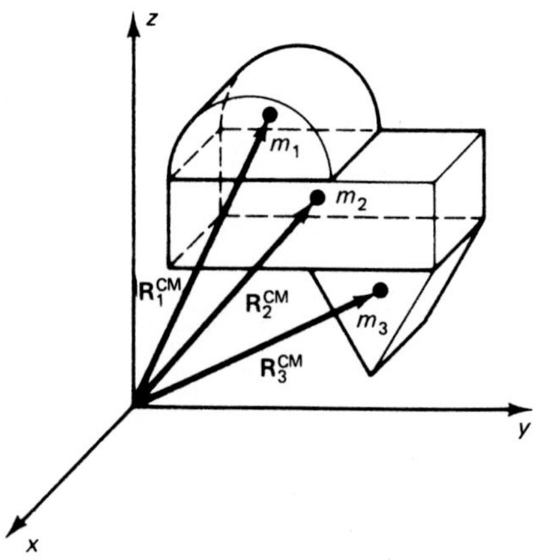

$$\mathbf{R}_{\text{CM}} = \frac{\displaystyle\sum_{i=1}^{N} m_i \mathbf{R}_i^{\text{CM}}}{\displaystyle\sum_{i=1}^{N} m_i}, \tag{12.10}$$

where the m_i's are the masses of the individual parts and the $\mathbf{R}_i^{\text{CM}}$, are the position vectors of the center of mass of these parts. The proof of this theorem is left as a problem.

12.3 Newton's Laws for Center-of-Mass Motion

THE CONCEPT of center of mass was introduced in Chapter 10 when we studied conservation principles and collision processes. The center-of-mass coordinate system was defined as that frame of reference where the total linear momentum of the system of particles was zero. We also showed that for such an isolated system of particles the center-of-mass velocity was a constant of the motion.

But what happens when an external force acts on the system? How does the system respond? We limit our discussion of these questions to low-velocity Newtonian systems, although some of the results are applicable to relativistic particles. We are free to choose the boundary of our system of interest, but once we have made the choice, we must be consistent about which particles are included in the system and which are not. We begin with the equation for the position vector of the center of mass of a collection of particles Eq. 12.1:

$$\mathbf{R}_{\text{CM}} = \frac{1}{M}(m_1\mathbf{r}_1 + m_2\mathbf{r}_2 + \cdots + m_n\mathbf{r}_n) = \frac{1}{M}\sum_{i=1}^{N} m_i\mathbf{r}_i.$$

Differentiating it with respect to time yields

$$\frac{d\mathbf{R}_{\text{CM}}}{dt} = \frac{1}{M}\left(m_1\frac{d\mathbf{r}_1}{dt} + m_2\frac{d\mathbf{r}_2}{dt} + \cdots + m_n\frac{d\mathbf{r}_n}{dt}\right)$$

$$= \frac{1}{M}\sum_{i=1}^{N} m_i\frac{d\mathbf{r}_i}{dt} = \mathbf{V}_{\text{CM}}. \tag{12.11}$$

We recognize Eq. 12.11 as the generalization of Eq. 10.16, the definition of the velocity of the center of mass. Differentiating Eq. 12.11 again with respect to time gives the acceleration of the center of mass:

$$\mathbf{a}_{\text{CM}} \equiv \frac{d\mathbf{V}_{\text{CM}}}{dt} = \frac{1}{M}\left(m_1\frac{d^2\mathbf{r}_1}{dt^2} + m_2\frac{d^2\mathbf{r}_2}{dt^2} + \cdots + m_N\frac{d^2\mathbf{r}_N}{dt^2}\right)$$

$$= \frac{1}{M}\sum_{i=1}^{N} m_i\frac{d^2\mathbf{r}_i}{dt^2} \tag{12.12}$$

According to Newton's second law of motion, the total external force on a point particle is equal to the time rate of change of the linear momentum, or

$$\mathbf{F}_i = \frac{d}{dt}(m_i\mathbf{v}_i). \tag{12.13}$$

But if the mass is constant, the second law becomes

$$\mathbf{F}_i = m_i\frac{d\mathbf{v}_i}{dt} = m_i\frac{d^2\mathbf{r}_i}{dt^2}. \tag{12.14}$$

The total force on a point mass is the vector sum of all the forces that act on that body. These forces can be divided into two catego-ries: those that are *external* to the entire system of particles, and those that come from the other forces within the system. Mathematically, this can be expressed as

$$\mathbf{F}_i = \mathbf{F}_i^{\text{ext}} + \sum_{\substack{j=1 \\ j \neq i}}^{N} \mathbf{F}_{ij}, \tag{12.15}$$

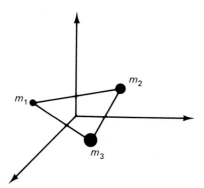

Fig. 12.9

where the $\mathbf{F}_{ij}$ are the forces *within* the system of interest.

Let's examine a three-body system (shown in Fig. 12.9) to understand Eq. 12.15. Imagine that our three particles are connected by "massless" rigid rods which only transmit the force between the particles. Since there are just three particles in our system, the summation in Eq. 12.15 is quite easy. For mass 1,

$$\mathbf{F}_1 = \mathbf{F}_1^{\text{ext}} + \mathbf{F}_{12} + \mathbf{F}_{13}, \tag{12.16}$$

where $\mathbf{F}_{12}$ is the force *on* mass 1 *due to* mass 2, $\mathbf{F}_{13}$ is the force *on* mass 1 *due to* mass 3. Note that there is *no* term in the summation $\mathbf{F}_{11}$ since the sum has the explicit restriction $j \neq 1$.*

Substituting Eq. 12.15 into Eq. 12.14, we have

$$\mathbf{F}_i^{\text{ext}} + \sum_{\substack{j=1 \\ j \neq i}}^{N} \mathbf{F}_{ij} = m_i \frac{d^2 \mathbf{r}_i}{dt^2}. \tag{12.17}$$

Summing over all particles of interest in the system and substituting Eq. 12.17 into the equation for the acceleration of the center of mass (Eq. 12.12), we arrive at

$$\mathbf{a}_{\text{CM}} = \frac{1}{M} \sum_{i=1}^{N} \left(\mathbf{F}_i^{\text{ext}} + \sum_{\substack{j=1 \\ j \neq i}}^{N} \mathbf{F}_{ij} \right). \tag{12.18}$$

Our expression for the acceleration of the center of mass has two terms. The first term is easy to interpret. The vector sum of all the *external* forces acting on all the particles of the system is just the *total* force on the system. The second term is not as obvious. It is a double summation: first a sum over j, *not* including $j \neq i$, and then a second summation over i. This is also written

$$\sum_{i} \sum_{i \neq j} \mathbf{F}_{ij}.$$

Returning to our example of a three-mass system and examining this double summation, we find that

$$\sum_{i=1}^{3} \sum_{\substack{j=1 \\ j \neq i}}^{3} \mathbf{F}_{ij} = \mathbf{F}_{12} + \mathbf{F}_{13} + \mathbf{F}_{23} + \mathbf{F}_{21} + \mathbf{F}_{31} + \mathbf{F}_{32}. \tag{12.19}$$

By Newton's *third* law, the forces between any two particles are equal and opposite and act on a line connecting these particles.

---

* Physically, this means that the particle does not exert a force on itself.

Then the sum equation 12.19 is equal to zero, since

$$\mathbf{F}_{21} = -\mathbf{F}_{12}, \mathbf{F}_{32} = -\mathbf{F}_{23}, \text{ and } \mathbf{F}_{13} = -\mathbf{F}_{31}.$$

The generalization of Eq. 12.19 to N particles is obvious, even if N becomes a large number. Thus Eq. 12.17 collapses to

$$\mathbf{a}_{CM} = \frac{1}{M} \sum_{i=1}^{N} \mathbf{F}_i^{ext} \equiv \frac{\mathbf{F}_{ext}}{M}$$

or

$$\mathbf{F}_{ext} = M\mathbf{a}_{CM}. \qquad (12.20)$$

That equation sure looks familiar! If you had just opened the book to this page, you might mistake Eq. 12.20 for Newton's second law; it looks like $\mathbf{F} = ma$. Indeed, Eq. 12.20 means that the center of mass of a composite body *acts as if* a point particle of mass $M = \Sigma\, m_i$ is subjected to a single force, $\mathbf{F}_{ext}$. Remember: The center of mass is a mathematical construct—*not* a physical parameter! The motion of a complex extended body may appear complicated, but a single point in space, on or near that body, acts in a rather ordinary way, moving like a point particle subjected to a single external force. What could be easier? By the way, although we set out to study rigid extended objects, *nothing* in the derivation of Eq. 12.20 requires that the relative positions of the composite mass particle be constant. The rigid rods only assured us that the forces between the component parts obeyed Newton's third law. Equation 12.20 is a general result, applicable to exploding, oscillating, vibrating, stretching, and rotating collections of particles.

Equation 12.20 does not tell the "whole story" of how an extended object moves. It only describes the translational acceleration of a point, a particular point: the center of mass. That's important to remember!

EXAMPLE 1

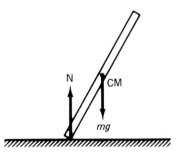

Suppose that one wished to describe the motion of a falling meterstick on a frictionless surface, as shown in Fig. 12.10. Two external forces act on the meterstick: the gravitational force on all the mass points of the stick and the normal force from the surface, which acts on one end. No external forces act in the horizontal direction, so there is no acceleration of the center of mass in that direction. Equation 12.20 makes it clear that the center of mass will fall directly down, following a vertical line, while the meterstick itself rotates. A full description of the motion in this problem must wait until we have discussed rotational motion of extended bodies.

Fig. 12.10

EXAMPLE 2

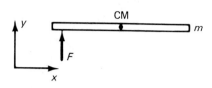

Fig. 12.11

A meterstick of mass m rests on a smooth frictionless surface, such as a sheet of ice, as shown in Fig. 12.11. Describe the motions of the meterstick's center of mass if a constant force **F** acts on it in the y-direction.

The answer to this question is deceptively simple. Equation 12.20 tells us that the center of mass accelerates in the same direction as the net external force and with a magnitude F^{ext}/m. Of course, this is not all there is to the motion of the meterstick. It tells us about the motion of only *one* point, the center of mass. The stick will execute a combination of rotational and translational motion due to this force, but detailed analysis of this complicated motion will be considered later.

Because of the human bodie's flexibility, a high jumper can clear the crossbar without her center of mass reaching the height of the bar. Think about it!

EXAMPLE 3

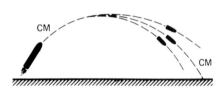

Fig. 12.12

A ballistic missile with multiple warheads is shot into space. While in flight it divides into three 1-megaton hydrogen bonds. If we consider the missile and its warheads as our system, then even after the warheads have been ejected by small rocket engines inside the missile, the center of mass of the system continues to move in its parabolic path (neglecting air resistance). The only external force is Earth's gravitational field, which we take as constant (not a very good approximation for a ballistic missile). The motion of a point mass in such a field is a parabola (Fig. 12.12), as we showed in Section 9.8.

Even if all the warheads were to explode in space, the center of mass would continue on its uninterrupted parabolic path.

EXAMPLE 4   The Mexican jumping bean, a wonderful creature of nature with its own peculiar life-style, demonstrates an important physics principle. If you are not familiar with this marvelous animal, let me describe it. The jumping bean appears from the outside to be a rather large kidney bean, but inside resides a worm that is free to move around. We will analyze an oversimplified model of this shell–worm system, shown in Fig. 12.13. If the bean is placed on a smooth frictionless horizontal surface, it appears to move around. Yet there are no *horizontal external* forces acting on the bean. How can the bean move in the horizontal direction when it was initially at rest and has no horizontal forces acting on it?

If the net external force on a system is zero, Eq. 12.20 says that the center of mass cannot accelerate. Indeed, although the bean does move in the horizontal direction, *its center of mass remains fixed*. What happens, of course, is that the worm moves inside the shell and the shell moves in the opposite direction to keep the center of mass at its original position. That's the qualitative explanation. Now let's work out a quantitative calculation using our idealized model.

Let the shell be exactly symmetrical, with a length l and mass m_2, and let the worm be a very small compact object of mass m_1. By symmetry, the center of mass of the shell alone is at its geometric center. In its initial position, the worm is on the left side of the shell at $x = \dfrac{l}{2}$ and the shell's geometric center is located at the origin. The initial position of the center of mass is then

$$X_{\mathrm{CM}} = \frac{1}{m_1 + m_2}\left(-m_1 \cdot \frac{l}{2} + m_2 \cdot 0\right) = -\frac{m_1 l}{2(m_1 + m_2)}. \quad (12.21)$$

If the worm crawls to the right side of the shell and the shell slides to

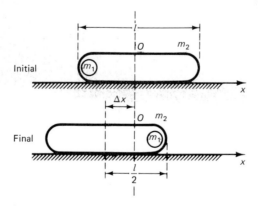

Fig. 12.13

the left a distance Δx, with the center of mass remaining fixed, we have

$$X_{\text{CM}} = \frac{-m_1 l}{2(m_1 + m_2)} = \frac{1}{m_1 + m_2}\left[-m_2\,\Delta x + m_1\left(\frac{l}{2} - \Delta x\right)\right], \quad (12.22)$$

which reduces to

$$\Delta x = \frac{m_1 l}{m_1 + m_2}. \quad (12.23)$$

If the worm has twice the mass of the shell ($m_1 = 2m_2$), the shell will slide a distance

$$\Delta x = \frac{2}{3}l.$$

12.4 Einstein's Gedanken Experiment

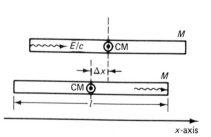

Fig. 12.14

EINSTEIN was certainly not thinking about Mexican jumping beans when he derived the relationship between mass and energy, but the analysis he followed is analogous to that discussion. Like so many of Einstein's contributions to our understanding of nature, his reasoning was deceptively simple and the consequences profound.

Einstein asked what would happen to a long enclosed tube if a photon left one side and traveled to the other end of the tube, where it is absorbed. The photon acts just like the worm in our jumping bean. Let's see what happens.

A photon emitted from the left side of the tube with energy E and momentum E/c travels in the positive x-direction toward the end of the tube (Fig. 12.14). Conservation of linear momentum requires the tube (of mass M) to recoil with a velocity (V) in the negative x-direction, or

$$-MV + \frac{E}{c} = 0, \quad (12.24)$$

since the initial momentum of the system was zero. Rearranging and solving for the velocity of the tube during the photon's flight gives us

$$V = \frac{E}{Mc}. \quad (12.25)$$

The light will travel a distance l (actually, $l - \Delta x$, but if l is large, we can correctly neglect Δx) in a time t, or

$$t = \frac{l}{c} \quad (12.26)$$

where c is the velocity of light. The tube will move a distance Δx in the time t,

$$\Delta x = Vt = \frac{E}{Mc}\frac{l}{c} = \frac{lE}{Mc^2} \qquad (12.27)$$

before it stops. The tube stops when the photon strikes its right sides.

Einstein believed that the center of mass of an object could not accelerate without an external force. But in his experiment the system was initially at rest, and clearly, there were no external forces. Thus he assumed that the position of the center of mass, originally at the origin $x = 0$, would remain at the origin. For the center of mass to remain at the origin (and the tube to move a distance Δx), the photon, like our worm, *must carry mass from one side of the tube to the other.* If we define that mass to be m, we can use Eq. 12.23 to calculate the distance the tube moves due to the mass transport. For this experiment, where $M \gg m$, Eq. 12.23 can be written as

$$\Delta x = \frac{ml}{M + m} \qquad \text{or} \qquad \Delta x = \frac{ml}{M}. \qquad (12.28)$$

Combining Eqs. 12.28 and 12.27, we obtain

$$\Delta x = \frac{ml}{M} = \frac{lE}{Mc^2}$$

or

$$\boxed{E = mc^2.}$$

Recognize this equation? The photon of energy E carried an equivalent mass m from one side of the tube to the other. Einstein understood that this relationship had much broader implications than the motion of a tube with a photon moving inside it. He realized that he had uncovered a fundamental connection between mass and energy. No longer would these two physical parameters be considered as separate entities; they would now be recognized as two different manifestations of the same phenomenon. Einstein made only one serious mistake when he discussed his new discovery: He initially thought it would be of little or no practical value! How wrong can you be?

12.5 **Static Equilibrium of Extended Bodies**

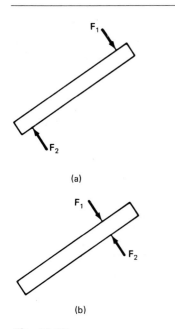

(a)

(b)

Fig. 12.15

FOR A *point* particle to be in static equilibrium (i.e., if it is at rest, it remains at rest), the vector sum of all the forces acting on the point particle must be zero. But that condition is *not* sufficient to describe the equilibrium of *extended* bodies. For example, consider the metersticks shown in Fig. 12.15, subjected to the two external forces, $\mathbf{F}_1$ and $\mathbf{F}_2$, which are equal in magnitude and opposite in direction. If the two forces do not act along the same line, as in (a), the stick will rotate in the clockwise direction. But if the *same* two forces act along a common line, as in (b), the stick will not rotate. In both cases the sum of the forces on the meterstick is zero. According to Eq. 12.20, the sum of the external forces on an extended body is equal to the acceleration of the center of mass of the body. In both (a) and (b), the center of mass of the meterstick remains stationary, but in case (a) the meterstick rotates about the center of mass.

When we are examining the effect of applied forces on bodies of finite size, the lines of action of the forces, as well as the directions and magnitudes of the forces, are important.

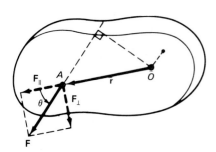

Fig. 12.16

Consider a body that pivots about a fixed point, such as the object drawn in Fig. 12.16, where the frictionless pivot is located at point O. A force $\mathbf{F}$ is applied to point A at an angle θ with respect to the position vector $\mathbf{r}$. This force can be resolved into two components, $\mathbf{F}_\perp$ and $\mathbf{F}_\parallel$, where $F_\perp = |\mathbf{F}| \sin \theta$ and $F_\parallel = |\mathbf{F}| \cos \theta$. The parallel component of the force $\mathbf{F}_\parallel$ produces *no* tendency to rotate about the pivot point. It does, however, cause the pivot point to exert a force equal and opposite to $\mathbf{F}_\parallel$ that acts along the same line as $\mathbf{F}_\parallel$. The component of the force perpendicular to $\mathbf{r}$, $\mathbf{F}_\perp$, will produce a counterclockwise rotation of the entire body about the pivot point. The effect of the perpendicular component depends not only on its magnitude and direction, but also on its *distance* from the pivot point. All this sounds somewhat complicated. To understand what causes an extended body to rotate, we must keep track of three things: the magnitude and direction of the perpendicular component of the force and the distance to the pivot point. Vector algebra once again comes to our rescue, since all of the foregoing properties are subsumed into a quantity called TORQUE, which is the cross product of two vectors.

The torque of the forces **F** about a point O (Fig. 12.17), sometimes called "the moment of the force," is defined as

Fig. 12.17

$$\boldsymbol{\tau}_0 \equiv \mathbf{r} \times \mathbf{F}. \tag{12.29}$$

This is the second example of the usefulness of the cross product of two vectors; the first example was the magnetic force on a moving charged particle. Since any two vectors define a plane, we can begin our discussion of torque by considering $\mathbf{r}$ and $\mathbf{F}$ to lie in the x–y plane of the paper. The magnitude of the torque about the origin O is given by

$$|\boldsymbol{\tau}_o| = |\mathbf{r}|\,|\mathbf{F}|\,\sin\theta \tag{12.30}$$

which is the same as

$$|\boldsymbol{\tau}_o| = F_\perp |\mathbf{r}|.$$

Equation 12.30 can be read: "The magnitude of the torque about the point O is the perpendicular component of the force times the length of the lever arm". The subscript on τ indicates the origin about which the torque is evaluated.

> *In calculating the torque of a given force, it is absolutely essential that the origin or pivot point is specified. In fact, it makes no sense to speak about the torque of a force without specifying the point about which the torque is acting.*

In summing torques, one must keep track of the sense of rotation of the force about the pivot point. A torque that produces a *counterclockwise rotation is defined as positive* and a torque that produces a *clockwise rotation* is defined as *negative*. When the *net* torque on an extended body is calculated about a particular origin, the sense of rotation must be taken into account.

EXAMPLE 1   Archimedes' principle of the lever provides a good example of how to sum torques on an extended body. Figure 12.18 shows a meterstick of mass M balanced about its geometric center (which is also its

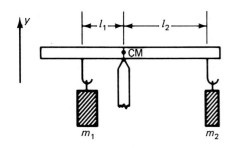

Fig. 12.18

center of mass) with two different weights hanging vertically at different distances l_1 and l_2 from the center.

(a) What is the ratio of l_1 to l_2?

(b) What is the force on the pivot point at the center of mass?

To solve these problems, we must formally state the *two* conditions for static equilibrium of extended rigid bodies.

---

1. *The vector sum of the force on the body must be zero:*

$$\sum_i \mathbf{F}_i = 0. \tag{12.31}$$

2. *The vector sum of the torques about* any *point must be zero:*

$$\sum_i \boldsymbol{\tau}_{oi} = 0. \tag{12.32}$$

Applying principle 2 to the lever yields

$$\boldsymbol{\tau}_{o2} + \boldsymbol{\tau}_{o1} = 0.$$

But $\tau_{o1} = +M_1 g l_1$ and $\tau_{o2} = -m_2 g l_2$, since both forces act at right angles to their lever arms. The gravitational force on the meterstick itself causes no torque about the center of mass since the entire force of gravity can be thought of as acting at the center of mass.

$$m_1 g l_1 = m_2 g l_2 \quad \text{or} \quad \frac{m_1}{m_2} = \frac{l_2}{l_1}, \tag{12.33}$$

which is another way of stating Archimedes' principle of the lever. The force on the pivot point can be calculated from Eq. 12.31, defining the positive y-axis to be up, vertical:

$$\sum_i \mathbf{F}_{yi} = -m_1 g - m_2 g - Mg + F_{\text{pivot}} = 0$$

$$F_{\text{pivot}} = (m_1 + m_2 + M)g \quad \text{in the positive } y\text{-direction.}$$

EXAMPLE 2 A ball of mass 3.5 kg is held in a forearm in the horizontal position as shown in Fig. 12.19a. The bicep muscle is attached 6 cm from the elbow at an angle of 70° and the forearm is .4 m long. If the arm itself has a mass of .5 kg and is considered symmetric so that its center of mass is at its geometric center $\frac{l}{2}$, find the force that the bicep exerts on the forearm.

To solve this problem we must first construct a working model of this anatomical system. Figure 12.19b represents a realistic model of such a forearm where m is the mass of the arm, $\mathbf{F}_p$ is the resultant

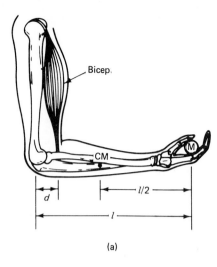

(a)

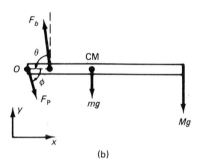

(b)

Fig. 12.19

force at the pivot, and F_b is the resultant bicep force. For equilibrium the vector sum of the forces on the extended body must be equal to zero. For each component,

$$\sum F_x \Rightarrow F_p \cos \phi - F_b \cos \theta = 0$$

$$\sum F_y \Rightarrow F_b \sin \theta - F_p \sin \phi - mg - Mg = 0.$$

Summing torques about the pivot point O, we obtain

$$(F_b \sin \theta)d - (mg)\frac{l}{2} - (Mg)l = 0$$

solving for F_b

$$F_b = \frac{mg\frac{l}{2} + Mgl}{d \sin \theta}$$

$$= \frac{.5 \text{ kg} \times 9.8 \text{ m/s}^2 \times .2 \text{ m} + 3.5 \text{ kg} \times 9.8 \text{ m/s}^2 \times .4 \text{ m}}{.06 \text{ m} \times \sin 70°}$$

$$= 260.7 \text{ N}.$$

Note: It was not necessary to sum forces. Only the sum of torques was needed to find F_b. However, if you wish to know the force on the pivot F_p, one may use these equations:

$$F_p \cos \phi = 260.7 \cos 70° = 89.16 \text{ N}$$

$$F_p \sin \phi = 260.7 \sin 70° - .5 \text{ kg} \times 9.8 \text{ m/s}^2 - 3.5 \text{ kg} \times 9.8 \text{ m/s}^2$$

$$= 205.77 \text{N}.$$

Squaring and adding the equations above eliminates the angle ϕ, and solving for the magnitude of F_p, one finds that

$$F_p = 224.24 \text{ N}$$

and

$$\cos \phi = \frac{89.16}{224.24} = .3976$$

$$\phi = 66.57°.$$

There are many ways to solve this problem using the same principles. Another strategy is to sum torques about the point of contact of the bicep muscle. That is,

$$(F_p \sin \phi)d - (mg)\left(\frac{l}{2} - d\right) - Mg(l - d) = 0$$

or

$$F_p \sin \phi = \frac{mg\left(\frac{l}{2} - d\right) + Mg(l - d)}{d}$$

Putting in the numbers

$$F_p \sin \phi = 205.8 \text{ N}.$$

This is the same expression that we obtained from summing the forces in the y-directions. Using the expression for the sum of the forces in the x-direction, one can again obtain both F_p and the angle ϕ.

Which way is the best? There is no single strategy that will always point you toward the path of least resistance. The freedom to choose the origin allows you to pick a point where you don't know the forces, such as the pivot in this problem. Then those forces do not enter into the torque equation. Sometimes that can be a helpful trick to remember.

Static equilibrium problems often involve the gravitational force that acts on all mass elements in an extended object. If the body is in a uniform gravitational field, such as near the Earth's surface, the following theorem is quite useful:

> *The sum of all the uniform gravitational forces acting on the system is equivalent to a single force acting on the center of the mass of the system.*

That is, this single force accounts for all the gravitational forces *and* torques that act on the system of particles. We already used this theorem in solving the lever problem, where the only gravitational force that was considered was the single force Mg acting at the center of mass. The theorem is equally applicable to rigid bodies as well as nonrigid systems of particles. The proof is straightforward.

An arbitrary collection of particles of masses $m_1, m_2, m_3, \ldots$ at positions $\mathbf{r}_1, \mathbf{r}_2, \mathbf{r}_3, \ldots$ experiences a uniform gravitational field, which produces gravitational forces in the negative z-direction, $-m_1 g\mathbf{k}, -m_2 g\mathbf{k}, -m_3 g\mathbf{k}, \ldots$ (Fig. 12.20). The total gravitational force is then

$$\mathbf{F}_G = \sum_i -m_i g\mathbf{k} = -g\mathbf{k} \sum_i m_i = -Mg\mathbf{k}. \qquad (12.34)$$

where M is the total mass of the system, $\sum_i m_i$. The torque about the origin due to the gravitational force on the mass m_i is

$$\boldsymbol{\tau}_{oi} = \mathbf{r}_i \times \mathbf{F}_i = \mathbf{r}_i \times (-m_i g\mathbf{k}) = -m_i g\mathbf{r}_i \times \mathbf{k}.$$

Summing all the torques about 0 yields

$$\boldsymbol{\tau}_{\text{Total}} = \sum_i \boldsymbol{\tau}_{oi} = \sum_i -m_i g\mathbf{r}_i \times \mathbf{k} = -g\left(\sum_i m_i \mathbf{r}_i\right) \times \mathbf{k}.$$

But we know from the definition of the center of mass of a system of particles that

$$\sum_i m_i \mathbf{r}_i = M\mathbf{r}_{\text{CM}}.$$

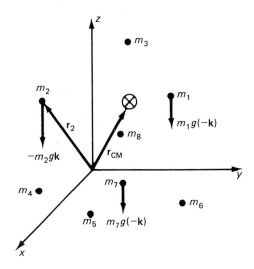

Fig. 12.20

Thus
$$\tau_{\text{Total}} = \mathbf{r}_{\text{CM}} \times (-gM\mathbf{k}) = \mathbf{r}_{\text{CM}} \times \mathbf{F}_G. \tag{12.35}$$

A single force $(-Mg\mathbf{k})$ acting on the center of mass does indeed give the correct total gravitational force (Eq. 12.34) and the correct total torque (Eq. 12.35) on our system of arbitrary masses. This theorem will be particularly useful when we examine the motion of spinning tops and the gyroscope. When the gravitational field is *not* uniform over the entire body (such as the sun's gravitational force on the Earth), the theorem cannot be applied.

EXAMPLE 3   It seems as if every student who has ever studied physics has to be able to solve the infamous "ladder on the wall" problem. Question: What is the farthest distance from the wall that one can place a ladder before it will fall, assuming that there exists a frictional force of the ladder with the ground and neglecting its friction with the vertical wall (Fig. 12.21)?

We can analyze this problem with the aid of a free-body diagram (Fig. 12.22) which graphically represents all the forces. There are four forces acting on the ladder: (1) the gravitational force, $mg\mathbf{j}$ on the center or mass; (2) the normal force of the wall, $\mathbf{F}_{N2}$; (3) the frictional force of the floor, $\mathbf{F}_f$; and (4) the normal force of the floor, $\mathbf{F}_{N1}$. Condition 1 for translational static equilibrium requires:

x-comp.: $F_{N2} - F_f = 0$ (a)

y-comp.: $-mg + F_{N1} = 0.$ (b)

Condition 2 for rotational equilibrium requires that the sum of the torques about any point is zero. We are free to choose the point at which we wish to sum the torques. If we pick point A at the top of the ladder,

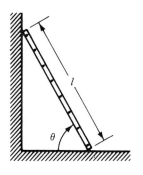

Fig. 12.21

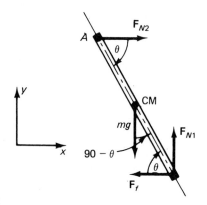

Fig. 12.22

$$-F_f l \sin \theta + F_{N1} l \sin (90 - \theta) - mg \left(\frac{l}{2}\right) \sin (90 - \theta) = 0$$

or

$$-F_f \sin \theta + F_{N1} \cos \theta - mg \frac{1}{2} \cos \theta = 0. \quad \text{(c)}$$

Finally, we can specify the maximum frictional force of the ladder with respect to the floor, in terms of the normal force and coefficient of frictions, as

$$F_f = \mu F_{N1}. \quad \text{(d)}$$

If Eq. (d) is used for the maximum frictional force and the four equations (a), (b), (c), and (d) are solved simultaneously, θ will be the angle at which slipping will begin. From (b),

$$F_{N1} = mg,$$

but from (a) and (d),

$$F_f = F_{N2} = \mu F_{N1} = \mu mg.$$

Then Eq. (c):

$$-\mu mg \sin \theta_s + mg \cos \theta_s - \frac{1}{2} mg \cos \theta_s = 0$$

or

$$\tan \theta_s = \frac{1}{2\mu}.$$

To get a feel for the slipping angle, assume that $\mu = .3$:

$$\tan \theta_s = \frac{1}{.6} \quad \text{or} \quad \theta_s = 59°.$$

Note that neither the mass nor the length of the ladder entered into the final result.

You may wonder what dictated our choice of reference point around which we summed the torques. The choice in any statics problem is arbitrary. One can show (see Problem 22) that for a body in translational equilibrium, that is, the net force on the body is zero if the resultant torque around any reference point is zero, the resultant torque about any other reference point must also be zero.*

EXAMPLE 4   Find the force **F** necessary to raise the cylinder of mass M up a stair of height h (Fig. 12.23). Assume that the cylinder does not slip and that the force is applied in the horizontal direction.

When the cylinder has enough force F exerted on it to just raise it, the contact force at point C becomes zero and thus all the forces on the cylinder are shown in Fig. 12.24. Summing forces, we have

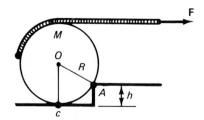

Fig. 12.23

---

* This is true for a single inertial reference frame.

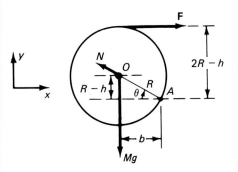

Fig. 12.24

$$\Sigma \, F_x: \qquad F - N \cos \theta = 0 \qquad \text{(a)}$$

$$\Sigma \, F_y: \qquad N \sin \theta - Mg = 0. \qquad \text{(b)}$$

Summing torques about A yields

$$Mgb - F(2R - h) = 0.$$

Note $(R - h)^2 + b^2 = R^2$, so $b = [R^2 - (R - h)^2]^{1/2} = (2Rh - h^2)^{1/2}$,

$$F = \frac{Mgb}{2R - h} = \frac{Mg(2Rh - h^2)^{1/2}}{2R - h}.$$

One can use the force equations to solve for the normal force N by squaring and adding (a) and (b):

$$N = (M^2g^2 + F^2)^{1/2} \qquad \text{and} \qquad \tan \theta = \frac{Mg}{F}.$$

12.6 Axial and Polar Vectors

In the last derivation, we treated torques as three-dimensional vectors. In our discussion of static equilibrium of two-dimensional planar objects the vector properties of torques was accounted for by assigning positive values to the magnitude of the torque if it produced counterclockwise rotation, and a negative value if it produced a clockwise rotation. For rigid two-dimensional or planar objects, the latter technique is sufficient. Remember, however, that torque is a vector quantity with both direction and magnitude. In an examination of the rotational properties of rigid bodies, the three-dimensional vector properties of torques become essential.

A close look at the vector torque reveals an interesting difference between its vector properties and the vector properties of velocity, force, acceleration, and linear momentum. There are, in fact, *two kinds* of vectors:

1. POLAR VECTORS, such as velocity, force, and acceleration
2. AXIAL VECTORS, such as torque and angular momentum

Axial vectors differ from polar vectors in one important property: namely, if the axes of a Cartesian coordinate system x, y, z are reversed, that is, $x \rightarrow -x$, $y \rightarrow -y$, and $z \rightarrow -z$ (called an inversion of coordinates), all the components of an axial vector *remain the same*, and all the components of a polar vector *change*.

The cross product of two polar vectors is an axial vector; for example, torque, $\boldsymbol{\tau} = \mathbf{r} \times \mathbf{F}$. Notice that all the components of *both* $\mathbf{r}$ and $\mathbf{F}$ would change sign if the Cartesian coordinates were reversed, but the torque does *not* change its components.

The existence of two types of vectors gives rise to another new quantity, the PSEUDOSCALAR. The scalar product of two axial vectors and the scalar product of two polar vectors is a true scalar, a numeric whose value is unchanged by any transformation of coordinates, including an inversion. However, the scalar product of an axial vector and a polar vector is called a "pseudoscalar" since its sign *is* changed upon inversion of coordinates.

One important consequence of this subtle difference in the vector quantities is that one can only equate polar vectors to polar vectors and axial vectors to axial vectors to maintain vectorial consistency in physics equations. The same is true for scalar equations. To familiarize yourself with this idea, try the following: Assume that **a** and **b** are polar vectors. Then:

1. $\mathbf{a} \cdot \mathbf{b}$, $\mathbf{a} \cdot \mathbf{a}$, and $\mathbf{b} \cdot \mathbf{b}$ are true scalars.
2. $\mathbf{a} \times \mathbf{b}$ is an axial vector.
3. $\mathbf{a} \cdot (\mathbf{a} \times \mathbf{b})$ is a pseudoscalar.
4. $\mathbf{a} \times (\mathbf{a} \times \mathbf{b})$ is a polar vector.

12.7 Principle of Virtual Work

There is another method of calculating forces on rigid bodies which is particularly useful for pulley systems, levers, hoists, and other devices that have what is called "mechanical advantage." This is called the **PRINCIPLE OF VIRTUAL WORK**. To use this method the system must satisfy three demands:

1. No stored or internal energy in any part
2. No friction or heat
3. No kinetic energy of any part

For such a system, or one that can reasonably approximate these criteria, the principle of conservation of mechanical energy (discussed in Chapter 11) reduces to

$$
\begin{array}{c}
\text{CHANGE IN WORK INPUT} = \text{CHANGE IN WORK OUTPUT} \\
\Delta W_{\text{input}} = \Delta W_{\text{output}}
\end{array}
\tag{12.36}
$$

Let's illustrate the use of this principle. Consider the apparatus shown in Fig. 12.25a. How much force must be exerted on the rope so that the massless platform will hold the mass m at the center?

We imagine that the rope over the pulley is pulled down a very small distance Δx with an unknown force **F**. Then the work input into the system is

$$\Delta W_{\text{input}} = F \, \Delta x.$$

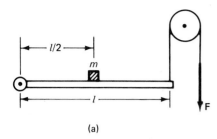

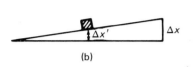

Fig. 12.25

What is the work "output"? The mass m is lifted an amount $\Delta x'$, so

$$\Delta W_{\text{output}} = mg\,\Delta x'.$$

From Fig. 12.22b, similar triangles

$$\frac{\Delta x'}{\Delta x} = \frac{l/2}{l} \qquad \text{or} \qquad \Delta x' = \frac{1}{2}\,\Delta x.$$

Thus, applying the principle of virtual work, we obtain

$$F\,\Delta x = mg\,\Delta x' = mg \cdot \frac{1}{2}\,\Delta x$$

or

$$F = \frac{1}{2}\,mg,$$

or half the force that it would take to hold the mass by itself.

This method is applicable to pulley systems and levers. The technique is always the same. One assumes an infinitesimal amount of work input (a constant force over an infinitesimal displacement) and equates that to the work output of the system. If the technique is done correctly, the infinitesimal displacement does *not* appear in the final expression for the force.

12.8 **Summary of Important Equations**

Position of center of mass, discrete masses:

$$\mathbf{R}_{\text{CM}} = \frac{1}{M} \sum_{i=1}^{N} m_i \mathbf{r}_i$$

Position of center of mass, continuous body:

$$\mathbf{R}_{\text{CM}} = \frac{1}{M} \int \mathbf{r}\, dm \qquad \text{or} \qquad \mathbf{R}_{\text{CM}} = \frac{\int \mathbf{r}\rho(x,\,y,\,z)\,dV}{\int \rho(x,\,y,\,z)\,dV}$$

Velocity and acceleration of the center of mass:

$$\mathbf{V}_{\text{CM}} = \frac{d\mathbf{R}_{\text{CM}}}{dt} \qquad \mathbf{a}_{\text{CM}} = \frac{d\mathbf{V}_{\text{CM}}}{dt}$$

Newton's second law for translational motion of the center of mass:

$$\sum_{i=1}^{N} \mathbf{F}_i^{\text{ext}} = M\mathbf{a}_{\text{CM}}$$

Definition of the torque of a force $\mathbf{F}$, acting on the point $\mathbf{r}$, about the origin of $\mathbf{r}$ at O:

$$\boldsymbol{\tau}_o \equiv \mathbf{r} \times \mathbf{F}$$

Conditions for static equilibrium for macroscopic extended bodies:

$$\sum_{i=1}^{N} \mathbf{F}_i = 0 \qquad \sum_{i=1}^{N} \tau_{oi} = 0$$

Principle of virtual work:

$$\Delta W_{\text{input}} = \Delta W_{\text{output}}$$

Chapter 12   QUESTIONS

1. Give several examples of objects whose center of mass does not lie within the object itself.

2. The center of mass of a planar object of arbitrary shape can be found in the following way: Hang the object by one point and scribe a line directly below the suspension point in the vertical direction (Fig. 12.26). This can be done using a plumb line. Now do the same thing, this time hanging the object from another point. The intersection of these two lines is the center of mass of the body. Explain how this works. Actually, this point is the center of gravity. Why is it an excellent approximation to claim that the center of gravity and the center of mass are at the same point?

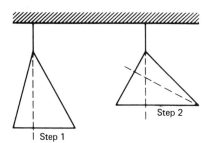

Fig. 12.26

3. A rocket propels itself with its own internal fuel and engines. What is meant by the statement that when we send a rocket to the moon, nothing really goes anywhere—the center of mass stays on the Earth's surface? The only external force on the rocket is the Earth's gravity. Explain this paradox.

4. Explain why high-wire performers in the circus carry long poles as they walk across the wire.

5. How is it possible for a high jumper to clear a 7-ft horizontal pole and not have his center of gravity go higher than 6 ft 8 in.?

6. Two sailors are becalmed on the sea without food or fresh water. Desperate for wind to fill their sails, they blow on them. Will this help? Explain.

7. Explain why a block-and-tackle system with many loops of rope requires less force to lift a given mass than one with only a few loops. Why don't you see such hoists with 50 loops?

8. Can one apply the principle of virtual work to a system that has friction? Explain.

Chapter 12   PROBLEMS

1. A water molecule has the shape of an isosceles triangle with an oxygen atom at the vertex and two hydrogen atoms at the base. The height of the triangle is 19 Å (angstroms) and the angle of the vertex is 105°. If the atomic mass of oxygen is 16 μ and of hydrogen is 1 μ (μ = atomic mass units):

    (a) Locate the center of mass of the molecule.
    (b) Locate the center of mass if the hydrogen is replaced by deuterium (heavy hydrogen) of mass 2 μ.

2. Find the center of mass of a collection of point masses that are located at the following coordinates:

m_1 of 2.8 kg at $x = 3$, $y = -7$, $z = 1$ (meters)

m_2 of 6.3 kg at $x = 0$, $y = 5$, $z = -3$ (meters)

m_3 of .25 kg at $x = -6$, $y = -1$, $z = 10$ (meters)

m_4 of 8.7 kg at $x = 2.6$, $y = -4.2$, $z = .15$ (meters)

3. An equilateral tetrahedron has three equal point masses at the corner of its base and a twice-larger point mass at its apex, as shown in Fig. 12.27. Find the location of the center of mass.

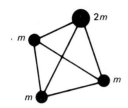

Fig. 12.27

4. Two solid spheres of masses 5 kg and 15 kg are connected together at the end of a thin uniform rod of mass .5 kg and length 1.5 m. If each sphere is .2 m in diameter, find the center of mass of the system.

5. Find the center of mass of the uniform planar object shown in Fig. 12.28.

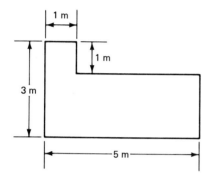

Fig. 12.28

6. Find the center of mass of the uniform planar structure shown in Fig. 12.29. Note the hole of 1.5 m

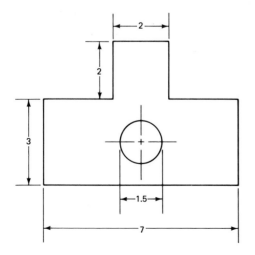

Fig. 12.29

diameter in the geometric center of the lower rectangle. All dimensions are in meters.

7. (a) Find the center of mass of this thick-walled uniform cylinder of length l and mass M_c if it is exactly half-filled with solid epoxy resin whose mass is M_E (Fig. 12.30).

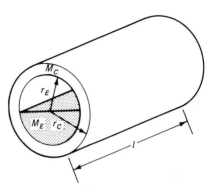

Fig. 12.30

(b) If the cylinder is rolled through an angle θ, find the increase in gravitational potential energy of the system.

(c) Plot the gravitational potential energy as a function of θ for this system.

8. Two identical bars of flat steel are welded together to form the L-shaped object shown in Fig. 12.31. When a hammer strikes a sudden blow in the x-direction at a point y_0, the object moves on a frictionless surface *without* rotating. Find the distance y_0 and explain why the object does not rotate. What happens when the hammer strikes below or above y_0.

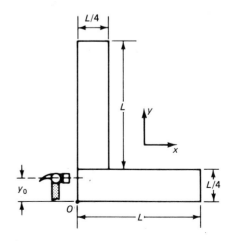

Fig. 12.31

9. Find the center of mass of two blocks with uniform square cross section which are welded together as shown in Fig. 12.32. Suppose that a hole is drilled down the axis of these blocks. If the diameter of the hole is $a/6$, does the center of mass shift its location? If so, how must it shift?

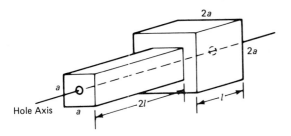

Fig. 12.32

10. Find the center of mass of a smooth, curved, semicircular, thin sheet of radius R shown in Fig. 12.33.

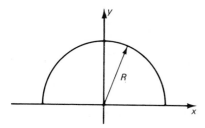

Fig. 12.33

11. Find the center of mass of the uniform triangular planar object, .01 m thick, shown in Fig. 12.34.

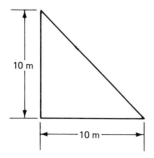

Fig. 12.34

12. Show that the center of mass of a solid homogeneous cone, shown in Fig. 12.35, is on the cone's z-axis at the point $z = h/4$.

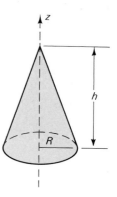

Fig. 12.35

13. Find the center of mass of a homogeneous uniform wedge shown in Fig. 12.36.

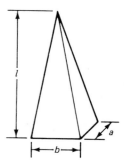

Fig. 12.36

14. Find the center of mass of a uniform spherical ball of radius R which has a large internal bubble of radius r almost at the surface (Fig. 12.37). (*Hint:* Think of the bubble as an object with negative mass.)

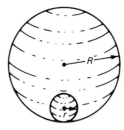

Fig. 12.37

15. A bar supports two strings that are wrapped around the axis of this special falling and rotating yo-yo (Fig. 12.38). Find the force **F** necessary to support the bar if the yo-yo's center of mass is observed to fall with an acceleration of 4.2 m/s². Find the tension T in each string. Assume that the mass of the yo-yo is 3.2 kg.

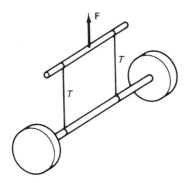

Fig. 12.38

16. Consider the 7 kg block shown in Fig. 12.39. We will assume that the center of mass is in the center of the block and that the coefficient of sliding friction at the two points of contact is $\mu = .3$. What force **F** is needed to accelerate this block at 5 m/s²?

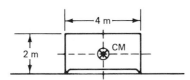

Fig. 12.39

17. Find the slipping angle in Example 3 of Sec. 12.5 if the friction of the ladder with the vertical wall is *not* neglected. Call this coefficient of friction μ_2 and the coefficient with the ground μ_1.

18. Derive the theorem expressed mathematically by Eq. 12.10.

19. A hiker dropped her 15-kg knapsack off a 150-m-high cliff into the ravine below. Being a bright physics student, she wondered about the fall and how it affected the Earth. How much did the Earth "come up" to meet the knapsack during the approximate 150-m fall? (*Hint:* Consider the earth–knapsack an isolated system and use $R_{Earth} = 6400$ km and $M_{Earth} = 6 \times 10^{24}$ kg. Does the center of mass of this system move?)

20. Bob and Jane decide to exchange places in their canoe. Assuming no friction of the canoe with the water, find the new location of the center of the canoe after they make the change. Note from Fig. 12.40 that

the canoe is L long, has mass M, and that Bob and Jane's masses are m_B and m_J, respectively.

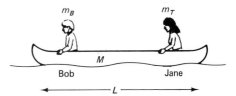

Fig. 12.40

21. There are several subtle things wrong with the analysis in Section 12.4 of Einstein's gedanken experiment that demonstrated the relationship between rest energy and mass. The errors concern the fact that the photon will travel to the opposite end of the tube *before* that end has had a chance to move, because the tube can only transmit mechanical forces at approximately the speed of sound in a metal (which is much, much slower than the velocity of light). However, the basic principles are correct. They are (1) conservation of linear momentum, and (2) laws of motion of the center of mass in the absence of external forces. Now use these two principles in an analysis of the problem anew. This time consider the motion of the center of mass of two separated blocks of ordinary matter m_1 and m_2 shown in Fig. 12.41, separated by a distance L. At $t = 0$, the block m_1 emits a photon of energy E and recoils with a velocity v_1. Later block m_2 absorbs the photon and recoils with a new mass m_2' with a velocity v_2. Carry out the analysis and show that $E = \Delta mc^2$. Reference: A. P. French, *Special Relativity* (New York: W. W. Norton, 1968), p. 27.

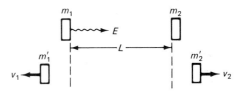

Fig. 12.41

22. Four identical blocks are stacked on top of one another as shown in Fig. 12.42. Show that the *largest* extension can be obtained by having the top block hanging over the one below it by $a/2$, the second by $a/4$, and the third by $a/6$.

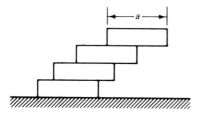

Fig. 12.42

23. A ladder 25 m long of 40-kg mass rests against a wall. The top of the ladder is 18 m above the ground. Find the minimum coefficient of friction between the ground and the ladder so that a person of 80-kg mass can climb 90% of the ladder without the ladder slipping. Neglect the friction of the ladder with the vertical surface.

24. A platform of mass 5 kg is hinged to the wall and supported by a string as shown in Fig. 12.43. A 3-kg block slowly slides along the platform away from the wall. At what point x will the string break if the maximum tension it can withstand is 120 N?

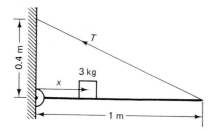

Fig. 12.43

25. Consider the symmetric y-shaped planar structure shown in Fig. 12.44 which is subject to *four*

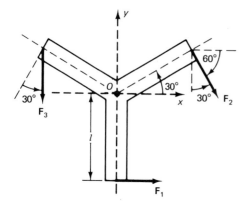

Fig. 12.44

forces, three at the ends of the arms and one at the geometric center. If $\mathbf{F}_1 = 5\mathbf{i}$ N and $\mathbf{F}_3 = -8\mathbf{j}$ N, what force $\mathbf{F}_2$ is needed to keep the body in static rotational equilibrium? What force must be exerted at the center O to maintain translational equilibrium?

26. The stability of a ladder resting on the ground and against a wall depends on friction. For this problem we ignore the frictional effects of the wall and concentrate on the ground. Show that the stability of the system actually *increases* as a person climbs up the ladder (Fig. 12.45). Do this by plotting the increase

Fig. 12.45

in the frictional force at the base of the ladder as the person ascends the ladder. You may assume that the person's mass is eight times the mass of the ladder. Find the smallest angle θ that the ladder can be placed to be sufficiently stable for the person to climb all the way to the top. Assume that the coefficient of friction between the ladder and the ground is $\mu = .6$.

27. Alexander Calder must have known some physics when he invented the "mobile" (Fig. 12.46). Suppose that you decide to build a fish mobile for your

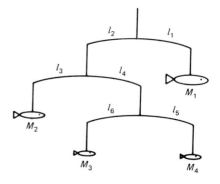

Fig. 12.46

newborn nephew. All the lengths of your lightweight rods $l_1, l_2, l_3, l_4, l_5,$ and l_6 have been chosen. What are the values of the masses M_1, M_2, M_3, M_4 which will assure that the mobile is in static equilibrium?

28. The pivoted platform of mass m_1 in Fig. 12.47 supports a block of mass m_2. Find the tension in the cord that is attached to both the block and the platform, and the force on the pivot.

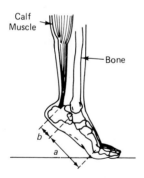

Fig. 12.47

29. The calf muscle is primarily responsible for allowing human beings to stand on their toes (Fig. 12.48). If an 85-kg man stands on tiptoe (with two identical feet) at an angle of 45°, find the force the calf muscle must exert on the bone if the dimensions of the feet are $a = 17$ cm and $b = 6.5$ cm.

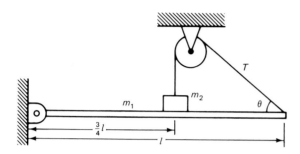

Fig. 12.48

30. An old English inn hung a new sign along the road to attract tourists (Fig. 12.49). The sign is uniform: Its dimensions are 2.5 m × 1.5 m, mass 45 kg. If it is hung on a 4-m-long rod of mass 4 kg:

   (a) Find the tension in the cable that supports the rod from a point 3 m above the pivot.

   (b) Find the resultant force exerted by the main support post on the pivot.

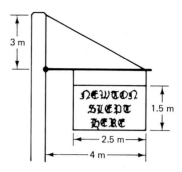

Fig. 12.49

31. Find the smallest value of the coefficient of static friction necessary to keep this cylinder of mass M and radius R from sliding down the slope inclined at an angle θ (Fig. 12.50). Find the tension in the rope.

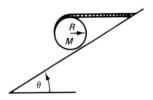

Fig. 12.50

32. A 7-m crane is used to lift a 25-kg mass as shown in Fig. 12.51. Find the tension in the cable T and the resultant force acting on the pivot point, neglecting the mass of the crane.

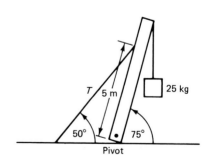

Fig. 12.51

33. A car of mass 4000 kg has traversed 75 m across a 100-m-long bridge (Fig. 12.52). If the bridge itself has a mass of 5×10^4 kg and is supported at each end, find the forces on each support.

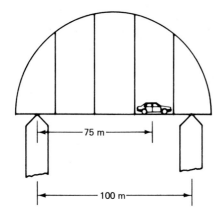

Fig. 12.52

34. A physician's scale has several remarkable properties that you may not have thought about. It is a balance that can weigh a large mass with a small mass, and its measurement is independent of where you stand on the platform. This is remarkable because the scale equilibrium condition is based on the torques as well as the forces summing to zero. Consider Fig. 12.53. The scale is built so that $\dfrac{A}{B} = \dfrac{D}{C}$. Under these

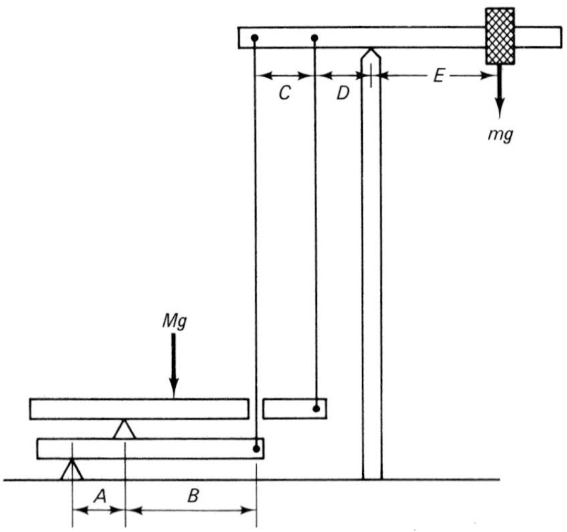

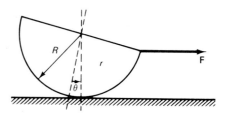

Fig. 12.53

conditions, show that $MgD = Emg$, independent of where the weight of the person (Mg) is located on the scale platform. (*Hint:* This scale has three major parts. Each of them must be in static equilibrium.

Identify the forces and the equilibrium condition on each of these parts.)

35. A horizontal force **F** is applied to a uniform solid hemisphere of mass m as shown in Fig. 12.54. If the coefficient of friction of the sphere with the floor is μ, find the maximum angle of tipping θ that will occur before the hemisphere begins to slide. (*Hint:* See the text for the location of the center of mass of a solid hemisphere.)

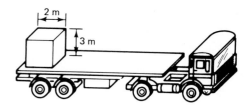

Fig. 12.54

36. A large box 2 m square by 3 m high rests on the back of a moving flatbed truck as shown in Fig. 12.55. The truck is moving at a speed of 20 m/s when

Fig. 12.55

the driver steps on his brakes and decelerates at a constant rate of 5 m/s². If the frictional coefficient between the box and the truck bed is $\mu = .3$, will the box slide, tip over, or remain at rest relative to the truck? Assume that the center of mass of the box is in its geometric center.

37. A large trunk of width W and height h is pushed across a rough floor by a shipping clerk. Find the maximum height above the floor that the clerk can exert a horizontal force on the trunk without tipping it. Assume that the mass inside is uniformly distributed throughout the volume. Note that neither the mass nor the depth of the trunk is needed to solve this problem. Explain this.

38. A uniform rectangular block shown in Fig. 12.56 is rotated about the edge AB.

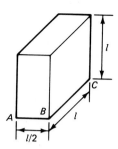

Fig. 12.56

(a) Find the maximum angle it can be tipped and still return to its original position after the tipping forces are released.

(b) Find this maximum angle if it is tipped about the BC edge.

39. A homogeneous solid sphere of radius r and mass M is suspended on a wall by a string of length L (Fig. 12.57). Assuming no friction:

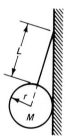

Fig. 12.57

(a) Find the tension in the string.

(b) Find the normal force the wall exerts on the sphere.

40. Two uniform planks, each of mass M, are hinged together by a string at the middle (Fig. 12.58). Find the tension in the string if the boards rest on a frictionless surface.

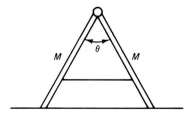

Fig. 12.58

41. The deltoid muscle is principally responsible for your ability to hold things in your hand in a horizontal outstretched position. A simplified mechanical model of your arm is shown in Fig. 12.59. If the weight of your arm is 8 lb and $\theta = 18°$, what is the tension T in the deltoid which is necessary just to support your arm?

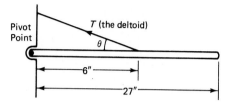

Fig. 12.59

42. A solid hemisphere of radius R rests on a rough flat surface (Fig. 12.60). The hemisphere rolls on the surface maintaining point contact friction. Show that this system is in stable equilibrium with respect to small rolling displacement from its equilibrium configuration.

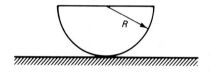

Fig. 12.60

43. Find the two tensions in the two ropes T_1 and T_2 for the machine shown in Fig. 12.61 using the principle of virtual work.

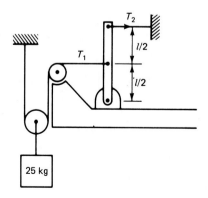

Fig. 12.61

44. Find the force necessary to lift the mass of 50 kg attached to the machine shown in Fig. 12.62.

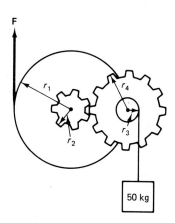

Fig. 12.62

45. Ignoring all friction in the system, calculate the tension in the rope necessary to lift the 500-kg load attached to the four-pulley machine shown in Fig. 12.63.

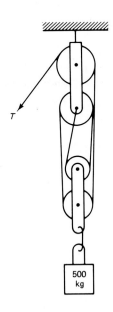

Fig. 12.63

46. What force must be applied to the handle in the differential windlass shown in Fig. 12.64 that will lift the attached mass *M*.

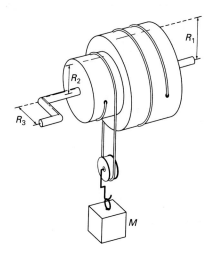

Fig. 12.64

47. Prove the theorems on axial and polar vectors that are given in the text. What kind of a scalar is $\mathbf{a} \cdot (\mathbf{b} \times \mathbf{c})$ if $\mathbf{a}$, $\mathbf{b}$, and $\mathbf{c}$ are polar vectors?

48. Prove that if the vector sum of the forces on an object is zero and if the vector sum of the torques is also zero about a particular point, then the sum of the torques about *any* point must also be zero.

13

ANGULAR MOMENTUM AND ROTATION ABOUT A FIXED AXIS

13.1 Introduction

IN THIS chapter we expand our discussion of the dynamics of rigid bodies to include rotational motion. We introduce new physical concepts that organize and clarify this particular kind of motion. In Chapter 12 you studied torques as they applied to static equilibrium, but now we will see their relevance to rotational motion, particularly their relationship to a new quantity, ANGULAR MOMENTUM. Angular momentum is the principal "actor" in this new scene, while MOMENT OF INERTIA is cast in the supporting role. You may draw on your knowledge of translational motion, where forces produce time rates of change of linear momentum, to guess that torques produce time rates of change of angular momentum and that moment of inertia takes the place of mass inertia for rotational motion. Newton's laws apply for all the systems we examine, but they are recast in a way that is appropriate to describe both general rotational motion and rotational motion of an ideal rigid body about a fixed axis. Vamanos!*

13.2 Angular Momentum

ANGULAR momentum is one of those concept in physics that has proven to be more fundamental, universal, and contemporary than its originators could ever have anticipated. Remember, you first encountered it in our discussion of elementary particles, where the intrinsic spin angular momentum of a particle turns out to be quantized in either integral or half-integral units of Planck's constant divided by 2π: $\hbar = h/2\pi$. Even particles without rest mass, such as

---

* Spanish for "Let's go!"

the photon and the neutrino, have intrinsic quantized angular momentum. But that's way ahead of the story; let's begin with the classical definition of angular momentum $\mathbf{L}_0$ of a point particle of mass m, which has linear momentum $\mathbf{p}$, and is located at the position $\mathbf{r}$ with respect to a given origin of coordinates O (Fig. 13.1). The angular momentum with respect to the origin O is defined as

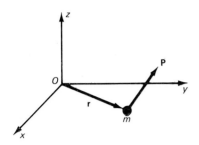

Fig. 13.1

$$\mathbf{L}_0 = \mathbf{r} \times \mathbf{p}. \tag{13.1}$$

The cross product of any two polar vectors is itself a vector, an axial vector. We have become used to working with polar vectors that can be "pushed around" as long as their magnitude and direction remain the same. That is *not* permitted with axial vectors such as torque and angular momentum. Both these quantities can be defined *only* when a particular origin is specified. The angular momentum of any object, be it a point mass or a massive spinning flywheel, can be specified only with respect to a specific origin. This means that *both* the magnitude and direction of the angular momentum depend on the choice of origin. For that reason, our notation includes a subscript on the $\mathbf{L}$, indicating the chosen origin. The unit of angular momentum in the SI system is kg-m²/s. There is no special name, such as watt or joule, for these units. Make one up; maybe it will catch on.

The direction of the angular momentum of either a point particle or an extended rigid body is not intuitive for the beginning student. Since angular momentum is the cross product of two familiar vector quantities, it will help you greatly to renew your acquaintance with the "right-hand" rule and the vector algebra associated with cross products in Section 6.2. Assuming this familiarity, we can examine a few examples of particle motion in two dimensions. Figure 13.2 depicts a particle of mass m moving in a straight line in the y-direction with velocity $v\mathbf{j}$ in the x–y plane. From Eq. 13.1 the angular momentum of this particle *with respect to the origin O* is

$$\mathbf{L}_0 = \mathbf{r} \times m\mathbf{v} = m|\mathbf{r}||\mathbf{v}| \sin\theta\mathbf{k} \tag{13.2}$$

and is directed along the positive z-axis. This result may also be arrived at by writing out both the momentum and position vectors in terms of the unit vectors:

$$\mathbf{r} = r_x\mathbf{i} + r_y\mathbf{j} \quad \text{and} \quad \mathbf{p} = mv\mathbf{j}.$$

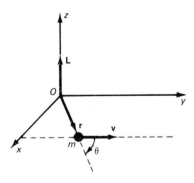

Fig. 13.2

Then

$$\mathbf{L}_0 = \mathbf{r} \times \mathbf{p} = (r_x\mathbf{i} + r_y\mathbf{j}) \times mv\mathbf{j} = mr_xv(\mathbf{i} \times \mathbf{j}) + mr_yv(\mathbf{j} \times \mathbf{j})$$

but

$$\mathbf{i} \times \mathbf{j} = \mathbf{k} \quad \text{and} \quad \mathbf{j} \times \mathbf{j} = 0.$$

Thus

$$\mathbf{L}_0 = mr_xv\mathbf{k}. \tag{13.3}$$

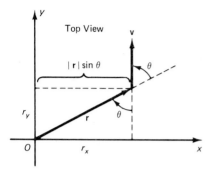

Fig. 13.3

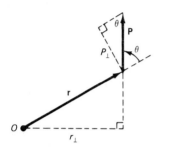

Fig. 13.4

Examine Fig. 13.3, the x–y plane from the top view. From this perspective it is clear that

$$r_x = |\mathbf{r}| \sin \theta, \tag{13.4}$$

so that Eqs. 13.3 and 13.2 are the same.

There is a third way to calculate angular momentum of a point particle about a specific origin. Any two nonparallel vectors define a plane. In the plane defined by $\mathbf{r}$ and $\mathbf{p}$, the quantity $|\mathbf{r}|\sin \theta$ is the same as $r_\perp$, the component of $\mathbf{r}$ perpendicular to $\mathbf{p}$ (as shown in Fig. 13.4). Thus the magnitude of the angular momentum can be written

$$|\mathbf{L_0}| = r_\perp p. \tag{13.5}$$

Since $p_\perp = p \sin \theta$, an alternative way to write Eq. 13.5 is

$$|\mathbf{L}| = r p_\perp, \tag{13.6}$$

where $p_\perp$ is the component of $\mathbf{p}$ that is perpendicular to the position vector, $\mathbf{r}$.

Finally, one can directly use the definition of the cross product in the determinant representation:

$$\mathbf{L_0} = \mathbf{r} \times \mathbf{p} = \begin{vmatrix} \mathbf{i} & \mathbf{j} & \mathbf{k} \\ r_x & r_y & r_z \\ p_x & p_y & p_z \end{vmatrix}$$

$$= (r_y p_z - r_z p_y)\mathbf{i} + (r_z p_x - r_x p_z)\mathbf{j} + (r_x p_y - r_y p_x)\mathbf{k}. \tag{13.7}$$

In this particular problem, $\mathbf{r}$ and $\mathbf{p}$ define a plane. We are free to choose this to be the x–y plane; thus

$$\mathbf{L_0} = \begin{vmatrix} \mathbf{i} & \mathbf{j} & \mathbf{k} \\ r_x & r_y & 0 \\ p_x & p_y & 0 \end{vmatrix} = (r_x p_y - r_y p_x)\mathbf{k} = r_x p_y \mathbf{k} \tag{13.8}$$

(since $p_x = 0$)

Why does the minus sign appear in this representation of angular momentum? What is its significance?

Of all the mathematical representations of angular momentum, Eq. 13.6 ($|\mathbf{L_0}| = r p_\perp$) is possibly the easiest to interpret. Angular momentum is a measurement of rotation. If there is no sense of rotation with respect to a given origin, there is no angular momentum. For example, straight-line motion toward or away from the origin has no angular momentum about that origin. A particle moving with a velocity transverse to the position vector has maximum tendency to rotate about that origin. That's what Eq. 13.6 is telling us. At the instant when $\mathbf{p}$ is perpendicular to $\mathbf{r}$, one has the largest rotation about the origin, because then $p_\perp$ and p are the same. When the particle is moving toward or away from the origin, $p_\perp$ is zero and there is no angular momentum or rotation about O. Between these two extreme cases, the component of the momentum perpendicular to **r** contributes to the rotation tendency, while the parallel component of the linear momentum does not. One easy way to test whether a particle has angular momentum with respect to a particular origin is to construct a straight line from the origin to the particle. If the line rotates about the origin as the particle moves, it has angular momentum, but if the line only shrinks or stretches, it does not.

EXAMPLE 1   Consider a small sliding puck of mass m moving along the y-axis with a constant velocity $\mathbf{v} = v\mathbf{j}$, as shown in Fig. 13.5. The angular momentum of the puck with respect to the origin O can easily be calculated. The position vector is $\mathbf{r} = r_y\mathbf{j}$ and the linear momentum is $\mathbf{p} = mv\mathbf{j}$. Thus the angular momentum is

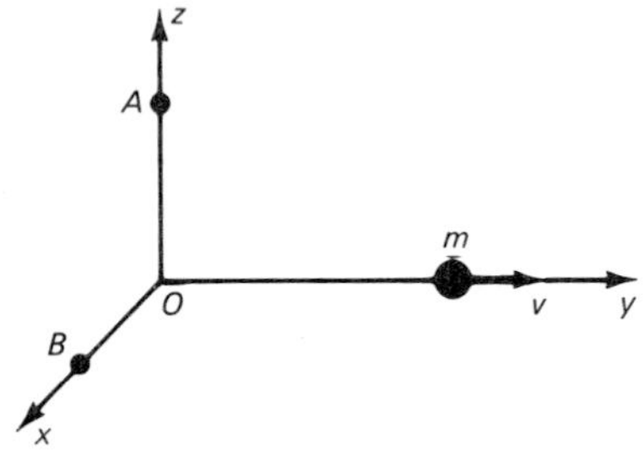

$$\mathbf{L}_0 = \mathbf{r} \times \mathbf{p} = r_y\mathbf{j} \times mv\mathbf{j} = r_y mv(\mathbf{j} \times \mathbf{j}) = 0, \qquad (13.9)$$

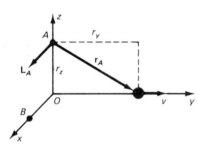

Fig. 13.5

since $\mathbf{j} \times \mathbf{j} = 0$. The puck has zero angular momentum with respect to the origin. Next we shall calculate the angular momentum of the same puck about point A on the z-axis (Fig. 13.6). Here $\mathbf{r}_A = r_y\mathbf{j} - r_z\mathbf{k}$. Thus

$$\mathbf{L}_A = \mathbf{r}_A \times m\mathbf{v} = (r_y\mathbf{j} - r_z\mathbf{k}) \times mv\mathbf{j}$$

$$= r_y mv(\mathbf{j} \times \mathbf{j}) - r_z mv(\mathbf{k} \times \mathbf{j})$$

$$= r_z mv\mathbf{i}, \qquad (13.10)$$

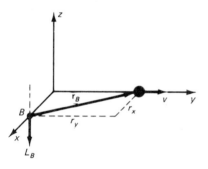

Fig. 13.6

(since $\mathbf{k} \times \mathbf{j} = -\mathbf{i}$). The puck has a counterclockwise sense of rotation about A.

Finally, we calculate its angular momentum about point B (Fig. 13.7) on the positive x-axis, where $\mathbf{r}_B = -r_x\mathbf{i} + r_y\mathbf{j}$:

$$\mathbf{L}_B = (-r_x\mathbf{i} + r_y\mathbf{j}) \times mv\mathbf{j}$$

$$= -r_x mv(\mathbf{i} \times \mathbf{j}) + 0$$

$$= -r_x mv\mathbf{k}. \qquad (13.11)$$

These examples demonstrate the dependence of the angular momentum of a point particle on the choice of reference point. But hidden in all these simple examples is the germ of an important principle of physics, another "absolute" conservation principle. The example assumed that the puck moves with a constant speed in a straight line along the x-axis. This means that there is no *net* external force acting on the puck and the linear momentum is a constant of the motion.

Please examine Eq. 13.10 carefully. The mass, the magnitude of the velocity $|\mathbf{v}|$, and r_z are also constant of the motion. Regardless of the position of the puck along the y-axis, r_z remains the same. Thus the magnitude of the angular momentum about A is a constant of the motion. It should be obvious that the direction of the angular momentum also remains constant. *Conclusion:* The angular momentum about A is a constant of the motion. The angular momentum for

Fig. 13.7

this particle is in fact a constant of the motion about any stationary reference point.

You should *not* quickly jump to the conclusion that conservation of linear momentum is the same as conservation of angular momentum or that angular momentum is conserved if there is no net force on a body. It turns out that the criteria for angular momentum to be conserved are in general *different* from those for linear momentum conservation. We shall see what they are in the next section.

EXAMPLE 2
Circular Motion

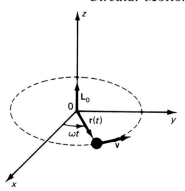

Fig. 13.8

Among the many examples of particle motion, uniform circular motion is one of the most important (Fig. 13.8). In Section 9.5 we studied both the kinematics and the dynamics of circular motion. Before proceeding with this discussion of angular momentum, you should review this material. Equations 9.15 and 9.16 give the position and velocity vectors for a particle executing uniform circular motion. Then angular momentum of such a particle with respect to the origin, taken at the center of the circle, is

$$\mathbf{L}_0 = \mathbf{r} \times \mathbf{p} = \mathbf{r} \times \left(m\, \frac{d\mathbf{r}}{dt} \right)$$

$$= [r \cos\omega t \; \mathbf{i} + r \sin\omega t \; \mathbf{j}] \times m[-r\omega \sin\omega t \; \mathbf{i} + r\omega \cos\omega t \; \mathbf{j}]$$

$$= r^2 \omega m [\cos^2\omega t \; (\mathbf{i} \times \mathbf{j}) - \sin^2\omega t \; (\mathbf{j} \times \mathbf{i})].$$

But since $\mathbf{i} \times \mathbf{j} = \mathbf{k}$ and $\mathbf{j} \times \mathbf{i} = -\mathbf{k}$ and $\sin^2\theta + \cos^2\theta = 1$,

$$\mathbf{L}_0 = r^2 \omega m \mathbf{k}. \tag{13.12}$$

However, Eq. 9.13 gives

$$v = r\omega,$$

and thus

$$\mathbf{L}_0 = rvm\mathbf{k}. \tag{13.13}$$

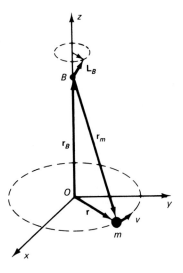

Fig. 13.9

We might have derived these two expressions for $\mathbf{L}_0$ in a simpler way by recalling that for *circular motion* the velocity of the particle is always perpendicular to the position vector, and thus

$$|\mathbf{L}_0| = |\mathbf{r} \times \mathbf{p}| = |\mathbf{r}||\mathbf{p}| \sin\theta = |\mathbf{r}||\mathbf{p}| = rmv.$$

Equations 13.12 and 13.13 demonstrate that the angular momentum (with respect to the center of the circle) of a point mass moving with uniform circular motion is *constant* and directed along the axis perpendicular to the plane of the circle.

Now we will reexamine the angular momentum of this *same* particle but from a different origin, point B, along the positive z-axis (as shown in Fig. 13.9). From the diagram it is apparent that $\mathbf{r}_m$ and $\mathbf{v}$ are always perpendicular, so that $|\mathbf{L}_B| = |\mathbf{r}_m \times \mathbf{p}| = r_m mv$. It should also be obvious that although the magnitude of the angular momentum is constant, its direction in space rotates with the particle.

To obtain the complete vector expression for the angular momentum about B, we write the vector relationship between $\mathbf{r}$, $\mathbf{r}_B$, and $\mathbf{r}_m$. They are

$$\mathbf{r} = \mathbf{r}_B + \mathbf{r}_m \qquad \text{or} \qquad \mathbf{r}_m = \mathbf{r} - \mathbf{r}_B. \tag{13.14}$$

But $\qquad \mathbf{r}_B = r_B\mathbf{k} \qquad$ and $\qquad \mathbf{r} = r\cos\omega t\,\mathbf{i} + r\sin\omega t\,\mathbf{j}.$ (13.15)

Combining Eqs. 13.14 and 13.15, we obtain an expression for $\mathbf{r}_m$: namely,

$$\mathbf{r}_m = r\cos\omega t\,\mathbf{i} + r\sin\omega t\,\mathbf{j} - r_B\mathbf{k}. \tag{13.16}$$

Now we are ready to calculate $\mathbf{L}_B$:

$$
\begin{aligned}
\mathbf{L}_B = \mathbf{r}_m \times \mathbf{p} &= [r\cos\omega t\,\mathbf{i} + r\sin\omega t\,\mathbf{j} - r_B\mathbf{k}] \\
&\quad \times m\omega r[-\sin\omega t\,\mathbf{i} + \cos\omega t\,\mathbf{j}] \\
&= mr^2\omega[(\mathbf{i} \times \mathbf{j})\cos^2\omega t - (\mathbf{j} \times \mathbf{i})\sin^2\omega t] \\
&\quad + mr\omega r_B[\sin\omega t\,(\mathbf{k} \times \mathbf{i}) - \cos\omega t\,(\mathbf{k} \times \mathbf{j})] \\
&= mr^2\omega\mathbf{k} + \omega r_B mr[\cos\omega t\,\mathbf{i} + \sin\omega t\,\mathbf{j}]
\end{aligned}
$$

or

$$\mathbf{L}_B = mr^2\omega\mathbf{k} + m\omega r_B\mathbf{r}(t). \tag{13.17}$$

To find the magnitude of this angular momentum, we form the dot product:

$$
\begin{aligned}
|\mathbf{L}_B| = (\mathbf{L}_B \cdot \mathbf{L}_B)^{1/2} &= (m^2 r^4 \omega^2 + m^2 \omega^2 r_B^2 r^2)^{1/2} \\
&= mr\omega(r^2 + r_B^2)^{1/2} = mv(r^2 + r_B^2)^{1/2}. \tag{13.18}
\end{aligned}
$$

But

$$r^2 + r_B^2 = r_m^2 \qquad \text{so} \qquad |\mathbf{L}_B| = mvr_m,$$

which is what we had originally demonstrated in our first calculation. However, our efforts here were not wasted, since the formal solution Eq. 13.17 gave us the *direction, magnitude,* and *time dependence* of the angular momentum. Compare Eqs. 13.13 and 13.17, the angular momentum of the same particle about *different reference points.* Not only do we find the magnitude and direction of the angular momentum to be different, we also discover that for one reference point (the origin) the angular momentum is independent of time, while for reference point B the angular momentum is time dependent. The time dependence is not in the magnitude but in the direction of the angular momentum. From Eq. 13.17 we see that the angular momentum rotates about the z-axis with the same time dependence as $\mathbf{r}(t)$. The z-component of the angular momentum $(mr^2\omega\mathbf{k})$ is a constant of the motion, whereas the x–y component $[m\omega r_B\mathbf{r}(t)]$ is time dependent.

WHAT CAUSES the angular momentum to change with time? In translational motion it is the force on a particle that produces the time rate of change of its linear momentum. Let's examine the analogous law for rotational motion.

Consider the definition of angular momentum (Eq. 13.1):

$$\mathbf{L}_0 = \mathbf{r} \times \mathbf{p}.$$

The time rate of change of the angular momentum can be obtained by direct differentiation:

$$\frac{d\mathbf{L}_0}{dt} = \frac{d}{dt}(\mathbf{r} \times \mathbf{p}) = \frac{d\mathbf{r}}{dt} \times \mathbf{p} + \mathbf{r} \times \frac{d\mathbf{p}}{dt}. \qquad (13.19)$$

The first term in Eq. 13.19 is zero because

$$\frac{d\mathbf{r}}{dt} = \mathbf{v} \qquad \text{and} \qquad \mathbf{p} = \gamma m \mathbf{v}$$

and then

$$\frac{d\mathbf{r}}{dt} \times \mathbf{p} = \mathbf{v} \times \gamma m \mathbf{v} = \gamma m (\mathbf{v} \times \mathbf{v}) = 0$$

since the cross product of any vector with itself is always zero. The second term in Eq. 13.19 can be simplified since by Newton's second law,

$$\mathbf{F} = \frac{d\mathbf{p}}{dt},$$

then

$$\mathbf{r} \times \frac{d\mathbf{p}}{dt} = \mathbf{r} \times \mathbf{F} = \boldsymbol{\tau}_0,$$

the torque of the force $\mathbf{F}$. Finally, Eq. 13.19 reduces to

$$\boxed{\boldsymbol{\tau}_0 = \frac{d\mathbf{L}_0}{dt}.} \qquad (13.20)$$

Equation 13.20 is an elegant, compact, and extremely important law of physics that was derived using the *correct relativistic expressions* for both linear momentum and force. We expect this law to hold for relativistic systems. If the net torque on the body is zero, Eq. 13.20 states that the time rate of change of angular momentum of the body also equals zero. Under such circumstances the angular momentum is a constant of the motion. The conservation of angular momentum can be written as follows:

> *The angular momentum of a system is conserved if there is zero net external torque on the system.*

The law of angular momentum conservation holds for a single point mass, a collection of point masses, an extended body, or a collection of extended bodies. All the experimental evidence so far supports the validity of the law from the smallest dimensions inside the proton to the largest objects in the universe. Astrophysicists use the principle to understand everything from binary star systems and exploding supernovas to the condensation of gas clouds forming gigantic galaxies. Atomic physicists use it to explain the structure of atoms and molecules, while high-energy physics uses the same law to understand the quark structure of baryons and mesons. The conservation of angular momentum is one of the cornerstones of modern physics.

Equation 13.20, relating the time rate of change of the angular momentum to the torque, is analogous to Newton's second law for translational motion, but it has some important differences. To begin with, it is an equation of *axial vectors* rather than the polar vectors you are used to. Both L_0 and τ_0 are cross products of polar vectors, $\tau_0 = \mathbf{r} \times \mathbf{F}$ and $L_0 = \mathbf{r} \times \mathbf{p}$, and are thus axial vectors. In practical terms this means that the origin chosen to calculate the angular momentum *must be the same* as the origin chosen to calculate the torques of the body. The equation holds for *any* origin as long as the *same* origin is used for calculating *both* L_0 and τ_0. Like Newton's second law, this law of rotational motion is valid in inertial reference frames.

For linear momentum to be conserved, the system must be isolated from all external forces. For angular momentum to be conserved, Eq. 13.20 does *not* require the system to be isolated from external forces, only from external torques. There are many examples of systems that conserve angular momentum, yet are subjected to external forces (but no external torques). Our own planetary system is a good example, where the sun exerts an external force but no external torque on all the planets.

EXAMPLE 1   It is appropriate now to reexamine Example 2 of the preceding section: a point mass executing uniform circular motion, for the purpose of understanding the relationship between torque and angular momentum. We must, however, include the forces on the object that are responsible for circular motion. Suppose that our small mass was supported by a "massless" fiber in the configuration of the conical pendulum (Fig. 13.10). There are only two forces that act on m: the tension in the fiber $\mathbf{T}$ and the gravitational force $-mg\mathbf{k}$. Figure 13.11 is the free-body diagram showing these forces. Newton's second

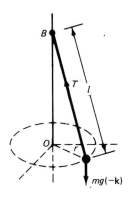

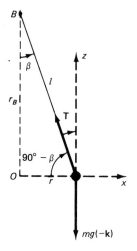

Fig. 13.10

Fig. 13.11

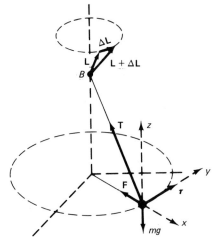

Fig. 13.12

law, in component form, for this point mass:

$$x\text{-comp.:} \qquad -T \sin \beta = ma_x = -m\omega^2 r \qquad (13.21)$$

$$z\text{-comp.:} \qquad T \cos \beta - mg = 0. \qquad (13.22)$$

The *resultant* instantaneous force on the mass m is then $-T \sin \beta\mathbf{i}$. The torque caused by this resultant force can be calculated with respect to *any* origin. First consider the origin at the center of the circle 0. Then

$$\boldsymbol{\tau}_0 = \mathbf{r} \times \mathbf{F} = (r\mathbf{i}) \times (-T \sin \beta\mathbf{i})$$

$$= -rT \sin \beta(\mathbf{i} \times \mathbf{i}) = 0,$$

since $\mathbf{i} \times \mathbf{i} = 0$. There is zero torque about the origin and thus no time rate of change of the angular momentum about that origin. This confirms what we had already calculated in Eq. 13.13, where $\mathbf{L}_0 = rvm\mathbf{k}$ is a constant of the motion. For the origin at 0, the position vector $\mathbf{r}$ and the resultant force $-T \sin \beta\mathbf{i}$ are parallel vectors, and therefore this force produces no torque about 0.

What about the torque of the same force about the reference point B? From Fig. 13.11 we can readily calculate the magnitude of the torque about B:

$$|\boldsymbol{\tau}_B| = \text{(radial distance)(net force)(sine of angle between)}$$

$$= (l)(T \sin \beta)[\sin (90 - \beta)]. \qquad (13.23)$$

But $\sin (90 - \beta) = \cos \beta$, and from Eq. 13.22, $T \cos \beta = mg$; then Eq. 13.23 reduces to

$$|\boldsymbol{\tau}_B| = lmg \sin \beta. \qquad (13.24)$$

Using the right-hand rule, it is clear that the torque acts in the $\mathbf{j}$ or y-direction. Actually, what we have calculated is the torque at the instant the mass is along the x-axis. As the mass moves around the circle, the torque also changes direction, always tangential to the circular path, as shown in Fig. 13.12. This is also the direction of the time rate of change of the angular momentum, as one can see from the same diagram.

Now that we have shown that a nonzero torque exists about point B and that there is indeed a time-varying angular momentum about B, let us demonstrate that they are equal. Starting with Eq. 13.17, we have

$$\mathbf{L}_B = mr^2\omega\mathbf{k} + \omega r_B mr[\cos\omega t \, \mathbf{i} + \sin\omega t \, \mathbf{j}] \qquad (13.25)$$

which shows that the angular momentum consists of a time-independent z-component and a time-varying x–y component. Differentiating, one obtains

$$\frac{d\mathbf{L}_B}{dt} = mr_B r\omega^2[-\sin\omega t \, \mathbf{i} + \cos\omega t \, \mathbf{j}]. \qquad (13.26)$$

The magnitude of this rate of change is

$$\left|\frac{d\mathbf{L}_B}{dt}\right| = mr_Br\omega^2, \qquad (13.27)$$

which should be equal to the magnitude of the torque, given in Eq. 13.24: $lmg \sin \beta$. To show this, we note from Eqs. 13.21 and 13.22 that

$$T \sin \beta = m\omega^2 r \qquad \text{and} \qquad T \cos \beta = mg.$$

Then $$|\tau| = lmg \sin \beta = lmg \frac{m\omega^2 r}{T} = lm\omega^2 r \cos \beta.$$

But $\cos \beta = r_B/l$ (from geometry), so that the expression above reduces to

$$|\tau| = lm\omega^2 r \frac{r_B}{l} = m\omega^2 r r_B,$$

which is the value calculated in Eq. 13.27. The magnitudes are equal.

The direction of the rate of change of the angular momentum is obtained by examination of Eq. 13.26. We have already seen that for uniform circular motion

$$v(t) = r\omega[-\sin\omega t\ \mathbf{i} + \cos\omega t\ \mathbf{j}]$$

(Chapter 9). Then Eq. 13.26 becomes

$$\frac{d\mathbf{L}_B}{dt} = mr_B\omega\mathbf{v}(t). \qquad (13.28)$$

Equation 13.28 tells us that the rate of change of the angular momentum is in the direction of the instantaneous velocity of the conical pendulum. This is also the direction of the instantaneous torque on the pendulum. All this long-winded calculation demonstrates that Eq. 13.20, $\tau = d\mathbf{L}/dt$, holds for this particular example. It's always good to check!

EXAMPLE 2

Two masses are connected by thin Mylar tape over a frictionless air bearing, shown in Fig. 13.13. The tape is connected to both blocks along the line of their geometric centers. Find the linear acceleration of m using the dynamics equation for rotational motion $\tau = d\mathbf{L}/dt$, neglecting the sliding friction of m_1 with the horizontal surface.

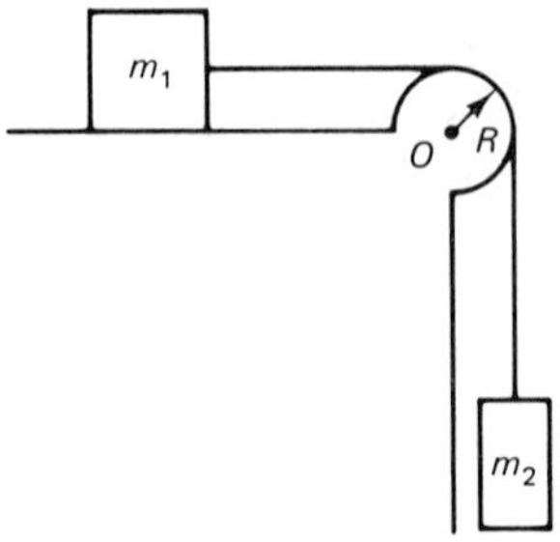

I am sure that all of you can solve this problem using the technique we developed in Chapter 9 for Newton's second law. This time we analyze the machine using rotational dynamics. But what's rotating? Where's the angular momentum?

The first order of business when you attack a rotational dynamics problem is to pick a reference point, since you will be working with *axial* vectors. The most convenient one in this machine is the center

Fig. 13.13

of the air bearing O. At some instant of time t, both m_1 and m_2 have a linear velocity $v(t)$. The angular momentum of the two blocks about the reference point O is

$$\mathbf{L}_0 = Rm_1v(t) + Rm_2v(t) = Rv(t)(m_1 + m_2) \quad \text{(into the paper).}$$

Differentiating with respect to time yields

$$\frac{d\mathbf{L}_0}{dt} = R(m_1 + m_2)\frac{dv(t)}{dt} = R(m_1 + m_2)a,$$

where a is the acceleration of both masses. What is the net external torque on this system about O? The only external force on the machine is gravity, and it produces a net force only on m_2 (since the normal force on m_1 cancels the gravitational force). The torque of this force on m_2 about O is

$$\boldsymbol{\tau}_0 = \mathbf{R} \times \mathbf{F}_{\text{gravity}} = Rm_2g \quad \text{(into the paper).}$$

Equating torque to the time rate of change of angular momentum, we have

$$\boldsymbol{\tau}_0 = \frac{d\mathbf{L}_0}{dt} \quad \text{or} \quad Rm_2g = R(m_1 + m_2)a \quad \text{or} \quad a = \frac{m_2g}{m_1 + m_2},$$

the result we expected from our past experience.

13.4 **Center-of-Mass Theorem**

SUPPOSE THAT we extend our discussion of angular momentum to a system consisting of many particles. It will now be advantageous for us to bring into the analysis the center-of-mass coordinates, for as you will see, the center of mass will greatly simplify the calculations. We begin by proving an important theorem:

> *If the center of mass of a system of particles is at rest in an inertial frame of reference, the angular momentum of these particles is independent of the reference point chosen in that frame.*

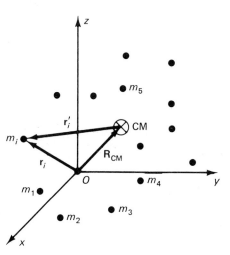

Fig. 13.14

To prove this theorem we consider the arbitrary collection of point masses m_i shown in Fig. 13.14 and define the following position vectors:

$\mathbf{r}_i$ = position of mass m_i with respect to the origin O

$\mathbf{r}_i'$ = position of mass m_i with respect to the center of mass

$\mathbf{R}_{\text{CM}}$ = position of center of mass with respect to the origin O

$\mathbf{v}_i$ = velocity of mass m_i, $d\mathbf{r}_i/dt$.

Then, by vector addition,

$$\mathbf{r}_i = \mathbf{R}_{\text{CM}} + \mathbf{r}_i'. \tag{13.29}$$

The total angular momentum with respect to the *arbitrary* origin of coordinates O is

$$\mathbf{L}_0 = \sum_i \mathbf{r}_i \times m_i \mathbf{v}_i = \sum_i m_i(\mathbf{R}_{\text{CM}} + \mathbf{r}_i') \times \mathbf{v}_i$$
$$= \sum_i m_i \mathbf{R}_{\text{CM}} \times \mathbf{v}_i + \sum_i m_i \mathbf{r}_i' \times \mathbf{v}_i. \tag{13.30}$$

The first term in Eq. 13.30 can be simplified by removing R_{CM} from the summation,

$$\sum_i m_i \mathbf{R}_{\text{CM}} \times \mathbf{v}_i = \mathbf{R}_{\text{CM}} \times \sum_i m_i \mathbf{v}_i = \mathbf{R}_{\text{CM}} \times M\mathbf{v}_{\text{CM}} \equiv \mathbf{L}_{\text{CM}}. \tag{13.31}$$

We recognize this term as the angular momentum of the center of mass of the system about the origin, that is, the angular momentum of a fictitious point particle of mass $\sum_i m_i = M$ moving with a velocity $\mathbf{v}_{\text{CM}}$ at a position $\mathbf{R}_{\text{CM}}$ about the origin. Equation 13.31 was derived using the definition of the velocity of the center of mass, Eq. 10.16.

What about the second term, $\sum_i m_i \mathbf{r}_i' \times \mathbf{v}_i$, in Eq. 13.30? Consider $\mathbf{v}_i$ first:

$$\mathbf{v}_i = \frac{d}{dt}(\mathbf{r}_i) = \frac{d}{dt}(R_{\text{CM}} + \mathbf{r}_i') = \mathbf{v}_{\text{CM}} + \mathbf{v}_i'. \tag{13.32}$$

Substituting this expression for $\mathbf{v}_i$ into the second term of Eq. 13.30, we have

$$\sum_i m_i \mathbf{r}_i' \times (\mathbf{v}_{\text{CM}} + \mathbf{v}_i') = \sum_i m_i(\mathbf{r}_i' \times \mathbf{v}_{\text{CM}}) + \sum_i m_i(\mathbf{r}_i' \times \mathbf{v}_i'). \tag{13.33}$$

From the definition of the center of mass, one can show that*

$$\sum_i m_i \mathbf{r}_i' = 0, \tag{13.34}$$

---

* Here is the proof:

$$\mathbf{R}_{\text{CM}} \equiv \frac{1}{M} \sum_i m_i \mathbf{r}_i.$$

But $\quad \mathbf{r}_i = \mathbf{R}_{\text{CM}} + \mathbf{r}_i', \quad$ thus $\quad \mathbf{R}_{\text{CM}} = \frac{1}{M} \sum_i m_i(\mathbf{R}_{\text{CM}} + \mathbf{r}_i')$

or $\quad \mathbf{R}_{\text{CM}} = \frac{1}{M} \sum_i m_i \mathbf{R}_{\text{CM}} + \frac{1}{M} \sum_i m_i \mathbf{r}_i' = \mathbf{R}_{\text{CM}} \frac{1}{M} \sum_i m_i + \frac{1}{M} \sum_i m_i \mathbf{r}_i'.$

However,

$$\sum_i m_i = M, \quad \text{thus} \quad \mathbf{R}_{\text{CM}} = \mathbf{R}_{\text{CM}} + \frac{1}{M} \sum_i m_i \mathbf{r}_i' \quad \text{or} \quad \sum_i m_i \mathbf{r}_i' = 0.$$

and since $\mathbf{v}_{CM}$ is a constant vector not included in the summation, the first term in Eq. 13.33 can be simplified:

$$\sum_i m_i \mathbf{r}_i' \times \mathbf{v}_{CM} = -\mathbf{v}_{CM} \times \sum_i m_i \mathbf{r}_i' = 0.$$

All this reduces Eq. 13.30 to

$$\mathbf{L}_0 = \mathbf{L}_{CM} + \sum_i \mathbf{r}_i' \times m_i \mathbf{v}_i'. \tag{13.35}$$

What has been accomplished with all this vector manipulation? We have found an expression for the angular momentum of an arbitrary collection of particles (which could be the mass elements of a rigid extended body) in terms of the angular momentum of the center-of-mass motion about the origin, and the motion of all the point mass elements *relative to the position of the center of mass*. In other words, we have decomposed arbitrary angular motion into: *motion of the imaginary point particle of mass M at a distance* $\mathbf{R}_{CM}$ *about the origin, and the motion of all the mass elements about the center of mass.* Even if you do not remember the rather long proof, you should not forget the result. It is important for simplifying complex rotational motions.

But we started out to prove a theorem; have we neglected our task? No, we have already proved it. If the center of mass of a system of particles is at rest in an inertial frame, $\mathbf{L}_{CM}$ must be zero. Should that be the case, Eq. 13.35 tells us that the total angular momentum about the arbitrary origin O is

$$\mathbf{L}_0 = \sum_i \mathbf{r}_i' \times m_i \mathbf{v}_i' \quad (\mathbf{L}_{CM} = 0). \tag{13.36}$$

But this equation contains only *relative coordinates*, that is, coordinates with respect to the center of mass of the system. This equation is *invariant* (unchanging) for any reference point. The vectors in Eq. 13.36 refer to the center of mass, but that is a property of the object. For such systems, the angular momentum *does not depend on the choice of origin*. We began the discussion of angular momentum with all sorts of warnings that you must always specify the origin about which you measure the angular momentum, and now we discover that for *one class of systems*, that is not necessary. One cannot apply this theorem to a "point" particle, for it makes no sense to speak of the motion of the mass elements of a point particle about its center of mass. All the mass elements of a classical point particle are located at the same point.*

---

* Here we are referring to classical point particles, not elementary particles such as electrons, which, although they appear to have no size or structure, do have intrinsic quantized spin angular momentum. Classical physics cannot predict or understand this property of nature.

Since the total angular momentum can be decomposed into two types of angular momentum, according to Eq. 13.35, we will assign to each type of angular momentum their commonly used names.

$$\mathbf{L}_{CM} \equiv \mathbf{L}_{\text{orbit}} \quad \text{(orbital motion of the center of mass about the origin)}$$

$$\mathbf{L}_{\text{relative to CM}} \equiv \sum_i \mathbf{r}'_i \times m_i \mathbf{v}_i \qquad (13.37)$$

$$\equiv \mathbf{L}_{\text{spin}} \quad \text{(angular momentum of the system about its center of mass)}$$

Then

$$\mathbf{L}_{\text{total}} = \mathbf{L}_{\text{orbit}} + \mathbf{L}_{\text{spin}}. \qquad (13.38)$$

EXAMPLE 1 Consider four point particles of equal mass constrained to rotate in a circle of radius r with the same speed v, as shown in Fig. 13.15. It is apparent from the symmetry that the center of mass is located at the center of the circle and is at rest. According to the theorem we just proved, the angular momentum of this system should be independent of the origin. Let's check the theorem by calculating the angular momentum of the system $\mathbf{L}$ for three different reference points.

(a) Origin at the center of mass:

$$\mathbf{L}_{CM} = \sum_{i=1}^{4} \mathbf{r}_i \times \mathbf{p}_i = \overset{(1)}{(r\mathbf{i})} \times (mv\mathbf{j}) + \overset{(2)}{(r\mathbf{j})} \times (mv(-\mathbf{i}))$$

$$+ \overset{(3)}{(r(-\mathbf{i}))} \times (mv(-\mathbf{j})) + \overset{(4)}{(r(-\mathbf{j}))} \times (mv(\mathbf{i}))$$

$$= 4mrv\mathbf{k}.$$

(b) Origin at B:

$$|\mathbf{L}_B| = \overset{(1)}{r_B mv \sin \theta'} + \overset{(2)}{2rmv} + \overset{(3)}{vr_B m \sin \theta'} + \overset{(4)}{0}.$$

But $\sin \theta' = r/r_B$; thus

$$|\mathbf{L}_B| = rmv + 2rmv + rmv = 4mrv\mathbf{k}$$

(direction determined by inspection).

(c) Origin at A:

$$|\mathbf{L}_A| = \overset{(1)}{r_A mv \sin \theta} + \overset{(2)}{[r + (r - x)]mv} + \overset{(3)}{r_A mv \sin \theta} + \overset{(4)}{xmv}.$$

But $\sin \theta = r/r_a$; thus

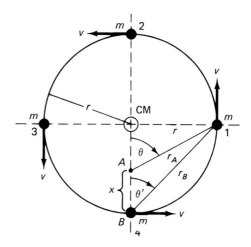

Fig. 13.15

$$|\mathbf{L}_A| = rmv + (2r - x)mv + rmv + xmv = 4mrv\mathbf{k}$$

(direction determined by inspection).

It works! Now you are at liberty to choose the most convenient origin to calculate $\mathbf{L}$ *if the center of mass is at rest.*

13.5 Rotation about a Fixed Axis

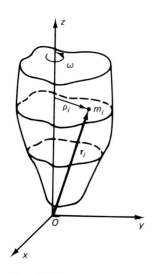

Fig. 13.16

Many practical problems that are concerned with rotational motion involve rotation about a fixed axis of nearly ideal rigid bodies. The examples are many, from gears and pulleys to automobile and bicycle wheels, rotating shafts, flywheels, spinning gyroscopes, and many others. Almost every piece of machinery has at least one such component. It is well worth our time to develop the special equations that govern the rotational motion of such objects. We will limit ourselves to nonrelativistic systems, because human-made rigid mechanical objects moving at relativistic speeds are still in the science fiction stage of development.

Imagine that we have an arbitrary rigid body rotating about a fixed axis, like the one in Fig. 13.16. If we divide it up into many small point masses m_i, we can calculate the total kinetic energy of the entire body simply by summing the kinetic energies of each point mass as

kinetic energy of rotation $= \frac{1}{2}m_1v_1^2 + \frac{1}{2}m_2v_2^2 + \cdots + \frac{1}{2}m_nv_n^2$

(13.39)

$$\mathrm{KE}_{\mathrm{rot}} = \frac{1}{2}\sum_i^n m_iv_i^2$$

> *For rotation about a fixed axis, every mass element executes* circular motion *about the axis.*

This fact greatly simplifies the analysis of these systems. We define the z-axis to be the axis of rotation. It is also convenient to define the distance of each mass element from the axis of rotation, ρ_i (Fig. 13.16). Now we can relate the linear speed of each mass element v_i to the angular speed of the body ω by Eq. 9.13, derived earlier:

$$v_i = \rho_i \omega_i. \tag{9.13}$$

All the mass elements, regardless of their distance from the axis of rotation, have the *same angular velocity* ($\boldsymbol{\omega}_i = \boldsymbol{\omega}_j = \boldsymbol{\omega}$). Substituting Eq. 9.13 into Eq. 13.39 and factoring ω out of the summation, we have

$$\text{KE}_{\text{rot}} = \tfrac{1}{2} \sum_{i=1} m_i \rho_i^2 \omega^2 = \tfrac{1}{2}\omega^2 \sum_i m_i \rho_i^2. \tag{13.40}$$

We define a new quantity, the MOMENT OF INERTIA about the z-axis, as

$$\text{MOMENT OF INERTIA} \equiv I_z \equiv \sum_i m_i \rho_i^2, \tag{13.41}$$

so the expression for the rotational kinetic energy of a rigid body about a fixed axis reduces to

$$\text{KE}_{\text{rot}} = \tfrac{1}{2} I_z \omega^2. \tag{13.42}$$

What about the *angular momentum* of our rigid body rotating about its fixed axis? To calculate the total angular momentum, we must add up the angular momentum of each mass element about the fixed axis of rotation. The angular momentum of m_i is

$$\mathbf{L}^i = \mathbf{r}_i \times m_i \mathbf{v}_i. \tag{13.43}$$

The addition of these angular momentum vectors is greatly simplified for fixed-axis rotation, for then we need only concern ourselves with the z-component of the angular momentum. All the other component are zero. From the definition of cross product,

$$\mathbf{L}^i = L_z^i = m_i(r_x v_y - r_y v_x)\mathbf{k}. \tag{13.44}$$

Figure 13.17 shows a cross-sectional view down the axis of the body with the mass element m_i moving with a speed v_i in the circular

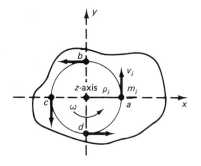

Fig. 13.17

path of radius ρ_i. Wherever Eq. 13.44 is evaluated for the mass elements anywhere on the circle they traverse, we will obtain the same result. To show this I have calculated $\mathbf{L}_z^i$ at four points a, b, c, and d.

At a: $\mathbf{L}_z^i = m_i r_x v_y = m_i \rho_i v_i \mathbf{k}$

At b: $\mathbf{L}_z^i = m_i r_y(-v_x) = m_i \rho_i v_i \mathbf{k}$

At c: $\mathbf{L}_z^i = m_i(-r_x)(-v_y) = m_i \rho_i v_i \mathbf{k}$

At d: $\mathbf{L}_z^i = m_i(-r_y)(v_x) = m_i \rho_i v_i \mathbf{k}$

or
$$L_z^i = m_i(\rho_i v_i) = m_i \rho_i^2 \omega \qquad (13.45)$$

(dropping the unit vector $\mathbf{k}$ and substituting for v_i from Eq. 9.8). Summing all the z-components of the angular momentum, we have

$$L_z = \sum_i L_z^i = \omega \sum_i m_i \rho_i^2 = \omega I_z. \qquad (13.46)$$

Once again the moment of inertia appears in our equation of rotational motion for rigid bodies. Finally, we examine Eq. 13.20 for the time rate of change of the angular momentum,

$$\frac{d\mathbf{L}}{dt} = \boldsymbol{\tau},$$

as it relates to the special case of rigid bodies executing rotation about a fixed axis. Considering the *z-component of the torque* about the z-axis,

$$\tau_z = \frac{d}{dt} L_z = \frac{d}{dt}(\omega I_z) = I_z \frac{d\omega}{dt}.$$

But if we then define the rate of change of the angular speed as the ANGULAR ACCELERATION, or

$$\frac{d\omega}{dt} \equiv \alpha, \qquad (13.47)$$

then

$$\tau_z = I_z \alpha. \qquad (13.48)$$

If there is more than one torque on the same body, one must sum the torques, keeping track of the sense of rotation. That is, a torque that produces a counterclockwise rotation is positive, and one that produces a clockwise rotation is negative, using the convention of the right-hand rule as we did in our discussion of static equilibrium of extended bodies in Section 12.5. Because we have derived so many new equations for rotational motion, it is worth listing them in Table 13.1 for your review. They are not as difficult as you might at first imagine, because they are analogous to the familiar equations for translational motion.

TABLE 13.1   Equations Describing Rotational and Translational Motion for Constant Forces and Torques, and Fixed Axis of Rotation About the z-Axis

Translational Motion	Rotational Motion, Fixed Axis
Inertia → mass m	Moment of inertia → I_z
Displacement → $\mathbf{r}$	Angular displacement → θ
Infinitesimal displacement $d\mathbf{r}$	Infinitesimal angular displacement $d\boldsymbol{\theta}$
Linear velocity → $\mathbf{v} = \dfrac{d\mathbf{r}}{dt}$	Angular velocity → $\boldsymbol{\omega} = \dfrac{d\boldsymbol{\theta}}{dt}$
Linear acceleration → $\mathbf{a} = \dfrac{d\mathbf{v}}{dt}$	Angular acceleration → $\boldsymbol{\alpha} = \dfrac{d\boldsymbol{\omega}}{dt}$
$d\mathbf{r} = \mathbf{v}\,dt$	$d\boldsymbol{\theta} = \boldsymbol{\omega}\,dt$
$\mathbf{v} = \mathbf{v}_0 + \mathbf{a}t$	$\omega = \omega_0 + \alpha t$
$\mathbf{r} = \mathbf{r}_0 + \mathbf{v}_0 t + \frac{1}{2}\mathbf{a}t^2$	$\theta = \theta_0 + \omega_0 t + \frac{1}{2}\alpha t^2$
$v_x^2 = v_{0x}^2 + 2a_x(x - x_0)$	$\omega^2 = \omega_0^2 + 2\alpha(\theta - \theta_0)$
Linear momentum → $m\mathbf{v}$	Angular momentum → $\mathbf{L}_z = \omega I_z\mathbf{k}$
Force → $\mathbf{F}$	Torque → τ_z
$\mathbf{F} = \dfrac{d}{dt}\mathbf{p}$	$\tau_z = \dfrac{d\mathbf{L}_z}{dt}$
Kinetic energy → $\frac{1}{2}mv^2$	Rotational kinetic energy → $\frac{1}{2}I_z\omega^2$
Work = $\int \mathbf{F} \cdot d\mathbf{s}$	Work = $\int \boldsymbol{\tau} \cdot d\boldsymbol{\theta}$
Power = $\mathbf{F} \cdot \mathbf{v}$	Power = $\tau\omega$

13.6 Moment of Inertia

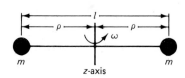

Fig. 13.18

IN THE preceding section I probably introduced too much new physics to swallow in one gulp. To become familiar with these new, powerful theoretical tools, let's first learn how to calculate the moment of inertia of a rigid body about a fixed axis of rotation. A symmetric dumbbell composed of two "point" masses, held together by a rigid massless rod of length l (Fig. 13.18), is a good problem to tackle. From the definition of moment of inertia about the z-axis, Eq. 13.41, we can see that

$$I_z = m\left(\frac{l}{2}\right)^2 + m\left(\frac{l}{2}\right)^2 = \frac{ml^2}{2}.$$

The same dumbbell rotating about a different fixed axis, through one

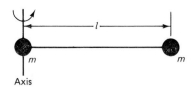

Fig. 13.19

of the masses (Fig. 13.19), has a moment of inertia:

$$I_z = m(0) + m(l)^2 = ml^2.$$

Actually, these two trivial calculations demonstrate an interesting general theorem of fixed-axis rotational motion. The moment of inertia about any axis that does *not* pass through the center of mass of the body is always *larger* than the moment of inertia about an axis that passes through the center of mass.*

For bodies that are not composed of discrete mass elements but are a continuous distribution of mass, Eq. 13.41 becomes an integral rather than a finite summation. That is,

$$I_z = \sum_i m_i \rho_i^2 \rightarrow \int_{\text{body}} \rho^2 \, dm, \tag{13.49}$$

where the integration is taken over the whole body. The mathematics is similar to the techniques we used in Chapter 12 to find the center of mass of a continuous body. For a body with uniform density σ, one can rewrite Eq. 13.49 as

$$dm = \sigma \, dV$$

and

$$I_z = \sigma \int_{\text{body}} \rho^2 \, dV. \tag{13.50}$$

EXAMPLE 1 Consider a uniform flat disk of radius R, thickness h, rotating around its geometric center as shown in Fig. 13.20. The infinitesimal ring of radius ρ and width $d\rho$ contributes an infinitesimal moment of inertia about the z-axis:

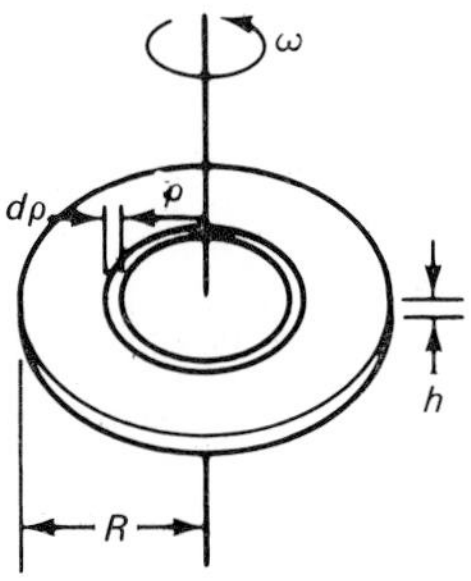

$$dI_z = \rho^2 \, dm.$$

But $dm = \sigma \, dV$, where σ is the density of the disk and dV is the volume of the infinitesimal ring. Thus

$$dm = \sigma(2\pi\rho \, d\rho)h \tag{13.51}$$

and

$$dI_z = 2\pi\sigma h \rho^3 \, d\rho. \tag{13.52}$$

Fig. 13.20

Substituting into Eq. 13.50 yields

$$I_z = 2\pi\sigma h \int_0^R \rho^3 \, d\rho = \frac{2\pi\sigma h R^4}{4}. \tag{13.53}$$

This can be simplified by noting that the density of a uniform disk is

$$\sigma = \frac{M}{\text{vol.}} = \frac{M}{\pi R^2 h}. \tag{13.54}$$

---

* This can be proven from the parallel axis theorem, discussed below.

Substituting into Eq. 13.53 gives us

$$I_z = \frac{2\pi h M}{\pi R^2 h} \frac{R^4}{4} = \frac{MR^2}{2}. \tag{13.55}$$

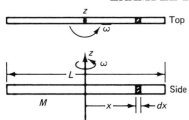

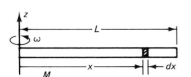

Fig. 13.21

EXAMPLE 2   Next let's calculate the moment of inertia of a uniform *thin* stick of mass M and length L (Fig. 13.21). In principle we should integrate over the thickness and the height. However, every mass element along the height is the same distance from the axis, and because the stick is thin, it is an excellent approximation to take all mass elements over the thickness also to be the same distance from the axis. Thus the problem is reduced to a one-dimensional integral. The linear mass density λ is defined as

$$\lambda = \frac{M}{L} \quad \text{and} \quad dm = \lambda \, dx$$

$$I_z = \int_{L/2}^{L/2} x^2 \, dm = \lambda \int_{-L/2}^{L/2} x^2 \, dx = \frac{Mx^3}{3L} \Big|_{-L/2}^{L/2} = \frac{ML^2}{12}.$$

The same stick, now rotating about a fixed axis at its *end* (as shown in Fig. 13.22), has a moment

$$I_z = \frac{M}{L} \int_0^L x^2 \, dx = \frac{ML^2}{3},$$

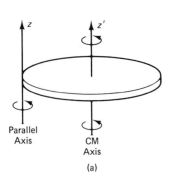

Fig. 13.22

once again demonstrating that the minimum moment of inertia of a rigid body occurs when the axis passes through the center of mass. Table 13.2 pictures a variety of symmetrical solid objects with their moments of inertia for the particular axis shown.

It is not always necessary to go through the integration process to calculate the moment of inertia of a body about a given fixed axis. A simple relationship exists between the moment of inertia of the body about an axis that passes through its center of mass and the moment of inertia of the *same* body about an axis that is parallel to the center-of-mass axis. It is called the **PARALLEL AXIS THEOREM**. Its proof is another demonstration of the power of vectors.

Consider the disk shown in Fig. 13.23 with two possible axes of rotation, one about the center of mass, the other parallel to it but along the edge of the disk. The z-component of the moment of inertia about the z-axis is, by definition,

$$I_z = \sum_i m_i r_i^2, \tag{13.56}$$

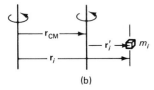

(b)

Fig. 13.23

where m_i are the mass elements of the disk. But from Fig. 13.22b,

$$\mathbf{r}_i = \mathbf{r}_{CM} + \mathbf{r}_i'. \tag{13.57}$$

TABLE 13.2 Moments of Inertia of Some Familiar Symmetrical Solid Objects about a Fixed Axis

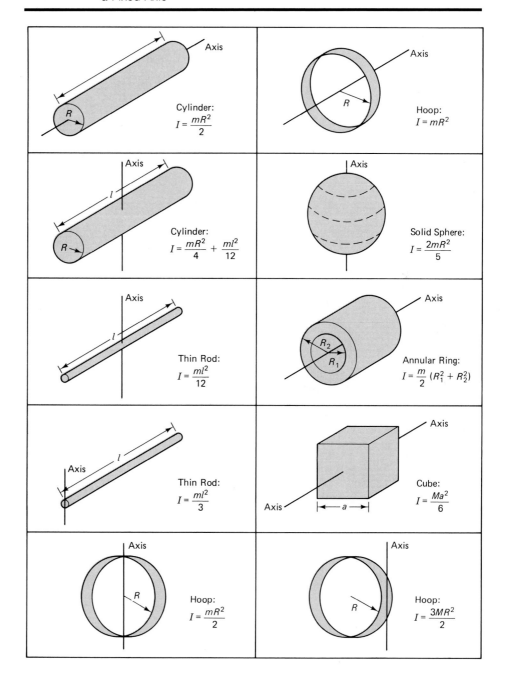

Substitution into Eq. 13.56 yields

$$I_z = \sum_i m_i (\mathbf{r}_{CM} + \mathbf{r}'_i)^2.$$

Squaring and writing out in three terms gives us

$$I_z = \overset{(1)}{\sum_i m_i r_{CM}^2} + \overset{(2)}{\sum_i m_i r_i'^2} + 2 \sum_i m_i \mathbf{r}_{CM} \cdot \mathbf{r}'_i. \qquad (13.58)$$

Let us evaluate Eq. 13.58 term by term:

(1) $\sum_i m_i r_{CM}^2 = r_{CM}^2 \sum_i m_i = M r_{CM}^2.$

(2) $\sum_i m_i r_i'^2 \equiv I_z^{CM},$ the moment of inertia of the body about the center of the mass axis.

(3) $2 \sum_i m_i \mathbf{r}_{CM} \cdot \mathbf{r}'_i = 2\mathbf{r}_{CM} \cdot \sum_i m_i \mathbf{r}'_i = 0.$

Since $\sum_i m_i \mathbf{r}'_i = 0$ (see Eq. 13.34 for proof), Eq. 13.58 becomes

$$I_z = I_z^{CM} + M r_{CM}^2. \qquad (13.59)$$

Equation 13.59 is the parallel axis theorem. A few examples will demonstrate how to use this theorem. In Example 2 we calculated the moment of inertia of a thin stick about its center of mass. The result was

$$I_z^{CM} = \frac{mL^2}{12}.$$

Using the parallel axis theorem to find the moment of inertia about an axis at the end of the stick,

$$I_z = \frac{1}{12} ML^2 + M \left(\frac{L}{2}\right)^2 = \frac{1}{3} ML^2$$

which agrees with the calculation done by straightforward integration. We can easily check some of the results in Table 13.2. For example, consider the moment of inertia of a hoop about any edge, Fig. 13.24.

$$I_z^{CM} = \frac{1}{2} MR^2.$$

Using the parallel axis theorem to find I_z, about an edge

$$I_{z'} = \frac{1}{2} MR^2 + MR^2 = \frac{3}{2} MR^2.$$

It checks!

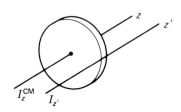

Fig. 13.24

NOW THAT you understand how to calculate the rotational inertial properties of extended bodies about fixed axes, it is time to summarize the equations of motions that govern the dynamics of this kind of rotation. Three essential equations were presented in Sec. 13.5:

$$L_z = \omega I_z \qquad \text{(angular momentum)}$$

$$\tau_z = I_z \alpha \qquad \text{(angular acceleration due to torque)}$$

$$\text{KE}_{\text{rot}} = \tfrac{1}{2} I_z \omega^2 \qquad \text{(kinetic energy of rotation)}.$$

For the special case of *constant angular acceleration*, you can readily derive a set of kinematic equations for rotational motion equivalent to the kinematic equations derived in Chapter 8 (Eqs. 8.30) for constant linear acceleration. They are:

$$\omega(t) = \omega_0 + \alpha t$$

$$\theta(t) = \theta_0 + \omega_0 t + \tfrac{1}{2}\alpha t^2 \qquad (13.60)$$

$$\omega^2(t) = \omega_0^2 + 2\alpha[\theta(t) - \theta_0].$$

Work can also be done by a torque when it produces a rotation of a body about a fixed axis. The expression for rotational work can be derived from the original definition of work Eq. 11.1:

$$dW = \mathbf{F} \cdot d\mathbf{s}. \qquad (11.1)$$

For rotational work the displacement is a rotation caused by a torque, so polar coordinates are the natural system to use. Figure 13.25 shows a flywheel mounted on a fixed bearing with a force **F** applied

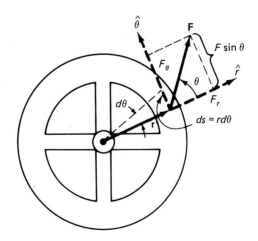

Fig. 13.25

at the point **r** at the angle θ. The force has two components,

$$\mathbf{F} = F_r\hat{\mathbf{r}} + F_\theta\hat{\boldsymbol{\theta}}, \tag{13.61}$$

and the displacement

$$d\mathbf{s} = r\, d\theta\, \hat{\boldsymbol{\theta}}. \tag{13.62}$$

Substituting Eqs. 13.61 and 13.62 into Eq. 11.1, we have

$$dW = \mathbf{F} \cdot d\mathbf{s} = (F_r\hat{\mathbf{r}} + F_\theta\hat{\boldsymbol{\theta}}) \cdot r\, d\theta\, \hat{\boldsymbol{\theta}} = F_\theta r\, d\theta.$$

But

$$F_\theta = F \sin \theta$$

so

$$dW = F \sin \theta\, r\, d\theta.$$

We recognize $(F \sin \theta)r = \tau_z$, the torque, so

$$dW = \tau_z\, d\theta. \tag{13.63}$$

Equation 13.63 expresses the differential of work done by the z-component of the torque acting on a rigid body rotating about fixed z-axis. The *rate* at which work is being done on a body executing this kind of motion is the power input, or

$$\frac{dW}{dt} = \frac{d}{dt}(\tau_z\, d\theta) = \tau_z \frac{d\theta}{dt}$$

$$\frac{dW}{dt} = \text{power} = \tau_z\omega. \tag{13.64}$$

EXAMPLE 1   A rope is wrapped around the rim of a solid uniform cylinder that is supported on a fixed axis through its geometric center (Figs. 13.26 and 13.27). The axis rotates without friction on stationary bearings. If the disk is initially at rest and a constant tangential force **F** is applied at $t = 0$, find (a) the angular acceleration of the disk at time t, (b) the angular velocity at time t, (c) the angular momentum at time t, (d) the angle θ through which it has rotated at time t, and (e) the power input to the system by the external force **F** at time t.

So that some numbers can be calculated, let us design the disk with:

$$M = 2 \text{ kg} \qquad R = 35 \text{ cm}$$
$$F = 8 \text{ N} \qquad t = 15 \text{ s}$$

(a) The angular acceleration can be calculated from $\tau_z = I_z\alpha$ if we know I_z and the z-component of the torque.

$$\tau_z = (\mathbf{r} \times \mathbf{F})_z = RF = (.35 \text{ m})(8 \text{ N}) = 2.8 \text{ N-m}$$

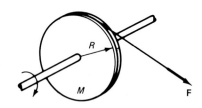

Fig. 13.26

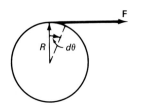

Fig. 13.27

$$I_z = \tfrac{1}{2}MR^2 = \tfrac{1}{2}(2 \text{ kg})(.35 \text{ m})^2 = .123 \text{ kg-m}^2$$

$$\alpha = \frac{\tau_z}{I_z} = (2.8 \text{ N-m})(.123 \text{ kg-m}^2)^{-1} = 22.9 \text{ rad/s}^2.$$

(b) The angular velocity can be calculated from the constant-torque equations 13.60:

$$\omega(t) = \omega_0 + \alpha t \qquad \text{but} \qquad \omega_0 = 0$$

$$= \alpha t = (22.9 \text{ rad/s}^2)(15 \text{ s}) = 342.9 \text{ rad/s}$$

(c) $L_z = I_z\omega = (.123 \text{ kg-m}^2)(342.9 \text{ rad/s}) = 42 \text{ kg-m}^2/\text{s}$

(d) $\theta(t) = \theta_0 + \omega_0 t + \tfrac{1}{2}\alpha t^2 \qquad \text{but} \qquad \theta_0 = 0, \quad \omega_0 = 0$

$$= \tfrac{1}{2}(22.9 \text{ rad/s}^2)(15 \text{ s})^2 = 2572 \text{ rad}$$

(e) $P = \tau_z\omega = (2.8 \text{ N-m})(342.9 \text{ rad/s}) = 960 \text{ W}.$

It should be noted that to calculate power in watts, the angular velocity must be measured in radians per second, not in some other units, such as revolutions per minute.

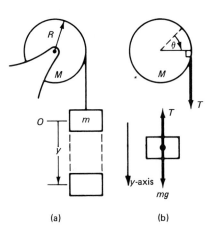

(a)

Fig. 13.28

EXAMPLE 2   A uniform cylinder of radius R and mass M is mounted with an axle through its geometric center on frictionless bearings (Fig. 13.28). A lightweight line is wrapped around the cylinder and attached to a hanging weight of mass m. Find (a) the acceleration of both the hanging mass and the cylinder, (b) the tension in the line, (c) the distance the weight falls in a time t, (d) the work done by the falling mass on the cylinder, and (e) the final kinetic energy of the rotating cylinder.

The physical parameters in this example are:

$$M = 1.5 \text{ kg} \qquad m = .5 \text{ kg}$$

$$R = .12 \text{ m} \qquad t = 7 \text{ s}$$

The tension in the line produces the torque that causes angular acceleration of the cylinder. But the tension depends on the acceleration of the falling mass. To solve this problem, we must deal with two simultaneous equations, which are coupled through the constraints that connect the rotational motion of the cylinder to linear motion of the falling mass. As we did in the analysis of linear motion, the machine will be dissected into its separate components with the forces and torques on each part. Free-body diagrams, coordinate systems, constraint equations, Newton's laws, and the law of rotational dynamics must now be employed.

Begin with the falling mass. Figure 13.28b is the free-body diagram. For linear motion in the y-direction,

$$mg - T = ma_y. \tag{b}$$

For our cylinder with the tension tangential to the external surface, the torque on the cylinder is

$$\tau_z = RT.$$

Substitution in Eq. 13.48 yields

$$\tau_z = I_z \alpha \qquad \text{or} \qquad RT = (\tfrac{1}{2}MR^2)\alpha. \qquad \text{(c)}$$

Now we are stuck. We have two equations, (b) and (c), and three unknowns, α, a_y, and T. There is a constraint equation, however, that relates the *angular acceleration* to the *linear acceleration*, which we can quickly derive. As the cylinder rotates through an angle θ, the length of line that comes off is $y = R\theta$, the same distance y that the mass falls. Differentiating this expression *twice* with respect to time, remembering that R is a constant, gives

$$\frac{d^2y}{dt^2} = \frac{d^2}{dt^2}(R\theta) = R\frac{d^2\theta}{dt^2}$$

or

$$a_y = R\alpha, \qquad\qquad\qquad (13.65)$$

and our Eqs. (b) and (c) can now be written as

$$RT = \frac{MR^2 a_y}{2R} \qquad \text{and} \qquad a_y = \frac{mg - T}{m}.$$

Solving for the linear acceleration a_y, we have

$$a_y = \frac{mg}{\frac{1}{2}M + m} = \frac{(.5\ \text{kg})(9.8\ \text{m/s}^2)}{(.5)(1.5\ \text{kg}) + .5\ \text{kg}} = 3.92\ \text{m/s}^2$$

and the tension

$$T = \frac{Ma_y}{2} = \frac{(1.5\ \text{kg})(3.92\ \text{m/s}^2)}{2} = 2.94\ \text{N}.$$

The angular acceleration

$$\alpha = \frac{a_y}{R} = \frac{3.92\ \text{m/s}^2}{.12\ \text{m}} = 32.67\ \text{rad/s}^2.$$

The distance the mass falls in a time t:

$$\theta(t) = \theta_0 + \omega_0 t + \tfrac{1}{2}\alpha t^2, \qquad \theta_0 = 0, \quad \omega_0 = 0$$

$$= \tfrac{1}{2}\alpha t = \tfrac{1}{2}(32.67\ \text{rad/s}^2)(7\text{s})^2 = 800.3\ \text{rad}$$

$$y = R\theta = (.12\ \text{m})(800.3\ \text{rad}) = 96.0\ \text{m}.$$

The work done by the tension in the line on the cylinder is

$$W = \tau\theta = RT\theta = (2.94\ \text{N})(.12\ \text{m})(800.3\ \text{rad}) = 282.3\ \text{J}.$$

The kinetic energy of the rotating cylinder at the time t can be calculated once the angular velocity $\omega(t)$ is determined.

$$\omega(t) = \omega_0 + \alpha t \qquad \omega_0 = 0$$

$$= \alpha t = (32.67 \text{ rad/s}^2)(7\text{s}) = 228.7 \text{ rad/s}$$

$$\text{KE}_{\text{rot}} = \tfrac{1}{2}I_z\omega^2 = \tfrac{1}{2}(\tfrac{1}{2})(1.5 \text{ kg})(.12 \text{ m})^2(228.7 \text{ rad/s})^2 = 282.4 \text{ J}.$$

Finally, let's calculate the change in the potential energy of the falling mass;

$$\Delta U = mgy = (.5 \text{ kg})(9.8 \text{ m/s}^2)(96.03 \text{ m}) = 470.5 \text{ J},$$

which does *not* equal the rotating kinetic energy of the cylinder. What is wrong?

Well, it shouldn't equal that rotational kinetic energy, because we have forgotten the linear kinetic energy acquired by the falling mass. Let's calculate it. First we need to know the linear velocity of the mass at time t.

$$v(t) = v_0 + at \qquad \text{but} \qquad v_0 = 0$$

$$= at = (3.92 \text{ m/s}^2)(7\text{s}) = 27.44 \text{ m/s}$$

$$\text{KE}_{\text{trans}} = \tfrac{1}{2}mv^2 = \tfrac{1}{2}(.5 \text{ kg})(27.44 \text{ m/s})^2 = 188.2 \text{ J}$$

$$\text{total kinetic energy of system} = \text{KE}_{\text{rot}} + \text{KE}_{\text{trans}}$$

$$\text{KE}_{\text{T}} = 282.4 + 188.2 = 470.6 \text{ J}.$$

Everything checks!

The work–energy theorem is applicable to rotational motion as well as to translational motion. It is essential to use the *net* torque acting on a body when calculating the change in rotational kinetic energy of that body. The work–energy theorem could have been invoked to calculate the rotational kinetic energy of the cylinder in Example 2, but we chose to demonstrate the theorem by calculating KE_{rot} and the work independently.

EXAMPLE 3   Now the tools are at hand to reconsider the analysis of the Atwood's machine (Fig. 13.29), this time including the effects of the moment of inertia of the pulley. (The rope will remain massless.) The task will be simplified by choosing two coordinate systems so that both masses will be accelerating in their respective positive directions. Figure 13.30 is the free-body diagram of the component parts.

For the hanging masses, Newton's second law:

$$m_1g - T_1 = m_1a_{1x} \tag{a}$$

$$T_2 - m_2g = m_2a_{2x'}. \tag{b}$$

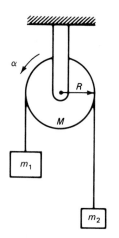

Fig. 13.29

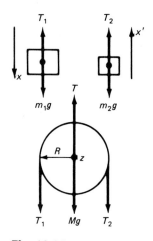

Fig. 13.30

For the pulley:

$$T - Mg - T_1 - T_2 = Ma_{\text{pulley}}. \tag{c}$$

Summing the torques about the axis of the pulley and using Eq. 13.48 gives

$$\tau_z = T_1R - T_2R = I_z\alpha. \tag{d}$$

The constraints of the machine are

$$a_{\text{pulley}} = 0 \qquad a_{1x} = a_{2x'} = a$$

$$a = \alpha R \qquad I_z = \tfrac{1}{2}MR^2 \qquad \text{(for a solid cylinder)}.$$

Solving (a) for T_1 and (b) for T_2 and substituting into (d), we arrive at

$$m_1(g - a)R - m_2(a + g)R = \frac{aMR^2}{2R}$$

or

$$a = \frac{g(m_1 - m_2)}{\tfrac{1}{2}M + m_1 + m_2}.$$

This value of the linear acceleration can be substituted in (a) and (b) and solved for the two tensions T_1 and T_2:

$$T_1 = m_1(g - a) = \frac{m_1g(\tfrac{1}{2}M + 2m_2)}{\tfrac{1}{2}M + m_1 + m_2}$$

and

$$T_2 = m_2(a + g) = \frac{m_2g(\tfrac{1}{2}M + 2m_1)}{\tfrac{1}{2}M + m_1 + m_2}.$$

Several things should be apparent from this calculation. The acceleration of the machine is reduced by the inertial mass of the pulley. The tension in the rope is *not* the same on both sides of the pulley. It is this difference in tension that is responsible for the net torque on the pulley, which causes the pulley to accelerate. No longer can the rope slide without friction over the pulley; there must be contact friction of the rope with the pulley for the pulley to accelerate.

13.8 Translational and Rotational Motion Combined

HOW DO we analyze the motion of a rolling rigid body that is *both* translating and rotating (without slipping) along a surface? One may theoretically understand such motions in two separate but equally correct ways: a combination of rotation and translation, and pure rotation. We will work out both approaches and demonstrate their equivalence. Consider the second approach first: *pure rotation*.

At any instant of time the cylinder shown in Fig. 13.31 has its bottom in contact with the horizontal surface. The point of contact P is *instantaneously at rest*, since it does not slide. Thus the *axis*

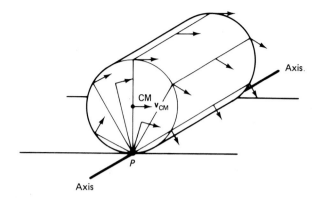

Fig. 13.31

through point P can be considered in INSTANTANEOUS AXIS OF ROTATION for the cylinder. At that instant, the linear velocity of *every* particle of the cylinder is directed at right angles to a line joining the particle and the axis at P. The magnitude of the velocity is proportional to the distance from the axis.

This is obviously the same as the cylinder rotating about a fixed axis at its edge, with an angular velocity ω *at that instant*. Thus we can write the total kinetic energy as

$$\text{KE} = \tfrac{1}{2}I_p\omega^2 \tag{13.66}$$

where $I_p \equiv$ moment of inertia of the cylinder about the axis P.

We can rewrite Eq. 13.66 in terms of the moment of inertia of the center of mass about P plus the moment of inertia of the body about the center of mass, using the parallel axis theorem (Eq. 13.59). The theorem tells us that

$$I_p = I_{\substack{\text{about the}\\\text{center of mass}}} + I_{\substack{\text{of the CM about}\\\text{the inst. axis } p}} = I_{\text{CM}} + MR^2.$$

Using this theorem, we rewrite the total kinetic energy in Eq. 13.66 as

$$\text{KE} = \tfrac{1}{2}I_{\text{CM}}\omega^2 + \tfrac{1}{2}MR^2\omega^2. \tag{13.67}$$

The constraint equation for this kind of motion (rolling without slipping) can be written by inspection (since the center of mass is executing circular motion about the instantaneous point of contact p) as

$$v_{\text{CM}} = R\omega, \tag{13.68}$$

where v_{CM} is the speed of the center of mass with respect to the point P. The speed of the center of mass with respect to P is the same as the speed of P with respect to the center of mass. Hence the angular speed ω of the center of mass about P, as observed by someone at P, is the same as the angular speed of a particle at P about CM (as seen by someone moving with the center of the cylinder). This is equivalent to saying that any reference line in the cylinder turns through the same *angle* in a given time whether it is observed from a reference frame fixed with respect to the surface or from a moving frame.

Substituting Eq. 13.68 into Eq. 13.67, our expression for the total kinetic energy becomes

$$\text{KE} = \tfrac{1}{2}I_{\text{CM}}\omega^2 + \tfrac{1}{2}Mv_{\text{CM}}^2. \tag{13.69}$$

Equation 13.69 suggests that the combination of translational motion of the center of mass $\tfrac{1}{2}Mv_{\text{CM}}^2$ and rotation about the center of mass $\tfrac{1}{2}I_{\text{CM}}\omega^2$ are equivalent to pure rotation with the same angular velocity about an axis through the point of contact of the rolling body.

To understand this better, consider the instantaneous velocities of various parts of our rotating cylinder with respect to a fixed reference frame. From the point of view of pure rotation about an instantaneous axis, a cylinder whose center of mass has velocity $\mathbf{v}_{\text{CM}}$, point A has $2\mathbf{v}_{\text{CM}}$ and zero velocity at point P (Fig. 13.32).

The following diagrams (Fig. 13.33) consider the motion as a combination of rotation about the center of mass CM and pure translation. If the cylinder was *only* translating, all points on it would have the same speed $\mathbf{v}_{\text{CM}}$. If it was *only* rotating about the center of mass, A would move with $+\omega R$ and P with $-\omega R$. When we combine these:

For point A:

$$v_A = v_{\text{CM}} + \omega R = v_{\text{CM}} + \frac{v_{\text{CM}}}{R} R = 2v_{\text{CM}}$$

For the CM:

$$v_c = v_{\text{CM}} + 0 = v_{\text{CM}}$$

For point P:

$$v_P = v_{\text{CM}} - \omega R = v_{\text{CM}} - \frac{v_{\text{CM}}}{R} R = 0.$$

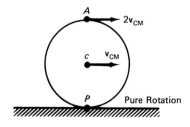

Fig. 13.32

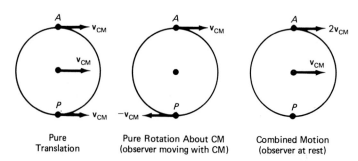

Pure
Translation

Pure Rotation About CM
(observer moving with CM)

Combined Motion
(observer at rest)

Fig. 13.33

> *Although our discussion has focused on one particular example of a cylinder's motion, the ideas presented are applicable to all types of rigid body motion involving simultaneous translation and rotation.*

EXAMPLE 1   A cylinder of mass M and radius R rolls without slipping down an inclined plane. Find the translational acceleration of the center of mass of the cylinder at any time.

We will use the coordinate system parallel and perpendicular to the inclined plane, as shown in Fig. 13.34. The motion of the center of mass is governed by Eq. 12.20: $\sum_i \mathbf{F}_i = m a^{\mathrm{CM}}$, and in component form:

x-comp.: $Mg \sin \theta - F_f = M a_x^{\mathrm{CM}}$

y-comp.: $F_N - Mg \cos \theta = M a_y^{\mathrm{CM}}.$

Rotational motion about the center of mass:

$$\text{torque} = \tau_z = I_z^{\mathrm{CM}} \alpha.$$

The constraints for this problem are

$$a_y = 0 \qquad \alpha = \frac{a_x^{\mathrm{CM}}}{R}.$$

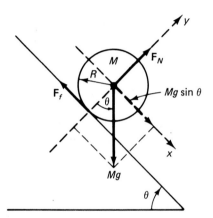

Fig. 13.34

Neither the normal force nor the gravitational force Mg can cause (directly) the rotation. It is the *frictional force* that produces the torque:

$$\tau_z = \mathbf{R} \times \mathbf{F}_f = R F_f = I_z^{\mathrm{CM}} \alpha.$$

But $I_z^{\mathrm{CM}} = \tfrac{1}{2} M R^2.$

Thus $R F_f = \dfrac{1}{2} M R^2 \dfrac{a_x^{\mathrm{CM}}}{R}$ or $F_f = \dfrac{M a_x^{\mathrm{CM}}}{2}.$

Substituting F_f into translational equations for x-axis motion of the center of mass yields

$$Mg \sin \theta - \tfrac{1}{2} M a_x^{\mathrm{CM}} = M a_x^{\mathrm{CM}}$$

or $a_x^{\mathrm{CM}} = \tfrac{2}{3} g \sin \theta.$

Neither the mass of the cylinder nor the frictional force enters the final result. Because there is no slippage, the frictional force does no work on the system for it only acts at a point, not over some displacement of the body. Observe that the rolling cylinder takes longer to go down the inclined plane than if it slides without friction. Why?

13.9 Summary of Important Equations

Definition of angular momentum of point particle with respect to the origin at O:

$$\mathbf{L}_0 = \mathbf{r} \times \mathbf{p}$$

Torque and time rate of change of angular momentum (note the same origin for calculating torque and angular momentum):

$$\sum_i \boldsymbol{\tau}_{oi} = \frac{d\mathbf{L}_0}{dt}$$

Center-of-mass theorem:

$$\mathbf{L}_0 = \mathbf{L}_{CM} + \sum_i m_i \mathbf{r}'_i \times \mathbf{v}'_i = \mathbf{L}_{orbit} + \mathbf{L}_{spin}$$

Moment of inertia
(z-component)

$$\equiv I_\gamma = \sum_{i=1}^{N} m_i \rho_i^2 \quad \text{(discrete point bodies)}$$

$$= \int_{body} \rho^2 \, dm = \sigma \int_{body} \rho^2 \, dV \quad \text{(uniform density)}$$

Kinetic energy of rotation about a fixed axis:

$$KE_{rot} = \tfrac{1}{2} I_z \omega^2$$

Angular momentum of rigid body about fixed z-axis:

$$L_z = \omega I_z$$

Fixed axis in z-direction:

$$\tau_z = I_z \frac{d\omega}{dt} = I_z \alpha$$

Fixed axis rotation motion:

$$\frac{d\omega}{dt} = \alpha \quad \frac{d\theta}{dt} = \omega \quad a = R\alpha$$

Parallel axis theorem:

$$I_z = I_z^{CM} + MR_{CM}^2$$

Rotational work:

$$dW = \tau_z \, d\theta$$

Rotational power:

$$power = \tau_z \omega.$$

Chapter 13 QUESTIONS

1. Why does the minus sign appear in the equation that represents angular momentum (Eq. 13.8).? What is its significance? There is no minus sign in Eq. 13.6. Explain.

2. If the resultant torque on a body is zero, is its angular momentum also zero? Explain.

3. Why does a helicopter have two rotors, a main rotor on top and a second rotor in the rear? What would happen to the helicopter if the rear rotor fell off while it was in flight? Describe the motion under such circumstances.

4. Can the spin angular momentum of a system be zero with respect to one reference frame and be non-zero with respect to another frame? Explain.

5. Does the axis of symmetry always pass through the center of mass of a rigid body? Explain.

6. Explain how you might experimentally determine the moment of inertia of a person standing on a turntable.

7. What axis would you choose to give this book the smallest moment of inertia? What axis would give it the largest?

8. How could you distinguish a lead sphere from an aluminum sphere if they were painted the same color and had the same mass and diameter? Explain.

9. A bicycle is ridden around a banked oval track. Explain all the angular momenta that exist in this bike–rider system. Is there spin and orbital angular momentum?

10. Describe the motion of a simple pendulum oscillating back and forth from the point of view of the equation of dynamics of rigid bodies $\tau_0 = d\mathbf{L}_0/dt$.

11. Why does a spiraling football wobble on its way to the receiver? What does this mean? Why do some throws not wobble?

12. Consider a ball rotating on a string with uniform circular motion. From the point of view of an observer on a rotating platform that rotates along with the ball, the ball is stationary and thus has *no* angular momentum. Thus for this reference frame, the angular momentum is zero. What's wrong with this analysis?

13. Why do divers tuck their bodies into a ball-like configuration when they want to rotate rapidly in flight?

14. Why do you suppose that artificial satellites are given a rotational motion before they are launched?

15. You can distinguish a raw egg from a hard-boiled egg in their shells by spinning both of them. Explain what happens and how it works. Try it.

Chapter 13 PROBLEMS

1. A particle of mass m, starting at the origin with an initial velocity $v_{0y}\mathbf{j}$, continues to move in the y-direction until it comes to rest (Fig. 13.35). The particle experiences a frictional force $-\mu m g\mathbf{j}$ during its motion.

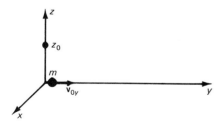

Fig. 13.35

(a) What is the angular momentum of the particle as a function of time with respect to both the origin O and the point Z_0? Give direction as well as magnitude.

(b) What is the torque on the particle with respect to O and Z_0?

(c) Show that the equation $\tau = d\mathbf{L}/dt$ holds for this system.

2. Show that if the net force on a system of particles is zero, the total torque on that system is the same for all reference points.

3. A small ball of mass m is allowed to free-fall from rest in Earth's gravitational field, as shown in Fig. 13.36.

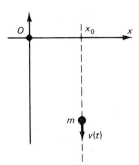

Fig. 13.36

(a) Calculate the angular momentum of the ball as a function of time about both the origin 0 and the point x_0.

(b) Calculate the torque on the ball exerted by the gravitational force about the same two reference points.

(c) Show that Eq. 13.20 holds for this system.

4. A particle of mass m moves in a region of space where it experiences a central force, that is, a force that can be written $\mathbf{F} = f(r)\hat{\mathbf{r}}$ (Fig. 13.37). Show that for such a force, the angular momentum of the body about the origin is a constant of the motion (i.e., the angular momentum does not change with time).

Fig. 13.37

5. An early model of the hydrogen atom proposed by Niels Bohr pictured the electron executing uniform circular motion about the positively charged proton in quantized orbits (Fig. 13.38). This model predicted

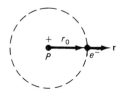

Fig. 13.38

some (not all) important properties of hydrogen and has remained a frequently used conceptual model of

the atom (even though it is fundamentally wrong!). Let's analyze this electron–proton system step by step. The electrostatic force between the proton and the electron was given by Eq. 2.2:

$$\mathbf{F} = -\frac{1}{4\pi\varepsilon_0}\frac{e^2}{r^2}\hat{\mathbf{r}},$$

where e is the magnitude of the charge on both the proton and electron and $\hat{\mathbf{r}}$ is the radial unit vector.

(a) Calculate the angular momentum of the electron about the origin if the radius of the oribit is $r_0 = .529$ Å $= .529 \times 10^{-10}$ m. Is the angular momentum conserved? Explain.

(b) Bohr proposed that the angular momentum was quantized, that is, that it existed only in discrete values. If the orbital angular momentum is $L = \hbar$, calculate the electron's velocity in terms of r_0, $\hbar$, and m.

(c) Calculate the kinetic energy of the electron.

(d) Calculate the work necessary to move a stationary electron from r_0 to infinity (ionization).

(e) Calculate the work done by the electrostatic force on the electron as it makes one circular orbit.

(f) Calculate the potential energy of the electron at r_0 if the potential is zero at $r = \infty$.

(g) Calculate the total mechanical energy of the system at $r = r_0$.

(h) Should the relativistic expressions for kinetic energy be used in this calculation? How much error is made by using the Newtonian equations?

6. An electron is injected into a storage ring of 150 m radius with a velocity of $.98c$ (Fig. 13.39). What is the angular momentum at injection with respect to points 0 and B? (Give direction and magnitude.) What is the angular momentum with respect to 0 when it arrives at point B, assuming that it has not gained kinetic energy?

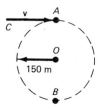

Fig. 13.39

7. Two identical particles are moving at the same speed in opposite directions along parallel lines that are separated by a distance a. Find the angular momentum of this two-particle system. Prove that it is independent of the origin taken.

8. A NASA satellite is placed into a highly eccentric elliptical orbit about the Earth, traveling at a speed of 800 m/s at 2800 km above the Earth. Find its speed when it is at its closest approach, which is 700 km above Earth's surface. (*Hint:* Consider the gravitational force to come from a single point at the Earth's center.)

9. Rutherford used α-particles to detect the concentration of mass in the nucleus of gold atoms. Let's take another look at the interaction of the α-particle with the electric field of the nucleus. If the α-particle has an initial velocity v_0 when it is far away from the nucleus (Fig. 13.40):

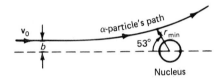

Fig. 13.40

(a) What is its angular momentum with respect to the center of the nucleus? Note the direction.

(b) What is the angular momentum at the point of closest approach?

(c) If $r_{min} = 6b$, what is the ratio of v_0/v, where v is the velocity at the point of closest approach?

10. Four particles of mass m are supported on a massless rigid rod as shown in Fig. 13.41. The rod rotates about O with a constant angular velocity ω.

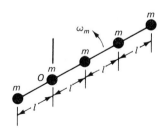

Fig. 13.41

(a) Find the linear speed of all four masses.

(b) Calculate the moment of inertia of the system about O.

(c) Calculate the angular momentum of each particle and the entire system about the axis at O.

11. A particle of mass m moves in a circle of radius r with constant angular velocity $\boldsymbol{\omega}$ about the origin at y_0 (Fig. 13.42). The particle moves in the x–y plane.

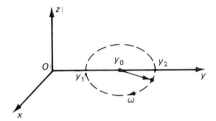

Fig. 13.42

(a) Calculate the angular momentum of the particle about 0 when it reaches points y_1 and y_2.

(b) Show that the difference in the angular momentum $\mathbf{L}_{y_1}$ and $\mathbf{L}_{y_2}$ for the two points calculated in part (a) is given by $\mathbf{L}_{y_1} - \mathbf{L}_{y_2} = mvy_0\mathbf{k}$.

12. Derive the equations of motion for a simple pendulum from the laws of rotational motion (i.e., $\boldsymbol{\tau} = d\mathbf{L}/dt$) rather than from Newton's laws for translational motion.

13. Compute the torque about the origin of coordinates for a force $\mathbf{F} = mg\mathbf{j}$ and a position vector $\mathbf{r} = r_x\mathbf{i} + r_y\mathbf{j}$. Show that this torque is independent of the coordinate y. What physics system might this exercise represent?

14. Calculate the moment of inertia of a thin flat sheet about the three possible rotational axes x, y, z shown in Fig. 13.43.

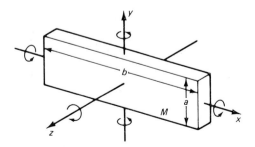

Fig. 13.43

(a) If the mass of the sheet is M, show that

$$I_x = M \frac{a^2}{12} \qquad I_y = M \frac{b^2}{12},$$

(b) For those students who have studied multiple integration, show that

$$I_z = M \frac{a^2 + b^2}{12}.$$

15. Show that the moment of inertia of a thin uniform disk of mass M and radius R, rotated about its vertical diameter (Fig. 13.44), is given by $I_z = \frac{1}{4} MR^2$.

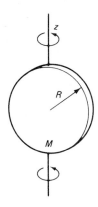

Fig. 13.44

16. Show that the moment of inertia of a thin spherical shell about an axis through its center of mass is

$$dI_{CM} = \frac{8}{3}\pi r^4 \rho \, dr$$

where dr is the thickness of the shell and ρ is the material's density. Use this result to show that the moment of inertia of a solid sphere is

$$I_{CM} = \frac{2}{5}MR^2,$$

where M is the total mass of the sphere of radius R.

17. A straight uniform wire of mass M and length L is bent at the midpoint and attached to a bearing for rotation (Fig. 13.45) What is the moment of inertia of the wire in this configuration as it rotates about an axis directed out of the paper?

Fig. 13.45

18. A thin circular metal disk is rotated about its center of mass as shown in Fig. 13.46. Find the moment of inertia about this axis if the disk has four circular holes of radius a cut out of its body. (*Hint:* The parallel axis theorem will simplify the problem.)

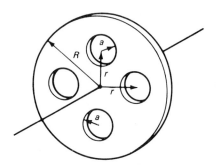

Fig. 13.46

19. For thin flat objects that lie in the x–y plane, it can be shown that the three moments of inertia about orthogonal axes are related by the equation

$$I_z = I_x + I_y.$$

This is called the *perpendicular axis theorem*. Prove it. (*Hint:* Start with the definition of I_z:

$$I_z = \int \rho R^2 \, dV,$$

where ρ is the density and R is the distance of the infinitesimal mass element to the z-axis of rotation.) For the flat object in the x–y plane, how is R related to the x and y coordinates of the mass elements?

20. A flywheel of radius .26 m is subjected to a constant tangential force of 287 N. The flywheel increases its rate of rotation from 3 to 48 Hz (cycles/second) in 1 minute.

(a) What is the moment of inertia of the flywheel about its axis?

(b) How much did the angular momentum change in that minute?

(c) How many radians did the flywheel turn in that minute?

(d) How much energy was put into the flywheel?

21. Show that the moment of inertia of a perfect cube of dimension a about an axis through the center of mass and perpendicular to a face is $I_{CM} = \frac{1}{6}Ma^2$.

22. What is the approximate rotational and translational kinetic energy of Earth?

23. A 5-kg mass slides on a frictionless table as it is pulled by the hanging 2-kg mass, shown in Fig. 13.47. The pulley in the system is a homogeneous cylinder of radius 8 cm and mass .5 kg.

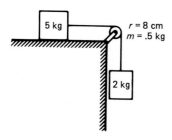

5 kg $r = 8$ cm $m = .5$ kg

2 kg

Fig. 13.47

(a) Calculate the external torque on the *entire system* about the center of the pulley.
(b) Calculate the angular momentum of the *system* about the center of the pulley. Assume that the center of mass of the 5-kg mass is at the height of the string.
(c) Find the acceleration of the system by the equation $\tau = d\mathbf{L}/dt$.
(d) Find the acceleration of the system using the equations of translational motion.
(e) Find the acceleration of the system using energy considerations.

24. A modified Atwood's machine is shown in Fig. 13.48. The two pulleys have lightweight line wrapped around their outer diameters, fixed friction-

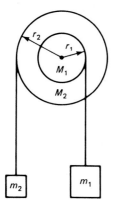

r_2 r_1 M_1 M_2 m_2 m_1

Fig. 13.48

less bearings, and are rigidly connected together. If

$$r_1 = .5 \text{ m} \quad m_1 = 5 \text{ kg} \quad M_1 = 1.5 \text{ kg}$$
$$r_2 = .7 \text{ m} \quad m_2 = 3 \text{ kg} \quad M_2 = 3.5 \text{ kg},$$

find the acceleration of m_2 and m_1.

25. A uniform cylinder of length .5 m and radius .05 m, whose mass is 2.5 kg, is supported by three cords wrapped around its body as shown in Fig. 13.49. If

Fig. 13.49

the cylinder is held stationary and then released to fall in the gravitational field of the Earth:

(a) Determine the linear acceleration of the center of mass of the cylinder as the cords unwind.
(b) Find the tension in *each* string.
(c) Find the tension in each string necessary to prevent the center of mass of the cylinder from accelerating.

26. Consider a race between two objects down an inclined plane. In position 1 we have a solid sphere of radius R and mass m_s; in position 2 we see a solid cylinder of mass $2m_s$ and radius R and length l. Both will roll *without* slipping. Which one would you put your money on? Don't guess—you might lose. This is a bet that any physicist can win!

27. A long slender rod of mass m and length l is pivoted at one end by a frictionless bearing. The free end is held almost vertically above the pivot and released from rest (Fig. 13.50).

(a) Find the angular acceleration of the rod when it is at the angle θ.
(b) Find the magnitude of the translational acceleration of the free end at this angle.
(c) What is the total kinetic energy of the rod when $\theta = \pi$.

Fig. 13.50

28. What fraction of the total kinetic energy is rotational kinetic energy in the case of (**a**) a rolling sphere, (**b**) a rolling cylinder, and (**c**) a rolling hoop?

29. A boy throws a softball at his girlfriend, who is standing at the end of a playground platform (Fig. 13.51). Find the angular speed of the platform after

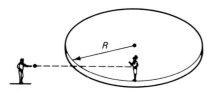

Fig. 13.51

she catches the ball, assuming that it was at rest before the ball was caught. The physical parameters of the system are:

   (1) Mass of the girl, 50 kg
   (2) Moment of inertia of the platform, 285 kg-m²
   (3) Initial speed of the ball, 12 m/s
   (4) Radius of the platform, 4.5 m
   (5) Mass of the ball, .2 kg

30. A thin, very light Mylar tape is wrapped around a solid cylinder of length l, mass M, and radius R (Fig. 13.52). The tape passes over a frictionless massless

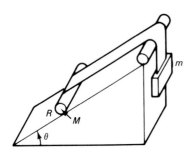

Fig. 13.52

pulley and is attached to a hanging mass m. If the cylinder rolls without slipping on a plane inclined at an angle of θ radians, find the linear acceleration of the center of mass of the cylinder and the tension in the tape.

31. A spool, shown in Fig. 13.53, consists of two wheels of radius r_1 and a bobbin of radius r_2. The bobbin has a lightweight string wound around it and the string emerges from below at an angle α with respect to the horizontal. A force **F** is applied to the string.

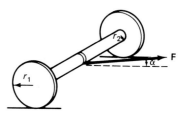

Fig. 13.53

   (a) Show that if $\cos \alpha < r_2/r_1$, the string will unwind.
   (b) Show that if $\cos \alpha > r_2/r_1$, the spool will wind up on the string.

32. The same spool as in Problem 31 (without the string) is placed on a narrow inclined plane so that the bobbin part rolls without slipping down the wedge-shaped incline. If the angle of the incline is θ, find the translational acceleration of the center of mass.

33. A meterstick is held in a vertical position with one end on a smooth frictionless horizontal surface. The top end of the stick is released and begins to fall while the bottom end slides across the horizontal surface. Find the speed of the top end of the stick when it strikes the horizontal surface. (*Hint:* Note that the meterstick performs rotational motion. Use energy considerations.

34. A particle (mass m) revolves with a uniform angular velocity ω_0 on a horizontal plane in a circle of radius a at about O held by a string passing through 0 (Fig. 13.54). Below O the string is pulled very slowly until m revolves in a circle of radius $a/2$.

   (a) What is the angular momentum about O at $r = a$?
   (b) What is the angular momentum about O at $r = a/2$?
   (c) Does the kinetic energy increase or decrease as m is pulled in? How much?

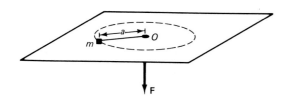

Fig. 13.54

(d) What work is done in pulling m in from a to $a/2$?

(e) Find the r dependence of the force applied at end of string.

35. A small solid ball of radius a rolls without slipping down a track shown in Fig. 13.55. How high up along the track (h) must the ball be released so that it will remain on the track over the entire loop-the-loop? Assume that $R \gg a$.

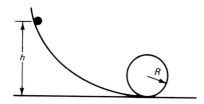

Fig. 13.55

36. A rigid body rotates about a fixed axis. At $t = 0$ its rotational energy is 5.0 J. During the time interval from $t = 0$ to $t = 4.0$ s, a constant torque is applied whose absolute amount is .20 N-m, reducing the rotational energy to 1.80 J. The rotation is counterclockwise during the entire time interval.

   (a) Calculate the initial and the final angular velocity.

   (b) Calculate the moment of inertia about the fixed axis.

37. A flywheel has friction in the bearings. Show the direction of the torque on the flywheel caused by the friction in the bearings (Fig. 13.56). What effect does that torque have on the spin angular momentum of the flywheel? Explain.

Fig. 13.56

38. Two equal masses are supported by massless rods of length r and are rotating about the origin at fixed angular separations θ (Fig. 13.57). Find the spin, orbit, and total angular momentum of the system. Discuss the two limits where $\phi = 0°$ and $\phi = 180°$.

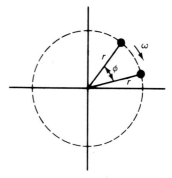

Fig. 13.57

39. A rod of length l is thrown vertically in the air in such a way that one end of the rod has exactly zero velocity as it leaves. The rod goes into the air and rotates, making N complete turns before being caught at the same position it left. Show that the maximum height at the center of mass of the rod traveled was

$$h_{max} = \frac{\pi}{4} Nl.$$

14 ROTATIONAL DYNAMICS

Introduction

IN THIS chapter we expand our discussion of rotational motion. No longer will we limit the analysis to the special case of rotation about a fixed axis. Now we examine rotation about *any* axis, including those not fixed in an inertial reference frame. The analysis can become extremely difficult and mathematically complex for objects that do not have a high degree of symmetry and that do not include some "special effects." We will steer clear of these complexities and emphasize the physics of freely rotating bodies. Unlike much of the discussion of linear motion, you may find this subject nonintuitive, even counterintuitive at times. To grasp the new concepts, you will have to use your knowledge of vectors: both vector addition and vector multiplication, particularly the vector cross product. A well-known advertisement for the telephone company's Yellow Pages reads: "Let your fingers do the walking." I suggest that a good strategy for dealing with rotational dynamics would be:

> *"Let the **vectors** do the talking."*

The vector nature of rotational motion is the key to its comprehension. When the axis of rotation is fixed, it is possible to use scalar equations such as $\tau_z = I_z\alpha$ or $L_z = I_z\omega$; that will *not* work with free-axis rotation! Perhaps the best known example of this type of motion is the gyroscope. The apparently mysterious behavior of this device can only be explained by using the correct *vector* equations. We will invest some time in studying and demonstrating the gyroscope. Finally, conservation of angular momentum will be examined again, as a law that can describe the motion of an Olympic high diver or ice skater as well as the decay of elementary particles and the trajectories of our planetary system. So stay tuned!

426

CONFESSION, they say, is good for the soul, and I want to confess and rectify a small deception. To this point we got away with writing angular velocity as a scalar quantity having only magnitude. Angular velocity is in fact a vector. For a quantity to be a vector, not only must it have magnitude and direction, but it must also add and subtract like a vector. This means that a vector quantity must obey the associative and distributive rules of addition, as, for example, linear displacement obviously does. The time rate of change of the linear displacement, the linear velocity, also has all the properties of a vector quantity.

What are the properties of angular displacement? The magnitude of the angular displacement θ is the angle through which a body rotates. The direction for this angular displacement might be chosen to be the axis around which the body rotates. For example, $\theta_x\mathbf{i}$ would be a rotation of θ_x radians around the x-axis. If angular displacement is a vector quantity, it must obey the commutation laws of vector addition. That is, the *order* in which the angular displacement is carried out does not matter. Mathematically, this is expressed as

$$\boldsymbol{\theta}_1 + \boldsymbol{\theta}_2 = \boldsymbol{\theta}_2 + \boldsymbol{\theta}_1. \qquad (14.1)$$

In Chapter 9 we demonstrated the commutative properties for translational vectors, but Eq. 14.1, in fact, does *not* hold for finite rotation. Figure 14.1 illustrates the failure of two finite rotations of $\pi/2$ radians to commute.* Consider an ordinary solvent container that is subjected to two rotations: a $\pi/2$ rotation about the z-axis and a second $\pi/2$ rotation about the y-axis. In Fig. 14.1a, we rotate the container $\theta_z + \theta_y$; in Fig. 14.1b, the order is reversed; $\theta_y + \theta_z$. Clearly, these two rotation sequences do not result in the same *net* rotation. This example alone shows that finite angular displacements do not commute. Therefore, they are *not* vectors.

However, if we perform rotations of the same body by only $10°$ on each axis, we notice something surprising. (Take a book and try this experiment.) For two operations, each with two $10°$ rotations, the object is not left in exactly the same orientation, but it is much closer. If the two rotations are only $1°$, the results are almost indistinguishable. It can be proven mathematically that *infinitesimal* angular displacements do indeed commute. That is,

$$d\theta_z\,\mathbf{k} + d\theta_y\,\mathbf{j} = d\theta_y\,\mathbf{j} + d\theta_z\,\mathbf{k}. \qquad (14.2)$$

Similarly, $d\theta_x\,\mathbf{i}$ commutes with $d\theta_y\,\mathbf{j}$ and $d\theta_z\,\mathbf{k}$, the other two infinitesimal displacements. The infinitesimal angular displacements meet all the criteria for vectors: they have magnitude and direction

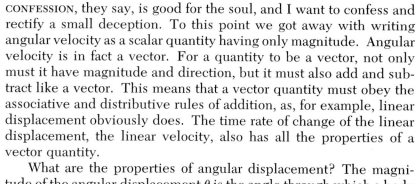

(a)

(b)

Fig. 14.1

---

* An asymmetric container is used in this illustration to keep track of the orientation of the rotated object.

and obey all the laws of vector addition and multiplication (although we have not proved it). From now on they will be treated with all the "rights and privileges accorded those who have attained the title of vectors."

The derivatives of these vectors with respect to the scalar quantity time are obviously also vectors: namely,

$$d\boldsymbol{\theta} = d\theta_x \, \mathbf{i} + d\theta_y \, \mathbf{j} + d\theta_z \, \mathbf{k} \tag{14.3}$$

$$\frac{d\boldsymbol{\theta}}{dt} \equiv \boldsymbol{\omega} = \frac{d\theta_x}{dt} \mathbf{i} + \frac{d\theta_y}{dt} \mathbf{j} + \frac{d\theta_z}{dt} \mathbf{k} \tag{14.4}$$

or $$\boldsymbol{\omega} = \omega_x \mathbf{i} + \omega_y \mathbf{j} + \omega_z \mathbf{k}, \tag{14.5}$$

whose magnitude is

$$|\boldsymbol{\omega}| = (\omega_x^2 + \omega_y^2 + \omega_z^2)^{1/2}. \tag{14.6}$$

EXAMPLE 1   A small battery-powered electric motor rotates on a horizontal turntable (Fig. 14.2). Find the resultant angular velocity of its shaft if $\omega_1 = 5$ rad/s about the vertical axis of the turntable and $\omega_2 = 8$ rad/s about the axis of the motor.

At time t, when the axis of the motor is aligned at an angle θ with respect to the x-axis (Fig. 14.3), the magnitude and direction of the resultant angular velocity vector $\boldsymbol{\omega}$ can easily be calculated.

$$\boldsymbol{\omega} = 8 \cos\theta \, \mathbf{i} + 8 \sin\theta \, \mathbf{j} + 5\mathbf{k} \quad \text{(in rad/s).}$$

But since $\theta = \omega_1 t$, the above becomes

$$\boldsymbol{\omega} = 8 \cos\omega_1 t \, \mathbf{i} + 8 \sin\omega_1 t \, \mathbf{j} + 5\mathbf{k},$$

and its magnitude

$$|\boldsymbol{\omega}| = [8^2(\cos^2 \omega_1 t + \sin^2 \omega_1 t) + 5^2]^{1/2} = 9.43 \text{ rad/s}.$$

Another way to calculate the magnitude of $\boldsymbol{\omega}$, with these two perpendicular components follows:

$$|\boldsymbol{\omega}| = (\omega_1^2 + \omega_2^2)^{1/2} = (64 + 25)^{1/2} = 9.43 \text{ rad/s}$$

and $$\alpha = \tan^{-1} \frac{\omega_1}{\omega_2} = \tan^{-1} \frac{5 \text{ rad/s}}{8 \text{ rad/s}} = 32°.$$

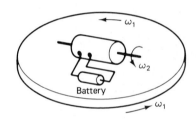

Fig. 14.2

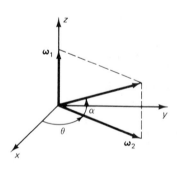

Fig. 14.3

Some fundamental vector properties of the "new" vector, angular velocity, can be seen by carefully examining a specific example of rotational motion: a spinning top rotating about an arbitrary instantaneous axis (Fig. 14.4). The instantaneous axis of rotation of the top is defined as the z-axis. The origin O is taken at the point of contact of the top with the horizontal surface. As the top rotates with an angular velocity $\boldsymbol{\omega}$, every particle of the top executes circular motion about the instantaneous axis of rotation $\mathbf{k}$.

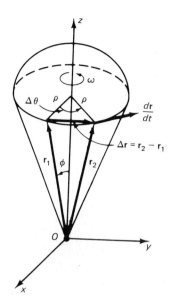

Fig. 14.4

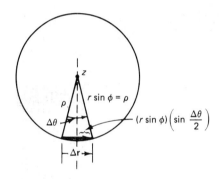

Fig. 14.5

With this example we intend to demonstrate that a vector relationship exists between the angular velocity $\boldsymbol{\omega}$, the linear velocity $\mathbf{v}$, and the position vector $\mathbf{r}$ of a point on the top, namely that

$$\frac{d\mathbf{r}(t)}{dt} = \mathbf{v}(t) = \boldsymbol{\omega}(t) \times \mathbf{r}(t). \qquad (14.7)$$

To show this, consider a view from above, looking down the z-axis at this rotating body. From Figs. 14.4 and 14.5 one can see that

$$|\mathbf{r}_2 - \mathbf{r}_1| = |\Delta\mathbf{r}| = 2(r \sin \phi)\left(\sin \frac{\Delta\theta}{2}\right).$$

The magnitude of the rate of change of the position vector with respect to time is

$$\left|\frac{\Delta\mathbf{r}}{\Delta t}\right| = \frac{2r \sin \phi \sin (\Delta\theta/2)}{\Delta t}. \qquad (14.8)$$

In the limit of $\Delta t \to 0$, and $\Delta\theta \to d\theta$, the sine of an infinitesimally small angle (in radians) is

$$\sin \frac{d\theta}{2} = \frac{d\theta}{2},$$

and thus

$$\lim_{\Delta t \to 0} \left|\frac{\Delta\mathbf{r}}{\Delta t}\right| = \left|\frac{d\mathbf{r}}{dt}\right| \equiv |\mathbf{v}| = r \sin \phi \frac{d\theta}{dt} = (r \sin \phi)\omega. \qquad (14.9)$$

Equation 14.7 (our original assumption) predicts

$$|\mathbf{v}(t)| = |\boldsymbol{\omega} \times \mathbf{r}| = \omega r \sin \phi$$

(since ϕ is the angle between $\mathbf{r}$ and $\boldsymbol{\omega}$).

From Fig. 14.4 it is apparent that the direction of the linear velocity is tangent to the circle and perpendicular to the vector $\mathbf{r}$. The angular velocity is directed along the axis of rotation $\mathbf{k}$. The vector cross product $\boldsymbol{\omega} \times \mathbf{r}$ is directed perpendicular to *both* $\boldsymbol{\omega}$ and $\mathbf{r}$ and is tangent to the circle at the point $\mathbf{r}$ in the direction given by the right-hand rule. We have satisfied all the requirements of Eq. 14.7 for both magnitude and direction for all three vectors. This is an important relationship that you will need in the analysis of rigid-body rotation.

14.3 Angular Velocity and Angular Momentum

NOW THAT angular velocity has been recognized for what it really is, a vector quantity, we can examine the relationship between it and other vector quantities, such as torque and angular momentum.

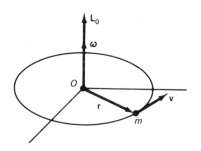

Fig. 14.6

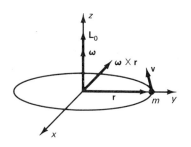

Fig. 14.7

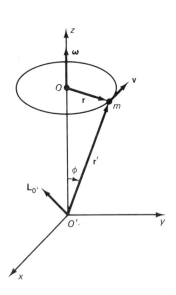

Fig. 14.8

Consider, for example, a particle of mass m that executes uniform circular motion in the x–y plane with constant angular velocity $\boldsymbol{\omega}$, linear velocity $\mathbf{v}$, and radial distance $\mathbf{r}$ (Fig. 14.6). The angular momentum about the origin can readily be calculated:

$$\mathbf{L}_0 = \mathbf{r} \times \mathbf{p} = \mathbf{r} \times m\mathbf{v}, \qquad (14.10)$$

and since both $\mathbf{r}$ and $\mathbf{v}$ lie in the x–y plane, $\mathbf{L}_0$ (according to the right-hand rule) is directed along the positive z-axis. For circular motion, $\mathbf{r}$ and $\mathbf{v}$ are always mutually perpendicular, so

$$|\mathbf{L}_0| = |\mathbf{r} \times m\mathbf{v}| = rmv.$$

For uniform circular motion, the angular momentum about the origin $\mathbf{L}_0$ is constant in both magnitude and direction.

The angular momentum about the origin can also be obtained using the vector properties of angular velocity. Examine Fig. 14.7. The particle executing circular motion is drawn at a time when it lies along the y-axis, so $\mathbf{r} = r\mathbf{j}$. The angular velocity is always directed along the z-axis, $\boldsymbol{\omega} = \omega\mathbf{k}$. The angular momentum is then

$$\mathbf{L}_0 = \mathbf{r} \times m\mathbf{v} = mr\mathbf{j} \times (\boldsymbol{\omega} \times \mathbf{r}) = mr\mathbf{j} \times (\omega\mathbf{k} \times r\mathbf{j}).$$

But $\qquad \mathbf{k} \times \mathbf{j} = -\mathbf{i} \qquad$ so $\qquad \omega\mathbf{k} \times r\mathbf{j} = -\omega r\mathbf{i}$

and $\qquad mr\mathbf{j} \times (-\omega r\mathbf{i}) = mr^2\omega\mathbf{k} \qquad$ since $\qquad \mathbf{j} \times (-\mathbf{i}) = \mathbf{k}.$

Thus $\qquad\qquad\qquad\qquad \mathbf{L}_0 = m\omega r^2\mathbf{k}. \qquad (14.11)$

Taking the magnitude of Eq. 14.7,

$$|\mathbf{v}| = |\boldsymbol{\omega} \times \mathbf{r}| \qquad \text{or} \qquad v = \omega r,$$

Eq. 14.11 becomes

$$\mathbf{L}_0 = mvr\mathbf{k},$$

as we demonstrated previously by direct examination using the definition of angular momentum (Eq. 14.10). We arbitrarily chose to calculate $\mathbf{L}_0$ when the particle was located along the y-axis. You might want to convince yourself that the same result holds for the particle located anywhere in the x–y plane (Problem 6).

Now consider the same rotating particle from the point of view of a different origin O', Fig. 14.8. The angular momentum with respect to O' is

$$\mathbf{L}_{0'} = \mathbf{r}' \times m\mathbf{v}. \qquad (14.12)$$

As you can see from the diagram, $\mathbf{L}_{0'}$ is a *constant-magnitude* rotating vector that is pointed at an angle of $(90 - \phi)$ from the angular velocity ($\boldsymbol{\omega}$) axis. Because the angular momentum is time dependent (in its direction), there must be a torque acting on the particle. In this case, the torque is generated by the force that causes the particle to move in a circle: the centripetal force. We have not specified this force in this example, but in Example 1 of Section 13.3, where the same system is analyzed, the force is supplied by the tension in the string of a conical pendulum. We will analyze this motion by the

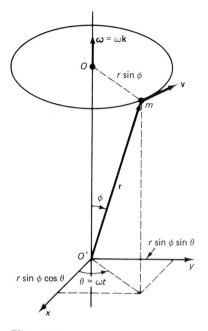

Fig. 14.9

straightforward application of the vector properties of angular velocity.

From Fig. 14.9 you can see that the position vector $\mathbf{r}'$ can be written in terms of its Cartesian coordinates as

$$\mathbf{r}' = r(\sin \phi \cos \theta \, \mathbf{i} + \sin \phi \sin \theta \, \mathbf{j} + \cos \phi \, \mathbf{k}) \qquad (14.13)$$

and $\boldsymbol{\omega} = \omega \mathbf{k}$. Now the linear velocity is

$$\mathbf{v} = \boldsymbol{\omega} \times \mathbf{r}' = \omega r[\sin \phi \cos \theta (\mathbf{k} \times \mathbf{i}) + \sin \phi \sin \theta (\mathbf{k} \times \mathbf{j})]$$

or

$$\mathbf{v} = \omega r[\sin \phi \cos \theta \, \mathbf{j} + \sin \phi \sin \theta (-\mathbf{i})]. \qquad (14.14)$$

Substituting Eqs. 14.14 and 14.13 into Eq. 14.12, we arrive at

$$\mathbf{L}_{0'} = m\omega r^2[(\sin \phi \cos \theta \, \mathbf{i} + \sin \phi \sin \theta \, \mathbf{j} + \cos \phi \, \mathbf{k})$$

$$\times (\sin \phi \cos \theta \, \mathbf{j} + \sin \phi \sin \theta (-\mathbf{i}))]$$

$$= m\omega r^2[\sin^2 \phi \cos^2 \theta \, \mathbf{k} + \sin^2 \phi \sin^2 \theta \, \mathbf{k} + \sin \phi \cos \theta \cos \phi (-\mathbf{i})$$

$$+ \sin \phi \sin \theta \cos \phi (-\mathbf{j})]$$

and since $\theta = \omega t$,

$$\mathbf{L}_{0'} = m\omega r^2 \{\sin^2 \phi \, \mathbf{k} - \sin \phi \cos \phi [\cos \omega t \, \mathbf{i} + \sin \omega t \, \mathbf{j}]\}. \qquad (14.15)$$

The bicycle and its rider are a complicated collection of rotating components with various torques, friction, angular velocities, angular momentum, and moments of inertia all interconnected. Make a diagram of this racer rounding the corner, showing some of these quantities.

Here we have derived an explicit expression for the angular momentum about the origin O', which shows the time-dependent x–y component and the time-independent z-component.

EXAMPLE 1
Rotation of a
Tilt-Axis Dumbbell

How should you have your automobile tires balanced when you buy a new set? (Or, how should you advise your parents?) Should you spend less money and have the simple static or bubble balance done, or should you invest in what tire dealers call spin or dynamic balancing? This book is hardly *Consumer's Report,* but it may be helpful in answering this particular question.

The prototype system we examine consists of two equal masses rigidly supported by two massless rods of length l (Fig. 14.10). The dumbbell is mounted on an axle of negligible mass and is supported by a pair of excellent bearings that allow it to rotate with its angular velocity in the z-direction. The system is statically balanced; that is, it consists of two equal masses supported on the axle at their center of mass. When the dumbbell rotates, the angular momentum does *not* lie along the angular velocity axis. This means that the angular momentum executes rotational motion and, as we have seen, time-dependent angular momentum requires a torque. But we are getting ahead of ourselves. Let's do the mathematics.

The angular momentum of this tilt-axis dumbbell can be calculated from the definition of angular momentum:

$$\mathbf{L}_0 = \sum_i \mathbf{r}_i \times \mathbf{p}_i = \mathbf{r}_1 \times \mathbf{p}_1 + \mathbf{r}_2 \times \mathbf{p}_2. \qquad (14.16)$$

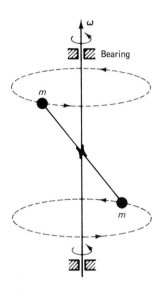

Fig. 14.10

Each mass moves in a circle of radius $l \cos \phi$ with an angular speed ω, so the magnitude of the linear velocity is $|\mathbf{v}| = |\boldsymbol{\omega} \times \mathbf{r}| = \omega l \cos \phi$ and the magnitude of the linear momentum of each particle $|\mathbf{p}_1| = |\mathbf{p}_2| = m\omega l \cos \phi$ (Fig. 14.11). Each mass executes circular motion. The magnitude of the total angular momentum is then the algebraic sum (since they point in the same direction):

$$|\mathbf{L}_0| = m\omega l^2 \cos \phi + m\omega l^2 \cos \phi = 2m\omega l^2 \cos \phi. \qquad (14.17)$$

You can see from the diagram that the angular momentum is directed perpendicular to the connecting rod and rotates in a circle about the z-axis.

The vector properties of the angular velocity can also be used to calculate the angular momentum of this object. Imagine a stop-action photograph of the dumbbell shown in Fig. 14.12. The angular velocity is decomposed into two components, one parallel to the support rod (and the radial vector), $\omega_\parallel$, and one perpendicular to the rod, $\omega_\perp$. Mathematically, this can be written

$$\boldsymbol{\omega} = \omega_\parallel \hat{\parallel} + \omega_\perp \hat{\perp} \quad \text{or} \quad \boldsymbol{\omega} = \omega \sin \phi \hat{\parallel} + \omega \cos \phi \hat{\perp}, \qquad (14.18)$$

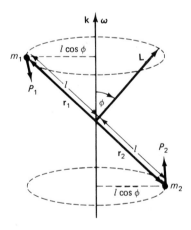

Fig. 14.11

where I am using two new unit vectors, $\hat{\parallel}$ and $\hat{\perp}$, only for this example. The angular momentum about the origin O' can be written

$$\mathbf{L}_0 = \sum \mathbf{r}_i \times m(\boldsymbol{\omega} \times \mathbf{r}_i)$$

$$= m\mathbf{r}_1 \times (\boldsymbol{\omega} \times \mathbf{r}_1) + m\mathbf{r}_2 \times (\boldsymbol{\omega} \times \mathbf{r}_2). \qquad (14.19)$$

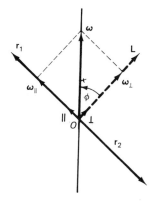

Fig. 14.12

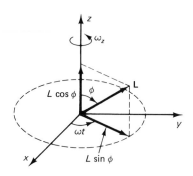

Fig. 14.13

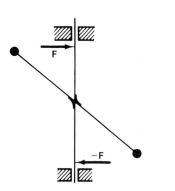

Fig. 14.14

The component of the angular velocity parallel to the radial vectors $\omega_\parallel$ does not contribute to the angular momentum since $\omega_\parallel \hat{\parallel} \times \mathbf{r} = 0$ (because $\mathbf{r}$ and $\hat{\parallel}$ are parallel vectors). Only the perpendicular component of $\boldsymbol{\omega}$ is needed to calculate the linear momentum,

$$m\omega_\perp \hat{\perp} \times \mathbf{r}_1 = m\omega r_1 \cos\phi = m\omega l \cos\phi,$$

directed out of the paper, and

$$m\omega_\perp \hat{\perp} \times \mathbf{r}_2 = m\omega l \cos\phi,$$

directed into the paper. To complete the calculation of angular momentum indicated in Eq. 14.19, we must take a second cross product with the respective $\mathbf{r}_1$ and $\mathbf{r}_2$ and arrive at

$$\mathbf{L}_0 = 2ml^2\omega\cos\phi\,\hat{\perp} \qquad (14.20)$$

in the direction of the perpendicular component of the angular velocity. These results are exactly the same as we obtained in Eq. 14.17. Equation 14.17 is only an expression for the instantaneous value of the angular momentum; the full mathematical expression would contain its time dependence. The angular momentum executes circular motion at an angular speed ω. This vector can be decompose into two components, a time-independent part along the z-axis and a time-dependent part in the x–y plane. From Fig. 14.13 we see that

$$\mathbf{L}_0 = L\sin\phi\cos\omega t\,\mathbf{i} + L\sin\phi\sin\omega t\,\mathbf{j} + L\cos\phi\,\mathbf{k}. \qquad (14.21)$$

Equation 14.21 is the expression for the angular momentum in its full glory! The net external torque on the system can easily be calculated using the master equation relating torque to the *time* rate of change of the angular momentum. Differentiating Eq. 14.21 with respect to time, we have

$$\frac{d\mathbf{L}_0}{dt} = \boldsymbol{\tau}_0 = -L\omega\sin\phi\sin\omega t\,\mathbf{i} + L\omega\sin\phi\cos\omega t\,\mathbf{j}, \qquad (14.22)$$

and the magnitude of this torque is

$$|\boldsymbol{\tau}_0| = [L^2\omega^2\sin^2\phi(\sin^2\omega t + \cos^2\omega t)]^{1/2} = L\omega\sin\phi. \qquad (14.23)$$

The torque on the dumbbell is supplied by the bearings (Fig. 14.14). It is constant in magnitude, $L\omega\sin\phi$, but varying in direction. In effect, the torque "follows" the dumbbell around as it rotates.

So what does all this have to do with balancing your tires? Remember, this dumbbell is a system whose center of mass is located at the center, on the axle, so it is statically balanced. But a torque, supplied by the bearings, is required to keep it rotating about its fixed axis. The bearings in your car will experience unbalanced forces, which eventually will damage or destroy them as well as produce vibrations of the wheel as it rotates. The angular momentum and angular velocity of a dynamically balanced wheel *are in the* *same direction.* Such a system requires no torque to keep it rotating at a constant angular velocity.*

One way to eliminate the torque on our dumbbell is to set $\phi = 0$. Then the dumbbell is rotating in the horizontal plane. But that is not an option for our automobile tire. The imbalance in the tire can be caused by a mass m of extra rubber that is part of the tire, a consequence of poor quality control in the manufacturing process (Fig. 14.15). The mechanic could, in principle, shave it off at the risk of damaging the tire's structural cords. Instead, weights are added. These weights create angular momentum when the tire rotates that cancels out the x–y components of the angular momentum. The mass and location of the weights can be determined *only* by spinning the tire. The tire, mounted on its rim (because the rim may not be perfectly balanced), is placed on the axle of the "spin balance" machine. The machine is designed to sense the torque on its axle while the tire is rotating and to compute the location and mass of the add-on "weights." These will compensate for the time-dependent angular momentum. This cannot be done by a static balancing system; the directions of the angular momentum can be obtained only when the tire is rotating.

One last piece of practical advice: *Don't* have your brand new tires balanced. For the best, most enduring balancing, have your new tires mounted and installed on the car and then drive the car a few hundred miles, keeping the speeds below 50 mi/hr. This allows a tire to "seat" on the rim in a nearly permanent location. A new tire may shift on the rim after a few hundred miles and negate your expensive dynamic balancing job. This means two trips to the shop—but it is worth the trouble, although few places will tell you this. I suspect they fear you will not return to have the work done. However, that is not a physics consideration.

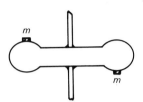

Fig. 14.15

14.4 **Angular Momentum of Rigid Bodies about a Free Axis**

THE PROBLEM of calculating the angular momentum for a point particle about a particular origin, using the vector properties of the angular velocity, becomes a problem in manipulating the double cross product, $\mathbf{r} \times (\boldsymbol{\omega} \times \mathbf{r})$. Calculating the angular momentum of a *rigid body* executing rotational motion (with the center of mass at rest) involves the vector summation of the angular momentum of all the point masses of the body, as

$$\mathbf{L}^{\text{CM}} = \sum_i \mathbf{r}_i \times m_i \mathbf{v}_i \tag{14.24}$$

---

* Friction in the bearings and contact friction with the road (which propels the car) also produce torques on the tires, but they change the magnitude of the angular momentum, not its direction.

or
$$\mathbf{L}^{\mathrm{CM}} = \sum_i \mathbf{r}_i \times m_i(\boldsymbol{\omega} \times \mathbf{r}_i). \qquad (14.25)$$

Although we have restricted ourselves to the case where the center of mass is at rest in the inertial frame, we have not specified any special direction for the angular velocity $\boldsymbol{\omega}$ (Fig. 14.16). The most general expression for the angular velocity in Cartesian coordinates is

$$\boldsymbol{\omega} = \omega_x \mathbf{i} + \omega_y \mathbf{j} + \omega_z \mathbf{k} \qquad (14.26)$$

and
$$\mathbf{r}_i = x_i \mathbf{i} + y_i \mathbf{j} + z_i \mathbf{k}.$$

Thus

$$\boldsymbol{\omega} \times \mathbf{r}_i = (z_i\omega_y - y_i\omega_z)\mathbf{i} + (x_i\omega_z - z_i\omega_x)\mathbf{j} + (y_i\omega_x - x_i\omega_y)\mathbf{k}. \quad 14.27)$$

The direct calculation of $\mathbf{r}_i \times (\boldsymbol{\omega} \times \mathbf{r}_i)$ gives a very long and complicated looking expression. It can, however, be written as the determinant

$$\mathbf{r}_i \times (\boldsymbol{\omega} \times \mathbf{r}_i) = \begin{Vmatrix} \mathbf{i} & \mathbf{j} & \mathbf{k} \\ x_i & y_i & z_i \\ z_i\omega_y - y_i\omega_z & x_i\omega_z - z_i\omega_x & y_i\omega_x - x_i\omega_y \end{Vmatrix} \qquad (14.28)$$

and thus

$$\mathbf{L}^{\mathrm{CM}} = \sum_i m_i \|\det\|. \qquad (14.29)$$

Let's examine only one component of $\mathbf{L}_{\mathrm{CM}}$, the z-component:

$$L_z^{\mathrm{CM}} = \sum_i m_i[x_i(x_i\omega_z - z_i\omega_x) - y_i(z_i\omega_y - y_i\omega_z)], \qquad (14.30)$$

or, expanding into three sums,

$$L_z^{\mathrm{CM}} = \sum_i m_i(x_i^2 + y_i^2)\omega_z - \sum_i m_i z_i\, x_i\omega_x - \sum_i m_i y_i\, z_i\omega_y. \qquad (14.31)$$

If we define the following quantities:

$$I_{zz} \equiv \sum_i m_i(x_i^2 + y_i^2) \qquad (14.32)$$

$$I_{zx} \equiv -\sum_i m_i z_i x_i \qquad (14.33)$$

$$I_{zy} \equiv -\sum_i m_i y_i z_i, \qquad (14.34)$$

Eq. 14.31 becomes

$$L_z^{\mathrm{CM}} = I_{zx}\omega_x + I_{zy}\omega_y + I_{zz}\omega_z. \qquad (14.35)$$

Calculating the other components of the angular momenta, L_x^{CM} and L_y^{CM}, is a straightforward exercise in determinant manipulation. We can also define another two sets of these analogous I quantities and arrive at

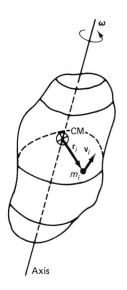

Fig. 14.16

$$L_x^{CM} = I_{xx}\omega_x + I_{xy}\omega_y + I_{xz}\omega_z$$
$$L_y^{CM} = I_{yx}\omega_x + I_{yy}\omega_y + I_{yz}\omega_z. \qquad (14.36)$$

These certainly are complicated-looking equations. Each Cartesian component of the angular momentum depends on *three* terms, and each term has a different component of the angular velocity. What a mess!

Equations 14.35 and 14.36 include a collection of nine elements of the I's. They can be arranged in an ordered array, where they are called the elements of the moment-of-inertia TENSOR. In the mathematics of tensors, they are written

$$\mathbf{I} \equiv \begin{bmatrix} I_{xx} & I_{xy} & I_{xz} \\ I_{yx} & I_{yy} & I_{yz} \\ I_{zx} & I_{zy} & I_{zz} \end{bmatrix}. \qquad (14.37)$$

Equations 14.35 and 14.36 can be written in the extremely compact form

$$\mathbf{L} = \tilde{\mathbf{I}}\boldsymbol{\omega}. \qquad (14.38)$$

Don't panic! We will not actually do any problems with these tensors; that is properly left for a more advanced mechanics course. We can, however, get some idea of the physical significance of these moment-of-inertia elements by reexamining Eq. 14.35, the z-component of the angular momentum.

In Chapter 13 we considered rotation about a fixed axis where the angular velocity had only *one* component, $\boldsymbol{\omega} = \omega_z\mathbf{k}$. For this special case Eq. 14.31 collapses to

$$L_z = I_{zz}\omega_z = m_i(x_i^2 + y_i^2)\omega_z. \qquad (14.39)$$

But $\rho_i^2 = x_i^2 + y_i^2$ (see Eq. 13.45), so that as I_{zz} is the same as the quantity I_z, the moment of inertia about the z-axis for fixed z-axis rotation. For rotational motion about any one of the three Cartesian directions x, y, and z, only the diagonal elements of the moment of inertia tensor are needed to calculate the angular momentum about the respective axis. Thus for *fixed* x-or y-axis rotation, we have

$$L_x = I_{xx}\omega_x \quad \text{and} \quad L_y = I_{yy}\omega_y \quad \text{(fixed axis)}. \quad (14.40)$$

All this was derived from the vector nature of angular velocity and its relation to the angular momentum of extended rigid bodies.

14.5 **Gyroscopes**

THE VECTOR nature of angular momentum can be dramatically demonstrated by observing a gyroscope. During World War II the gyroscope was developed into a highly accurate and sophisticated instrument for navigation. Today the gyroscope remains one of the most important navigational instruments in common use on surface ships, submarines, and aircraft. The world's most sensitive and sophisticated gyroscope has been under development for over 15 years at Stanford University. Eventually, it will be placed in orbit around the Earth to test Einstein's general theory of relativity. Einstein's theory predicts an extremely small "extra" motion of the gyroscope, which is not allowed according to Newton's theory of gravity.

A detailed mathematical theory of the motion of a gyroscope is beyond the scope of this book. Our presentation will not explain all the interesting types of gyroscopic behavior that one can observe with an ordinary toy gyroscope. We limit our discussion to the major features of its motion and how they are related to the fundamental laws of rotational motion.

Most physics departments have in their collection of demonstration apparatus a special bicycle wheel with a heavy solid rubber rim and a pair of handles on its axis that can be held by a student or the instructor. It can be used for many interesting demonstrations of the vector properties of angular momentum. Consider first the wheel at rest (not spinning) with one handle mounted on a low-friction support and the other supported by the demonstrator (Fig. 14.17). If the demonstrator removes his support at S, the gravitational force $mg(-\mathbf{k})$, acting on the center of mass of the bicycle wheel, exerts a torque about the point O,

$$\boldsymbol{\tau}_0 = r\mathbf{j} \times mg(-\mathbf{k}) = rmg(-\mathbf{i})$$

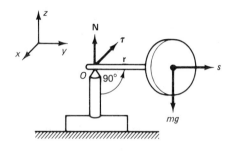

Fig. 14.17

in the negative x-direction. The normal force $\mathbf{N}$, exerted by the stand at the pivot point O, cannot contribute to the torque about O since the lever arm of this force is zero. The gravitational torque must produce a time rate of change of the angular momentum about O. But initially there was *no* angular momentum of the system; all parts of the wheel were at rest. Of course, what happens is the wheel begins to fall in the y–z plane (Fig. 14.18.) The falling wheel is actually rotating about the pivot O, *creating* an angular momentum about O in the same direction as the torque (the negative x-direction). As the wheel falls, the direction of the angular momentum remains the same, but its *magnitude* increases with time.

Nothing so far should surprise you; this is merely a description of the motion of a falling extended object pivoting about one point. But what happens if the experiment is repeated, this time spinning the bicycle wheel so that it has substantial spin angular momentum about its y-axis *before the support handle is released?* Because it is designed with a large amount of mass uniformly distributed around its rim, it has a substantial moment of inertia about the spin axis. When the wheel is spinning about its axle with an angular velocity $\boldsymbol{\omega}_s$, it acquires a large angular momentum directed along the same axis as $\boldsymbol{\omega}_s$. Initially, both $\boldsymbol{\omega}_s$ and $\mathbf{L}_s$ are directed along the same axis, as shown in Fig. 14.19. In the second experiment the motion of the system is dominated by this large initial spin angular momentum.

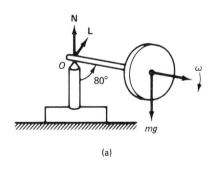

(a)

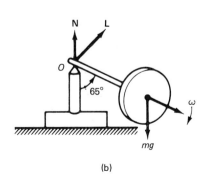

(b)

Fig. 14.18

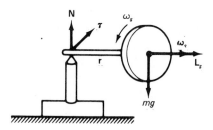

Fig. 14.19

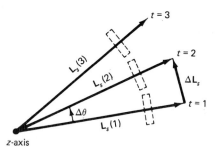

Fig. 14.20

The gravitational force acting on the center of mass of the wheel produces a time rate of change of the *direction* of the large spin angular momentum. This can best be "seen" by an aerial view of the apparatus looking down the z-axis (Fig. 14.20). The magnitude of the spin angular momentum remains constant, but the direction changes due to the torque. The net result is a PRECESSIONAL motion in the x–y plane.

The angular velocity of the precessional motion can be calculated easily. The magnitude of the torque on the wheel is $|\tau_0| = rmg$. For a short time Δt the wheel precesses through a small angle $\Delta\theta$ such that

$$\Delta L_s = \Delta\theta L_s.$$

Dividing both sides by the time interval Δt,

$$\frac{\Delta L_s}{\Delta t} = \frac{\Delta\theta}{\Delta t} L_s. \tag{14.41}$$

In the limit of very small time intervals, Eq. 14.41 becomes

$$\frac{dL_s}{dt} = \Omega_p L_s \tag{14.42}$$

where $\Omega_p \equiv \dfrac{d\theta}{dt}$ is the PRECESSIONAL ANGULAR VELOCITY. Now we can equate this time rate of change of the angular momentum to the external torque:

$$|\tau_0| = \left|\frac{d\mathbf{L}_0}{dt}\right| \qquad \text{or} \qquad rmg = \Omega_p|\mathbf{L}_s|$$

$$\Omega_p = \frac{rmg}{|\mathbf{L}_s|}. \tag{14.43}$$

The magnitude of the spin angular momentum is related to the moment of inertia of the bicycle wheel about the spin axis by $|\mathbf{L}_s| = \omega_s I_s$. Then an alternative way to express the precession angular velocity is

$$\Omega_p = \frac{rmg}{\omega_s I_s}. \tag{14.44}$$

Equation 14.44 is an interesting and perhaps unexpected result. The horizontal precession rate is *inversely* proportional to the spin angular velocity. As the wheel slows down because of friction in its bearings and drag in the air (something we have neglected in this simplified analysis), the precession rate *increases*. This effect is readily demonstrated.

What happens when $\omega_s \to 0$? If you watch a gyroscope, you will see that as $\omega_s \to 0$, the motion of the wheel becomes erratic. Nothing in our analysis predicts this. However, we have not done a complete analysis of the motion but have only examined the case where the spin angular momentum is large and dominates the dynamics. As $\omega_s \to 0$, these assumptions are violated, so we should not expect our theoretical model to explain the motion under these conditions.

A gyroscope is a device that utilizes the vector properties of the large spin angular momentum of its flywheel. In a gyroscope the flywheel is mounted on supports, called *gimbals*, which limit the possible torques that can act on the spin angular momentum of its flywheel. A gyroscope can act as a compass—the gyrocompass—if it is mounted in a special configuration. Again, we shall not work out the details of motion; we examine the physical ideas.

The following experiment demonstrates the basic principles of the gyrocompass. Consider the apparatus shown in Fig. 14.21, a flywheel supported by a special mounting system standing on a rotating platform. Begin by "spinning up" the flywheel so that it has a large spin angular momentum. This is usually done by attaching a string to the shaft of the flywheel and pulling hard on the string for a few seconds while it unwinds (exerting a large torque), keeping the platform stationary. After the spin angular momentum has been established (in the negative x-direction), slowly rotate the platform, giving it angular velocity $\boldsymbol{\omega}_p$ in the z-direction. You will notice that the flywheel will rotate about the C–D axis so that the spin angular momentum L_s will align itself along the angular velocity vector of the rotating platform $\boldsymbol{\omega}_p$. If one reverses the direction of rotation of the platform, making $\boldsymbol{\omega}_p$ point in the negative z-direction, the gyroscope will rotate again about C–D in such a way that $\mathbf{L}$ and $\boldsymbol{\omega}_p$ are parallel. The gyrocompass will attempt to make its spin angular momentum align itself with the rotational angular velocity of its environment. If that environment is the Earth, the gyrocompass will point its spin

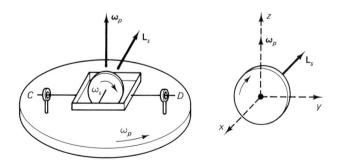

Fig. 14.21

angular momentum along the spin angular velocity of the Earth. It acts as a true compass (unlike the magnetic compass, which points toward the magnetic pole) and point in the direction of the true rotational spin axis of the Earth.

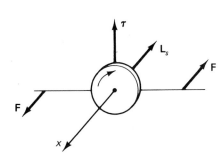

Fig. 14.22

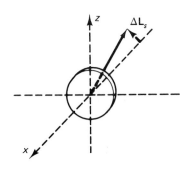

Fig. 14.23

How can we understand this interesting behavior? The first thing to notice is the mounting structure of the gyrocompass. Because of the pivot at C–D (the y-axis), *no torque can be exerted along this axis*. When the platform is rotated in the counterclockwise direction, the supports exert sideward forces $\mathbf{F}$ and $-\mathbf{F}$ on the pivots at C and D. These forces produce a torque in the *z-direction* on the flywheel (Fig. 14.22). The torque produces a time rate of change of the angular momentum in the z-direction, described by our now-familiar master equation. If the flywheel were *not* spinning, it could produce this changing angular momentum in the z-direction by rotating the stationary flywheel about the z-axis, creating an $L_z = \omega_p I_z$, where I_z is the moment of inertia of the flywheel about the z-axis. However, if the flywheel already has a large spin angular momentum in the negative x-direction $\mathbf{L}_s$, the same torque can create a time-varying angular momentum by *rotating* the flywheel about the y-axis and tilting the spin angular momentum toward the z-axis, as shown in Fig. 14.23. It is easy to feel the force necessary to change the large spin angular momentum of the flywheel when one attempts to rotate the platform, but once the spin angular momentum has aligned itself along the z-axis, it no longer requires a torque to maintain the alignment of $\boldsymbol{\omega}_p$ and $\boldsymbol{\omega}_s$. Try it!

14.6 Spinning Tops

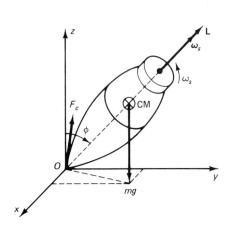

Fig. 14.24

THE ANALYSIS of a spinning top is essentially the same as the one we used for the motion of the gyroscope. Consider the symmetric top spinning with an angular velocity $\boldsymbol{\omega}_s$ along its own symmetry axis, tilted at the angle ϕ, with respect to the z-axis (Fig. 14.24). Two external forces act on the top: the contact force acting at the origin $\mathbf{F}_c$ (which involves friction as well as a normal force), and an action-at-a-distance force $mg(-\mathbf{k})$, which we can treat as acting on its center of mass. Since we do not know the magnitude and direction of the contact force $\mathbf{F}_c$, we will avoid using it in the analysis by examining both the torque and angular momentum of the body about the origin O. Remember, you must use the *same origin* for $\boldsymbol{\tau}$ and for $\mathbf{L}$ when using the master equation

$$\boldsymbol{\tau}_0 = \frac{d\mathbf{L}_0}{dt}$$

The entire gravitational force on all parts of the body can be treated *as if* it acted at one point (Fig. 14.25), the center of mass of the top. The torque of this force at the instant that the center of mass

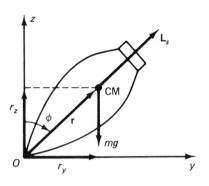

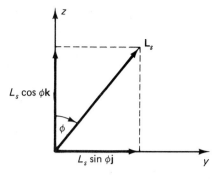

Fig. 14.25   **Fig. 14.26**

of the top lies in the y–z plane (Fig. 14.26) is

$$\tau_0 \left(\begin{array}{c} \text{center of mass} \\ \text{in } y\text{–}z \text{ plane} \end{array} \right) = \mathbf{r} \times mg(-\mathbf{k})$$

$$= (r_y\mathbf{j} + r_z\mathbf{k}) \times mg(-\mathbf{k}) = r_y mg(-\mathbf{i})$$

$$= rmg \sin \phi(-\mathbf{i}).$$

This torque produces a time rate of change of the angular momentum in the negative x-direction. The torque acts on the horizontal component of the spin angular momentum (the z-component remains constant), causing it to precess in the x–y plane, just like the gyroscope. From Eq. 14.43 we know that

$$\left| \frac{d\mathbf{L}_s}{dt} \right| = \Omega_p L_s \sin \phi,$$

and equating this to the applied torque,

$$rmg \sin \phi = \Omega_p L_s \sin \phi,$$

we arrive at
$$\Omega_p = \frac{rmg}{L_s},$$

the same result as we obtained for the horizontal gyroscope. The precession angular velocity Ω_p is *independent* of the angle of tilt of the top. Both the applied torque and the horizontal component of the spin angular momentum depend in the same way on the tilt angle ϕ. The angular dependence cancels out.

14.7 **Conservation of Angular Momentum, Revisited**

ONCE AGAIN we examine some consequences of the conservation of angular momentum for a rotationally isolated system. The master equation of rotational dynamics $\tau_0 = d\mathbf{L}_0/dt$ specifies the criterion for rotational isolation. If the net external torque on a system is zero, the time rate of change of the total angular momentum must be zero. The total angular momentum is a constant of the motion. The system does *not* have to be isolated from external *forces* for its angular momentum to be conserved, *only from external torques*. Since both torque and angular momentum are vector quantities, if there is no net torque in one particular direction, the angular momentum *in that direction* is a constant of motion. Now we are ready to apply this principle to a variety of systems, including extended bodies.

EXAMPLE 1   I should like to return to our discussion of elementary particles, where many of the conservation principles were used except this very important one. Most elementary particles have intrinsic spin angular momentum. That is, for these particles, the spin angular momentum is as "natural" a property of the particle as is its rest mass, electrical charge, family membership, strangeness, and so on. The electron, for example, cannot exist without its $\hbar/2$ of spin angular momentum. All the intrinsic angular momenta are quantized in units of $\hbar$. We can see how this conservation principle applies to elementary particles by examining the decay of the K^+-meson.

Consider a K^+-meson at rest in an inertial reference frame. The K^+-meson has zero intrinsic spin angular momentum and several decay modes:

$$K^+ \rightarrow \mu^+ + \nu_\mu$$

$$K^+ \rightarrow \pi^0 + \pi^+$$

$$K^+ \rightarrow \pi^+ + \pi^+ + \pi^-.$$

The last two modes also involve particles with zero intrinsic spin angular momentum, and thus are not relevant to this discussion. The first decay mode involving the muon and the neutrino (each lepton having intrinsic spin angular momentum $\hbar/2$) must conserve angular momentum.* Since the initial angular momentum of the system is zero, the final total angular momentum (of the decay products) must also be zero. That means that the two product particles not only have equal and opposite linear momentum, they also have equal and opposite spin angular momentum. If you imagine a classical "picture" of this decay, it might look like Fig. 14.27. Don't get carried away by the drawing; no one has ever observed the structure of any lepton, so that showing it as a rotating sphere is pure fantasy!

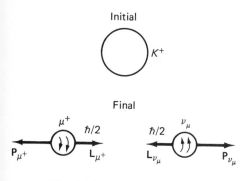

Fig. 14.27

For a two-body collision of elementary particles the problem becomes more complicated. A particle moving in a straight line with

---

* There are no external torques on the system that can cause measurable changes in the angular momentum of elementary particles since the torque due to the gravitational force on an elementary particle is extremely small. We can also neglect the effect of magnetic fields for this discussion.

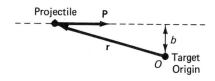

Fig. 14.28

linear momentum **p** has angular momentum with respect to an origin displaced from the path by a distance b (as shown in Fig. 14.28) whose magnitude is written $L_0 = b\mathbf{p}$. This relative orbital angular momentum, as well as the intrinsic spin angular momentum, must be taken into account when one calculates the initial and final angular momenta of an interacting system.

EXAMPLE 2 One of the most interesting types of systems is extended bodies rotating about a fixed axis, whose moments of inertia can be varied by *internal* forces. The ice skater, the high diver, and the gymnast demonstrate this kind of motion. The ice skater is about to go into one of those spectacular spins. She begins by extending her arms as far from her body as possible and executes a rotational motion by pushing off on one skate against the ice. Then she brings her arms in as close to her axis of rotation as possible, rapidly increasing her angular speed.

The dominant features of her motion can be analyzed by examining a simplified mechanical model of the skater. Figure 14.29 shows two equal masses supported by massless arms mounted on an axle fixed in frictionless bearings. From Chapter 13 we know that the initial angular momentum of this machine is

$$L_z^i = (I_{\text{axle}} + 2ml^2)\omega_i = I_z^i\omega_i.$$

If some device *within* the machine moves the arms closer to the axle, the new moment of inertia of the system becomes

$$I_z^f = I_{\text{axle}} + 2m(l \sin \theta_f)^2.$$

Since *no external* torque about the axis of rotation acted on the system, its angular momentum must be conserved, which means that

$$L_z^i = L_z^f \qquad \text{or} \qquad (I_{\text{axle}} + 2ml^2)\omega_i = [I_{\text{axle}} + 2m(l \sin \theta_f)^2]\omega_f$$

or

$$\frac{\omega_f}{\omega_i} = \frac{I_z^i}{I_z^f} = \frac{I_{\text{axle}} + 2ml^2}{I_{\text{axle}} + 2ml^2 \sin^2 \theta_f}.$$

This expression makes it clear that $\omega_f > \omega_i$; the angular velocity must increase to keep the angular momentum constant. The rotational kinetic energy of the skater is given by Eq. 13.42, $\text{KE} = \frac{1}{2}I_z\omega^2$. Substituting in our expressions for initial and final moments of inertia and angular velocity, we obtain

$$\frac{\text{KE}_{\text{final}}}{\text{KE}_{\text{initial}}} = \frac{\frac{1}{2}I^f\omega_f^2}{\frac{1}{2}I^i\omega_i^2} = \frac{\frac{1}{2}I^f(I_z^i)^2}{\frac{1}{2}I^i(I_z^f)^2} = \frac{I_z^i}{I_z^f}.$$

Since $I_z^i > I_z^f$, the final kinetic energy of the skater is greater than her initial kinetic energy.

How can this happen? There were no external torques on the system. What caused the rotational kinetic energy to increase?

If you were the ice skater and carried extra weights in each hand to increase your ability to *change* your moment of inertia, you would

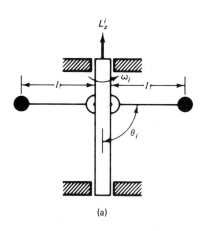

(a)

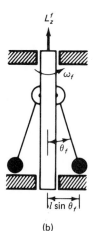

(b)

Fig. 14.29

The high diver clasps his legs and arms tight against his body to decrease his moment of inertia and thereby increase his angular velocity. To complete the dive he will straighten out his body and almost stop rotating.

know the answer. You would remember the effort it took to bring these weights in from their extended positions to a place near the axis of rotation. In fact, you did *work* (the physics kind of work) on the system by exerting an additional *internal* force on the weights over the distance they moved. The work you did was exactly equal to the change in kinetic energy achieved in the final configuration. All this was derived from the principle of the conservation of angular momentum, but it is also consistent with the conservation of total energy.

EXAMPLE 3   There are situations where an external force is applied to a system, but there are no external torques. In Example 2, the bearings supply forces that constrain the axle to rotate around the z-axis, but to the

Fig. 14.30

degree that they are free of friction, they do not exert a torque about the z-axis. Another example, a well-known classroom demonstration, consists of a small ball attached to a fishline which passes through a hollow tube (Fig. 14.30). The tube and line are held by the demonstrator and the ball is spun around in the horizontal plane. Since the string exerts a radial force as the ball executes horizontal circular motion, the string cannot exert a torque on the ball.

If an addition force is exerted on the line, decreasing the radius of the ball, the ball must speed up. How can we calculate the new velocity of the ball? The additional force transmitted by the string does *not* exert a torque on the ball, so the angular momentum of the ball cannot change. Thus

$$\mathbf{L}_i = \mathbf{L}_f \qquad \text{or} \qquad r_i m v_i = r_f m v_f$$

$$\frac{v_i}{v_f} = \frac{r_f}{r_i}.$$

Did the kinetic energy increase? How is this possible? Is there work done on the system? How much, and by whom? All these questions are left as an exercise for you.

EXAMPLE 4
Law of Equal Areas

Had Kepler known about angular momentum conservation, there is no doubt that he would have been able to give a physical explanation of his "law of equal areas." Kepler worked out several important empirical relationships which were in agreement with Tycho de Brahe's measurements of planetary motion. Among these empirical rules was the law of equal areas. It states that if a line is constructed from the force center, the sun, to the planet and the area in the plane swept out by this line in a given unit of time Δt is measured, this area is a constant of the motion. That is, the *rate* at which the area is swept out does not depend on the position of the planet in its elliptical orbit. This empirical rule, it turns out, is a direct consequence of the conservation of angular momentum. Let's prove it.

Consider the motion of the planets over a sufficiently short time period Δt so that the path over which the planets traveled Δs can be reasonably approximated by a straight line. Then

$$\Delta s = v \, \Delta t.$$

From an examination of Fig. 14.31, the area of the triangle OAB is half the height times the base r, or

$$\text{area} = \tfrac{1}{2}(\Delta h)r \qquad \text{but} \qquad \Delta h = (\Delta s) \sin \theta$$

so $\qquad \text{area} = \Delta A = \tfrac{1}{2}r(\Delta s) \sin \theta = \tfrac{1}{2}rv \, \Delta t \sin \theta.$

Dividing by Δt, to estimate the rate of change of the area with respect to time, we obtain

$$\frac{\Delta A}{\Delta t} = \frac{1}{2} rv \sin \theta.$$

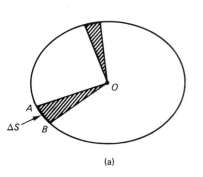

(a)

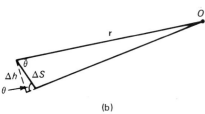

(b)

Fig. 14.31

But the magnitude of the angular momentum of the planet with respect to the origin O is $|\mathbf{L_0}| = |\mathbf{r} \times \mathbf{p}| = rmv \sin \theta$, so

$$\frac{\Delta A}{\Delta t} = \frac{|\mathbf{L_0}|}{2m}.$$

The last equation is still an approximation, since Δs is not a straight line. In the limit, as $\Delta t \to 0$, it is no longer an approximation; it becomes exactly true and the ratio $\Delta A/\Delta t$ becomes the time derivative:

$$\frac{dA}{dt} = \frac{|\mathbf{L_0}|}{2m} \qquad \text{(exactly)}.$$

The rate of change of the area is proportional to the angular momentum. Since the angular momentum is a constant of the motion, the rate at which the area is swept out must also be a constant. This discussion is *not* restricted to uniform circular motion; it is valid for the motion of any object that experiences only radial forces, where the angular momentum is conserved.

Tycho de Brahe's observations of six planets (Mercury, Venus, Earth, Mars, Jupiter, and Saturn) were accurate enough for Kepler to check his phenomenological relation for all these planets. You might be interested to know that Tycho did all these measurements *without* a single lens in his instruments. His apparatus consisted of line-of-sight devices, with very precise angle-measuring protractors. It is a monumental tragedy that none of his original instruments survived.

After Tycho, more precise measurements with optical instruments discovered some small deviations from the "rule" of equal area in equal time. Astronomers and physicists were not about to give up on such a fundamental notion as angular momentum conservation, so they began to search for a source of an external torque on the planet to account for the deviation. The search ultimately led to the discovery of two additional planets, Neptune and Pluto, whose gravitational fields produce small but measurable perturbing torques on the other planets. These torques were indeed responsible for the "violation" of Kepler's law.

EXAMPLE 5
Classroom Demonstration of the Vector Nature of Angular Momentum

Students may find torque, angular momentum, and moment of inertia difficult concepts to master, but they love to participate in the classroom demonstration of these phenomena. For these demonstrations the most important part of apparatus is a stable, horizontal platform with excellent bearings which allows rotation with very little friction around a vertical axis. The other basic tool is a rim-loaded bicycle wheel with good axial bearings, a handle on each side of the axle, and considerable mass, uniformly distributed around the perimeter of the rim. A student is usually asked to sit on a stool, which has been placed on the axis of the platform, and hold the bicycle wheel in his or her hands (Fig. 14.32a).

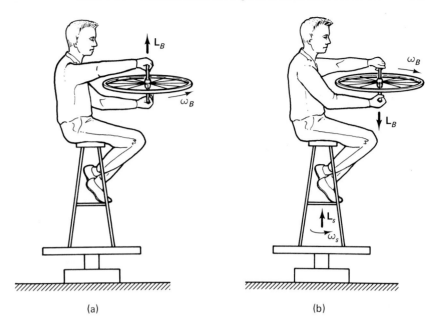

(a)   (b)

Fig. 14.32

Now we have a system consisting of a student plus bicycle wheel, both of which can rotate, one about the z-axis only and the other about any axis the student chooses. As a first experiment, the bicycle wheel is given an initial angular velocity in the **k** direction by the instructor and the student is kept at rest. The instructor then removes himself, leaving the student plus the spinning wheel isolated on the platform. Now the student is told to turn the wheel upside down, so that its angular velocity and angular momentum point in the $-\mathbf{k}$ direction (Fig. 14.32b). What happens?

The student begins to rotate in the counterclockwise direction. If the student again rotates the wheel 180°, back to its original position, he or she will stop rotating. If the wheel is given an initial angular velocity in the opposite direction (clockwise from the top) and the student again rotates its axis 180°, the student will start rotating, this time in the clockwise direction.

There are no external torques on the system, so the *total* (the vector sum) of the angular momentum of the *system* of the student plus wheel must remain constant (about the z-axis). When the wheel is rotated 180°, what happens to *its* angular momentum? How can its angular momentum change? If you ask the student, "Was it difficult to turn the wheel 180°?", the answer is definitely *yes*! Is a torque applied to the bicycle wheel? What does this torque do? What do you think will happen if the wheel is tilted only 50° or 30°? Remember that angular momentum, angular velocity, and torque are all vector quantities. Conservation of angular momentum requires *both* the magnitude and the direction of the vector to be a constant of the motion; that is, *both do not change with time.* All these questions are left as exercises. Or better yet, have your instructor do the demonstrations.

14.8 **Summary of Important Equations**

$$\boldsymbol{\omega} = \omega_x \mathbf{i} + \omega_y \mathbf{j} + \omega_z \mathbf{k}$$

$$\mathbf{v}(t) = \boldsymbol{\omega}(t) \times \mathbf{r}(t)$$

$$L_x = I_{xx}\omega_x + I_{xy}\omega_y + I_{xz}\omega_z$$

$$L_y = I_{yx}\omega_x + I_{yy}\omega_y + I_{yz}\omega_z$$

$$L_z = I_{zx}\omega_x + I_{zy}\omega_y + I_{zz}\omega_z$$

$$\mathbf{L} = \tilde{\mathbf{I}}\boldsymbol{\omega} \quad \text{(tensor equation)}$$

$$\Omega_p = \frac{rmg}{|\mathbf{L}_s|} = \frac{rmg}{\omega_s I_s} \quad \text{(gyroscope in Earth's gravity)}$$

Chapter 14   QUESTIONS

1. Figure 14.33 shows two kinds of Cartesian coordinate systems: right-handed and left-handed coordinates. Explain why these two systems were given their names. Why is it important to distinguish between the two systems?

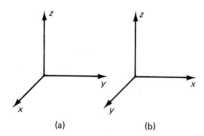

(a)   (b)

Fig. 14.33

2. In what direction does the *total* angular velocity vector point for the two rotating shafts in a helicopter? Draw a picture.

3. What configuration of masses could be added to the tilt-axis dumbbell that would eliminate the hori-

zontal torques on the bearings? Explain your additions and draw a diagram.

4. Explain the motion of the horizontal gyroscope when the spin angular velocity is in the opposite direction from that given in Fig. 14.18a.

5. A massive flywheel mounted in the bottom of a ship can be used as a "gyrostabilizer" to reduce the sideways roll of the ship in turbulent seas. Design a gyrostabilizer, explaining how you would mount the flywheel in the ship and what parameters you would use to make it an effective device.

6. In some Scandinavian countries buses are run on "flywheel power"; that is, a large flywheel mounted on the bus is set in motion at a starting station by an electric motor. The stored energy in the flywheel is used for the entire round trip. Can you see any problems that the mechanical engineers had to solve in designing such a system? What happens when the bus turns a corner or goes up a hill? Explain.

7. A common gyroscope demonstration apparatus is shown in Fig. 14.34. The mass M is movable on the shaft on the opposite side of the support pivot. Describe and explain the motion of the gyroscope, which has a large $\mathbf{L}_s$, as the mass M is moved closer to and farther away from the pivot point.

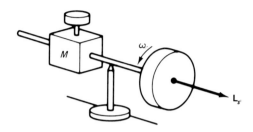

Fig. 14.34

8. Answer all the questions in Example 3, Section 14.7.

9. Answer all the questions in Example 5, Section 14.7.

10. A wheel and an axle have a combined mass M. The system is suspended by a rope attached to the end of the axle, has an angular momentum $I\omega$ lying in a horizontal plane, and is precessing counterclockwise about the point of suspension.

(a) What is the vertical force exerted by the wheel and axle on the rope at the point of suspension?

(b) Give the direction and magnitude of the force applied to the free end of the axle that will stop all precession.

11. A professor sits on a rotatable stool. He holds a bicycle wheel by its axle over his head so that it rotates in a horizontal plane clockwise as he looks up at it. He has no rotational velocity, but the wheel has angular momentum $I\omega$.

(a) What is the change of angular momentum if he stops the wheel, and which way will he rotate when the wheel stops?

(b) What is the change of angular momentum if he turns the wheel over so that it rotates counterclockwise as he looks up at it. Which way will he rotate?

12. Flywheels as practical energy storage devices and motors are discussed in an article in *Scientific American*, Dec. 1973, by Richard Post and Stephen Post. Look it up—you will find it fascinating.

Chapter 14 PROBLEMS

1. A motorcycle rides around the inside of a circular vertical track of radius 15 m, Fig. 14.35. If the linear speed of the cycle is a constant 25 m/s and the cycle

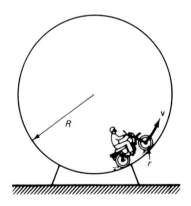

Fig. 14.35

traverses the track, without slipping, on wheels of 45-cm diameter:

(a) Find the angular velocity of the driver.

(b) Find the angular velocity of the cycle's wheels as measured by the driver.

(c) Find the total angular velocity of the wheels as measured by an observer on the ground watching the stunt.

2. A particle rotates in a vertical plane executing uniform circular motion with an angular velocity $\boldsymbol{\omega}$, Fig. 14.36. The angular velocity lies in the x–y plane at an angle of 30° with respect to the x-axis.

(a) Starting with the expression for the position vector $\mathbf{r}(t)$, find the linear velocity $\mathbf{v}$ from the relation $\mathbf{v} = d\mathbf{r}/dt$.

(b) Using the vector relation $\mathbf{v} = \boldsymbol{\omega} \times \mathbf{r}$, find the expression for $\mathbf{v}(t)$. Is it the same as you get for part (a)?

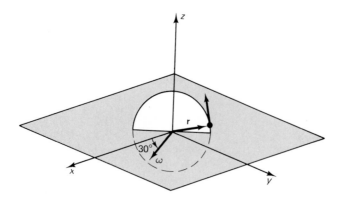

Fig. 14.36

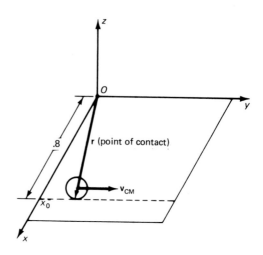

Fig. 14.38

3. An axle of length l is attached to a thin hoop of radius R and mass M, which rolls without slipping on a horizontal surface (Fig. 14.37). The hoop moves around the z-axis with a constant angular velocity $\boldsymbol{\Omega}$.

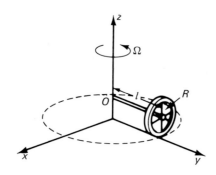

Fig. 14.37

(a) Find the instantaneous angular velocity of the hoop.

(b) Calculate the angular momentum of the hoop about point O. (*Hint:* The total angular momentum can always be written $\mathbf{L} = \mathbf{L}_{\text{spin}} + \mathbf{L}_{\text{orbit}}$.)

(c) Are the angular momentum and the angular velocity in the same direction? Explain.

4. A thin hoop rolls in a straight line along a horizontal plane with its center of mass moving with a velocity $\mathbf{v}_{\text{CM}} = .05\mathbf{j}$ m/s (Fig. 14.38). Its straight-line path intersects the x-axis at .8 m; the hoop has a radius of 25 cm and a mass of 0.3 kg.

(a) Find the general expression for the angular momentum of the hoop with respect to the origin O.

(b) Is the angular momentum time dependent?

(c) If the answer to part (b) is yes, what is the source of the torque that causes this time-dependent angular momentum?

5. Calculate the *total* angular momentum of the Earth–Sun system, assuming that the sun is at the center of a circle around which the Earth revolves and the Earth is executing uniform circular motion at a radius of 1.5×10^8 km. Assume that the Earth is a uniform sphere and its rotation axis is perpendicular to its plane of rotation. (Use Appendix A.IV for astronomical data.)

6. Show that the angular momentum of a particle moving in the x–y plane with uniform circular motion about the origin at constant $\boldsymbol{\omega}$, v, and radial distance r is given by

$$\mathbf{L}_0 = mvr\mathbf{k}.$$

Show this for the particle at an arbitrary position in the x–y plane using the vector relationship

$$\mathbf{v} = \boldsymbol{\omega} \times \mathbf{r}.$$

7. Equations 14.34 and 14.36 express the components of the moment of inertia tensor in terms of the x_i, y_i, z_i coordinates of the mass elements. Find the elements of the moment of inertia tensor that appears in the expression for L_x: namely, I_{xx}, I_{xy}, I_{xz} in terms of the x_i, y_i, z_i coordinates of the mass elements.

8. Use Eqs. 14.35 and 14.36 to calculate the angular momentum of the symmetric tilt-axis dumbbell shown in Fig. 14.11.

9. Two point masses M_1 and M_2 are fastened on the ends of a massless rigid rod of length l. The rod is connected to a rigid axle perpendicular to it, as shown in Fig. 14.39. The axle rotates with an angular velocity ω. At what position x should the axle be located so that the work required to set the system into rotation will be minimized?

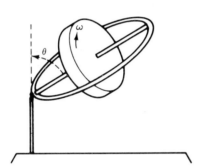

Fig. 14.39

10. Calculate the precession frequency of the gyroscope shown in Fig. 14.40 with the following physical parameters:

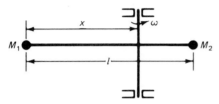

Fig. 14.40

(1) Disk mass, .017 kg

(2) $\theta = \dfrac{\pi}{6}$ rad

(3) Disk diameter, 4.8 cm

(4) Distance from support to disk center of mass, 3.0 cm

(5) Angular speed of disk, 310 revolutions per second

11. A cylinder with rotational inertia of 0.5 kg-m² is rotating about its axis of symmetry with an angular speed of 4 rad/s. Its angular momentum is in the positive y-direction.

(a) Indicate whether Fig. 14.41a or b is correct.

(b) What is the magnitude of the angular momentum?

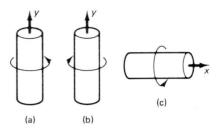

Fig. 14.41

(c) A torque is applied for 0.1 s. At the end of the 0.1-s interval the cylinder is rotating, as seen in Fig. 14.41c, with the same angular speed as at the beginning. What was the average torque applied? Give both magnitude and direction.

12. Magnetic fields can produce torques on elementary particles which have intrinsic angular momenta. Calculate the torque produced by such a field if it caused a 1.2×10^7-rad/s angular precession of a neutron whose quantized spin angular momentum is $\frac{1}{2}\hbar$.

13. Two ice skaters, moving at speeds of 3 m/s in opposite directions along parallel paths separated by .85 m, join hands when they are at the point of closest approach and move in a circular path. If each skater has a mass of 85 kg and they rotate about their center of mass,

(a) Find the initial angular momentum with respect to the center of rotation for each skater before they join. Is it constant?

(b) What is their final angular momentum?

14. A puck of mass m_2 slides along a frictionless horizontal surface at a constant velocity $\mathbf{v}$ toward a uniform stick of length l at rest on the surface (Fig. 14.42). The puck makes an elastic collision with the stick (of mass m_1) at a distance d from the center in such a way that it remains at rest and the stick moves off on the frictionless surface.

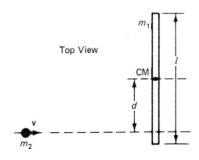

Fig. 14.42

(a) List all the conservation principles that are applicable for this problem.

(b) Using these principles, find the mass of the puck m_2, in terms of m_1, v, d, and l.

15. A disk with a moment of inertia I_z^1 is connected to a massless axle that is rotating with an angular velocity ω_i (Fig. 14.43). A second disk at rest (with moment

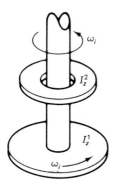

Fig. 14.43

of inertia I_z^2) is dropped onto the first, sliding at first and then stopping, so that there is no relative motion *between* the disks. Is mechanical energy, or angular momentum, or both, conserved? Calculate the final angular speed of the two-disk system.

16. A small ball is given a velocity $\mathbf{v}_0$ in the horizontal plane on the rim of a hemispherical cup (Fig. 14.44).

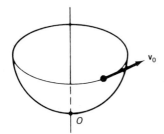

Fig. 14.44

If the ball rolls without friction, prove that it can never follow a path that would take it through the bottom of the cup. (*Hint:* Use conservation of angular momentum.) What is the torque on the ball at an arbitrary position on the hemisphere?

17. A bowling ball of mass M and radius R is thrown down the alley with an initial velocity $\mathbf{v}_0$. Initially, the ball begins to slide on the surface with a coeffi-

cient of friction μ. How fast will the ball be moving when it begins to roll *without* slipping? At what distance will the rolling without slipping occur? (*Hint:* Use angular momentum conservation about a point where there is zero torque.)

18. A mouse of mass m runs around the perimeter of a disk of radius a and moment of inertia I_z (Fig. 14.45).

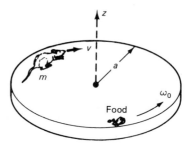

Fig. 14.45

The disk is supported by frictionless bearings which allow it to rotate freely about the z-axis. The mouse runs with a velocity v with respect to disk, and the disk has an angular speed of ω_0 with respect to the ground. If the mouse stops to eat food on the disk, what is the angular speed of the disk? What quantities are conserved?

19. A uniform circular hoop of mass m and radius R rolls without slipping down a vertical trough starting at a distance h above the bottom, as shown in Fig. 14.46.

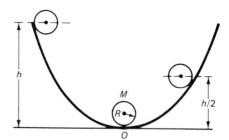

Fig. 14.46

(a) Calculate the speed of the hoop's center (1) when it is at the bottom, and (2) when its center reaches $h/2$.

(b) Calculate the total angular momentum of the hoop when it reaches the bottom of the trough at point O.

20. A turntable is mounted on frictionless bearings about a vertical axis with a moment of inertia I. It is designed with a special hollow axle and a slot that confines a small mass m to slide in the horizontal plane without friction. A string is attached to the mass, which runs down the center of the axle as shown in Fig. 14.47. Initially, the turntable and mass are set in motion with an angular speed ω_i and the mass is at position r_0. Then the mass is pulled into a new radius r by a force on the string.

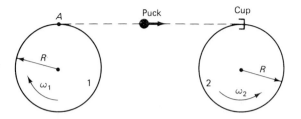

Fig. 14.48

Write down expressions for the following quantities in terms of R, m_1, m_2, ω_1, ω_2, and m, and calculate their numerical values.

(b) The angular velocity ω_1' (in rad/s) of disk 1 after the puck is released.

(c) The speed v (in m/s) of the puck when it is in the air.

(d) The angular velocity ω_2' (in rad/s, including direction), of disk 2 after the puck is caught.

(e) The ratio of the final kinetic energy of *the puck and disk 2* to their initial kinetic energy. Write this ratio in terms of ω_2' and v (as well as other known quantities). Where did the energy go?

22. A long thin rod of mass m and length l rest on a horizontal frictionless surface when a small piece of sticky mud of equal mass m and velocity v strikes it and adheres to it (Fig. 14.49). The collision is completely inelastic and lasts for an extremely short duration.

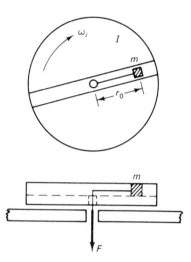

Fig. 14.47

(a) Find the initial and final angular momenta of the system.

(b) Show that the work done by the force on the string is equal to the energy difference between the initial and final energy states of the system.

(c) If the string breaks, with what radial velocity $dr/dt\ \hat{\mathbf{r}}$ will the mass pass the initial point r_0?

21. Two horizontal disks, both of radius $R = 20$ cm, spin about vertical shafts through their centers, the shafts being fixed to the earth. The first has mass $m_1 = 1$ kg and angular velocity $\omega_1 = 6$ rad/s; the second has $m_2 = 2$ kg and $\omega_2 = 3$ rad/s. (Note the directions as shown in Fig. 14.48.) On the rim of disk 1 is a small puck of mass $m = \frac{1}{4}$ kg, held to the disk by a (massless) pin. When the puck is at point A, the pin is pulled out so that the puck flies off disk 1, is caught, and is held by a (massless) cup on the rim of disk 2. (Neglect the gravitational force.)

(a) Name two quantities of the system that are conserved before the puck hits disk 2.

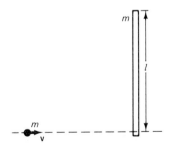

Fig. 14.49

(a) Calculate the velocity of the center of mass before and after the collision.

(b) Find the angular momentum of the system, about its center of mass, just before the collision.

(c) Calculate the angular velocity of the system about the center of mass just after the collision.

(d) How much kinetic energy is lost in the collision?

23. Two friction gears whose moments of inertia are I_1 and I_2, respectively, are supported on horizontal axes and separated a small distance as shown in Fig. 14.50. Gear 1 is set in motion with an angular speed

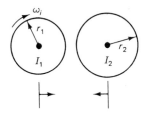

Fig. 14.50

ω_i and the second gear is kept at rest. The gears are slowly brought together until they touch, causing slippage at first until the gear system settles down so that the two gears are moving at a constant angular velocity.

(a) Is angular momentum conserved in this system? Be careful here—think about your answer.

(b) Using dynamics principles, $\tau = d\mathbf{L}/dt$, the torque coming from the friction between the two gears, find the final angular velocities of the two gears.

(c) Assuming that angular momentum is conserved, find the final angular velocity of the two gears.

(d) Are your answers the same for parts (b) and (c)? What does this tell you about your answer to part (a)?

24. A popular lecture demonstration of angular motion consists of a small toy electric train which runs on a circular track mounted on the rim of a bicycle wheel. For this problem the wheel is a thin hoop of mass M and radius R, and the train is a small mass m. The wheel rotates on a good bearing in the horizontal plane. If the system (wheel plus train) was initially at rest and then electric power was sent to the train so that the train reaches a steady linear velocity v *with respect to the tracks*, find the angular velocity of the circular track.

25. A horizontal racetrack is used to confine the motion of two small masses m_1 and m_2 so that they move in a circular path without friction (Fig. 14.51). The

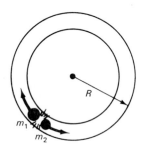

Fig. 14.51

two masses are in close proximity and at rest when a small explosive charge between them, with energy E, is set off, propelling each mass in opposite directions. The balls collide somewhere on the opposite side of the track.

(a) Find the location of the collision. (*Hint:* Use conservation principles.)

(b) Find the time it takes for the collision to occur.

15
HARMONIC OSCILLATION

15.1 Introduction

RICHARD FEYNMAN, one of the great scientists of the twentieth century, asked himself the following question: "If, in some cataclysm, all scientific knowledge were to be destroyed, and only one sentence passed on to the next generation of creatures, what statement would contain the most information in the fewest words?" His answer was, "All things are made of atoms—little particles that move around in perpetual motion attracting each other when they are a little distance apart but repelling each other upon being squeezed into one another."* Feynman never quite makes it clear why he believes this one piece of information to be so essential. Maybe the reason is obvious. Casual observations of most objects lead us to believe that things are at rest, quiet, motionless, and that only large-scale motion exists. But nature is much more lively, active, and interesting than she lets on.

In this chapter we examine one type of motion, harmonic oscillation, that is apparent in large-scale systems such as pendulums and vibrators, but also exists on the atomic, molecular, and nuclear scales. Our treatment of this oscillatory motion will be based on Newton's laws of motion and therefore clearly applicable to everyday objects. But what is so remarkable about the formalism we are about to develop is how much of it is also useful in the study of electromagnetism, electronic circuits, quantum optics, and atomic and molecular physics.

Repeated motion of an object in equal time intervals is defined as PERIODIC MOTION. The motion of a child on a playground swing has much in common with the vibrations of a violin string, or the oscillations of pendulums on old grandfather clocks. An unbalanced tire,

---

* R. P. Feynman, R. B. Leighton, and M. L. Sands, *Feynman Lectures on Physics*, Vol. I, Reading, Mass.: Addison-Wesley, 1964, pp. 1–2.

running at high speed on a car with worn shock absorbers, produces a familiar and very dangerous vibrational motion. Even tall sky-scrapers sway back and forth when they are subjected to high-velocity winds; sometimes you can feel the motion on the upper floors.

Many of the most interesting oscillatory motions are not readily visible. Microwave ovens cook primarily by causing the water molecules in the food to rotate back and forth about once every 10^{-12} s. Your digital watch has a tiny (but visible) piece of specially cut quartz crystal which is mounted in such a way that it vibrates about once every 10^{-6} s in response to an imposed electrical signal. That's the master clock that keeps the time in your watch. Stereo receivers have many electrically oscillatory components that help select the station you want to hear and reject all the other signals that are present. Even giant stars show oscillatory behavior, and their cycles can take from days to months to complete. We know about these distant suns because of observations of their regularly varying light intensity, as shown for Cephid stars in Fig. 15.1. No one will ever notice, but when a nuclear bomb explodes, the uranium or plutonium nuclei go through an oscillatory dance after capturing a neutron, before they break apart and begin the chain reaction of destruction. Oscillatory motion is everywhere!

Periodic motion can always be described mathematically in terms of sine and cosine functions. By suitable combinations of these functions, using techniques developed in mathematics called Fourier analysis, one can represent any periodic motion. We will not analyze generalized periodic motion, but only harmonic motion, that is, motion that can be described by a *single* sine or cosine function. Harmonic motion is not only periodic but it is bounded, and the mathematical functions that can describe a bounded harmonic motion are the sine and cosine functions.

Although in this chapter we limit our discussion to the motions of mechanical systems, the physical and mathematical analyses of these

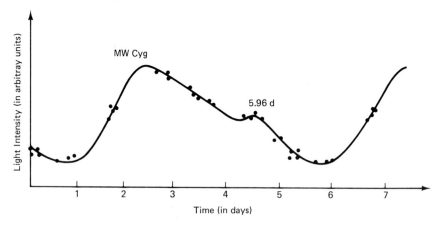

Fig. 15.1

systems are directly applicable to electromagnetic oscillations and radiation. Parts are even applicable to the quantum mechanical descriptions of mater on the microscopic atomic scale. In short, the material covered is widely applicable to many areas of physics.

15.2 Mass on an Ideal Spring

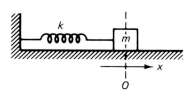

Fig. 15.2

IN Chapter 11 we have already considered an important example of SIMPLE HARMONIC MOTION (SHM) in one dimension: namely, a block attached to an ideal spring sliding without friction (Section 11.10). The zero of the x-coordinate is taken to be at the center of mass of the block when the spring is neither compressed nor extended and therefore exerts no force on the block (Fig. 15.2). This is the equilibrium position. The ideal spring obeys Hooke's law; it produces a force directly proportional to its displacement from equilibrium, and in the opposite direction to the displacement. Mathematically, this is expressed as

$$F_{\text{spring}} = -kx, \tag{15.1}$$

where F_{spring} is the force exerted *by* the spring, x is the displacement from equilibrium, and k is the spring constant characteristic of the particular spring. The minus sign tells us that this is a "restoring" force always acting to accelerate the object back in the direction of the equilibrium position. For positive displacement, the force acts in the negative x-direction, and for negative displacement, the force is positive always pushing the object back toward $x = 0$.

Since this is the only force on the block, Newton's second law becomes

$$-kx = m\,\frac{d^2x}{dt^2} \quad \text{or} \quad \frac{d^2x}{dt^2} = -\frac{k}{m}\,x, \tag{15.2}$$

whose solution is

$$x(t) = B\cos\omega_0 t + C\sin\omega_0 t \tag{15.3}$$

where

$$\omega_0 = \left(\frac{k}{m}\right)^{1/2} \tag{15.4}$$

and the constants B and C must be determined by the position and velocity of the block at $t = 0$ (initial conditions).

Direct substitution of Eqs. 15.3 and 15.4 into Eq. 15.2 verifies the solution. However, I should like to cast the solution into a slightly different, more convenient mathematical form:

$$x(t) = A \cos (\omega_0 t + \phi). \tag{15.5}$$

To show that Eqs. 15.5 and 15.3 are equivalent, let's make use of the trigonometric identity

$$\cos (\alpha + \beta) = \cos \alpha \cos \beta - \sin \alpha \sin \beta \tag{15.6}$$

or $A \cos (\omega_0 t + \phi) = A \cos\omega_0 t \cos \phi - A \sin\omega_0 t \sin \phi$

and identify in Eq. 15.3:

$$B = A \cos \phi \quad \text{and} \quad C = -A \sin \phi.$$

Since the two expressions are mathematically equivalent, we are free to choose either. We elect to use the expression

$$x(t) = A \cos (\omega_0 t + \phi), \tag{15.7}$$

where $x(t)$ is the instantaneous displacement of the mass m from its equilibrium position. What is the physical significance of the various parameters A, ω_0, and ϕ in this expression?

The cosine function itself varies between a maximum value of $+1$ and a minimum value of -1. Thus the constant A, called the AMPLITUDE, represents the maximum value of the displacement of the mass from equilibrium. Since the cosine is dimensionless, A must have the same units as $x(t)$. A full cycle of the cosine function occurs when its argument increases by 2π. The time T it takes to undergo a full cycle is called the PERIOD OF OSCILLATION (see Fig. 15.3). Since the argument of the cosine is $(\omega_0 t + \phi)$, when the time t increases by a full period T, then

$$\omega_0(t + T) + \phi = \omega_0 t + \phi + 2\pi$$

or

$$\omega_0 T = 2\pi \quad \text{or} \quad T = \frac{2\pi}{\omega_0}. \tag{15.8}$$

The FREQUENCY OF OSCILLATION is the reciprocal of the time necessary for one complete oscillation, or

$$f = \frac{1}{T} = \frac{\omega_0}{2\pi}, \tag{15.9}$$

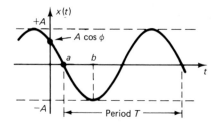

Fig. 15.3

where ω_0 is the ANGULAR FREQUENCY in units of radians per second and f is in the units of cycles per second or hertz (Hz).* We choose to write this oscillatory function in terms of an angular frequency, not only because it affords us a simple mathematical description, but also because of the close relationship between rotational and harmonic motion. That relationship will be explored shortly.

Finally, there is ϕ, the PHASE CONSTANT. It is related to the choice of starting times. If we start our clock at the time marked (a) in Fig. 15.3, ϕ will have the value $-\pi/2$, because $\cos(\omega_0 t - \pi/2) = -\sin \omega_0 t$; and if we start our clock at (b), then $\phi = -\pi$, because $\cos(\omega_0 t - \pi) = -\cos \omega_0 t$. If the clock's starting time is fixed, ϕ is related to the initial displacement of the mass, since this displacement is equal to $A \cos \phi$. Of course, we could just as easily and correctly have chosen to use $A \sin(\omega_0 t + \phi)$ as the mathematical representation of the oscillatory motion. The sine and cosine function are the same except for a phase shift of 90° [i.e., $\sin \omega_0 t = \cos(\omega_0 t + \pi/2)$].

It is essential that you become familiar with the parameters of SHM; amplitude, angular frequency, period, and phase. Figure 15.4 shows various solutions for simple harmonic motion with different amplitudes, frequencies, and phases. In Fig. 15.4a the two solutions have the same frequency and phase, but the amplitude in case I is twice as large as that in case II. In (b) the amplitudes are the same, but the angular frequency of I has twice the magnitude of II, and in (c), the amplitudes and frequencies of I and II are the same, but there is a 100° phase difference between the two solutions.

So far we have discussed only the displacement of the mass as a function of time for simple harmonic motion. We can readily obtain the velocity and the acceleration as functions of time by differentiating Eq. 15.5:

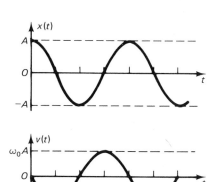

Fig. 15.4

$$x(t) = A \cos(\omega_0 t + \phi)$$
$$\frac{dx(t)}{dt} = v(t) = -\omega_0 A \sin(\omega_0 t + \phi)$$
$$\frac{dv(t)}{dt} = a(t) = -\omega_0^2 A \cos(\omega_0 t + \phi).$$

(15.10)

Figure 15.5 is a plot of these three dynamical variables on the same time scale for $\phi = 0$. The velocity is a maximum when the displacement is minimum and the acceleration is greatest when the block is at rest at it maximum displacement.

At $t = 0$, the spring has its maximum extension; the block is at rest, yet the spring exerts its maximum force on the block, producing

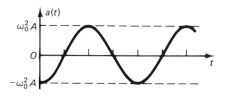

Fig. 15.5

---

* Hertz (Hz), the units of cycles per second, are named after the person who first observed the propagations of electromagnetic waves, Heinrich Hertz.

the maximum acceleration. The block moves toward the equilibrium position, reaching it a quarter of a period later and obtaining its maximum velocity. At equilibrium there is no extension or compression of the spring, zero force on the block, and zero acceleration. The block passes through the equilibrium position, moving in the negative x-direction. As it proceeds, it slows down and stops at its maximum negative displacement, $x = -A$. Here it reverses direction, has its maximum positive acceleration, but is instantaneously at rest. The mass again starts moving back toward the equilibrium position, passes it, and returns to its original position of maximum extension in the positive x-direction. The cycle begins again and repeats itself. For an ideal, frictionless, and lossless system, the block will oscillate forever.

The frequency of oscillation for simple harmonic motion is strictly independent of the amplitude of the motion. That may seem surprising; the greater the amplitude, the farther the mass has to travel, yet the time it takes to travel through one full cycle remains the same. A glance at Eq. 15.10 provides the explanation; the instantaneous velocity at any time *does* depend on the amplitude. As the amplitude increases, the velocity increases proportionally, allowing the frequency to remain the same. The *independence* of amplitude and frequency is an important characteristic of SHM.

Fig. 15.6

EXAMPLE 1   A student constructs a crude harmonic oscillator out of a metal spring and a fishing weight. He pulls the weight down from equilibrium, lets go, and observes that it travels a distance of 5 cm (minimum to maximum), executing 14 complete cycles in 9.8 s. What are the amplitude, frequency, period, maximum velocity, and maximum acceleration of this system, assuming that the weight executes simple harmonic motion?

If the weight traverses 14 complete cycles in 9.8 s, one cycle would be completed in

$$T = \frac{9.8 \text{ s}}{14 \text{ cycles}} = 0.70 \text{ s}.$$

The frequency of oscillations is

$$f = \frac{1}{T} = \frac{1}{.70} = 1.43 \text{ Hz}.$$

A system traverses 2π radians in one complete cycle, so the angular frequency is

$$\omega = 2\pi f = 2\pi(\text{rad/cycle}) \times 1.43 \text{ cycle/s} = 8.98 \text{ rad/s}.$$

Since we defined the amplitude A as the maximum excursion *from the equilibrium position*, the amplitude of this oscillator is 5 cm/2 = 2.5 cm. The position of the weight can be described by the equation

$$x(t) = (2.5) \cos (8.98t + \phi) \qquad \text{(in centimeters)}.$$

The maximum velocity and maximum acceleration are obtained from Eqs. 15.10:

$$v_{max} = \omega_0 A = 8.98(\text{rad/s}) \times 2.5 \text{ cm} = 22.44 \text{ cm/s}$$

$$a_{max} = \omega_0^2 A = (8.98 \text{ rad/s})^2 \times 2.5 \text{ cm} = 201.4 \text{ cm/s}^2.$$

EXAMPLE 2   When a free-hanging spring has a .02-kg ball attached to it, it extends 12 cm to a new equilibrium position (Fig. 15.7). Find the frequency, the maximum velocity, and acceleration of the ball if the ball is pulled down an additional 3 cm and let go.

The spring constant can be determined from the static extension of the spring before it is set into oscillation. From Hooke's law,

$$F = mg = kx \qquad \text{or} \qquad k = \frac{mg}{x}$$

$$k = \frac{.02 \text{ kg} \times 9.8 \text{ m/s}^2}{.12 \text{ m}} = 1.63 \text{ N/m}.$$

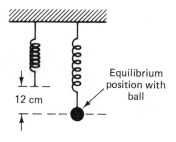

Equilibrium position with ball

12 cm

Fig. 15.7

The angular frequency is given by Eq. 15.4:

$$\omega_0 = \left(\frac{k}{m}\right)^{1/2} = \left(\frac{1.63 \text{ N/m}}{.02 \text{ kg}}\right)^{1/2} = 9.04 \text{ rad/s}.$$

The maximum velocity

$$v_{max} = \omega_0 A = 9.04 \text{ rad/s} \times .03 \text{ m} = .271 \text{ m/s}$$

and the maximum acceleration

$$a_{max} = \omega_0^2 A = (9.04 \text{ rad/s})^2 \times .03 \text{ m} = 2.45 \text{ m/s}^2.$$

15.3 Relationship to Uniform Circular Motion

THINK BACK to when we first introduced the mathematical description of a particle executing uniform circular motion. Equation 9.15 gives the position of the particle moving in the x–y plane, assuming that at time $t = 0$ the particle lies along the x-axis and $\theta = 0$. This expression is easily generalized so that at $t = 0$, θ can have any value between 0 and 2π. Such an expression is

$$\mathbf{r}(t) = r \cos(\omega t + \phi)\mathbf{i} + r \sin(\omega t + \phi)\mathbf{j}. \qquad (15.11)$$

Let's examine Eq. 15.11. It appears that uniform circular motion is the *sum* of *two* simple harmonic motions of the same amplitude but out of phase by $\pi/2$ radians. Circular motion is a vector sum of two components with one component along the x-axis, $\cos(\omega t + \phi)$, and the other along the y-axis and 90° out of phase,

$$\sin(\omega t + \phi) = \cos\left(\omega t + \phi + \frac{\pi}{2}\right),$$

represented pictorially in Fig. 15.8.

Conversely, think of simple harmonic motion as the projection along a diameter of a point executing uniform circular motion. The angular speed of the rotating point Q is the same as the angular frequency of the SHM. This analysis is not just a mathematical trick; it has a physical reality. One type of electric motor that uses alternating electric current creates an internal rotating magnetic field that the magnetic armature follows. This internal rotating magnetic field is formed by dividing the alternating current, phase shifting one part, and using both currents suitably placed at right angles to create a rotating magnetic field.

EXAMPLE 1   What is the period of rotation of an alternating current electric motor that uses household electric power?

Household current alternates at 60 Hz (cycles/second), so a rotating magnetic field will make 60 complete rotations every second. One rotation will then take

$$T = \frac{1}{f} = \frac{1}{60 \text{ cy/s}} = .0166 \text{ s.}$$

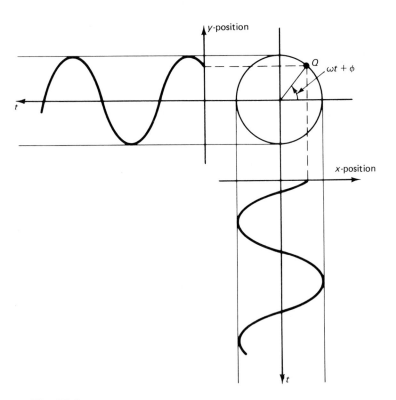

Fig. 15.8

OUR ANALYSIS of circular motion so far has been focused on a very special case, where the amplitude and frequency of the two components are the same and the phase difference is $-\pi/2$. Equation 15.11 can be generalized so that the x and y components have different amplitudes and phases. This can be written as

$$\mathbf{r}(t) = r_x \cos(\omega t + \phi_x)\mathbf{i} + r_y \cos(\omega t + \phi_y)\mathbf{j}, \qquad (15.12)$$

where r_x is the amplitude and ϕ_x the phase of the x-component, and r_y and ϕ_y correspond to the y-component. There are many physically relevant cases, but we will examine those where $r_x = r_y = 1$ and ϕ_x is set equal to zero. Setting $\phi_x = 0$ causes no real loss of generality, for it is only the *relative* phase of the two components that prescribes the type of motion created by the combination of harmonic oscillations. Figure 15.9 shows the resultant motion created by the combined harmonic oscillations with increments of $\pi/4$ radians for ϕ_y. It is possible to "create" everything, from straight line to elliptical to circular motion (rotating in either the clockwise or counterclockwise direction) with these combinations.

This discussion is directly applicable to understanding the nature of POLARIZED LIGHT. Classical physics describes light as oscillating electromagnetic fields that propagate in space. Certain interest-

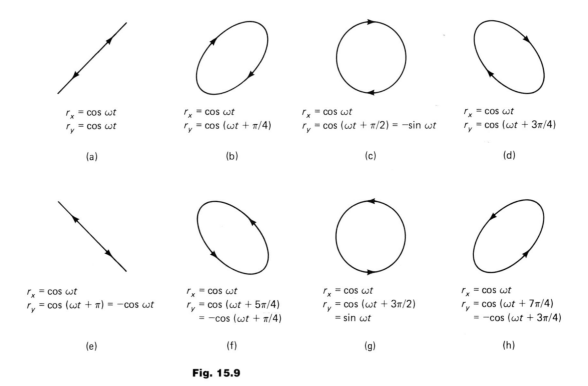

Fig. 15.9

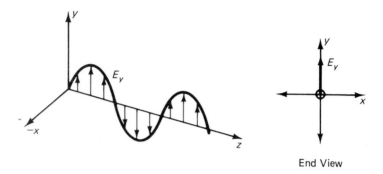

End View

Fig. 15.10

ing materials can create rotating polarized light by combining linearly polarized light at right angles with the proper phase shift. Even though we cannot explain all the details of the process, this phenomenon is worth a qualitative examination.

First some terminology. Light is propagating electric and magnetic fields. For this discussion we can ignore the magnetic fields and concentrate on the electric field part of the traveling fields. Light is called LINEARLY POLARIZED when its electric field oscillates in a plane, such as the y–z plane shown in Fig. 15.10. If the end of the electric field vector travels around in a circle (looking at it end on) as in Fig. 15.11b, the light is CIRCULARLY POLARIZED. Common light sources, such as an ordinary electric light bulb, produce unpolarized light, where the electric field randomly changes its direction of polarization over 2π radians, as depicted in Fig. 15.11c.

How is circularly polarized light created?* One way is through a phenomenon known as optical BIREFRINGENCE. A birefringent material is one in which light travels at different speeds in the material, depending on its direction of propagation. Suppose that a material consists of long cigar-shaped molecules that are all lined up so that the molecular axes are parallel, like sardines in a can. It should not surprise you that when light passes through such a material, the speed of propagation parallel to the molecular axis is different from the speed perpendicular to this axis. Common cellophane is such a substance, because it is made of long, fibrous molecules that are aligned in the manufacturing process.

Now we are ready to create circularly polarized light. All you need is an old pair of polaroid sunglasses and a sheet of cellophane. The polaroid "lenses" are another kind of special material that trans-

---

* Circularly polarized light is needed to cause certain electronic transitions in atomic or molecular systems because it carries the proper angular momentum. It is particularly important in optical pumping experiments (Arnold L. Bloom, *Scientific American*, Oct. 1960.

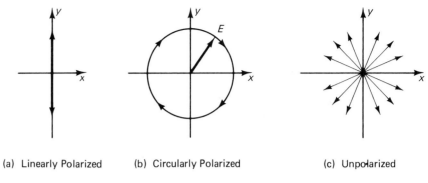

(a) Linearly Polarized   (b) Circularly Polarized   (c) Unpolarized

Fig. 15.11

mits light with only one direction of polarization (i.e., linearly polarized). For example, with proper orientation of the polaroid material, the electric field can be aligned at 45° in the x–y plane (Fig. 15.12). Place the sheet of cellophane in front of the polaroid filter with its optical axis in the y-direction and you have a source of circularly polarized light.

The linearly polarized light emerging from the polaroid material may be thought of as two electric fields, one in the x-direction (perpendicular to the optic axis of the cellophane) and the other in the y-direction (parallel to the optic axis of the cellophane), oscillating in phase. As the two components travel through the cellophane, they propagate with *different velocities*. They emerge at different times. Thus, although they were in *phase* when they entered, they leave the cellophane *out of phase*, because the time delay of a wave is the same as a phase shift of that wave.

By pure accident, common cellophane has a thickness for visible light such that the two components come out about $\pi/2$ radians out of phase. The electric field vectors of the light waves enter the cellophane as shown in Fig. 15.9a (where r_x and r_y should be thought of as E_x and E_y) and leave as depicted in Fig. 15.9c or g. The randomly

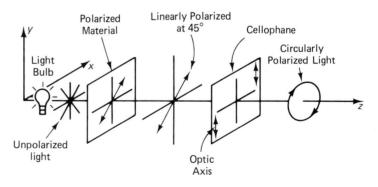

Fig. 15.12

polarized light source is turned into highly ordered circulating polarized light by its transmission through these two special materials.*

When the angular frequencies of the two harmonic solutions are different, the resulting motion is quite complicated, generally not even periodic. However, when the ratios of the two frequencies ω_x and ω_y are integers, the combined motion pattern produces LISSAJOUS FIGURES. These patterns can be used to determine an unknown frequency if one is known. Oscillation at *different* frequencies, but combined in the *same* direction, create "beat" patterns which are particularly important in the study of sound waves and nonlinear optical phenomena.

15.5 **Essential Conditions (Approximations) for SHM**

Robert Hooke (1635–1703), a contemporary of Isaac Newton, first proposed the empirical relationship between the restoring force and the extension from equilibrium produced by real springs. Although this is not a fundamental law of nature, it is a very good approximation for springs made of elastic materials, which are not extended beyond their ELASTIC LIMIT. The elastic limit is an extension, unique to a given spring, beyond which the spring will not return to its original shape after the external force has been removed. Exceeding the elastic limit permanently deforms the spring. Hooke's law is an especially good approximation when the spring is extended only a very small distance compared to its elastic limit. In fact, many objects other than springs obey Hooke's law for very small displacements.

In the discussion of energy diagrams (Chapter 11), we noted that any system that is subject to conservative forces, has a *potential minimum*, and executes small-amplitude oscillations about that minimum is describable by simple harmonic motion (see Section 11.10). For small excursions about the potential minimum where $\partial U/\partial r|_{r_0} = 0$, the potential energy function can be approximated using a Taylor series expansion (Eq. 11.92)

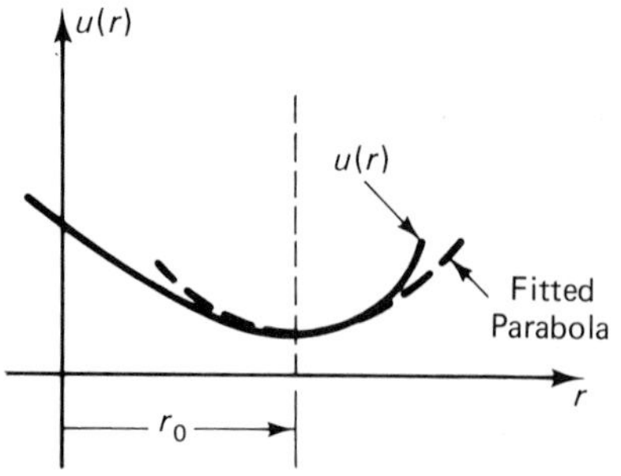

$$U(r) \approx U(r_0) + \frac{1}{2}(r - r_0)^2 \frac{d^2U}{dr^2}\bigg|_{r_0}$$

$$+ \text{ (higher-order terms which we neglect)} \qquad (15.13)$$

If the zero of coordinates is translated to the stable equilibrium point (Fig. 15.13) r_0 and the potential at r_0 is set equal to zero, $U(r_0) = 0$

Fig. 15.13

* You can do many interesting demonstrations with the two pieces of polaroid material and some cellophane. If you wish to read more about this, start with the *Feynman Lectures on Physics*, Vol. I (Reading, Mass.: Addison-Wesley, 1963), Chap. 33.

(since the absolute value of the potential is arbitrary), then Eq. 15.12 becomes

$$U(r) = \frac{1}{2} r^2 \frac{d^2U}{dr^2}\Big|_{r=0}. \tag{15.14}$$

Comparing this expression with the potential energy of an ideal Hooke's law spring,

$$U_{\text{spring}}(r) = -\int_0^r -kr\, dr = \frac{1}{2} kr^2,$$

one sees that the effective "spring constant" of the system is the second derivative of the potential energy curve, evaluated at the stable equilibrium point, or

$$k\ (\text{effective}) = \frac{d^2U}{dr^2}\Big|_{r=0}. \tag{15.15}$$

For small-amplitude excursions about the stable equilibrium position, the potential energy curve may be fitted, to a high degree of accuracy, by a parabola.

> *The quadratic dependence of the potential energy on particle displacement is essential for simple harmonic motion.*

As you may recall from our discussion of potential energy diagrams in Chapter 11 (section 11.11), a system subjected to a conservative force field with a parabolic potential energy can be described by the differential equation

$$E = \frac{1}{2} m \left(\frac{dx}{dt}\right)^2 + \frac{1}{2} kx^2. \tag{11.84}$$

This first-order differential equation describes simple harmonic motion as you can readily show. It is an alternative form of the master equation 15.2, which is a second-order equation. Sometimes it is easier to work with the first-order equation in solving a particular problem.

15.6 **Simple Pendulum**

THE SIMPLE pendulum is another of our "special effects," an idealization of a real physical system. Such a pendulum consists of a massless, perfectly rigid rod connected on one end to a frictionless bear-

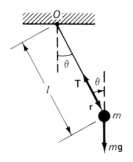

Fig. 15.14

ing and with a point mass attached to the other. Of course, no such device can be fabricated, but one can achieve a good practical approximation of this pendulum by obvious means.

When the pendulum is displaced an angle θ from equilibrium, the point mass experiences two forces on it—the tension **T** from the rod and the gravitational pull of the earth $m\mathbf{g}$, Fig. 15.14. We can analyze the device by considering the forces on the mass and its subsequent acceleration or by examining the torques acting on the mass about the pivot point o and the resulting change in angular momentum about the same point. We choose the latter method, because it is easily applied to the motion of other systems, namely the physical pendulum and the torsion oscillator.

What are the torques on m about the pivot o? The tension **T** cannot contribute, since $\mathbf{r} \times \mathbf{T} = 0$, where **r** is the vector from o to m. The gravitational force does produce a torque directed into the vertical plane swept out by the pendulum

$$\boldsymbol{\tau}_0 = \mathbf{r} \times m\mathbf{g} = lmg \sin \theta \qquad \text{(into the plane).} \qquad (15.16)$$

For fixed-axis rotation, this torque is related to the angular acceleration and the moment of inertia I_0 by Eq. 13.48, or

$$\tau_0 = I_0 \alpha = I_0 \frac{d^2\theta}{dt^2}.$$

Because the torque acts in a direction opposite to the angular displacement, the equations above combine to achieve a differential equation

$$I_0 \frac{d^2\theta}{dt^2} = -lmg \sin \theta. \qquad (15.17)$$

This is *not* an equation of simple harmonic motion because the angular displacement θ is *not* proportional to the angular acceleration. However, if we consider *only* small angular displacements, $\sin \theta \approx \theta$ *if* θ is in radians.* Accepting this restriction on the pendulum's amplitude, we can rewrite Eq. 15.17 as

$$\frac{d^2\theta}{dt^2} = -\frac{lmg}{I_0}\theta, \qquad (15.18)$$

which is mathematically identical to the master equation for SHM (Eq. 15.2).

The expression for the angular displacement and the angular

---

* Try it on your calculator and see how good this approximation really is. For a 10° or .17453 radian angle, there is only $\frac{1}{2}$% error.

frequency of oscillation is then

$$\theta = \theta_0 \cos{(\omega_0 t + \phi)} \quad \text{and} \quad \omega_0 = \left(\frac{lmg}{I_0}\right)^{1/2}. \quad (15.19)$$

The angular frequency expression can be further simplified by re-calling from Chapter 13 that the moment of inertia of a simple pendulum (a point mass a distance l from the pivot) is ml^2. Substituting for I_0 in Eq. 15.19 yields

$$\omega_0 \text{ (simple pendulum)} = \left(\frac{mgl}{ml^2}\right)^{1/2} = \left(\frac{g}{l}\right)^{1/2}. \quad (15.20)$$

This is a remarkable and important equation. What is most "surprising" is that the angular frequency does *not* depend on the mass. This seems counterintuitive. Galileo was the first to understand the reason and to recognize that the swinging pendulum represents the same physical phenomenon that is exhibited by all masses in free fall near the earth. A simple pendulum is nothing more than a *constrained* free-falling mass. Equation 15.20 also points out that the frequency is independent of amplitude (a characteristic of SHM) depending only on the length of the rod (for a fixed point near the earth).

Huygens (1629–1695) realized the use of the pendulum as a clock, since the period of oscillation

$$T = \frac{2\pi}{\omega} = 2\pi \left(\frac{l}{g}\right)^{1/2}$$

depends only on maintaining l constant and keeping the pendulum oscillating. He developed the escapement mechanism, which supplies the energy lost in the bearing friction and air resistance. This mechanism is necessary to keep the pendulum oscillating. For many years, until the development of electronics in the 1930s, pendulum clocks were the best human-made timekeepers. Today, geologists use a modern electronically calibrated pendulum to measure variations in the gravitational "constant" g at different parts of the earth's surface, by observing changes in the period of oscillation. It provides them with useful information about local density variations near the surface of the Earth's crust.

15.7 **Torsion Pendulum**

TWENTY-FIVE years ago, torsion oscillators were carried on the wrists of many people. The "balance wheel" inside their watches was a torsion oscillator with the restoring force provided by a hairspring.

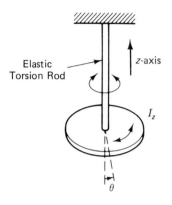

Elastic
Torsion Rod

z-axis

I_z

θ

Fig. 15.15

The period of oscillation provides the master timekeeping element. Figure 15.15 shows the essential elements of the torsion oscillator: an elastic rod (to provide the restoring torque) connected to the angularly oscillating member, in this case a simple disk. If the disk is rotated through an angle θ, the rod is twisted and provided that the rod obeys Hooke's law for twisting displacement, produces a torque as

$$\tau = -\kappa\theta, \tag{15.21}$$

where κ is the torsional constant and θ is the angular displacement from equilibrium. Following the identical procedure used to analyze the simple pendulum (Eq. 15.17), the differential equation that describes this angular oscillation is

$$I\frac{d^2\theta}{dt^2} = -\kappa\theta, \tag{15.22}$$

where I is the moment of inertia of the disk about the axis of rotation. The solution is of the form

$$\theta = \theta_0 \cos(\omega_0 t + \phi), \tag{15.23}$$

where θ_0 and ϕ must be determined from the initial conditions and

$$\omega_0 = \left(\frac{\kappa}{I}\right)^{1/2}.$$

Although balance wheels and mechanical watches are now almost extinct, the torsion oscillator has found at least one current application in low-temperature physics laboratories. At very low temperatures, helium becomes a SUPERFLUID, a truly unique liquid that flows without viscous drag. Such a fluid is not even imaginable in terms of classical physics, but classical physics does not describe nature on an atomic scale. Liquid helium is called a QUANTUM FLUID because quantum mechanics is needed to describe even its large-scale behavior. One model of this quantum fluid suggests that at very low temperatures helium is simultaneously a superfluid and a normal fluid, that is, a certain fraction of the liquid is in the normal state and flows like normal fluid with drag friction, and the other superfluid fraction flows without *any* friction or coupling to anything.*

Imagine a torsion oscillator immersed in this liquid. If it were all superfluid, the liquid would not couple to or affect in any way the motion of the torsion oscillator. The characteristic frequency of oscillation, $\omega_0 = (\kappa/I)^{1/2}$, would be the same when it was in the 100% superfluid as when it was in a vacuum. However, if we change the fraction of the fluid that is superfluid, making some of the fluid nor-

---

* A readable reference on superfluids is "Superfluid Turbulence," by Russell J. Donnelly, *Scientific American*, Nov. 1988, p. 100.

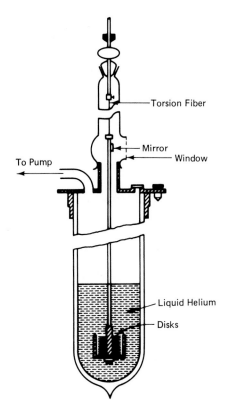

Fig. 15.16

mal liquid, this normal fluid will couple to the oscillating member and oscillate with it. The effect of the normal fluid will be to increase the mass of the disk, since some of the fluid would be carried along. That will increase its moment of inertia, decreasing the angular frequency, and increasing the period of oscillation. In fact, measurements of these parameters have been very useful in determining what is called the SUPERFLUID FRACTION of liquid helium for various physical situations.

The first measurements of the fraction of liquid helium in the superfluid state were made by the Soviet physicist E. L. Andronikashili in 1946. His apparatus consisted of 50 metal disks, each 35 mm in diameter and .13 mm thick, spaced apart by .21 mm. The period of oscillation of the system was about 20 s. The disks were held on a fine torsion suspension and the entire oscillator was enclosed in special vacuum-insulated devices (Thermos bottles), shown schematically in Fig. 15.16. The fraction of helium that is superfluid depends on the temperature of the helium, which can be varied by pumping on the liquid. Adronikashili's data on the superfluid fraction as a function of temperature (ρ_{normal}/ρ), where ρ is the total density, are shown in Fig. 15.17. In this diagram he compares measurements obtained from two independent techniques: one using the oscillating disks, the other using the velocity of "second sound" (the propagation of a temperature wave through the liquid). The agreement is excellent.

The superfluid state continues to intrigue physicists. In my own department, Frank Gasparini has pioneered the study of superfluid helium in confined spaces, where this remarkable fluid has even more remarkable properties. One important area of current interest to physicists is the study of the physical properties (mechanical,

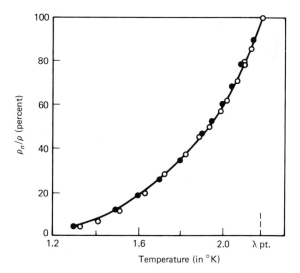

Fig. 15.17

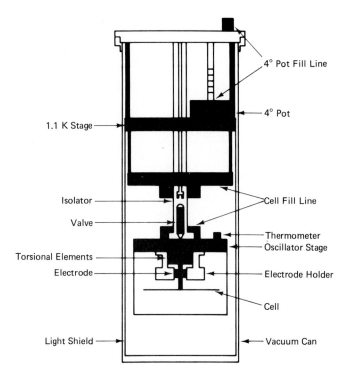

4° Pot Fill Line

4° Pot

1.1 K Stage

Cell Fill Line

Isolator

Valve

Thermometer
Oscillator Stage

Torsional Elements

Electrode

Electrode Holder

Cell

Light Shield

Vacuum Can

Fig. 15.18

electrical, magnetic, etc.) of various materials when they are formed in REDUCED DIMENSIONALITY. Reduced dimensionality is just our fancy way of saying that these materials are confined in a very small hole (zero dimensions), long and very skinny tubes (one-dimensional) or in thin films (two-dimensional). Of course, it is not possible to construct anything that really has zero, one, or two dimensions, but it turns out that some systems can be made to *behave* like the reduced dimensional models that the theoretical physicists calculate.

Figure 15.18 is a schematic diagram of a torsion oscillator cell constructed by Gasparini and Rhee out of silicon (using the latest high-tech methods developed for the semiconductor microminiature chips) which confines liquid helium into an effective two-dimensional material.* The entire apparatus is suspended in a special low-temperature refrigerator, which can reduce the temperature of the cell to 1.2 K above absolute zero. The period of this oscillator is about 2.7×10^{-3} s and it has an operating Q with helium in the cell of 750,000.† The entire oscillator has a maximum angular displacement of only 10^{-3} degree. Gasparini can observe *changes* in the period of oscillation as small as 4 parts in 10^9 (.0000004%) of the

---

* I. Rhee, F. Gasparini, and D. Bishop, *Physical Review Letters*, **63**, 410 (1989).
† Q is defined later in Eq. 15.47.

Fig. 15.19 Low temperature apparatus, showing the cell that holds the two-dimensional helium fluid.

period and thus can detect even a small quantity of superfluid helium in this confined geometry.

Some representative data using the apparatus shown in Figs. 15.18 and 15.19 are plotted in Fig. 15.20. The superfluid fraction

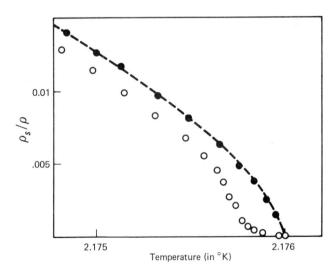

Fig. 15.20

ρ_s/ρ for both bulk (three-dimensional) helium (dashed line) and two-dimensional helium open circles, measured with this apparatus, is shown. Clearly, helium behaves differently as a two-dimensional liquid than it does as bulk liquid. These lower-dimensional systems are still not completely understood, despite extensive theoretical work. Even the lowly torsion oscillator helps push back the frontiers of human knowledge.

15.8 Physical Pendulum

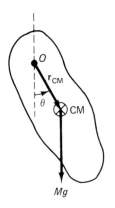

Fig. 15.21

Any rigid object of arbitrary shape that is free to rotate in a vertical plane about an arbitrary axis passing through it is called a physical pendulum. Figure 15.21 is an example of such a pendulum with mass M, pivoted at o, the center of mass at a distance $\mathbf{r}_{CM}$ from the pivot and rotated from equilibrium by an angle θ. In Chapter 12 we showed that the effect of a uniform gravitational field on an extended object can be accounted for by considering the entire gravitational force to act on a single point, the center of mass. Thus the magnitude of the torque about o due to gravity is simply

$$\tau_0 = r_{CM}Mg \sin \theta. \qquad (15.24)$$

The analysis of the physical pendulum is identical to the simple pendulum except that the moment of inertia of the physical pendulum is *not* Mr^2 but must be calculated for each particular pendulum. The moment of inertia depends on the location of the axis as well as the shape and density of each pendulum. Using our results from the simple pendulum, Eq. 15.19, the characteristic frequency is

$$\omega_0 = \left(\frac{r_{CM}Mg}{I_0}\right)^{1/2} \quad \text{(physical pendulum)}, \qquad (15.25)$$

where I_0 is the moment of inertia *about the pivot point*.

EXAMPLE 1

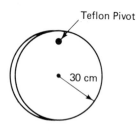

Fig. 15.22

A steel disk 30 cm in diameter and 5 cm thick is supported at its edge by a Teflon pin so that it can oscillate in the vertical plane (Fig. 15.22). Find the period of oscillation for small-amplitude motion.

Before you go looking up the density of steel and calculating the mass of the disk, let's examine the system to find out what information is actually needed to calculate the period. From Eq. 15.25 we can calculate the angular velocity and with Eq. 15.8 convert that into an expression for the period of oscillation: namely,

$$T = \frac{2\pi}{\omega_0} = 2\pi \left(\frac{I_0}{rMg}\right)^{1/2} \qquad (15.26)$$

But how do we calculate I_0, the moment of inertia about the pivot o? We can look up the moment of inertia of a solid disk about the *center of mass* (the geometric center) in Table 13.2; $I_{\mathrm{CM}} = \frac{1}{2}Mr^2$, but we will need to use the parallel axis theorem (Eq. 13.59) to calculate the moment of inertia about point o on the edge.

$$I_0 = I_z^{\mathrm{CM}} + Mr_{\mathrm{CM}}^2 = \tfrac{1}{2}Mr^2 + Mr^2 = \tfrac{3}{2}Mr^2. \qquad (15.27)$$

Substituting Eq. 15.27 into Eq. 15.26, we obtain

$$T = 2\pi \left(\frac{\frac{3}{2}Mr^2}{rMg}\right)^{1/2} = 2\pi \left(\frac{3r}{2g}\right)^{1/2}$$

$$= 2\pi \left(\frac{3 \times .3 \text{ m}}{2 \times 9.8 \text{ m/s}^2}\right)^{1/2} = 1.35 \text{ s.}$$

We don't have to know that the disk is 5 cm thick or that it is made of steel. The period depends only on the shape and the diameter of the uniform disk.

15.9 **Two-Body Oscillator**

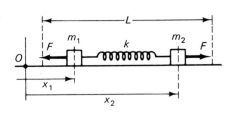

Fig. 15.23

TWO MASSES sliding on a frictionless surface and connected by a single spring (Fig. 15.23) is a mechanical system worth special attention. The analysis requires a coordinate transformation and provides an excellent model for the vibrational behavior of well-known diatomic molecules such as KCl, KBr, H_2, and CO. If the spring has an unextended length L, any net extension or compression x is given by the expression

$$x = (x_2 - x_1) - L. \qquad (15.28)$$

When the spring is compressed, as in Fig. 15.23, $L > x_2 - x_1$ and x is negative, and if the spring is stretched $L < x_2 - x_1$, then x is positive. The spring applies equal and oppositely directed forces on each block, whether it is extended or compressed. The spring is assumed to obey Hooke's law producing forces of magnitude kx.

For block m_1 (under compression when x *is negative*), we write (from Newton's second law)

$$kx = m_1 \frac{d^2x_1}{dt^2}$$

and for m_2,
$$-kx = m_2 \frac{d^2x_2}{dt^2}. \qquad (15.29)$$

These two equations can be combined by multiplying the top one by m_2, the bottom one by m_1, and subtracting them, yielding

$$-(m_1 + m_2)kx = m_1m_2 \frac{d^2x_2}{dt^2} - m_1m_2 \frac{d^2x_1}{dt^2}$$

or
$$-(m_1 + m_2)kx = m_1 m_2 \frac{d^2}{dt^2}(x_2 - x_1) = m_1 m_2 \frac{d^2 x}{dt^2}$$

since $x_2 - x_1$ differs from x by the constant L. Rearranging the equation above yields

$$\frac{d^2 x}{dt^2} = \frac{m_1 + m_2}{m_1 m_2}(-kx). \qquad (15.30)$$

Now we define a quantity called the reduced mass m_μ:

$$m_\mu = \frac{m_1 m_2}{m_1 + m_2}, \qquad (15.31)$$

so that Eq. 15.30 can be put in the standard form of a differentiated equation describing the SHM of a single object,

$$\frac{d^2 x}{dt^2} = -\frac{k}{m_\mu} x, \qquad (15.32)$$

where the mass of this nonexistent "object" is m_μ. The frequency of this system can be determined by inspection as

$$\omega_\mu = \left(\frac{k}{m_\mu}\right)^{1/2} \qquad (15.33)$$

If one of the masses is considerably smaller than the other, for example $m_1 \gg m_2$, then $m_\mu \approx m_2$; the lighter mass is the principal oscillating part of the system, with the larger mass remaining nearly at rest. If the two masses are equal, the frequency of oscillation is reduced by a factor of $2^{-1/2}$ from what it would have been if one of the masses had been confined to a fixed position.

15.10 Energy Considerations

LET US review the energy flow of a simple harmonic oscillator, a mass connected to an ideal spring. The potential energy $U(x)$ is located in the spring and is equal to the negative of the work done on the spring. Taking the zero of the potential at the equilibrium position $x = 0$, we have

$$U(x) = -\int_0^\infty -kx \, dx = \frac{1}{2} kx^2 = \frac{1}{2} kA^2 \cos^2(\omega_0 t + \phi), \quad (15.34)$$

and the kinetic energy at any point x is

$$KE = \frac{1}{2} mv^2 = \frac{1}{2} m\omega_0^2 A^2 \sin^2(\omega_0 t + \phi). \qquad (15.35)$$

The total mechanical energy of the system is then the sum of Eqs. 15.35 and 15.34:

$$E_{\text{total}} = \frac{1}{2} kA^2 \cos^2 (\omega_0 t + \phi) + \frac{1}{2} m\omega_0^2 A^2 \sin^2 (\omega_0 t + \phi).$$

But $\omega_0^2 = k/m$, so

$$E_{\text{total}} = \frac{1}{2} kA^2 [\cos^2 (\omega_0 t + \phi) + \sin^2 (\omega_0 t + \phi)]$$

$$E_{\text{total}} = \frac{1}{2} kA^2. \qquad (15.36)$$

The total mechanical energy is conserved. Notice that the total energy depends on the *square* of the maximum amplitude of oscillation. Electromagnetic and sound waves share this property.

It is instructive to see what form this energy takes at each part of the oscillator's cycle. Because the potential energy is only a function of the extension (or compression) of the spring, it reaches its maximum value when the extension is maximum, but the kinetic energy, which is only a function of the velocity, is zero at this point because there the particle is instantaneously at rest. When the extension is zero, at the equilibrium position, the potential energy is zero, but the velocity and hence the kinetic energy have their maximum values. Of course, the flow of energy from potential to kinetic to potential, and so on, is smooth and continuous as the particle moves back and forth over its complete cycle.

This flow is represented diagrammatically in Fig. 15.24 (for $\phi =$

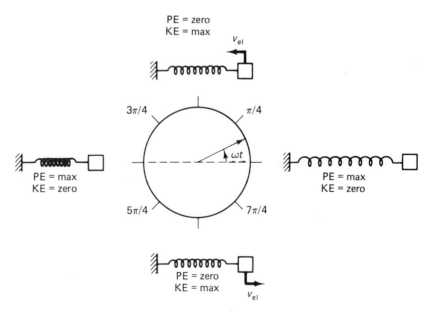

Fig. 15.24

0), where the circular motion representation of harmonic oscillation is used to depict the various stages. At the four intermediate positions

$$\frac{\pi}{4}, \quad \frac{3\pi}{4}, \quad \frac{5\pi}{4}, \quad \frac{7\pi}{4} \quad \text{radians,}$$

the energy of the system is stored equally as kinetic and potential. In these positions the spring is either compressed or extended to some degree, but the block is also moving with nonzero velocity.

15.11 Damped Harmonic Oscillation

Real mechanical systems do not conserve mechanical energy, since dissipative frictional forces are always present. Our mass connected to a spring will eventually stop oscillating, the pendulum will gradually come to rest, and our homemade fishing weight attached to a metal spring will quickly damp out after only a handful of excursions up and down. Our job is to develop a realistic physical model for this damping process that is also mathematically manageable.

Earlier, we introduced the basic physical ideas that we will now apply. In Chapter 8 we discussed drag forces that are velocity dependent and which always act to oppose the velocity of a body. Equation 8.37 expresses these forces as

$$\mathbf{F}_{\text{drag}} = -\alpha \mathbf{v},$$

where $\mathbf{v}$ is the particle velocity and α is a coefficient that depends on the size and shape of the moving object, as well as the medium through which it travels. The units of α are newton-seconds/meter. If this drag force is included in the model, the differential equation that describes the one-dimensional DAMPED harmonic oscillation is

$$-kx - \alpha \frac{dx}{dt} = m \frac{d^2x}{dt^2} \qquad (15.37)$$

or

$$\frac{d^2x}{dt^2} + \beta \frac{dx}{dt} + \omega_0^2 x = 0, \qquad (15.38)$$

where $\beta \equiv \dfrac{\alpha}{m}$ and, as before, $\omega_0^2 = \dfrac{k}{m}$.

Equation 15.38 is a new type of differential equation. This is not a book on differential equations, so we will not go into the general methods that are used to find the solutions to this kind of equation. But we may use our physics intuition to guess what the solution must look like, at least for the case where the damping is small. One would expect the mass to oscillate very nearly at its undamped characteristic frequency ω_0, but the amplitude will decay with a time characteristic of the damping coefficient β. A solution of the form

$$x(t) = Ae^{-(\beta/2)t} \cos(\omega_1 t + \phi), \qquad (15.39)$$

where
$$\omega_1 = \left(\omega_0^2 - \frac{\beta^2}{4}\right)^{1/2}$$

certainly seems reasonable. You should check, by direct substitution, to see whether Eq. 15.39 is a solution to Eq. 15.38. This solution is valid when $\omega_0 > \beta/2$, so that the angular frequencies are real numbers.

Velocity-dependent damping has changed the character of the motion. Figure 15.25 shows two cases: for lightly damped and heavily damped motion. The solution Eq. 15.39 appears almost identical to the undamped case except for the term $e^{-(\beta/2)t}$, which exponentially attenuates the amplitude of oscillation as time increases (see the envelopes in Fig. 15.25). The rate of decrease depends on the drag coefficients $\beta = \alpha/m$. The viscous drag has also lowered the "frequency" of oscillation of the motion. A new angular frequency ω_1 appears in the oscillating part of the solution. One must be careful in using the term "frequency" for such a solution, since we are no longer dealing with a pure sine or cosine function. In fact, a careful examination of this function shows that although the zero crossings occur at equal time intervals, separated by $2\pi/\omega_1$, the peak values are not located halfway in between. Formally speaking, this *cannot* be considered single frequency or simple harmonic motion, yet it is commonly referred to as oscillating at the frequency ω_1. Real springs with weights, torsion oscillators, pendulums, and other vibrating systems are experimentally found to obey Eq. 15.39, justifying our initial assumption of a dominant velocity-dependent dissipative force.

From energy considerations it is understandable that an oscillating mechanical system must eventually die down and come to rest. In time, any system will dissipate its mechanical energy by the frictional forces and turn it into internal energy of the materials involved. It is worth going through the quantitative analysis of this process to show that the rate of mechanical energy loss is also a simple exponential function.

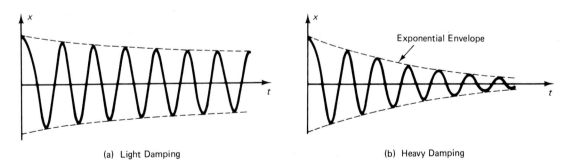

(a) Light Damping   (b) Heavy Damping

Fig. 15.25

The work–energy theorem tells us that the total mechanical energy at any time t, $E(t)$, is equal to the initial energy of the system E_0 plus the work done by the frictional forces. This is expressed as

$$E(t) = E_0 + W_{\text{friction}}(t). \qquad (15.40)$$

At any time t, the energy of the system is the sum of the kinetic energy $\frac{1}{2}mv^2(t)$ and the potential energy $\frac{1}{2}kx^2(t)$. We already have an expression for $x(t)$ (Eq. 15.39), so calculating the potential energy is obvious, but we must differentiate Eq. 15.39 with respect to time to obtain $v(t)$, which is needed to calculate the kinetic energy. Carrying out the differentiation yields

$$v(t) = -A\omega_1 e^{-(\beta/2)t} \left[\sin\left(\omega_1 t + \phi\right) + \frac{1}{2}\left(\frac{\beta}{\omega_1}\right) \cos\left(\omega_1 t + \phi\right) \right]. \qquad (15.41)$$

If we limit our discussion to the small damping case where $\beta/\omega_1 \ll 1$, we may neglect the second term in brackets in Eq. 15.41. Then the expression for the kinetic energy at any time t becomes

$$\text{KE}(t) = \frac{1}{2}\,mv^2(t) \approx \frac{1}{2}\,m\omega_1^2 A^2 e^{-\beta t} \sin^2\left(\omega_1 t + \phi\right) \qquad (15.42a)$$

and the potential energy at time t is

$$\text{PE}(t) = \frac{1}{2}\,kx^2 = \frac{1}{2}\,kA^2 e^{-\beta t} \cos^2\left(\omega_1 t + \phi\right). \qquad (15.42b)$$

The total mechanical energy is the sum, or

$$E(t) = \frac{1}{2}\,A^2 e^{-\beta t}\left[k\,\cos^2\left(\omega_1 t + \phi\right) + m\omega_1^2\,\sin^2\left(\omega_1 t + \phi\right)\right]. \qquad (15.43)$$

A further simplification can be made using the small damping approximation, replacing ω_1^2 by ω_0^2 (the undamped frequency), and recalling that $\omega_0^2 = k/m$. Under these conditions Eq. 15.43 collapses to the extremely simple expression,

$$E(t) = \frac{1}{2}\,kA^2 e^{-\beta t}, \qquad (15.44)$$

because $\cos^2\theta + \sin^2\theta = 1$.

At $t = 0$, the total mechanical energy is $E_0 = \frac{1}{2}kA^2$, which is in agreement with Eq. 15.36, derived in our discussion of undamped harmonic motions. Equation 15.44 can then be further simplified as

$$E(t) = E_0 e^{-\beta t}. \qquad (15.45)$$

The characteristic time τ associated with the decay of the mechanical energy to $\frac{1}{e}$ of its initial value is called the TIME CONSTANT of the system. That decrease will occur at the time when

$$\beta \tau = 1 \quad \text{or} \quad \tau = \frac{1}{\beta} = \frac{m}{\alpha} \quad \text{seconds.} \tag{15.46}$$

Do these expressions make physical sense? The larger the α (the drag coefficient), the shorter the time constant and the quicker the oscillation dies out. The larger the mass, the longer the time constant and the longer the oscillation continues. A system with a larger mass takes longer to dissipate the larger energy stored in its inertia, its kinetic energy. The analysis appears reasonable.

EXAMPLE 1   The amplitude of a 4-g harmonically oscillating mass is plotted as a function of time in Fig. 15.26, where the amplitude is measured in centimeters and the time in seconds. If this damped harmonic motion is described by Eq. 15.39, find A, ω_1, ϕ, β, k, and α.
Equation 15.39 is

$$x(t) = Ae^{-(\beta/2)t} \cos(\omega_1 t + \phi).$$

It is clear from Fig. 15.26 that $\phi = 0$, since at $t = 0$, $x(t)$ has its maximum value of 3 cm. Thus $A = 3$ cm.
The frequency ω_1 can be determined from zero crossings of $x(t)$, which occurs when

$$\omega_1 t = \frac{\pi}{2}, \quad \frac{5\pi}{2}, \quad \frac{9\pi}{2}, \quad \frac{13\pi}{2}.$$

From Fig. 15.26 the times for these zero crossing are .6, 3.1, 5.4, and

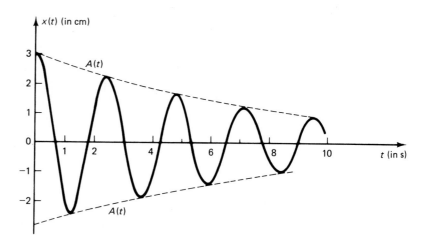

Fig. 15.26

7.7 s. The average *period* from these data is 2.45 s. Therefore,

$$\omega_1 = 2\pi f_1 = \frac{2\pi}{T} = \frac{2\pi \text{ rad}}{2.45 \text{ s}} = 2.64 \text{ rad/s}.$$

The exponential damping envelope function $A(t) = Ae^{-(\beta/2)t}$ is shown in the diagram as a dashed line. We shall determine the damping constant β from this curve by two independent methods. If we measure the amplitude of the envelope function at a time when $t = 2/\beta$, then

$$A(2/\beta) = Ae^{-(\beta/2)(2/\beta)} = Ae^{-1} = .368\ A.$$

The function will be reduced by a factor .368 from what it was at $t = 0$. For our oscillator

$$A = 3 \text{ cm} \quad \text{so} \quad (.368)\ 3 \text{ cm} = 1.104 \text{ cm}.$$

The envelope function has a value of 1.104 cm at $t = 6.9$ s. Thus

$$t = \frac{2}{\beta} = 6.9 \text{ s} \quad \text{or} \quad \beta = .29 \text{ s}^{-1}.$$

There is a more accurate way to extract β from the analysis of these data. Taking the natural logarithm of the envelope function $A(t) = Ae^{-(\beta/2)t}$, one gets

$$\ln A(t) = \ln A - \frac{\beta}{2}\ t,$$

which is the equation of a straight line whose slope is $\beta/2$. A plot of the natural log of the envelope function (using only the maximum positive excursions) versus time is shown in Fig. 15.27. The slope is

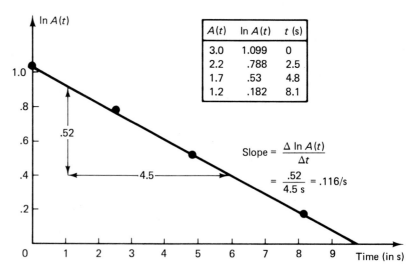

$A(t)$	$\ln A(t)$	t (s)
3.0	1.099	0
2.2	.788	2.5
1.7	.53	4.8
1.2	.182	8.1

$$\text{Slope} = \frac{\Delta \ln A(t)}{\Delta t}$$

$$= \frac{.52}{4.5 \text{ s}} = .116/\text{s}$$

Fig. 15.27

determined from the plot

$$\text{SLOPE} = .116 \text{ s}^{-1} = \frac{\beta}{2} \quad \text{or} \quad \beta = .232 \text{ s}^{-1}.$$

This is a more accurate value of β because it averages the data from several excursions rather than just one. If these were real experimental data, we would use both the positive and negative excursions as well as a statistical analysis of the data to determine the damping coefficient.*

To determine α we use $\beta = \dfrac{\alpha}{m}$, so

$$\alpha = \beta m = .232 \text{ s}^{-1} \times .04 \text{ kg} = 9.28 \times 10^{-3} \text{ kg/s},$$

and from Eq. 15.39, $\omega_1^2 = \omega_0^2 - \dfrac{\beta^2}{4}$, so

$$\omega_0^2 = \omega_1^2 + \frac{\beta^2}{4} = (2.64 \text{ rad/s})^2 + \frac{(.232/s)^2}{4}$$

or
$$\omega_0^2 = 6.98 \text{ rad}^2/\text{s}^2.$$

But $\omega_0^2 = \dfrac{k}{m}$, so

$$k = \omega_0^2 m = (6.98/s^2)(.04 \text{ kg}) = .279 \text{ kg/s}^2(\text{N/m}).$$

15.12 Q-Factor

THE Q of an oscillating system is a dimensionless parameter, commonly used in electronics and electrical engineering, to describe the amount of damping relative to the energy stored. The Q-factor immediately reveals how many *cycles* the system will traverse before it reduces its amplitude to an insignificant level. The precise definition of Q is

$$Q \equiv \frac{\text{energy stored in a system}}{\text{energy dissipated during } \textit{one radian} \text{ of oscillation}}. \tag{15.47}$$

A full cycle or 2π radians is traversed in one period $T = 2\pi/\omega_1$; a system will go through *one radian* of oscillation in $T/2\pi$ seconds, or

$$\frac{2\pi}{2\pi\omega_1} = \frac{1}{\omega_1} \quad \text{seconds.}$$

---

* The statistical analysis is usually a least-squares fit to the straight line.

The energy dissipated during a short time Δt is equal to

$$\Delta E = \frac{dE}{dt}\,\Delta t. \tag{15.48}$$

We can calculate $\frac{dE}{dt}$ from Eq. 15.45 by direct differentiation:

$$\frac{dE}{dt} = \frac{d}{dt}\,(E_0 e^{-\beta t}) = -\beta E_0 e^{-\beta t} = -\beta E(t). \tag{15.49}$$

Substituting into Eq. 15.48 yields

$$\Delta E = -\beta E(t)\,\Delta t.$$

The time interval is short for one radian, $\Delta t = 1/\omega_1$, and thus

$$\Delta E = -\frac{\beta}{\omega_1}\,E(t).$$

All this can now be substituted into our expression for Q (Eq. 15.47):

$$Q = \frac{E(t)}{\dfrac{\beta}{\omega_1}\,E(t)} = \frac{\omega_1}{\beta} \approx \frac{\omega_0}{\beta} \tag{15.50}$$

for the lightly damped oscillator where $\omega_1 \approx \omega_0$. A good spring oscillator might have a Q of 100, but a low-temperature torsion oscillator for superfluid helium measurements can be designed to have Q values on the order of 10^6. Specially fabricated superconducting microwave cavities have electrical Q factors as high as 10^{12}.

EXAMPLE 1   The Q of the oscillating system shown in Fig. 15.26 can easily be calculated using Eq. 15.50:

$$Q = \frac{\omega_1}{\beta} = \frac{2.64 \text{ s}^{-1}}{.232 \text{ s}^{-1}} = 11.4.$$

EXAMPLE 2   A spring attached to a fishing weight makes a crude damped harmonic oscillator. The damping mechanism for this system is primarily the internal friction of the spring. Assume that we measured the amplitude of oscillation of this system to decrease by about a factor of 2 in 5 cycles and the period of oscillation to be 1.4 s. Calculate the Q of the oscillator.

At any time t, the amplitude is given by Eq. 15.39. The ratio of the amplitude at $t = 0$, to that 5 cycles later (or $5 \times 1.4 = 7$ seconds) is 2, or

$$\frac{\text{amplitude at } t = 0}{\text{amplitude at } t = 7} = \frac{Ae^0}{Ae^{-(\beta/2)7}} = 2 \qquad \text{or} \qquad e^{7\beta/2} = 2.$$

Taking the natural logarithm of both sides gives us

$$\frac{7}{2}\beta = \ln 2 \qquad \text{or} \qquad \beta = \frac{2}{7}\ln 2 = .198 \text{ s}^{-1}.$$

The angular frequency of oscillation ω_1 or ω_0 is

$$\omega_0 \approx \omega_1 = \frac{2\pi}{T} = \frac{2\pi}{1.4\text{ s}} = 4.49\text{ s}^{-1}.$$

The Q can readily be calculated from Eq. 15.50:

$$Q = \frac{\omega_0}{\beta} = \frac{4.49\text{ s}^{-1}}{.198\text{ s}^{-1}} = 23.$$

15.13 Undamped Forced Oscillation

WE HAVE only analyzed free-standing oscillating machines, that is, machines that have been given some initial displacement or velocity and then allowed to move free of external influences. Interesting phenomena occur when the oscillators are coupled to harmonically oscillating driving forces, such as the one shown for the spring–mass system in Fig. 15.28. To be specific, we will assume that the driving

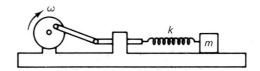

Fig. 15.28

force can be represented by the equation

$$F_{\text{driving}} = F_0 \cos \omega t, \tag{15.51}$$

where F_0 (in newtons) is the maximum driving force applied during the cycle and ω is the angular frequency of the applied force. Where there is *no damping* mechanisms, the differential equation describing the motion is

$$-kx + F_0 \cos \omega t = m\frac{d^2x}{dt^2}$$

or

$$\frac{d^2x}{dt^2} + \frac{k}{m}x = \frac{F_0}{m}\cos \omega t. \tag{15.52}$$

Once again we will use physical arguments to guess the solution and then check to see if it is correct. If we assume a harmonic solution of the form $x(t) = A \cos \omega t$ and substitute it into Eq. 15.52, both terms on the left side will have $\cos \omega t$ dependence and that is the same time dependence as the term on the right. That solution appears to be a good bet; let's try it.

$$-A\omega^2 \cos \omega t + \frac{k}{m}A \cos \omega t \stackrel{?}{=} \frac{F_0}{m}\cos \omega t, \tag{15.53}$$

which is a valid solution providing that

$$-A\omega^2 + \frac{k}{m} A = \frac{F_0}{m}$$

or

$$A = \frac{F_0}{k - m\omega^2} = \frac{F_0}{m} \frac{1}{(\omega_0^2 - \omega^2)}. \qquad (15.54)$$

The solution is then

$$x(t) = \frac{F_0}{m} \frac{1}{(\omega_0^2 - \omega^2)} \cos \omega t. \qquad (15.55)$$

What we have done is correct but not complete, because the solution obtained has no arbitrary constants—no way to allow for particular initial conditions.

The *complete* solution contains an additional term. The solution is

$$x(t) = \frac{F_0}{m} \frac{1}{(\omega_0^2 - \omega^2)} \cos \omega t + B \cos (\omega_0 t + \phi), \qquad (15.56)$$

in which the initial conditions will stipulate the proper values for B and ϕ. The second term in Eq. 15.56 you recognize as the solution for the undamped, undriven, simple harmonic oscillator. In real mechanical or electrical systems, which have some damping, this new term would decay to zero with time (B would not be a constant but an exponentially decreasing function of time). Since no damping term was included in the differential equation, the solution is that of a harmonic oscillator at the system's characteristic frequency, which never damps out. We disregard this term because in real systems it decays away and focus our attention on the solution given in Eq. 15.55.

Even a quick look at this equation tells us that something very new is going on. If we vary the motor speed on the drive mechanism so that its angular frequency ω approaches the characteristic frequency of the spring–mass oscillator ω_0, the amplitude of oscillation of the mass $x(t)$ becomes large without limit. This is clearly not physical. If we slowly vary the angular frequency of the driving force and measure the steady-state amplitude of the motion at each frequency, our theory predicts that it should behave as shown in Fig. 15.29.

The condition when the drive frequency ω is equal to the characteristic frequency ω_0 is called RESONANCE. The characteristic or natural frequency is often referred to as the RESONANCE FREQUENCY of the system. Our theoretical prediction that the amplitude of the motion will become large without limit cannot represent the real mechanical system. Nevertheless, our simple theory is telling us something important. A small but steady application of a driving force, at the right frequency, can produce large-amplitude oscillations of a simple harmonic oscillator. Infinite amplitudes appear in the solution because we did not put any dissipative terms in the original differential

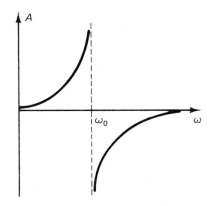

Fig. 15.29

equation. We will correct that defect in the theory in the next section.

Figure 15.29 also shows that the amplitude switches sign when the drive frequency ω exceeds the resonance frequency ω_0. How should we interpret this? Positive amplitudes mean that the displacement $x(t)$ varies as cos ωt, as does the applied force. Negative amplitudes implies that the displacement responds as $-$cos ωt even when the drive force is applied as cos ωt. That is, the driving force and the mass displacement are 180° or π radians *out of phase*. [Note that cos ($\omega t + \pi$) = $-$cos ωt.] This is an essential element of all harmonically driven oscillating systems. The response of the system undergoes a 180° phase shift as the driving forces' frequency is swept through the resonant frequency.

15.14 Damped Forced Oscillation

NOW AT LAST we can put everything together: a harmonic oscillating system, a harmonic driving force, and a velocity-dependent damping mechanism. This is mathematically the most complicated situation we have encountered so far, yet it is also the model that best represents real machines. Indeed, this model has extremely wide applications, including the vibrations of automobiles with poor shock absorbers, the devastation of the Tacoma Narrows Bridge due to a mild wind (see Fig. 15.30), vibration isolation of an optical bench for laser

Fig. 15.30 The Tacoma Narrows Bridge twisting in the wind

physics experiments, as well as the design of electrical antenna, seismic verification of nuclear test ban agreements, stimulated emission of radiation in lasers, infrared absorptions of hydrogen ions trapped in KCl (Fig. 15.31), and many of the fundamental mechanisms in spectroscopy. The list could go on, but you get the point. This is an important phenomenon, worth understanding.

Most of the work necessary to deal with this new model has already been done, so what we must do now is put it all together and see what emerges. If we add a velocity-dependent viscous force, $F_{\text{drag}} = -\alpha v$ to Eq. 15.52, we get

$$-kx - \alpha v + F_0 \cos \omega t = m \frac{d^2x}{dt^2}. \qquad (15.57)$$

Dividing by m, rearranging terms, and recalling that $\beta = \alpha/m$ and $\omega_0^2 = k/m$, the equation takes the form

$$\frac{d^2x}{dt^2} + \beta \frac{dx}{dt} + \omega_0^2 x = \frac{F_0}{m} \cos \omega t. \qquad (15.58)$$

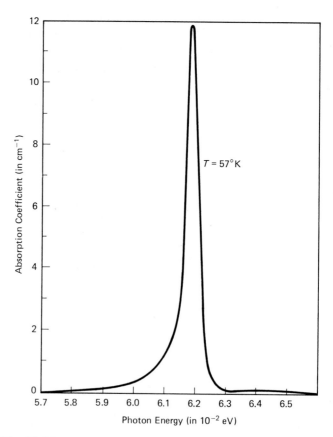

Fig. 15.31

This equation has a fundamental difference from the differential Eq. 15.52 for undamped motion, in that it has a term in dx/dt on the left side. A solution of the form of $x(t) = A \cos \omega t$ would clearly be inappropriate, since it would be differentiated into a sine function by the damping term while all others vary as $\cos \omega t$. The correct guess is

$$x(t) = A \cos (\omega t + \phi),$$ (15.59)

where one can show by direct substitution that it is a solution provided that

$$A = \frac{F_0}{m} \left[\frac{1}{(\omega^2 - \omega_0^2)^2 - \omega^2 \beta^2} \right]^{1/2} \quad \text{and} \quad \phi = \tan^{-1} \frac{\beta \omega}{\omega^2 - \omega_0^2}.$$ (15.60) (15.61)

Maybe the best way for you to get a handle on these solutions is to carefully examine the plots of the amplitude A and phase ϕ as a function of the driving frequency ω (Fig. 15.32). Here I have plotted two cases: I) heavy damping, where $\beta/\omega_0 \approx 1$, and II) light damping, where $\beta/\omega_0 \ll 1$. Both cases clearly show the resonance phenomenon, but the lightly damped system has the most dramatic change of amplitude at or near the natural resonance frequency. The relative

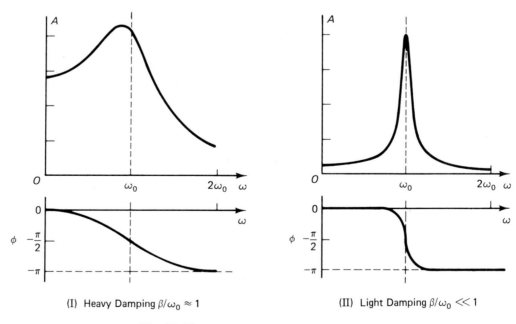

(I) Heavy Damping $\beta/\omega_0 \approx 1$ (II) Light Damping $\beta/\omega_0 \ll 1$

Fig. 15.32

phase of the driving force to the response of the mass changes a total of 180° over the entire frequency range, but for light damping most of the phase shift occurs over a small frequency interval.

The resonance phenomenon is quite remarkable. A small (but persistent) driving force can create an enormous response. How can this happen? How is it possible, for example, for the amplitude of vibration of our mass on the spring to acquire large-amplitude excursions with a small-amplitude driving force?

The answer, of course, lies in the very nature of resonant systems. These systems are capable of storing energy. When they are subjected to a steady oscillating driving force, during each cycle energy not only flows back and forth between kinetic and potential, but additional energy is added to the system by the driver. Without the driver, the system's energy decays away; with the driving force, the energy of the system can increase. The increase is reflected in the increase in amplitude. The amplitude increase is limited only by the damping mechanism. As the amplitude increases for a given frequency, the velocity of the mass must also increase, since the mass goes through one complete cycle in the same time, T, but must travel a greater distance. The increase in velocity produces an increase in dissipation, because the damping force is directly proportional to the velocity. Ultimately, the energy added equals the energy dissipated during each cycle. That is why this model does *not* give infinite amplitude, but does predict large increases in the response amplitude when the system is driven at or near resonance.

You might suspect that the shape of the resonance response curve is somehow related to the Q of the oscillator, since the Q factor is the ratio of the energy stored to the energy dissipated per radian. In fact, the Q of the system can be obtained from the response curve. The energy averaged over one complete cycle of the oscillating systems is given by the expression (see the appendix to this chapter)

$$\langle E \rangle = \frac{F_0^2}{8m} \left[\frac{1}{(\omega - \omega_0)^2 + \frac{\beta^2}{4}} \right], \tag{15.62}$$

where the quantity in square brackets is called the Lorentzian line shape function. The Lorentzian is characteristic of lightly damped harmonically driven oscillating systems of all types, such as electrically tuned filters, piezoelectric quartz oscillators, diatomic molecules, and many others. In Fig. 15.33 we have plotted the Lorentzian function for several values of $\beta = \alpha/m$.

The maximum height of the Lorentzian curve occurs when $\omega = \omega_0$ and has the value $\frac{4}{\beta^2}$. Its amplitude is reduced to one-half of the

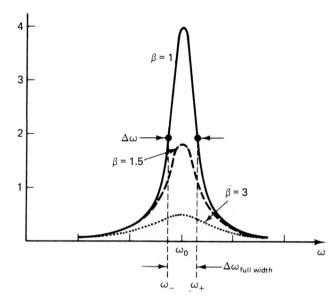

Fig. 15.33

maximum value when the two terms in the denominator are equal, or

$$(\omega - \omega_0)^2 = \frac{\beta^2}{4} \qquad \text{or} \qquad \omega - \omega_0 = \pm\frac{\beta}{2}. \qquad (15.63)$$

The *full width* at *half-maximum* amplitude for the average energy is exactly equal to the damping coefficient,

$$\text{FWHM} = \Delta\omega_{\text{full width}} = \beta = \frac{\alpha}{m}. \qquad (15.64)$$

From Eq. 15.50 we know that $Q = \omega_0/\beta$, so combining this with the measurement of the line width, we get

$$Q = \frac{\omega_0}{\Delta\omega} = \frac{\text{resonant frequency}}{\text{full-frequency line width at}} . \qquad (15.65)$$
$$\text{half maximum of average energy}$$

This method of measuring Q is particularly useful in electrical circuits. High-Q filters are often used to eliminate extraneous signals in electronic instrumentation, since the filter has a significant response only when it is driven at or very near the resonance frequency. Atomic systems can be made whose Q is as high as 10^9. They define a time standard which is much more precise than time as defined by either the earth's rotation about its axis or any mechanical pendulum.

IF I had to pick the one experimental technique that has made the most important contributions to all of science, including physics, chemistry, geology, astronomy, and biology, I would say, without hesitation, "spectroscopy." What is spectroscopy? The best way to explain this technique is to examine a diagram of a "generic" spectroscopy experiment, such as the one shown in Fig. 15.34.

There are three basic components in any spectroscopy experiment: (1) a source of electromagnetic radiation, (2) a sample to be studied, and (3) a detector of the transmitted (sometimes reflected, scattered, or emitted) electromagnetic radiation. A typical experiment might involve a source of infrared radiation, whose frequency can be varied, focused on a diatomic gas sample, such as HCl or CO. The transmitted light is detected by a semiconducting device. The detector produces a current proportional to the incident light intensity. The infrared "light" is in reality propagating electromagnetic oscillating fields, and when the frequencies of these electric fields are at or near the resonance frequency of the stretch modes of these diatomic molecules, the molecules strongly absorb the radiation.* This reduces the radiation incident on the detector for that particular frequency and the experimenter observes a decrease in current output of the detector. That is what is observed in sodium chloride films as shown in Fig. 15.35. Both the frequency and the shape of the absorption signal give valuable information about the molecules and the structure of the sample.

Spectroscopy is certainly not limited to a small range of frequencies such as the visible or infrared frequencies, but is carried out over the entire electromagnetic spectrum. Magnetic resonance spectroscopy is done from the low radio-frequency range of a few kilohertz to high microwave frequencies of 100 GHz (10^{11} Hz), far-infrared spectroscopy studies semiconductors, superconductors, and other solid materials; visible spectroscopy of every kind of material is used in many areas of science. Vacuum ultraviolet through x-ray spectroscopy is also important. Astronomy would not exist, as we know it today without the use of all kinds of spectroscopy. We know the chemical composition of our star (the sun) and the other stars in our galaxy and universe, as well as the giant gas clouds that fill the space in between the stars, because of the resonances observed in the spectra of the electromagnetic radiation coming from these or passing through these objects.

An unusual form of spectroscopy developed by Rudolph Mössbauer, and named after him, is capable of frequency selections of about 1 part in 10^{15} or Q values of 10^{15}! Mössbauer spectroscopy

---

* The frequency range of infrared radiation is 10^{11} to 10^{14} Hz. The human eye cannot detect light in this range.

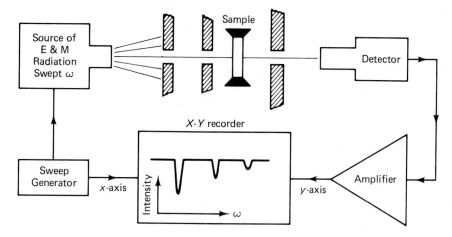

Fig. 15.34

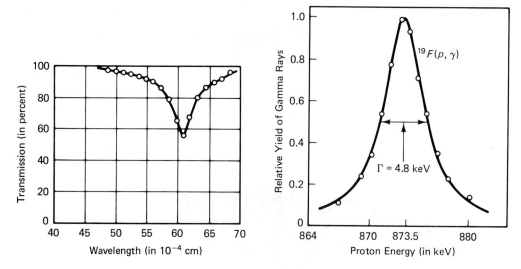

Fig. 15.35

Fig. 15.36

uses the same material, in effect, for both the source and the sample, and achieves this extremely high resolution, or narrow line, in the x-ray frequency range (Fig. 15.36). This is such a sharp resonance that R. V. Pound and G. Rebka were able to measure the predicted small frequency shifts of light as it "falls" three stories (30 m) in the earth's gravitational field.* This remarkable experiment verified one of the predictions of Albert Einstein's theory of general relativity.

---

* R. V. Pound and G. Rebka, Jr., *Physical Review Letters*, **4**, 337 (1960).

15.16 Summary of Important Equations

Simple harmonic oscillator:

$$\omega_0 = \left(\frac{k}{m}\right)^{1/2} \qquad \text{(independent of amplitude)}$$

$$T = \frac{2\pi}{\omega_0} \qquad f = \frac{1}{T}$$

$$x(t) = A \cos(\omega_0 t + \phi)$$

$$v(t) = -\omega_0 A \sin(\omega t + \phi)$$

$$a(t) = -\omega_0^2 A \cos(\omega t + \phi)$$

Effective "spring" constant:

$$k = \frac{d^2 U}{dr^2}\bigg|_{r=0}$$

Simple pendulum:

$$\omega_0 = \left(\frac{g}{l}\right)^{1/2}$$

Torsion pendulum:

$$\omega_0 = \left(\frac{k}{I_z}\right)^{1/2}$$

Physical pendulum:

$$\omega_0 = \left(\frac{r_{\text{CM}} Mg}{I_0}\right)^{1/2}$$

Two-body oscillator:

$$\omega_\mu = \left(\frac{k}{m_\mu}\right)^{1/2} \qquad \text{where } m_\mu \equiv \frac{m_1 m_2}{m_1 + m_2} \quad \text{(reduced mass)}$$

Total energy:

$$E_{\text{total}} = \frac{1}{2} kA^2 = \frac{1}{2} m\omega^2 A^2 \qquad \text{(SHM)}$$

Damped harmonic motion:

$$x(t) = Ae^{-(\beta/2)t} \cos(\omega_1 t + \phi) \quad \text{where } \omega_1 = \left[\omega_0^2 - \frac{\beta^2}{4}\right]^{1/2} \text{ and } \beta = \frac{\alpha}{m}$$

$$E(t) = \frac{1}{2} kA^2 e^{-\beta t} = E_0 e^{-\beta t}$$

$$Q = \frac{\omega_1}{\beta} \approx \frac{\omega_0}{\beta}$$

Forced oscillations:

$$x(t) = \frac{F_0}{m} \frac{1}{\omega_0^2 - \omega^2} \cos \omega t$$

Forced damped oscillations:

$$x(t)\, A \cos(\omega t + \phi) \quad \text{where } A = \frac{F_0}{m}\left[\frac{1}{(\omega^2 - \omega_0^2)^2 - \omega^2\beta^2}\right]^{1/2}$$

and

$$\phi = \tan^{-1}\frac{\beta\omega}{\omega^2 - \omega_0^2}$$

$$\langle E \rangle = \frac{F_0}{8m}\left[\frac{1}{(\omega - \omega_0)^2 + \frac{\beta^2}{4}}\right] \quad \text{(for light damping, Lorentzian)}$$

$$Q = \frac{\omega_0}{\Delta\omega}$$

$$\Delta\omega_{\text{full width}} = \beta$$

Appendix: **Lightly Damped Driven Harmonic Oscillation**

WE WILL demonstrate that the average energy of a lightly damped driven harmonic oscillator can be well represented by a Lorentzian line shape function, as given in Eq. 15.62. The steady-state solutions for the mass displacement of this system are given in Eq. 15.59, $x = A \cos(\omega t + \phi)$, and the velocity can be calculated by direct differentiation as

$$v = -A\omega \sin(\omega t + \phi).$$

We can calculate both the kinetic energy and the potential energy at any time t as

$$PE(t) = \frac{1}{2}kx^2 = \frac{1}{2}kA^2 \cos^2(\omega t + \phi)$$

$$KE(t) = \frac{1}{2}mv^2 = \frac{1}{2}mA^2\omega^2 \sin^2(\omega t + \phi),$$

and the total energy as the sum

$$E(t) = \frac{1}{2}A^2[m\omega^2 \sin^2(\omega t + \phi) + k \cos^2(\omega t + \phi)].$$

The average value of the square of either the sine or the cosine function *over one complete cycle* is $\frac{1}{2}$, no matter what the argument of the function. You can check this by carrying out the integration necessary to calculate this average value:

$$\langle \cos^2 \theta \rangle = \frac{1}{2\pi} \int_0^{2\pi} \cos^2 \theta \, d\theta,$$

or by making a plot of $\cos^2 \theta$ and examining it. Taking the average over one complete cycle of the expressions for the energy $E(t)$, we obtain

$$\langle E(t) \rangle = \frac{1}{2} A^2 \left(m \frac{\omega^2}{2} + \frac{k}{2} \right) = \frac{1}{4} mA^2(\omega^2 + \omega_0^2).$$

Substituting our expression for the amplitude equation 15.60, we can see how the average energy depends on the driving frequency ω:

$$\langle E(\omega) \rangle = \frac{F_0^2}{4m} \frac{\omega^2 + \omega_0^2}{(\omega^2 - \omega_0^2) - \omega^2 \beta^2}.$$

This equation is correct but not in the form we promised. We need to make an additional approximation. For light damping where $\beta \ll \omega_0$, all the "action" occurs at resonance and the average energy is very small except at resonance. It is then a reasonable approximation to replace ω by ω_0 in every term *except* in the resonant denominator $(\omega^2 - \omega_0^2)$ since this term varies rapidly at resonance. Thus we can simplify the expression above:

$$\langle E(\omega) \rangle = \frac{F_0^2}{2m} \frac{\omega_0^2}{(\omega^2 - \omega_0^2)^2 - \omega_0^2 \beta^2}.$$

This can be further simplified by the following approximation:

$$\omega^2 - \omega_0^2 = (\omega - \omega_0)(\omega + \omega_0) \approx 2\omega_0(\omega - \omega_0),$$

where $\omega = \omega_0$ has been substituted for in the sum term. Then

$$\langle E(\omega) \rangle = \frac{F_0^2}{2m} \frac{\omega_0^2}{4\omega_0^2(\omega - \omega_0)^2 - \omega_0^2 \beta^2} = \frac{F_0^2}{8m} \left[\frac{1}{(\omega - \omega_0)^2 - \frac{\beta^2}{4}} \right].$$

The term in brackets is the Lorentzian line shape function.

Chapter 15   QUESTIONS

1. How much does the frequency of a simple harmonic oscillator change when the spring constant doubles; when the mass doubles?

2. What happens to the period of oscillation of a simple harmonic oscillator when the amplitude doubles?

3. Imagine a spring whose spring constant you do not know attached to an unknown mass. Is it possible to predict the frequency of oscillation simply by measuring the extension of the spring when the mass is attached?

4. How far will a simple harmonic oscillator travel in two complete periods if its amplitude is A?

5. How much does the frequency of a simple pendulum change when the mass is doubled?

6. Assume that the amplitude of a simple harmonic oscillator changes by a factor of 2. How much do the following quantities change:

 (a) Frequency?

 (b) Period?

 (c) Maximum velocity?

 (d) Maximum acceleration?

 (e) Total mechanical energy?

7. Is it possible to specify uniquely the amplitude A and the phase ϕ for a simple harmonic oscillator if only the position at $t = 0$ is known? Explain.

8. Are the displacement and velocity, the displacement and acceleration, the acceleration and velocity, ever in the same direction in simple harmonic motion?

9. Give a qualitative description of the motion of a mass–spring oscillator assuming that the spring obeys Hooke's law but has a nonnegligible mass.

10. Why do cars have shock absorbers? Estimate α from what you recall about how a car responds to an external displacement in the vertical direction.

11. A particle's motion is described by the equation $x(t) = -A \sin \omega t$. What is the position of the particle at $t = 0, \pi/\omega, \pi/2\omega$? What is the phase constant?

12. A simple pendulum is suspended in an elevator. What happens to the period of oscillation when the elevator is

 (a) Accelerating upward?

 (b) Accelerating downward?

 (c) Moving at a relativistic velocity $\frac{2}{3} c$?

13. What happens to the frequency of oscillation of a spring–mass oscillator if the spring is cut in half and reattached to the same mass?

14. Can you construct a simple gadget to trace out sinusoidal traces using a pendulum? Draw it.

15. The potential energy of a spring–mass harmonic oscillating system is 3 J when $x = -A$. Find the total and potential energy of the system when $x = A$, $x = 0$. Find the kinetic energy at $x = 0$.

16. Explain why the frequency of a simple harmonic oscillating pendulum does not depend on its mass, but does for a spring–mass oscillator.

17. If the displacement of a harmonic oscillator can be written in the form $x(t) = A \cos (\omega t + \phi)$, find the phase constants ϕ for each of the three motions graphed in Fig. 15.37.

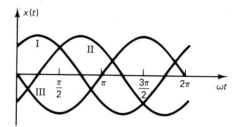

Fig. 15.37

18. Is a bouncing ball executing harmonic or periodic motion?

19. Describe what happens to the period of a simple pendulum when the small-angle approximation breaks down.

20. A hollow sphere is filled first with water, then with mercury, and then left empty. What is the approximate ratio of the frequencies for each case if the sphere is suspended by a fine wire and used as a simple harmonic oscillator?

21. How is it possible for an opera singer to break glass with her voice?

22. Describe what will happen to the frequency of the following simple harmonic oscillators if they were taken to the moon:

 (a) Simple pendulum

 (b) physical pendulum

 (c) mass–spring

 (d) torsion pendulum

23. How would you use a simple harmonic oscillating system to determine the mass of an object if you were deep in outer space?

24. Why is it dangerous to drive an automobile with worn-out shock absorbers?

1. Show that

$$A \sin \omega t + B \cos \omega t = (A^2 + B^2)^{1/2} \cos (\omega t - \phi)$$

and find the value of ϕ in terms of A and B.

2. A mass–spring harmonic oscillating system with spring constant $k = 250$ N/m and mass $m = 1.75$ kg is displaced 7 cm from equilibrium and given an initial velocity in the positive x-direction of 20 m/s. Find the position of the mass 3.5 s later.

3. The frequency of an ideal mass–spring harmonic oscillator is 1.7 Hz. What is the mass of the oscillating block if the spring constant is 22 N/m?

4. A block of mass 1.8 kg executes simple harmonic motion with a maximum amplitude of .22 m and moves with a maximum velocity of 12 cm/s. Find the spring constant k.

5. The period of a mass–spring simple harmonic oscillator is .95 s and its mass is 1.35 kg. Find the spring constant of the system.

6. Calculate the total mechanical energy of the mass–spring harmonic oscillator whose maximum amplitude of oscillation is 4.4 cm with a spring constant of 3.5 N/m.

7. Show that the amplitude of oscillation can be written in terms of the initial displacement of the system x_0 and the initial velocity as

$$A = \left[x_0^2 + \left(\frac{v_0}{\omega_0} \right)^2 \right]^{1/2}.$$

8. An object of mass .025 kg is attached to a spring and set into oscillating harmonic motion. If it is at rest at $t = 0$ but is displaced 55 cm from its equilibrium position and has an acceleration of $-.156$ m/s^2, find the following values.

(a) The spring constant of the system.

(b) The angular velocity of the system.

(c) The velocity of the mass at time $t = .167$ s.

(d) The position of the mass at $t = .167$ s.

9. Two particles α and β, moving with SHM on separate parallel tracks very close to one another, find themselves exactly at $x = 0$ at a time $t = 0$ moving in the positive x-direction (Fig. 15.38). They have the same maximum excursion from equilibrium of 5 cm but move with different angular frequencies $\omega_\alpha = 16$ s^{-1} and $\omega_\beta = 15$ s^{-1}.

(a) Find their *relative* velocities at $t = 0$.

(b) How far apart will they be separated (in x-direction) at the time $t = .46$ s, and what is their *relative* velocity at that time?

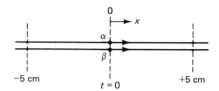

Fig. 15.38

10. A spring with a length L obeys Hookes' law and hangs freely in the vertical direction. A mass m is attached which stretches the spring to a length $L + B$ (Fig. 15.39). A second mass m is dropped from a height h onto the attached mass and sticks. Find the period, amplitude, and maximum displacement from the original equilibrium position of the $2m$ oscillator. What is the new equilibrium position of the oscillator?

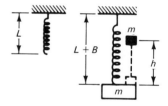

Fig. 15.39

11. A block of mass m slides without friction on an inclined surface shown in Fig. 15.40.

(a) When the mass is attached to the unstretched spring, how much does it extend?

(b) What is the expression for the frequency of oscillation of the system?

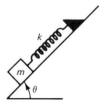

Fig. 15.40

12. A small .6-kg block is placed on top of a piston that is executing simple harmonic motion in the vertical direction. If the frequency of this motion is 6 Hz, find the maximum amplitude of the piston's motion that will permit the block to remain in contact at all times.

13. A bullet of mass m strikes a block of mass M and quickly comes to rest embedded in it (Fig. 15.41). If the frequency of vibration of the block is ω after the collision and the amplitude is A, find the velocity of the bullet before the collision, in terms of A, ω, m, and M.

Fig. 15.41

14. Our idealized spring–mass horizonal oscillator assumed, among other things, that the mass of the spring could be neglected. A more realistic model of this system would consider the spring's mass and its effect on the angular frequency of oscillation. If the spring is uniform over its length L, it would have a uniform mass density m_s/L, where m_s is the total mass of the spring. When the block has a velocity v, the element dx of the spring at position x will have a velocity

$$v_x = \frac{x}{L} v \qquad \text{(Fig. 15.42)}.$$

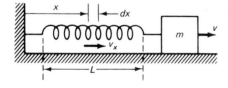

Fig. 15.42

(a) Show that the kinetic energy of the *spring* is

$$KE_s = \frac{m_s v^2}{6}.$$

(b) Show that the angular frequency of the system is given by

$$\omega = \left(\frac{k}{m_s/3 + m}\right)^{1/2},$$

where k is the spring constant. (*Hint:* Compare the expression for the total mechanical energy of an idealized system to the model when the spring's kinetic energy is taken into account.*)

15. Compare the periods of oscillation of a block of mass m connected to two identical ideal springs as shown in Fig. 15.43a and b to the period of oscillation of the same mass hung from one of the springs (Fig. 15.43c).

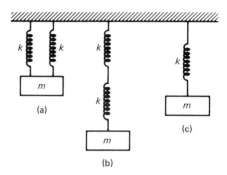

Fig. 15.43

16. Carbon dioxide (CO_2) can be modeled as a simple linear chain of three atoms connected together by springs shown in Fig. 15.44. The springs are repre-

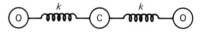

Fig. 15.44

sentations of the atomic bonds that hold the molecule together. It is observed that the system absorbs infrared electromagnetic energy whose frequency is about 3.9×10^{13} Hz. If this absorption corresponds to the motion of the two oxygen atoms on the ends while the carbon atoms remain stationary, what is the effective spring constant k of these atomic bonds?

17. A spring attached to a plate of mass m_1 is compressed by a mass m_2 until it has been shifted from equilibrium by an amount b (Fig. 15.45). The system is then released.

(a) How long will it take for the two masses to separate?

---

* See A. P. French, *Vibrations and Waves* (New York: W.W. Norton, 1971, p. 60, for a complete discussion of this problem.

(b) Calculate the energy of m_2 at the moment of separation.

(c) Write down the equation of motion for the spring–mass system for all times *after* the separation.

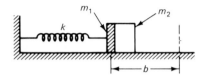

Fig. 15.45

18. A mass M slides along the x-axis on a frictionless surface under the influence of two identical springs of spring constant k (Fig. 15.46). A small piece of putty of mass m is dropped straight down onto M, just as M passes through its equilibrium position, and quickly sticks to it.

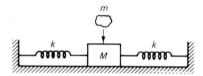

Fig. 15.46

(a) Calculate the new period, amplitude, and total energy of the system in terms of the original values of these quantities and m and M. If the energy has changed, where did it come from or go to?

(b) Repeat the calculation for the case in which the putty is dropped on M just as M reaches its maximum displacement.

19. A block of mass $m = .37$ kg has two 70-cm-long ideal springs attached as shown in Fig. 15.47.

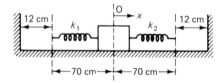

Fig. 15.47

(a) If $k_1 = 4$ N/m and $k_2 = 7$ N/m, find the new equilibrium position of the mass after the springs have been stretched and attached to the side walls.
If the mass is displaced from this new equilibrium position:

(b) Will it execute simple harmonic motion?

(c) What is the period of oscillation of the motion?

20. A block of mass m slides on a frictionless surface and is connected by two springs whose constants are k_1 and k_2 as shown in Fig. 15.48. At equilibrium both springs are in their unstretched condition. Find the frequency of simple harmonic oscillation of the system.

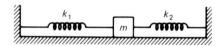

Fig. 15.48

21. An oscilloscope visually displays the electrical signals connected to its two inputs x and y (Fig. 15.49). It does this by deflecting an electron beam

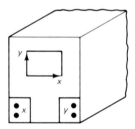

Fig. 15.49

inside its cathode ray tube on two mutually perpendicular axes in such a way that the time dependence of the electron deflection is given by the following equation:

$$x(t) = A_x \cos(\omega t + \phi)$$

$$y(t) = A_y \cos \omega t.$$

Neatly draw the electron path (which is observed on the screen) for the following conditions:

(a) $A_x = 2A_y$; $\phi = 0$, $\phi = \dfrac{\pi}{2}$, $\phi = -\dfrac{\pi}{2}$

(b) $A_x = A_y = A$; $\phi = 30°$, $\phi = -30°$

22. Two college students watch a bottle bobbing up and down in a pool of muddy water on a dark night (Fig. 15.50). One of the students, a physics major,

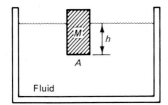

Fig. 15.50

bets his friend that he can calculate how much of the bottle is under water using only his watch. The skeptical friend doesn't believe him and so challenges him with a $5 bet. The physics student times the bottle moving up and down 10 times; it took 7.8 s. After a brief calculation on his hand calculator, he tells his friend that a 15-cm cylindrical bottle is submerged in the water. Who wins the $5? [*Hint:* Archimedes' principle states that a floating object experiences a buoyant (upward) force equal to the weight of the fluid that the floating object displaced. The weight of the fluid is equal to its density ρ times the volume of the fluid.]

23. Show that if the fluid in a vertical U-shaped tube is displaced a small amount, it will oscillate with the same period as that of an ideal pendulum of length equal to one-half of the total length of the water column in the tube.

24. A small block of mass m slides without friction on a slope whose curvature is given by the equation $y = \alpha x^2$ (Fig. 15.51). The mass is displaced a small distance y_0 and let go.

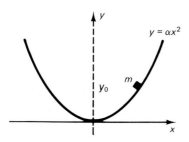

Fig. 15.51

(a) What are the units of α?
(b) What is the period of motion?

25. Suppose you wish to make a pendulum clock that keeps time to an accuracy of better than 1 s per day. The major source of timing error comes from the change in the length of the metal pendulum due to variations in the temperature of the environment. Metals change length by thermal expansion and contraction according to the equation $\Delta l = \alpha l \, \Delta T$, where Δl is the change in length, l is the length, ΔT is the change in temperature, and α is the coefficient of linear expansion. If you take $\alpha = 1.5 \times 10^{-5} \, °C^{-1}$, find the maximum allowed temperature variation that will permit the clock to keep the acceptable accuracy.

26. A disk with moment of inertia I_1 (about the axis of rotation) is suspended from a light torsion rod as shown in Fig. 15.52. The system oscillates with SHM with a period T_1. A second object with moment of inertia I_2 is placed on the first. If the period of the new combined system is T_2, find the value of I_2 in terms of I_1, T_1, and T_2.

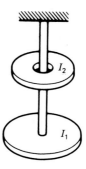

Fig. 15.52

27. A uniform disk of radius R is supported by a pivot at point o, a distance a from its center (Fig. 15.53).

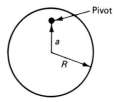

Fig. 15.53

(a) If the disk is set into small-amplitude oscillation, show that the frequency of oscillation is given by

$$\omega = \frac{ga}{\frac{1}{2}R^2 + a^2}.$$

(b) For what value of a does one obtain the shortest period of oscillation? What is the value of
a for the longest period?

28. Find the frequency of oscillation of a meterstick
that has been made into a physical pendulum by pivoting the stick through a hole placed at the 15-cm
mark (Fig. 15.54).

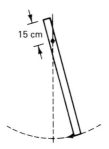

Fig. 15.54

29. Show that the period of oscillation of a long, very
thin cone suspended by its apex on a frictionless bear-
ing is given by the expression $2\pi\left(\dfrac{4l}{5g}\right)^{1/2}$ independent of the base of the cone (as long as it is very
small compared to the length; Fig. 15.55).

Fig. 15.55

30. The pendulum in Fig. 15.56 is a uniform bar of
length l and mass m attached to a uniform sphere of
mass M and radius R. The pendulum is supported by
a bearing at the end of the rod. Find the expression
for the period of small-amplitude oscillation of this
pendulum.

31. One end of a uniform rod of mass m and length L
is attached to a wall by a frictionless pivot (Fig.
15.57). The other end is attached to a spring of force
constant k. This spring is attached to another wall,
which is at right angles to the first. The rod and

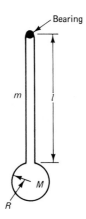

Fig. 15.56

spring lie on a frictionless *horizontal* table. In the
equilibrium position (when the spring is unstretched), the rod and spring are at right angles.

(a) A string is attached to the center of mass of the
rod and a force F_1 is applied by pulling on the
string in the positive x-direction until the end
of the rod attached to the spring is displaced a
small distance x. Find the force F_1 and the
force F_2 at the pivot.

(b) The string is now cut and the rod executes
simple harmonic motion about its equilibrium position. What is the torque about the
axis through the rod's pivot? Express your
answer in terms of θ, the angle the rod makes
with its equilibrium position. Use the small-
angle approximation. Include the direction
of the torque.

(c) Write down the torque equation of motion
and find the period of the rotational motion of
the rod about its pivot.

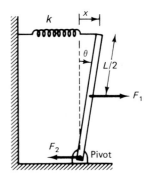

Fig. 15.57

32. A meterstick of mass .3 kg is pivoted about its center of mass and attached to an ideal spring ($k = 4$ N/m) at the 25-cm position (Fig. 15.58). The system is at equilibrium when the meterstick is in the horizontal position. Find the frequency of oscillation for small-amplitude oscillation.

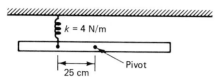

Fig. 15.58

33. A solid cylindrical gear attached to a spring of constant k executes rotational as well as translational motion as it travels back and forth from its equilibrium position (Fig. 15.59).

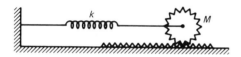

Fig. 15.59

(a) Show that the period of oscillation of the system is given by

$$T = 2\pi\left(\frac{3M}{2k}\right)^{1/2}.$$

(b) If the spring constant is 1.8 N/m and the spring is compressed .06 m, calculate the rotational and translational kinetic energy of the system at the instant it passes through the equilibrium position. Explain why it does not depend on the mass of the gear.

34. A stiff piece of uniform cross section and uniform-density wire is shaped into a semicircular arc structure and supported by a frictionless pivot at point O (Fig. 15.60). If the structure is displaced a small angle

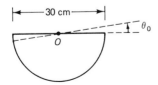

Fig. 15.60

θ_0 from its equilibrium position and allowed to oscillate in the vertical plane, find the frequency of the oscillating motion.

35. Show that the expression

$$x(t) = Ae^{-(\beta/2)t} \cos(\omega_1 t + \phi),$$

where $\omega_1 = \left(\omega_0^2 - \frac{\beta^2}{4}\right)^{1/2}$,

is a solution to the damped harmonic oscillator differential equation

$$\frac{d^2x}{dt^2} + \beta\frac{dx}{dt} + \omega_0^2 x + 0$$

when $\omega_0 > \beta/2$.

36. A pendulum reduces its angular amplitude of oscillation from 12° to 4° in 15 minutes. If the pendulum is .85 m long, find the value of β for this system.

37. A grandfather clock has a pendulum of mass 1.15 kg whose length is *equivalent* to a simple pendulum of length .993 m. The various frictional forces reduce its amplitude of oscillation by a factor of 2 in 16.5 minutes.

(a) Calculate the value of the damping factor α.

(b) How much power is needed to keep the pendulum oscillating with an amplitude of 9° to make up for the frictional losses?

38. A .16-kg block is attached to a spring whose constant is 65 N/m but is subjected to a velocity dependent resistive force $-\alpha v$.

(a) If the damped angular frequency is .87 times the undamped frequency, find the value of the damping coefficient α.

(b) What is the Q of the system?

(c) What fraction of the original amplitude of the system remains after 10 cycles?

39. The amplitude of oscillation decreases 2% during each cycle of a damped harmonic oscillator. What fraction of the energy is lost in each cycle?

40. A child's pendulum clock uses a .275-kg falling "weight" to supply the energy dissipated in the friction of the system. The pendulum has a mass of 15 g, a length of 25 cm, and a period of oscillation of 1 s. The angular amplitude of oscillation is 7° and the weight falls 0.7 m each day. Calculate the Q of the pendulum. If the weights were replaced by a battery whose total energy output was 2 J, how long would the clock run?

41. The natural logarithm of the ratio of successive maximum amplitudes of displacement (over one full period) is defined as the logarithmic decrement δ.

 (a) Show that $\delta Q = \pi$.

 (b) For a particular oscillator where $\delta = .018$, $m = 3.5$ kg, and $f = 1.2$ Hz, find the spring constant k and the damping constant α.

42. A particle of mass .003 kg executes simple harmonic motion when it is attached to a Hooke's law spring. At $t = 0$ the following parameters are observed for the system $x(0) = .13$ m, $v_x(0) = -4.5$ m/s, and $a_x(0) = -12$ m/s^2.

 (a) What is the angular frequency of the particle?

 (b) Calculate the amplitude of the particle.

 (c) What is the total energy of the oscillator?

 (d) Write the expression for $x(t)$ and $a(t)$.

43. Verify by direct substitution that the solution $x(t) = A \cos(\omega t + \phi)$ is indeed the solution to the driven damped harmonic oscillation differential equation 15.59 if

$$A = \frac{F_0}{m} \left[\frac{1}{(\omega^2 - \omega_0^2)^2 - \omega^2 \beta^2} \right]^{1/2}$$

$$\phi = \tan^{-1} \frac{\beta \omega}{\omega^2 - \omega_0^2}.$$

44. A 3.5-g simple pendulum is observed to slowly decrease its amplitude of oscillation. The amplitude *along the arc* is plotted as function of time in Fig. 15.61. Calculate the following values.

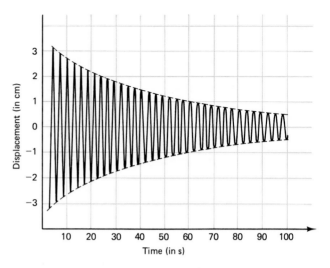

Fig. 15.61

 (a) The total energy of this system at $t = 10$ s.

 (b) The angular "frequency" of oscillation.

 (c) The Q of the oscillator.

 (d) The velocity-dependent damping coefficient α.

 (e) The total energy and the amplitude of the oscillator at $t = 250$ s.

 (f) How much energy would have to be put in every cycle so as to maintain the level of oscillation?

APPENDIX A Symbols, Constants, Units, Conversion Factors, Formulas

I. Mathematical Symbols

A	vector		$\cdot$	dot product
$\equiv$	is defined as		Δx	change in x
$\neq$	is not equal		$n!$	$n(n-1)(n-2)\cdots 1$
$\approx$	approximately equal		lim	limit
$\overset{?}{=}$	is the left side equal to the right side?		$\Delta t \to 0$	Δt approaches zero
$>$	is greater than		$\dfrac{dx}{dt}$	derivative of x with respect to t
$<$	is less than		$\dfrac{\partial x}{\partial t}$	partial derivative of x with respect to t
$\geq$	is equal to or greater than		$\gg$	much greater than
$\leq$	is equal to or less than		$\ll$	much less than
$\pm$	plus or minus		α	proportional to
ijk	unit vectors		ln	logarithm to base e
$\langle \; \rangle$	average value		$\hat{\boldsymbol{\theta}}, \hat{\mathbf{r}}$	polar unit vectors
$\times$	cross product			

II. The Greek Alphabet

Alpha	A	α		Pi	Π	π
Beta	B	β		Rho	P	ρ
Gamma	Γ	γ		Sigma	Σ	σ
Delta	Δ	δ		Tau	T	τ
Epsilon	E	ε		Upsilon	Y	υ
Zeta	Z	ζ		Phi	Φ	ϕ
Eta	H	η		Chi	X	χ
Theta	Θ	θ		Psi	Ψ	ψ
Iota	I	ι		Kappa	K	κ
Nu	N	ν		Lambda	Λ	λ
Xi	Ξ	ξ		Mu	M	μ
Omicron	O	o		Omega	Ω	ω

III. Fundamental Constants

Constant	Symbol	Computational Value	Best Value*
Speed of light (vacuum)	c	3.00×10^8 m/s	2.99792458×10^8 m/s
Gravitational constant	G	6.67×10^{-11} m³/s²-kg	6.6726×10^{-11} m³/s²-kg
Electron charge	e	1.60×10^{-19} C	1.602177×10^{-19} C
Electron mass	m_e	9.11×10^{-31} kg	9.109389×10^{-31} kg
Proton mass	m_p	1.67×10^{-27} kg	1.672623×10^{-27} kg
Planck's constant	h	6.63×10^{-34} J-s	6.626076×10^{-34} J-s
Permitivity constant	ε_0	8.85×10^{-12} F/m	$8.85418781762 \times 10^{-12}$ F/m
Boltzmann constant	k	1.38×10^{-23} J/K	1.380658×10^{-23} J/K
Avogadro constant	N_A	6.02×10^{23} mol^{-1}	6.022137×10^{23} mol^{-1}
Mass of neutron	m_n	1.67×10^{-27} kg	1.674928×10^{-27} kg
Mass of muon	m_μ	1.88×10^{-28} kg	1.883532×10^{-28} kg

* *1986 Adjustment of the Fundamental Physical Constants*, CODATA, Bulletin 86, E. R. Cohen and B. N. Taylor.

IV. Astronomical Data

Mass of Earth	5.98×10^{24} kg
Equatorial radius of Earth	6.378×10^6 m
Acceleration of gravity at the Earth's surface	9.807 m/s²
Mass of moon	7.36×10^{22} kg
Mean Earth–moon distance	3.8×10^5 km
Mass of sun	1.99×10^{30} kg
Radius of sun	6.96×10^8 m
Period of moon	2.36×10^6 s
Mean Earth–sun distance	1.5×10^8 km

V. Units of Measure

Prefix	Abbreviation	Multiple
tera-	T	10^{12}
giga-	G	10^{9}
mega-	M	10^{6}
kilo-	k	10^{3}
centi-	c	10^{-2}
milli-	m	10^{-3}
micro-	μ	10^{-6}
nano-	n	10^{-9}
pico-	p	10^{-12}
femto-	f	10^{-15}

VI. Basic and Derived Units

Quantity	Unit	Symbol	In Terms of Basic SI Units
Time	Second	s	
Length	Meter	m	
	Foot	ft	
Mass	Kilogram	kg	
	Slug	slug	
Temperature	Kelvin	K	
	Fahrenheit	F	
Force	Newton	N	m-kg/s^2
	Pound	lb	
Energy	Joule	J	kg-m^2/s^2
	Foot-pound	ft-lb	
	Electron-volt	eV	
Charge	Coulomb	C	
Power	Watt	W	
	Foot-pound/second	ft-lb/sec	
Speed	Meter/second	m/s	
	Foot/second	ft/s	
Frequency	Cycles/second	Hz	s^{-1}
Volume	Liter	l	1000.028 cm^3

VII. Conversion Factors

A. *Length*

Unit	Abbreviation	Value in Meters
1 fermi	1 F	10^{-15} m
1 angstrom	1 Å	10^{-10} m
1 centimeter	1 cm	10^{-2} m
1 inch	1 in.	2.54×10^{-2} m
1 foot	1 ft	.305 m
1 mile	1 mi	1.61×10^{3} m
1 light-year	1 Ly	9.46×10^{15} m
1 parsec	1 pc	3.08×10^{16} m

B. *Mass*

Unit	Abbreviation	Value in Kilograms
1 slug		14.59 kg
1 atomic mass unit	AMU	1.66×10^{-27} kg
1 metric ton		1000 kg
1 kilogram		1000 g

C. *Speed*

Unit	Value in m/s
1 kilometer/hour	.278 m/s
1 mile/hour	.447 m/s
1 foot/second	.305 m/s

D. *Angle*

1 degree = .01745 rad

1 radian = 57.30 degrees

π radian = 180 degrees

1 rev/minute = .1047 rad/s

E. Energy

$$1 \text{ J} = 10^7 \text{ ergs} = 6.25 \times 10^{18} \text{ eV}$$
$$1 \text{ eV} = 1.602 \times 10^{-19} \text{ J}$$
$$1 \text{ Cal} = 4.186 \text{ J}$$
$$1 \text{ Btu} = 1.054 \times 10^3 \text{ J}$$

F. Force

$$1 \text{ N} = 10^5 \text{ dynes} = .2248 \text{ lb}$$
$$1 \text{ lb} = 4.448 \text{ N}$$

G. Area

$$1 \text{ m}^2 = 10^4 \text{ cm}^2 = 10.76 \text{ ft}^2 = 1550 \text{ in}^2$$
$$1 \text{ ft}^2 = .0929 \text{ m}^2 = 144 \text{ m}^2$$
$$1 \text{ in}^2 = 6.452 \text{ cm}^2$$

H. Volume

$$1 \text{ m}^3 = 10^6 \text{ cm} = 6.102 \times 10^4 \text{ in}^3$$
$$1 \text{ liter} = 1000 \text{ cm}^3 = 1.0576 \text{ qt} = .0353 \text{ ft}^3$$
$$1 \text{ gal} = 3.786 \text{ liters} = 231 \text{ in}^3$$

I. Power

$$1 \text{ hp} = 550 \text{ ft-lb/s} = .746 \text{ kW}$$
$$1 \text{ W} = 1 \text{ J/s} = .738 \text{ ft-lb/s} = 1.341 \times 10^{-3} \text{ hp}$$
$$1 \text{ Btu/hr} = .293 \text{ W}$$

J. Pressure

$$1 \text{ Pascal (Pa)} = 1 \text{ N/m}^2 = 10 \text{ dyn/cm}^2 = 7.50 \times 10^{-3} \text{ mm Hg (torr)}$$
$$1 \text{ atmosphere (At)} = 1.013 \times 10^5 \text{ Pa} = 760 \text{ mm Hg (torr)}$$
$$1 \text{ lb/in}^2 \text{ (psi)} = 6.895 \times 10^3 \text{ Pa}$$
$$1 \text{ bar} = 100 \text{ kPa}$$

VIII. Mathematical Formulas

A. Quadratic Formula

If $ax^2 + bx + c = 0$,   then $x = \dfrac{-b \pm \sqrt{b^2 - 4ac}}{2a}$

B. Trigonometric Functions of Angle θ

$\sin \theta = \dfrac{y}{r}$ $\cos \theta = \dfrac{x}{r}$

$\tan \theta = \dfrac{y}{x}$ $\cot \theta = \dfrac{x}{y}$

$\sec \theta = \dfrac{r}{x}$ $\csc \theta = \dfrac{r}{y}$

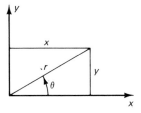

C. Properties of a Triangle

$\alpha + \beta + \gamma = \pi$

$a^2 = b^2 + c^2 - 2bc \cos \alpha$

$b^2 = c^2 + a^2 - 2ca \cos \beta$

$c^2 = a^2 + b^2 - 2ab \cos \gamma$

$\dfrac{a}{\sin \alpha} = \dfrac{b}{\sin \beta} = \dfrac{c}{\sin \gamma}$

For a right triangle $\gamma = \dfrac{\pi}{2}$, $a^2 + b^2 = c^2$

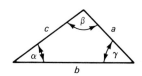

D. Trigonometric Identities

$\sin^2 \theta + \cos^2 \theta = 1$

$\sin 2\theta = 2 \sin \theta \cos \theta$

$\cos 2\theta = \cos^2 \theta - \sin^2 \theta = 2 \cos^2 \theta - 1 = 1 - 2 \sin^2 \theta$

$\sin \theta = \dfrac{e^{i\theta} - e^{-i\theta}}{2i}$ $\cos \theta = \dfrac{e^{i\theta} + e^{-i\theta}}{2}$ where $i = \sqrt{-1}$

$e^{\pm i\theta} = \cos \theta \pm i \sin \theta$

$\sin (\alpha \pm \beta) = \sin \alpha \cos \beta \pm \cos \alpha \sin \beta$

$\cos (\alpha \pm \beta) = \cos \alpha \cos \beta \mp \sin \alpha \sin \beta$

$\tan (\alpha \pm \beta) = \dfrac{\tan \alpha \pm \tan \beta}{1 \mp \tan \alpha \tan \beta}$

$\sin \alpha \pm \sin \beta = 2 \sin \tfrac{1}{2}(\alpha \pm \beta) \cos \tfrac{1}{2}(\alpha \mp \beta)$

E. Taylor's Series

$$f(x_0 + x) = f(x_0) + f'(x_0)x + f''(x_0)\frac{x^2}{2!} + f'''(x_0)\frac{x^3}{3!} + \cdots$$

F. Binomial Expansion

$$(1 + x)^n = 1 + \frac{nx}{1!} + \frac{n(n-1)}{2!}x^2 + \cdots$$

G. Exponential Expansion

$$e^x = 1 + x + \frac{x^2}{2!} + \frac{x^3}{3!} + \cdots$$

H. Logarithmic Expansion

$$\ln(1 + x) = x - \tfrac{1}{2}x^2 + \tfrac{1}{3}x^3 - \cdots$$

I. Trigonometric Expansions (θ in radians)

$$\sin\theta = \theta - \frac{\theta^3}{3!} + \frac{\theta^5}{5!} - \cdots$$

$$\cos\theta = 1 - \frac{\theta^2}{2!} + \frac{\theta^4}{4!} - \cdots$$

$$\tan\theta = \theta - \frac{\theta^3}{3} + \frac{2\theta^5}{15} + \cdots$$

J. Geometry

Circle of radius r:

Circumference $2\pi r$
Area πr^2

Sphere of radius r:

Area $4\pi r^2$
Volume $\tfrac{4}{3}\pi r^3$

Right circular cylinder of length l and radius r:

Area $2\pi r^2 + 2\pi rl$
Volume $\pi r^2 l$

K.   *Determination of Radius of Curvature from Arc*

Construct cord $2l$; then perpendicular bisector h. For the right triangle,

$$R^2 = (R - h)^2 + l^2$$

or
$$R = \frac{l^2 + h^2}{2h}$$

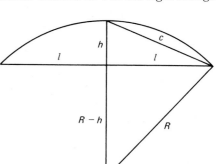

or $c^2 = h^2 + l^2$

so $R = \dfrac{c^2}{2h}$

IX.   Derivatives

1. $\dfrac{dx}{dx} = 1$

2. $\dfrac{d}{dx}(au) = a\dfrac{du}{dx}$

3. $\dfrac{d}{dx}(u + v) = \dfrac{du}{dx} + \dfrac{dv}{dx}$

4. $\dfrac{d}{dx}x^m = mx^{m-1}$

5. $\dfrac{d}{dx}\ln x = \dfrac{1}{x}$

6. $\dfrac{d}{dx}(uv) = u\dfrac{dv}{dx} + v\dfrac{du}{dx}$

7. $\dfrac{d}{dx}e^x = e^x$

8. $\dfrac{d}{dx}\sin x = \cos x$

9. $\dfrac{d}{dx}\cos x = -\sin x$

10. $\dfrac{d}{dx}\tan x = \sec^2 x$

11. $\dfrac{d}{dx}\cot x = -\csc^2 x$

12. $\dfrac{d}{dx}\sec x = \tan x \sec x$

13. $\dfrac{d}{dx}\csc x = -\cot x \csc x$

14. $\dfrac{d}{dx}\tan^{-1} x = \dfrac{1}{1 + x^2}$

15. $\dfrac{d}{dx}\sin^{-1} x = \dfrac{1}{\sqrt{1 - x^2}}$

16. $\dfrac{d}{dx}\sec^{-1} x = \dfrac{1}{x\sqrt{x^2 - 1}}$

APPENDIX B "Elementary" Particle Properties

In the following tables, the masses of the particles are actually mc^2, in units of MeV. The spin is the spin angular momentum in units of $\hbar$. The electric charge is in units of the electron's charge. EM, electromagnetic interactions; W, weak interactions; S, strong interactions. For the properties of the quark, see the end of Chapter 7. Only a few mesons and resonances have been tabulated. For data on more particles, see *Concepts of Particle Physics*, K. Gottfried and V. Weisskopf, Oxford University Press, Vol. I, Appendix I.

I. Leptons

Name	Symbol	Anti-Particle	Mass (MeV)	Electrical Charge	Spin	Average Lifetime (seconds)
Electron	e^-	e^+	.511	-1	1/2	Infinite
Electron's neutrino	ν_e	$\bar{\nu}_e$	0?	0	1/2	Infinite
Muon	μ^-	μ^+	105.66	-1	1/2	2.197×10^{-6}
Muon's neutrino	ν_μ	$\bar{\nu}_\mu$	0?	0	1/2	Infinite
Tau	τ	$\bar{\tau}$	1784	-1	1/2	3.2×10^{-13}
Tau's neutrino	ν_τ	$\bar{\nu}_\tau$	0?	0	1/2	

I. Leptons

Principal Decay Modes			Discovery		Name
Modes	Type	Fraction (%)	Date	By Whom	
			1897	J. J. Thomson, positron by C. Anderson in 1932	Electron
			1956	E. Cowan and F. Reines, using a nuclear reactor	Electron's neutrino
$e^- + \nu_\mu + \bar{\nu}_e$ $e^- + \nu_\mu + \bar{\nu}_e + \gamma$	W W	99%	1937	S. Neddermeyer and C. Anderson in cosmic radiation	Muon
			1962	M. Schwart's team at Brookhaven accelerator	Muon's neutrino
$\mu^- + \nu_\tau + \bar{\nu}_\mu$ $e^- + \nu_\tau + \bar{\nu}_e$ $\pi^- + \nu_\tau$ $\rho + \nu_\tau$	W W W W	19% 16% 11% 22%	1975	M. Perls's team at SLAC	Tau
					Tau's neutrino

II. Baryons

Name	Symbol	Anti-Particle	Mass (MeV)	Electrical Charge	Spin	Average Lifetime (seconds)
Proton	p	$\bar{\text{p}}$	938.26	+1	1/2	Infinite
Neutron	n	$\bar{\text{n}}$	939.55	0	1/2	918
Lambda	Λ^0	$\bar{\Lambda}^0$	1115.6	0	1/2	2.62×10^{-10}
Sigma: Sigma-plus	Σ^+	$\bar{\Sigma}^+$	1189.4	+1	1/2	0.8×10^{-10}
Sigma-minus	Σ^-	$\bar{\Sigma}^-$	1197.3	−1	1/2	1.48×10^{-10}
Sigma-zero	Σ^0	$\bar{\Sigma}^0$	1192.5	0	1/2	6×10^{-20}
Xi Xi-zero	Ξ^0	$\bar{\Xi}^0$	1315.	0	1/2	2.9×10^{-10}
Xi-minus	Ξ^-	$\bar{\Xi}^-$	1321.	−1	1/2	1.6×10^{-10}
Omega-minus	Ω^-	$\bar{\Omega}^-$	1672.5	−1	3/2	$.82 \times 10^{-10}$
Lambda-charm	Λ_c	$\bar{\Lambda}_c$	2282	+1	1/2	2.2×10^{-13}
Resonances	$\Delta(1232)$	$\bar{\Delta}(1232)$	1232	0, ±1, 2	3/2	10^{-23}
	$\Sigma(1660)$	$\bar{\Sigma}(1660)$	1660	0 ± 1	3/2	10^{-22}
	$\Lambda(1815)$	$\bar{\Lambda}(1815)$	1815	0	5/2	10^{-21}
	N(2200)	$\bar{\text{N}}(2200)$	1900–2230	1	5/2	10^{-23}

II. Baryons

| Quark Content | Principal Decay Modes | | | Discovery | | Name |
	Modes	Type	Fraction (%)	Date	By Whom	
uud				1911	E. Rutherford, α-particle scattering off atomic nuclei	Proton
ddu	$p + e^- + \bar{\nu}_e$	W	100	1932	J. Chadwick, beryllium bombarded by α-particles	Neutron
uds	$p + \pi^-$ $n + \pi^0$	W W	65 35	1951	C. Butler's group in cosmic radiation	Lambda
uus	$p + \pi^0$ $n + \pi^+$	W	52 48	1953	G. Tomasini's group in cosmic radiation	Sigma: Sigma-plus
dds	$n + \pi^-$	W	100	1953	W. Fowler's group at Brookhaven	Sigma-minus
uds	$\Lambda^0 + \gamma$	EM	100	1956	R. Plano's group at Brookhaven	Sigma-zero
uss	$\Lambda^0 + \pi^0$	W	100	1959	L. Alverez's group at Berkeley	Xi Xi-zero
dss	$\Lambda^0 + \pi^-$	W	100	1952	R. Armenteros's group, cosmic rays	Xi-minus
sss	$\Lambda^0 + K^-$ $\Xi^0 + \pi^-$	W W	69 23	1964	V. Barnes's group at Brookhaven	Omega-minus
uds	$p + K^- + \pi^+$ $\Delta^{++} + K^-$ $p + \bar{K}^0$	W	2	1975	Team at Brookhaven	Lambda-charm
ddd, ddu uud, uuu	$N + \pi$	S				Resonances
dds, dus uus	$N + \bar{K}$	S				
	$\Sigma^0 + \pi^0$	S				
	$N + \pi$	S				

III.   Mesons

Name	Symbol	Anti-Particle	Mass (MeV)	Electrical Charge	Spin	Average Lifetime (seconds)
Pion: Pi-zero	π^0	π^0	134.97	0	0	$.84 \times 10^{-16}$
Pi-minus	π^-	π^+	139.58	-1	0	2.6×10^{-8}
Pi-plus	π^+	π^-	139.58	$+1$	0	2.6×10^{-8}
Kaon K-zero	K^0	$\bar{K}^0$	497.8	0	0	$.86 \times 10^{-8}$ $.52 \times 10^{-8}$
K-plus	K^+	$\bar{K}^+$ or K^-	493.8	$+1$	0	1.24×10^{-8}
Eta	η	η	548.8	0	0	2.6×10^{-19}
Rho	ρ^+	ρ^-	765	$+1$	1	3×10^{-23}
D D-zero	D^0	$\bar{D}^0$	1870	0	0	10^{-12}
D-plus	D^+	$\bar{D}^+$	1870	$+1$	0	4×10^{-13}
J/PSI	J/ψ		3100	0	1	10^{-20}

III. Mesons

| Quark Content | Principal Decay Modes | | | Discovery | | Name |
	Modes	Type	Fraction (%)	Date	By Whom	
u$\bar{\text{u}}$ or d$\bar{\text{d}}$	$\gamma + \gamma$ $\gamma + e^+ + e^-$	EM EM	98.8 1.2	1949	R. Bjorkland's group at Berkeley	Pion: Pi-zero
d$\bar{\text{u}}$ u$\bar{\text{d}}$	$\mu^- + \bar{\nu}_\mu$ $\mu^+ + \nu_\mu$	W W	100 100	1947	C. Powell in the cosmic radiation	Pi-minus Pi-plus
d$\bar{\text{s}}$	$\pi^+ + \pi^-$ $\pi^0 + \pi^0$ $\pi^0 + \pi^0 + \pi^0$ $\pi^- + e^+ + \bar{\nu}_e$	W W W W	68.7 31.3 21.5 38.8	1947	G. Rochester and C. Butler in the cosmic radiation	Kaon K-zero
u$\bar{\text{s}}$	$\mu^+ + \nu_\mu$ $\pi^0 + \pi^+$ $\pi^+ + \pi^+ + \pi^-$	W W W	63.87 20.9 5.6			K-plus
	$\gamma + \gamma$ $\pi^0 + \pi^0 + \pi^0$ $\pi^+ + \pi^- + \pi^0$	EM EM EM	39 32 23			Eta
	$\pi^+ + \pi^0$	S	100			Rho
c$\bar{\text{u}}$ c$\bar{\text{d}}$	Numerous			1976	G. Goldhaber at Berkeley and SLAC	D D-zero D-plus
c$\bar{\text{c}}$	Numerous			1974	B. Richter, SLAC and S. Ting, Brookhaven	J/PSI

IV. Field Propagators

Name	Symbol	Anti-Particle	Mass (MeV)	Electrical Charge	Spin	Average Lifetime (seconds)
Photon	γ	Same	0	0	1	Infinite

W
  W-plus
  W-minus
	
	W^+
	W^-
		
		Same
		Same
			
			83,000
			
				+1
				−1
					
					1
					1
						
						10^{-25}

Z	Z	Same	93,000	0	1	10^{-25}
Gluon	g	Same	0?	0	1	Infinite?
Graviton		Same	?	0?	2?	Infinite?

IV.   Field Propagators

Principal Decay Modes			Discovery		Name
Modes	Type	Fraction	Date	By Whom	
			1905 1923	Einstein photoelectric effect, Compton γ + e scattering	Photon
			1983	UA1 and UA2 teams at CERN	W W-plus W-minus
			1983	UA1 and UA2 teams at CERN	Z
			1979	TASSO team at DESY	Gluon
				Never observed	Graviton

Discovery of the PSI Particles:
One Researcher's Personal Account

By Gerson Goldhaber

(This article first appeared in SLAC *Beamline*, November 1974.)

The circumstances which led to our discovery of the ψ particles began in 1973. That was the year we initiated our study of the cross section for hadron production from positron–electron (e^+e^-) annihilations. I will not attempt to describe here the gargantuan efforts and events that preceded and made the experiment possible—the construction of the SPEAR ring, the design and construction of the SLAC/LBL magnetic detector, the trigger logic, and the writing of software programs for track reconstruction.

Our aim was to find the behavior of the total cross section for e^+e^- annihilation as a function of energy. We began by taking data 100 MeV apart in energy of e^+ and e^- colliding beams; i.e., 200 MeV steps in total energy. We took a few hundred hadrons at each of these energies.

In those days, when we were working to understand the properties and systematics of our brand-new magnetic detector, we followed the philosophy of actually looking at the events. Having recently emerged from bubble-chamber physics, the most natural procedure to us was to visually scan these new kinds of pictures: the computer reconstructed topography of the events. Gerald (Gerry) Abrams was largely responsible for getting a "Berkeley" version of the program on the air which recorded on microfiche the computer reconstruction of the events from the SLAC/LBL SPEAR detector. At that point I was working on so-called "scanning programs" to identify the events automatically. This, of course, involved a large amount of cross-checking between the visual identification of the events and what those programs were saying. We finally convinced ourselves that we really could identify cosmic rays, electron-positron scattering (Bhabhas), mu-pairs, and hadronic events

Vera Lüth of the SLAC–LBL experimental group took this photo during the runs of November 9–10, 1974, in which the first psi particle was discovered. From the left are Willy Chinowsky of LBL; Martin Perl of SLAC; Francois Vannucci, at SLAC on leave from Orsay, France; and Gerson Goldhaber of LBL.

which were observed in the magnetic detector. In conjunction with Charles Morehouse from SLAC, we laid the groundwork for a "first filtering" program which eliminated cosmic rays and other obvious junk, which at that time was amounting to 90% or more of our triggers.

As the runs at these series of energies from 2.4 to 4.8 GeV became available, Gerry Abrams, John Kadyk and I started scanning them at Berkeley. Similar efforts were proceeding simultaneously at SLAC.

An Anomaly at 3.2 GeV

It was John Kadyk who first noted an anomaly at 1.6 GeV per beam, or 3.2 GeV total energy. He felt that there was a definite increase in the cross section by about 30% at that energy. In view of all the excitement of our early work, such as the much-larger-than-expected cross section confirming the earlier Cambridge Electron Accelerator (CEA) results, and the time required to get all the details of the interactions sorted out, this observation was not followed up for quite a while. In June 1974, Martin (Marty) Briedenbach at SLAC finally did follow up the observation and took additional data at energy points of 1.55, 1.6, and 1.65 GeV per beam respectively, as well as in the neighborhood of 2.0 GeV, where some hints of increased cross sections had also been observed. I was away for two weeks at the Finnish Summer School in Ekenäs at that time. When I came back, I heard that the additional measurements had been taken but that they showed nothing remarkable. Thus at Berkeley we did not pursue the matter any further at that time. The reason for our fallacious interpretation was a minor misunderstanding in the application of the analysis program which had been tuned up for running with an iron converter in an attempt to study gamma-ray energies. That's where the matter rested for awhile, and on July 2, the SPEAR ring was shut down for a three-month period for conversion to higher energy.

In the meantime, we were developing a bad conscience that while our cross section points were being reported at every meeting around the world, we had not really written a definitive paper on the subject. Roy Schwitters of SLAC undertook the job of writing the paper and, in the course of this effort, he went through each of the measured energies with a fine-tooth comb. He was assisted by my graduate student, Scott Whitaker, who had just decided to make the total cross section measurement his thesis topic.

Upon reexamination of the data they found that the new measurements at 1.6 GeV were indeed confirming a 30% higher cross section value, that the point at 1.65 GeV was back at the normal cross section level, but that the point at 1.55 GeV (a total energy of 3.1 GeV) behaved in a very peculiar fashion: the cross section at 1.55 GeV looked considerably higher than the average value in this energy region ($\approx$25 nb).

At a meeting at SLAC in mid-October, Schwitters showed me his result and asked us to double-check it at Berkeley—a procedure of cross-checking that we had followed throughout this experiment. Gerry

Abrams looked back at the entire series of runs and confirmed that the cross section indeed was larger at 1.55 GeV. But what was most peculiar, the study revealed, was that the cross section for 6 of the 8 runs looked perfectly normal! However, in two runs there were considerable discrepancies. Run 1380 showed an increase in hadronic cross section by about a factor of three, while Run 1383 showed an increase by about a factor of five! Here were a series of runs, all taken at the same energy, as far as we could ascertain, and yet they showed considerable internal inconsistencies! This was an obvious clue that something was going on that we didn't understand. And we could not publish our results until we understood what was causing the "fishy" results.

One obvious and simple cause which we considered was that something had gone wrong with our detector. But when all the events had been examined in great detail, we could find absolutely nothing wrong with them. Yet the glaring inconsistency persisted! We were further helped by two independent "red herrings." Scott Whitaker looked at charged K meson production and found a remarkable increase in Run 1383. I looked at K^0 production and also found a remarkable increase in Run 1383. That was the last straw: an increase in the cross section and an increase in strange particle production—just what the "charm

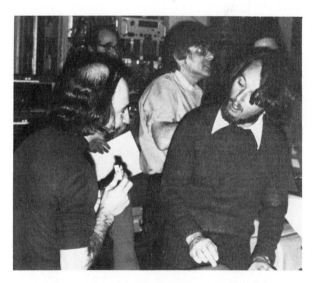

Left to right in the foreground are Marty Briedenbach and Harvey Lynch of SLAC, with Gerson Goldhaber and Willy Chinowsky of LBL to the rear. This is another of the photos taken by experimentalist Vera Lüth during the now-famous weekend of November 9–10, 1974.

advocates" had been talking about! As it turned out later, the apparent increase in strange particle production was part misidentification and part a statistical fluctuation; but it certainly was a fortunate one. While it was clear to us that we had to look at this energy region again before we could publish a total cross-section paper, the apparent strange particle increase pushed us to do it with great speed.

The Fateful Weekend

On Monday, November 4, I called Burton (Burt) Richter to discuss our problem. Was it conceivable that the energy of the SPEAR ring had drifted by a few MeV and that at the two anomalous runs I mentioned, a slightly different energy, presumably higher than 1.55 GeV, had been set? We determined that this matter needed clarification and, if possible, that we should do some more running in this region. This was not a trivial question, since in the conversion to SPEAR II, all changes had been made to allow the attainment of *higher* energies in the 2.5 to 4.5 GeV per beam region. Thanks to the efforts of Burt Richter, Ewan Paterson and the members of the SPEAR operating crew, however, the crew re-learned how to run in the 1.5 GeV region.

During the course of that week, there were many discussions between all the various experimenters. Finally, on Friday, Burt called to let me know that he had looked into the possibility of returning to low energies and that we were planning for the weekend to run in the 1.5 to 1.6 GeV region again. I called SPEAR in the early afternoon of Saturday, November 9, and talked with Vera Lüth. She told me that there had been some running at 1.57 GeV per beam and that an indication of a repeat of the higher cross section had been observed that morning. I went down to SLAC that same afternoon. When I arrived we were running at 1.5 GeV per beam, and at that energy the hadron cross section looked normal. No excess was seen. However, by evening we had clear proof that we were seeing a higher cross section!

Our data taking and analysis proceeded through three separate "channels": all triggers were recorded on magnetic tape for off-line analysis; Marty Briedenbach was just beginning to get results from his "on-line" cross-section determination; and at the same time we obtained the "one-event display" which showed a rough reconstruction of each event, as it occurred, on a CRT [cathode-ray tube] screen. We had sufficient information there to decide between cosmic rays, Bhabhas, hadrons, etc., and the data rate was low enough to allow this visual identification to be made. As I arrived, Rudolph Larsen was "keeping score" of the events as they came in on the CRT screen. I took over for him and continued the score-keeping as we changed the energy to 1.56 GeV per beam at 11:15 PM. As we did so, it became clear that the cross section was *definitely rising*. We observed a cross section about twice as large as at 1.5 GeV. While at 1.5 GeV we had observed 10 hadrons and 61 Bhabhas of more than two prongs, at 1.56 GeV we observed 55 hadrons for 170 Bhabhas. There was thus a clear increase of the hadron cross section by a factor of two! There was no more doubt in any of our minds that we had indeed observed a rise in the cross section or what might be a resonance in e^+e^- annihilation. At 3 AM, with this happy thought, I went to a local motel to catch a few hours of sleep.

Later that morning when I returned to SPEAR, I learned that a number of extra runs had been taken at various energies below and above 1.56 GeV, all confirming that the excess we had seen was very localized, since no additional increases were observed in the points further out. At about 9:00 AM we began to zero in on trying to find out where the exact peak might be. Suddenly, at about noon, we found an energy where the cross section had risen by a factor 7! By this time our excitement had risen to a fever pitch. Having been nurtured on strong interactions, I had hardly even seen an elementary-particle cross section that rose by a factor of 7 above the background level. At this point the results were so convincing that I suggested to some of my colleagues (Burt Richter, Roy Schwitters, William Chinowsky, and others) that we should sit down and write a paper. They agreed with me and Burt led me to a side room where I could quietly start the task. I picked up a piece of computer output from the floor to write on, and in about one hour I had written the first draft for our paper—an activity, incidentally, I had never done before: writing a paper on-line, so to speak.

A Further Increase Observed

At that point Willy Chinowsky burst into the room where I was working and told me very excitedly—I have rarely seen him so excited—that the cross section had risen by another factor of 10! With this news I jumped up and ran to the one-event display in the control room, and there indeed we were seeing hadron events coming in one after the other.

By now the hadrons were way outnumbering the Bhabhas and even the cosmic rays. It was absolutely unbelievable.

> "We were running along—one event in an interval, then zero, then one. Then 20 events all of a sudden. I said 'That's a fluctuation.' The guy with me on shift said 'That's *not* a fluctuation.' The next interval had 18 events. He said 'Do you believe me that it's not a fluctuation?' The next one was 15, then back to zero again. I believed him."
> —Chuck Morehouse, quoted by Robert Walgate in *New Scientist*, 11 March 1976

Were I to try to express my feelings as I saw the data coming in on the cathode-ray tube, I would say that this was perhaps the highest level of excitement I have ever experienced in a laboratory. In racking my memory, the only experience that was perhaps comparable in excitement to me occurred when Gosta Ekspong and I discovered the first event proving the antiproton annihilation process in photographic emulsions back in 1956.

I was obliged to go and delete all the numbers I had tentatively written in the paper and up everything by a factor of 10! By that time W. K. H. Panofsky had arrived and was pacing up and down the control room with a happy smile. I called George Trilling and John Kadyk, as well as Andrew Sessler (Director of LBL) and Robert Birge (Associate Director of Physics at LBL). Kadyk called J. D. Jackson who talked to Mahiko Suzuki and on it went. The news spread like wildfire.

A typical conversation went something like this:

"There really is a peak in the e^+e^- cross section. How high above background do you think it goes?"

Answers that came back were:

"50%?"
"A factor of 2?"
"A factor of 5?"

I would finally say:

"A factor of 70!" And then always had to add, "That is, seven-zero!!"

I wish I had been able to record the expressions of amazement that resounded in my ears.

The next few hours were spent talking and speculating what new quantum number or what new selection rule must be involved to inhibit the decay of our resonance and make it so narrow. I recall, in particular, discussions with Gary Feldman, George Trilling and Martin Perl, both on the first draft of our paper and on the properties of the new resonance. In the early afternoon, Burt took my first draft and started writing a second draft which finally resulted in the completion of our paper by that evening (Sunday, November 10, 1974).

Other Decay Channels of the Resonance

The only other activity that stands out in my mind occurred that afternoon when amidst the champagne and other goodies, we were also taking data and drawing graphs. Adam Boyarski pointed out a quantity he was routinely calculating with each run: the ratio of mu pairs to electron pairs over the QED [quantum electrodynamics] value expected for this ratio. This ratio was also suddenly showing deviations from unity for various points on the resonance curve. Then we all suddenly realized that a resonance which is produced via a photon must also decay via a photon. Hence one would expect large deviations from QED, which would appear as an increase in the $\mu^+\mu^-$ channel and for that matter in the e^+e^- channel as well.

Our graph which was then being plotted on a linear scale showed an enormous peak; but whatever scale allowed one to see the peak did not allow one to see the baseline. It was obvious that what we really needed was a logarithmic scale. Overcoming the problem of not having three-cycle semilog paper available, I pieced together two sheets of semilog paper and, with Harvey Lynch dictating the data points, I plotted them on that. Now, for the first time, it became clear how we had managed to observe this resonance. The radiative tail characteristic of e^+e^- annihilation, which is normally an experimenter's nightmare, had saved the day for us! This very long and well-defined tail toward the high energy side did indeed give rise to the 30% increase in the cross section at 1.6 GeV which had first drawn our attention to the entire phenomenon! Then I also plotted Adam Boyarski's muon-pair to electron-pair ratio on the same scale and it also showed a clear peaking coincident with the hadronic peak. Carl Friedberg later plotted a quantity proportional to electron-pair production and, sure enough, it also showed a clear peaking in the region of our resonance. When I finally left for home that evening, I agreed with Burt Richter that we would each announce this discovery at our respective laboratories on the following day, Monday, November 11. As there was no copying machine available, I made by hand a copy of my drawing to take with me to Berkeley and left the original in the log book.

Monday Morning

On Monday morning, George Trilling and I invited some of our colleagues at LBL to have champagne with us in our office area. Dave Jackson was one of the early arrivals. It was obvious that he had been working until quite late the previous evening. His time had been well-spent, for he was able to show—by a fairly straightforward application of the Breit-Wigner resonance formula—that not only had we discovered a new resonance, but also that its true width could be readily deduced from our measurements!

Dave's result was sufficiently straightforward—except for some details on radiative corrections—that many of us could have written it down independently and some did so, given some time of quiet thinking away from SPEAR. While Dave was explaining his calculations to a group of enthusiastic and champagne-sipping physicists in my office, I heard what to me was an astonishing piece of news. Somebody told me that he had just heard from SLAC that Sam Ting and his group working at Brookhaven had discovered the same effect we had found! While my first reaction was one of stunned disbelief, we all know now that it was obviously true.

At noon I presented our results, with the aid of view graphs prepared that morning, to a large and enthusiastic crowd at the LBL auditorium. All in all it was a good weekend's work.

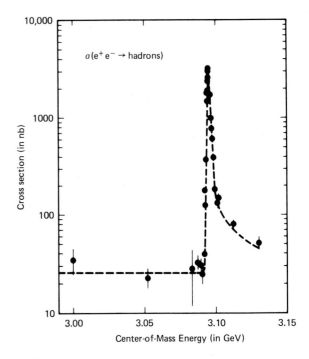

This figure illustrates the first psi resonance, $\psi(3095)$, as measured at SPEAR. The cross section (or probability of occurrence) for the process $e^+e^- \rightarrow$ hadrons rises by a factor of 70 to 100 as resonance is reached—from the "normal" level of about 25 nb to roughly 2300 nb at the peak of the resonance. So large a change requires the use of a logarithmic vertical scale to show both the low and high cross section values. The e^+e^- annihilation process produces a "radiative tail," which appears in the figure as a more gradual drop off on the high-energy side of the peak. It was the fact that this tail extends far enough out to affect the data taken at an energy of 3.2 GeV that provided the first hint that something unusual was happening nearby.

ANSWERS TO ODD-NUMBERED PROBLEMS

Chapter 2

1. 2×10^{15} N
3. $F_{\text{grav}} \approx 1.9 \times 10^8$ N, $F_{\text{elec}} \approx 2.3 \times 10^{44}$ N, will not hold.
5. 4.15×10^{42} for electrons, 1.24×10^{36} for protons
7. 4.4×10^{-13} kg

Chapter 3

1. Diameter 2.5×10^6 ft (370 miles), 10,000 TeV, $\$5.5 \times 10^{10}$

Chapter 4

5. Range $\approx 2.5 \times 10^{-18}$ m

Chapter 5

3. (a) 207.5 MeV (b) $\left(\dfrac{3}{4}\right)^{1/2} c$ (c) Ξ^0 (d) 1.04×10^{-27} kg
   (e) 529.2 J (f) 24.49 GeV (g) 2.507 (h) 1.04×10^8 m/s
7. (a) $-e$ (b) 212.9 MeV (c) π^- (d) .753 c
9. D^0 meson
13. (a) 65.82 MeV (b) $E_{\text{KE}}(\Lambda^0) = 8.6$ MeV, $E_{\text{KE}}(\pi^-) = 57.2$ MeV
15. .2193 c and 4,583.4 MeV
17. .706 c and 1.44 MeV
19. (a) .0308 c (b) 3.0919 m
21. $v_{\text{Nd}} = .000886\ c$, $E_{\text{R}}(\text{Sm}) = 141,512.15$ MeV;
    use *non*-relativistic expression for momentum because of large Nd mass
23. 13.33 cm/s
25. (b) $E_{\text{KE}}(i)/E_{\text{KE}(15)} = 1.012$

Chapter 6

1. $\mathbf{a} = 4\mathbf{i} + 5\mathbf{j}$, $\mathbf{b} = \mathbf{i} - 3\mathbf{j}$
3. (a) $5\mathbf{i} + 3\mathbf{j} + 7\mathbf{k}$ (e) -28 (i) $11\mathbf{i} - 11\mathbf{j} - 3\mathbf{k}$
   (b) $6\mathbf{i} + 8\mathbf{j} + 12\mathbf{k}$ (f) 2 (j) $55\mathbf{i} + 4\mathbf{j} - 15\mathbf{k}$
   (c) $\mathbf{i} + 13\mathbf{j} - 5\mathbf{k}$ (g) $5\mathbf{i} + \mathbf{k}$ (k) 57.57
   (d) 4 (h) $-49\mathbf{i} - 196\mathbf{j} - 147\mathbf{k}$
5. (a) $|\mathbf{S}| = 14.21$, $\theta = 50.7°$ (b) $\mathbf{A} \cdot \mathbf{C} = 82$

7. $\mathbf{u} = \pm \left(\dfrac{1}{17}\right)^{1/2} \mathbf{i} \mp \left(\dfrac{16}{17}\right)^{1/2} \mathbf{j}$

9. 456.09 mi/hr

15. $\dfrac{17}{\sqrt{5}}$

23. (a) π^+ (b) $p + \pi^0$ (c) γ (d) μ^+ (e) π^+

25. $.0718\,c$

27. (a) Cons. of mom. in transverse direction
   (b) 84.36 MeV (c) 53.13°

29. (a) $\pi^+ + \pi^- + \pi^-$
   (b) $p_2 = p_3 = 105.48$ MeV/c, $p_1 = 54.9$ MeV/c (c) 83.74°

31. Particle is ν_μ, $E(\nu_\mu) = 499.43$ MeV, $p(\nu_\mu) = 499.43$ MeV/c, zero rest mass

41. (a) $E_{KE}(2) = .1357mv_1^2$, $\theta = 28.68°$ (b) No

43. (a) 1.905 m/s (b) Yes

Chapter 7   **3.** (a) 6.47×10^{-16} m (c) 2.4×10^{-15} m
   (b) 4.14×10^{-17} m (d) 1.58×10^{-10} m

5. $S = -1$

7. 3.2×10^{-24} s

9. (a) Strangeness, weak (f) Electron, never (j) ok
   (b) Charge, never   (g) Strangeness, weak (k) ok
   (c) Strangeness, weak (h) Strangeness, weak (l) Strangeness, weak
   (d) Strangeness, weak (i) Charge, never   (m) Strangeness, weak
   (e) Baryon, never

Chapter 8   **1.** (a) Sarah (b) same (c) .015 mi/s

9. (a) 10 s²/m, 1 m   (e) .2 m/s
   (c) $\langle v(0, 0)\rangle = 0$, $\langle v(0, 1)\rangle = .1$ m/s   (f) $v(t) = t/5$
   (d) $\langle v(1, 1)\rangle = .3$ m/s, $\langle v(1, 0)\rangle \approx .2$ m/s (g) $v(1) = .2$ m/s, $v(0) = 0$

11. (a) α m/s⁴, β m/s³, γ m/s (c) Yes, initial conditions
   (b) $x(t) = \left(\dfrac{\alpha}{4}\right)t^4 - \left(\dfrac{\beta}{3}\right)t^3 + \gamma t$

15. (a) 10 s (b) 800 m

17. (a) A-m, ω-rad/s (d) $v_{max} = A\omega$, $v_{min} = -A\omega$
   (b) $-A\omega \sin \omega t$ (e) $\dfrac{2\pi}{\omega}$

19. (a) α 1/s, ω rad/s
   (c) $v = -e^{-\alpha t}(\alpha \cos \omega t + \omega \sin \omega t)$
     $a = \alpha e^{-\alpha t}(\alpha \cos \omega t + \omega \sin \omega t) - e^{-\alpha t}(\omega^2 \cos \omega t - \alpha\omega \sin \omega t)$

21. $v(t) = \dfrac{F_x}{m}(t - t_0) + v_0$

   $x(t) = x_0 + v_0(t - t_0) + \dfrac{F_x}{2m}(t - t_0)^2$

23. (a) $3\left(\dfrac{h}{2g}\right)^{1/2}$ (b) 3/4 h

   (c) $v_1 = -\dfrac{1}{2}\sqrt{2gh}, \quad v_2 = \dfrac{1}{2}\sqrt{2gh}$

25. (a) 4.6×10^6 s (53 days)   (b) 1.03×10^{14} m

27. Does not matter

29. For a 60 kg person, $F \approx 60{,}000$ N

31. 37,007 km/hr^2

33. 1.14 km

35. (a) 15.36 m/s,   15.33 m/s   (b) .079 s

37. (a) $\frac{3}{20} t^2 - 3t + 15$ (b) 10 s   (d) 50 m

41. (a) $\dfrac{dv}{dt} = \dfrac{6}{m}\, e^{-\alpha t}$

   (b) $v(t) = \dfrac{6}{m\alpha}(1 - e^{-\alpha t})$ (c) $x(t) = \dfrac{6}{m\alpha^2}(e^{-\alpha t} - 1) + \dfrac{6}{m\alpha} t$

43. (a) 4 newtons,   2 newtons/second   (c) $v(t) = -\frac{1}{2}t^2 + 2t$

45. $\alpha = \dfrac{mg}{v_t} = .21$ kg/s $v(t) = \dfrac{mg}{\alpha}(1 - e^{-(\alpha t/m)})$

47. (b) $\dfrac{dv}{dt} = -\dfrac{k}{m}\, x, \quad v(t) = -C\left(\dfrac{k}{m}\right)^{1/2}\sin\left(\dfrac{k}{m}\right)^{1/2} t$

   (c) $x(t) = C\cos\left(\dfrac{k}{m}\right)^{1/2} t$ (e) Velocity would increase without limit

   (d) Pendulum

49. (a) $n(t) = Ne^{-\alpha t}$ (b) $N(1 - e^{-\alpha t})$

Chapter 9

1. $T_1 = 280.9$ N, $T_2 = 244.9$ N, $T_3 = 245$ N

5. $a = 1.625$ m/s, $T_a = 4.875$ N, $T_b = 14.625$ N

7. $a_2 = \dfrac{F}{4M + m}, \quad a_1 = 2a_2, \quad T_1 = \dfrac{4M}{4M + m} F, \quad T_2 = \dfrac{T_1}{2}$

9. $T = 5$ N, $a_{7\text{kg}} = 9.1$ m/s^2

11. $a = \dfrac{2F - g(M + m)}{M + m}, \quad T = 2F$

13. (a) $a = \dfrac{F\cos\theta - \mu\,(Mg - F\sin\theta)}{M}$ (b) $\tan\theta_{\text{max acc}} = \mu$

15. (a) 2.19 m/s^2 (b) 1.35 s

19. (a) $a = g\sin\theta - \left(\dfrac{\mu_2 - \mu_1}{2}\right) g\cos\theta$ (b) $T = \left(\dfrac{\mu_2 - \mu_1}{2}\right) mg\cos\theta$

   (c) $\theta_c = \tan^{-1}\left(\dfrac{\mu_1 + \mu_2}{2}\right)$

21. $a = g\tan\theta,$ put on brakes it tilts forward; const. vel $\to \theta = 0$

23. $\mu = .124$ **25.** $F = (M + m)g\tan\theta$ **27.** $\left(\dfrac{\mu g}{\alpha^2 r}\right)^{1/2}$ s

29. For a 50 kg person, $W_{\text{pole}} = 490.35$ N, $W_{\text{equat}} = 488.66$ N, $\Delta W \approx 1.7$ N

31. $A \to mg - m\dfrac{4\pi^2 R}{\tau^2},$ $B, D \to mg,$ $C \to mg + m\dfrac{4\pi^2 R}{\tau^2}$

33. $T_1 = \dfrac{m}{\sqrt{2}}\left(g + \dfrac{\omega^2 L}{\sqrt{2}}\right), \quad T_2 = \dfrac{m\omega^2 L}{2} - \dfrac{mg}{\sqrt{2}}$ **35.** 2.21 rad/s

37. $\theta = \cos^{-1}\dfrac{g}{\omega^2 r}$ but $\omega_{\min} = \left(\dfrac{g}{r}\right)^{1/2}$

39. (a) $B = \dfrac{mv_0}{eL}$ into the paper $(-\mathbf{k})$ (b) $\mathbf{E} = \dfrac{v_0^2 m}{2\,eL}(-\mathbf{i} + \mathbf{j})$

41. $v_0 = \left(\dfrac{gL}{2}\right)^{1/2}$ or any value larger

43. (a) 441.4 m   (b) $\mathbf{v}_f = 12.2$ m/s $\mathbf{i} - 180$ m/s $\mathbf{j}$ (c) $t = 36.18$ s

45. $v_x = 5 \times 10^5$ m/s, $v_y = 3.5 \times 10^5$ m/s, deflection 21 cm

47. $\Delta x = \left[\dfrac{8Ey_0}{qB^2}(m_2 - m_1)\right]^{1/2}$ **49.** $a = g - \dfrac{\rho_w g x}{\rho_s L}$

Chapter 10

1. (a) $\mathbf{r} = 250 \text{ m}\mathbf{i} + 65 \text{ m}\mathbf{j} - 3.8 \text{ m/s } t\mathbf{i} - 4.9 \text{ m/s}^2 \, t^2\mathbf{j}$
     (b) $\mathbf{v}(t) = -3.8$ m/s $\mathbf{i} - 9.8t$ m/s^2 $\mathbf{j}$
     (c) -3.8 m/s $\mathbf{i} - 25.2$ m/s $\mathbf{j}$ (d) -9.8 m/s^2 $\mathbf{j}$

3. (a) $2\mathbf{i} - \mathbf{j}$ m/s   (b) 3/4 lost   (d) 90° deflection

7. $\mathbf{u}_{CM} = 3.96\mathbf{i} - 1.53\mathbf{j}$ m/s   (b) $\mathbf{P}_T = 27.72\mathbf{i} - 10.72\mathbf{j}$ kg m/s
     (c) $\mathbf{v}_1^{CM} = .16\mathbf{i} + 1.53\mathbf{j}$ m/s, $\mathbf{v}_2^{CM} = .54\mathbf{i} - 5.17\mathbf{j}$ m/s

9. (a) 1.56×10^{-12} J   (b) 40 MeV   (c) 1.07×10^{-20} J

11. $90° - \sin^{-1}\dfrac{b}{2a}$

15. $\mathbf{p}' = (p_x - mu)\mathbf{i} + p_y\mathbf{j} + p_z\mathbf{k}, \quad KE' = \dfrac{(p_x - mu)^2 + p_y^2 + p_z^2}{2m}$

17. $u^{CM} = -\frac{1}{3}v, \quad v'(1) = \frac{4}{3}v, \quad v'(2) = -\frac{2}{3}v$

21. (a) $v^{CM} = .367\,c, \quad v^{lab} = .644\,c$ (b) 289 MeV

23. (a) -2.45 m/s^2 (b) $g/2$ (c) 34.3 N

25. (a) 1.324 m/s^2 (b) -1.176 m/s^2 (c) 140.5 m, yes
     (d) 13.24 m/s   (e) The box's mass

27. (a) 5×10^6 m/s^2 (gravity too small to be important)
     (b) .00338 m/s

Chapter 11

1. (a) 4,009.52 J   (b) 3,732.02 J

3. (a) 2173 J   (b) -542 J   (c) -571 J   (d) 12.05 m/s

5. $v_{10} = \sqrt{2}$ m/s, $v_{20} = \sqrt{6}$ m/s

7. $v_f = \left(\dfrac{2m_2 g l}{m_1 + m_2}\right)^{1/2}$ **9.** $\Delta y = 14.84$ m

11. (a) $v = \sqrt{gR}$ (b) $v = \left(\dfrac{2gR}{1 + m/M}\right)^{1/2}$

13. $\left[\dfrac{-\alpha + (\alpha^2 + 2\beta mv_0^2)^{1/2}}{\beta}\right]^{1/2}$

15. (a) $-mgr(1 - \cos\theta)$ (b) $a_t = g\sin\theta, \quad a_r = 2g(1 - \cos\theta)$
     (c) $\theta = 48.19°$ (d) longer

17. (a) 2.94×10^6 W   (b) 9.27×10^{13} J

19. 8.83 m/s

21. (a) 5×10^4 J   (b) 10^7 N/m   (c) 5×10^6 W

23. $88g$

25. (a) $\frac{5}{2}R$ (b) $v_A = (3\ gR)^{1/2} = v_c,\quad v_B = \sqrt{Rg}$

29. .124

31. (a) 18.66 m/s   (b) 3.58 m   (c) 18.64 m from the bottom
   (d) $\sim$ 163 m

33. $u(r) = eE_x x$ for $\mathbf{E} = E_x \mathbf{i}$ zero at $x = 0$

35. $x_0 y_0^2$; the field is conservative

39. (a) $-\dfrac{G'm_1 m_2}{2r^2}$ (b) $\dfrac{\sqrt{G'm_E}}{R_E}$

41. (a) $F_x = -\alpha 2x\mathbf{i},\quad F_y = -\alpha 2y\mathbf{j}$ (b) $F_r = -\alpha 2r\hat{\mathbf{r}},\quad F_\theta = 0$

43. (a) $F = -\dfrac{6A}{r^7} + \dfrac{12B}{r^{13}}$ (b) $r_{eq} = \left(\dfrac{2B}{A}\right)^{1/6}$ (c) $-\dfrac{A^2}{4B}$

45. $u(x) = \dfrac{kx^2}{2}\left(1 - \dfrac{\alpha x^2}{2}\right)$

47. (a) $\Delta v = \left(v_i^2 + \dfrac{3 \times 10^4\ \text{eV}}{m}\right)^{1/2} - v_i$ (not unique)

   (b) $v_f = 1.7 \times 10^6$ m/s; $r = 1.1 \times 10^{-3}$ m
   (c) $v_m = \sqrt{n}\ (1.7 \times 10^6)$ m/s   (d) $r_n = \sqrt{n}\ (1.107 \times 10^{-3})$ m
   (e) $v_f = 5.37 \times 10^7$ m/s, small relativistic effects
   (f) 3.5 cm   (g) 2.05×10^{-6} s

Chapter 12

1. (a) 2.1 Å below oxygen, along perpendicular bisector
   (b) 3.8 Å, along the same line

3. $\frac{2}{5}h$ above the base, where h = height of tetrahedron along the axis

5. $\mathbf{R}_{CM} = -.182\mathbf{i} + .136\mathbf{j}$ (meters); origin at CM of large object

7. (a) $\dfrac{M_E}{M_E + M_c}\left(\dfrac{4r_1}{3\pi}\right)$ below the center of symmetry of the empty cylinder

   (b) $U(\theta) = \dfrac{M_E g 4 r_1}{3\pi}\ (1 - \cos\theta)$

9. (a) Along the axis at the point of intersection of the two blocks, or $2l$ from the long end
   (b) CM moves $.0055l$ towards the larger block.

11. 2.357 m; along the perpendicular bisector from the long edge

13. $\dfrac{\sqrt{l^2 - b^2/4}}{3}$ from the bottom of the base

15. $T = 8.96$ N, $F = 17.9$ N   **17.** $\theta = \tan^{-1}\left(\dfrac{1 - \mu_1\mu_2}{2\mu_1}\right)$

19. $\Delta x_{\text{earth}} = 3.75 \times 10^{-22}$ m   **23.** $\mu = .74$

25. $\mathbf{F}_2 = 5.93\ \text{N}\mathbf{i} - 10.3\ \text{N}\mathbf{j},\quad \mathbf{F}_4 = -10.97\ \text{N}\mathbf{i} + 18.3\ \text{N}\mathbf{j}$

27. Starting at the lowest part

$$x_{CM}^{(1)} = m_4 l_5 - m_3 l_6$$

$$x_{CM}^{(2)} = (m_3 + m_4) l_4 - l_3 m_2$$

$$x_{CM}^{(3)} = m_1 l_1 - (m_2 + m_3 + m_4) l_2$$

The CM of each unit must be at the supporting string.

29. 301 N   **31.** $T = \dfrac{mg \sin\theta}{1 + \cos\theta}, \quad \mu = \dfrac{\sin\theta}{1 + \cos\theta}$

33. 2.548×10^5 N, 2.744×10^5 N   **35.** $\theta = \sin^{-1}\dfrac{8\mu}{3 + 8\mu}$

37. $\dfrac{h}{\mu\sqrt{h^2 + W^2}}$ above the floor

39. (a) $T = \dfrac{mg(L + r)}{L\sqrt{1 - \dfrac{2r}{L}}}$ (b) $N = \dfrac{mgr(L + r)}{L\sqrt{1 - \dfrac{2r}{L}}}$

41. $T = 58.25$ lb   **43.** $T_2 = 61.25$ N, $T_1 = 122.5$ N   **45.** $T = 1.23 \times 10^3$ N

Chapter 13

1. (a) $\mathbf{L}_0 = 0$, $\mathbf{L}_{z0} = z_0 m(v_{0y} - \mu g t)\mathbf{i}$ (b) $\boldsymbol{\tau}_0 = 0$, $\boldsymbol{\tau}_{z0} = -z_0\mu mg\mathbf{i}$

3. (a) $\mathbf{L}_{x0} = 0$, $\mathbf{L}_0 = -mx_0 gt\mathbf{k}$ (b) $\boldsymbol{\tau}_0 = 0$, $\boldsymbol{\tau}_{x0} = -mgx_0\mathbf{k}$

5. (a) $|\mathbf{L}_0| = 1.05 \times 10^{-34}$ kg/m²s, $\mathbf{L}_0$ conserved $\boldsymbol{\tau}_0 = 0$

(b) 2.19×10^6 m/s   (d) 4.35×10^{-18} J or 27.2 eV

(c) 1.16×10^{-28} J   (e) None

(f) 27.2 eV   (g) -4.35×10^{-18} J or -27.2 eV   (h) Nonrelativistic, $\dfrac{v^2}{c^2} \sim 5 \times 10^{-5}$

7. $\mathbf{L}_0 = amv\mathbf{k}$

9. (a) $\mathbf{L}_0 = -bmv_{ox}\mathbf{k}$

(b) Same as (a); no torque due to radial force

(c) $v_0/v_{min} = 6$

11. (a) $\mathbf{L}_0(y_1) = m\omega r(y_0 - r)\mathbf{k}$, $\mathbf{L}_0(y_2) = -m\omega r(y_0 + r)\mathbf{k}$

13. $\boldsymbol{\tau} = mgr_x\mathbf{k}$

17. $\dfrac{mL^2}{12}$ independent of θ

23. (a) 1.568 kg m²/s²   (c) 2.70 m/s²   or $\dfrac{m_1 g}{\frac{1}{2}m_3 + m_1 + m_2} = a$

(b) $\mathbf{L}_0 = .58v$ kg m²/s

25. (a) 6.533 m/s², (b) 2.72 N, (c) 8.17 N

27. (a) $\dfrac{3g \sin\theta}{2l}$ rad/s², (b) $\dfrac{3g \sin\theta}{2}$ m/s², (c) mgl

29. .0083 rad/s   **33.** 5.42 m/s   **35.** $h = \frac{27}{10}R$

Chapter 14

1. (a) 1.67 rad/s out of paper

(b) 55.56 rad/s into paper

(c) 53.89 rad/s into paper

3. (a) $\omega = \Omega\left(1 - \dfrac{l^2}{R^2}\right)^{1/2}$ at angle $\theta = \tan^{-1}\dfrac{R}{l}$

(b) $\mathbf{L}_T = Ml\Omega(l\mathbf{k} - R\mathbf{j})$ at angle $\alpha = \tan^{-1}\dfrac{l}{R}$

(c) $\mathbf{L}$ and $\boldsymbol{\omega}$ same direction if $l = R$

5. 2.68×10^4 J-s $\mathbf{k}$ (orbital mom. dominates)

9. $x = \dfrac{lm_2}{m_1 + m_2}$

11. (a) a (b) 2.0 J-s   (c) $20(\mathbf{i} - \mathbf{j})$ N-m

13. (a) $L_1 = L_2 = 108.375$ J-s into paper

(b) 216.75 J-s

15. (a) Angular momentum conserved   (b) $\omega_f = \left(\dfrac{I_z, \omega_i}{I_z^1 + I_z^2}\right)$

17. (a) $v_f = \frac{5}{7}v_0$ (b) Distance $= \dfrac{12}{49}\dfrac{v_0^2}{\mu g}$

19. (a) 1. $v_{\text{CM}} = [g(h - R)]^{1/2}$ 2. $v_{\text{CM}} = \left(\dfrac{gh}{2}\right)^{1/2}$

(b) $\mathbf{L}_0 = 2Rm[g(h - R)]^{1/2}$ into paper

21. (a) Kinetic energy and angular momentum   (d) $\omega_2' = 1.2$ rad/s (out of paper)

(b) $\omega_1' = 9$ rad/s (into paper)

(c) $v = 1.2$ m/s   (e) $\dfrac{E_f}{E_i} = .10$

23. (a) No

(b) $\omega_{2f} = \left(\dfrac{I_1 R_1 R_2}{R_1^2 I_2 - R_2^2 I_1}\right)\omega_0$ $\omega_{1f} = \dfrac{I_1 R_2^2 \omega_0}{R_1^2 I_2 - R_2^2 I_1}$

(c) $\omega_{2f} = \dfrac{I_1 R_1 \omega_0}{I_1 R_2 - I_2 R_1}$ $\omega_{1f} = \dfrac{I_1 \omega_0 R_1 R_2}{I_1 R_2 R_1 - I_2 R_1^2}$

25. (a) m_1 travels $2\pi\left(\dfrac{m_2}{m_1 + m_2}\right)$ rad and m_2 travels $2\pi\left(\dfrac{m_1}{m_1 + m_2}\right)$ rad

(b) $\pi R\left[\dfrac{2m_1 m_2}{(m_1 + m_2)E}\right]^{1/2}$ seconds

Chapter 15   **3.** 0.193 kg   **5.** 59.05 N/m

9. (a) $v_\alpha - v_\beta = .05$ m/s

(b) $x_\alpha - x_\beta = .015$ m, $v_\alpha - v_\beta = -.233$ m/s

11. (a) $\dfrac{mg \sin \theta}{k}$ (b) $\left(\dfrac{k}{m}\right)^{1/2}$

13. $\dfrac{(m + M)A\omega}{m}$ **15.** (a) $T = \dfrac{T_0}{\sqrt{2}}$ (b) $T = \sqrt{2}\,T_0$

17. (a) One-quarter of a cycle or $\dfrac{\pi}{2}\left(\dfrac{m_1 + m_2}{k}\right)^{1/2}$

(b) $\dfrac{m_2 b^2 k}{2(m_1 + m_2)}$ (c) $x = b\left(\dfrac{m_1}{m_1 + m_2}\right)\cos\left(\dfrac{k}{m_1}\right)^{1/2}t$

19. (a) $x_{\text{eq}} = .0327$ m   (b) Yes   (c) 1.15 s

25. 1.54° Celsius

27. (b) $2\pi\left(\dfrac{\sqrt{2}\,R}{g}\right)$ shortest period

29. $2\pi\left(\dfrac{4l}{5g}\right)^{1/2}$

31. (a) $F_1 = 2\,kx$, $F_2 = kx$ (b) $\tau = -kL^2\theta$ (c) $T = 2\pi\left(\dfrac{m}{2k}\right)^{1/2}$

33. (b) $E_{\text{rot}} = 1.08 \times 10^{-3}$ J, $E_{\text{trans}} = 2.16 \times 10^{-3}$ J

37. (a) .00161 kg/s   (b) 3.48×10^{-5} W

39. 3.96×10^{-2}

41. (b) $\alpha = .15$ kg/s, $k = 198.97$ N/m

INDEX